A Text-Book of
EXPERIMENTAL CYTOLOGY

A Text-Book of
EXPERIMENTAL CYTOLOGY

by

J. GRAY, M.A., F.R.S.

*Fellow of King's College, Cambridge, and
Lecturer in Experimental Zoology,
University of Cambridge*

CAMBRIDGE
AT THE UNIVERSITY PRESS
1931

CAMBRIDGE UNIVERSITY PRESS
Cambridge, New York, Melbourne, Madrid, Cape Town,
Singapore, São Paulo, Delhi, Mexico City

Cambridge University Press
The Edinburgh Building, Cambridge CB2 8RU, UK

Published in the United States of America by Cambridge University Press, New York

www.cambridge.org
Information on this title: www.cambridge.org/9781107625662

© Cambridge University Press 1931

This publication is in copyright. Subject to statutory exception
and to the provisions of relevant collective licensing agreements,
no reproduction of any part may take place without the written
permission of Cambridge University Press.

First published 1931
First paperback edition 2013

A catalogue record for this publication is available from the British Library

ISBN 978-1-107-62566-2 Paperback

Cambridge University Press has no responsibility for the persistence or
accuracy of URLs for external or third-party internet websites referred to in
this publication, and does not guarantee that any content on such websites is,
or will remain, accurate or appropriate.

CONTENTS

	page
I: The Cell as a Unit of Life	1
II: The Cell as a Physical Unit	6
III: Cell Dynamics	18
IV: The Cell as a Colloidal System	33
V: The Physical State of Protoplasm	57
VI: Cell Membranes and Intercellular Matrices	102
VII: The Nucleus	120
VIII: Mitosis	141
IX: Cell Division	189
X: The Shape of Cells	248
XI: The Growth of Cells	267
XII: Cell Variability	307
XIII: The Equilibrium between a Living Cell and Water	324
XIV: The Permeability of the Cell Surface	349
XV: The Nature of the Cell Surface	382
XVI: The Germ Cells	408
XVII: Contractile Cells	451
XVIII: Phagocytosis	490
Index to Authors	507
Index to Subjects	512

PLATES

Figs. 9 A–D	*between pp.* 40 *and* 41
Fig. 10	*facing p.* 42

available for download in colour from www.cambridge.org/9781107625662

Fig. 184	„ *p.* 470

PREFACE

The present book represents the substance of a series of lectures delivered in Cambridge for some years past. It is intended to be an introduction to that aspect of biology which is commonly associated with the experimental method. The so-called 'experimental' outlook is really a misnomer, for there is no peculiar virtue in experiment as such. The crucial point is whether or not we feel that an analytical study of living processes is more useful, for the time being, than the inductive morphology of the past century.

Until recently the central theme in zoological thought has been Evolution and whatever be the mechanism whereby one species has given rise to others we may be quite sure that the processes involved are essentially of a dynamic nature; the potentialities of the organism and the nature of its environment have operated together to produce new and varied forms of life. Any real insight into the causes of Evolution involves a knowledge of function, and any conception of the organism as a dynamic unit is incomplete without a knowledge of its evolutionary history. It is just as illogical to restrict the study of animal life to morphological observations as it is to study an aeroplane as though it were a static and inert machine. Precise knowledge of the reactions between an organism and its environment is, however, very rare in those cases where the sequence of evolutionary forms is best known, and students of Evolution must either be content with relatively vague and indefinite speculation or must set to work to analyse environmental responses for themselves. Unfortunately, few students of zoology have had that training in the physical sciences which alone can make them realise the full significance of the analytical method. It is important to remember that those whose interests lie primarily in embryology or morphology undoubtedly find that a purely biological conception of the organism carries them much farther and much more securely than any physico-chemical conception can possibly hope to do. On the other hand, any attempt to analyse the relationship of the organism to its external environment without an accurate and precise knowledge of physico-chemical laws is doomed to failure. No branch of biological knowledge can be complete until the facts of form and of function have been subjected to quantitative treatment and the concepts of Evolution will continue to remain vague and indefinite until we have established units of measurement which will enable us to orientate the facts one to another on a quantitative basis, or until we can decide how far we may accept, as fundamental, those units which have proved to be satisfactory in the inanimate world. It is commonly stated that a study of function must inevitably rest on a knowledge of form. In a restricted sense this is true, but it is a travesty of general truth unless we realise that the instruments of morphological enquiry are more subtle and more significant than the microscope and scalpel. With our present lack of knowledge it is not easy to weld into one generalised hypothesis any considerable mass of morphological and analytical data, and it is safer to put before the student the alternative points of view and leave him to frame his own philosophy after following, as far as he can, each rather bewildering trail. Within the compass of cytology the advantages of the dual viewpoint are obvious, and for this reason the present work was planned as a convenient means of introducing to students some conception of the complexity of ideas which arise from a study of the simplest of morphological units.

Cytology may rightly claim to be the frontier state in the biological commonwealth, for within its borders biologists and chemists find common ground. The chemist is interested in the activities of living matter and the biologist is impressed by the orderly analysis to which the chemist can submit his facts. A biologist who attempts to give a generalised conception of the living cell inevitably invites active criticism, for the attempt to co-ordinate the relevant facts is fast becoming a task beyond the competence of any one individual. Sir William Hardy has recently described the ideal biological college: 'It should have three floors—a ground floor for molecular physics, a first floor for biophysics, and a top floor for cell mechanics'. The present book represents an honest, if pathetic, attempt to creep downstairs. It represents the impressions which the author, as a biologist, has gained by contact with such physical facts as appear to him to bear on the structure of living matter, and which can be imparted to others like himself, whose knowledge of inanimate matter is limited.

In following the analytical outlook in cytology it is often not easy to assess the relative value of related data. Almost unconsciously we adopt from time to time views which are dangerously near to

PREFACE

incompatibility. The need for a covering hypothesis, however frail, is often acute and we tend to frame such hypotheses in terms which are elastic or obscure; this usually ends by adopting a terminology with which we, ourselves, are least familiar. As biologists we tend to put ingenuous trust in the possibilities of the colloidal state—that Alsatia (to borrow again from Sir William Hardy) 'wherein difficult states of matter find refuge from a too exacting enquiry'. So with the chemist, we note from time to time the bones of ignorance draped in the tinsel of hypothetical biology. One may note, not without some element of humour, that those who have done most to resolve biological phenomena into physical terms are the most prone to lay their tribute before the elements of purpose and design, whereas Genetics has produced the vocal advocate of co-ordinated chemistry. Speculation is an essential element in all scientific enquiry, but it is well to define and not conceal the difficulties if we are to reap the full benefit of co-operative effort. It is important to realise that no single activity of the living cell has yet been analysed in physical terms, and although the debt of biology to the physical sciences is fast mounting up, we have not yet emerged from the depths of a very profound ignorance. It is also well to remember that biological units may, for some time to come, prove more valuable in cytology than those of physics—for the cell has already provided us with units (the chromosomes and the genes) which enable us to express the facts of heredity in numerical form as surely as the properties of matter can be expressed in units of molecules or of atoms.

In the following pages the outlook is frankly analytical; at the same time an attempt is made to impress upon the student the fundamental complexity of living matter. Without stirring too vigorously the embers of a sterile controversy, it is well to ask how far it is unreasonable to look upon the cell as an aggregate of matter which is unique in the material world, and over which laws of a specific type hold sway. If once this question has been faced, it is safe to find an inspiration in the thought that a purely mechanistic conception of living matter has led to a more orderly conception of biological data than that provided by any other hypothesis. Whatever be our natural outlook we can at least find common satisfaction in the belief that the established facts will remain long after our most cherished theories have passed away. Whatever viewpoint enables us to establish new facts, that viewpoint is the one we should

PREFACE

cherish; if it stimulates others to frame hypotheses more fruitful than our own we may rest content.

The reader is reminded that this book represents the materials of a course of lectures and for this reason much has been included which has already been incorporated into the works of more competent authors. To these authors the writer has endeavoured to acknowledge his indebtedness. In particular there should be mentioned the works of E. B. Wilson, T. H. Morgan, F. R. Lillie and E. V. Cowdry. Most of the material has been collected together, in the form of notes, from year to year and it is only too probable that I have failed to make due acknowledgment of all original and secondary sources. For many of the diagrams I have to thank both authors and publishers; in particular, I have to acknowledge the generosity of the Rockefeller Institute, New York, for permission to reproduce several figures from their publications.

J. GRAY

KING'S COLLEGE
July 1930

CHAPTER ONE

The Cell as a Unit of Life

The cellular structure of living organisms has been a demonstrable fact for nearly a century and its discovery represents a definite landmark in the history of biology. Hitherto, the only unit of life had been the entire organism and the change in biological thought which is represented in the original Cell Theory[1] is comparable to the change produced in the physical sciences by the concept of the molecule as a unit of chemical structure. In both cases the establishment of a definable unit led automatically to greater precision of observation and of thought.

As a unit of structure the significance of the cell is obvious, for it enables us to define with accuracy the morphological origin and the microscopic structure of most living tissues. Throughout life the cell remains the unit of organisation just as the bricks remain the units of which a house is built. Each cell-unit has a fairly well-defined structure and although this may vary from the normal we can, without much difficulty, form a reasonable conception of a 'typical' cell; it is bounded by a cell membrane and it contains a nucleus embedded in a cytoplasmic matrix of a heterogeneous nature. The problems of classical cytology are largely concerned with intracellular structure and not, primarily, with the parts which individual cells play in the economy of the whole organism; in embryology the significance of the cell is clearly associated with the structure of the whole organism and the facts of cell lineage owe their origin to the conception of the cell as a fundamental unit. For such purposes the cell is essentially a static unit, whose structure we investigate or whose position in the organism can be rigorously determined.

As a physiological unit the rôle of the individual cell is not quite so clear. In the first place the cell is not a fundamental unit in the sense that any process of subdivision leads to a complete disor-

[1] 'The body is composed entirely of cells and their products, the cell being the unit of structure and function and the primary agent of organisation' (Sharp, 1921, p. 7).

ganisation of its properties. An enucleated fragment of an amoeba, or of a sea urchin's egg is by no means dead, but retains for a considerable time typical properties of living matter; bacteria are alive, yet their structure is by no means similar to that of a typical cell; fungi are also alive, and yet they exhibit no cell structure in the strict sense of the word. Difficulties of this kind have perplexed cytologists for many years and are still real objections to the time-honoured Cell Theory. Further difficulties arise when it is suggested that a metazoön is to be regarded as a colony of cells in which each cell is a fundamental unit of function. Nearly every experimental biologist has declined to accept this view. Driesch's (1907) analysis of the part played by individual cells during the development of echinoderm eggs shows clearly that 'all attempts to depict the organism as an aggregate of cells have proved to be wrong'. The whole organism moulds the cells for its own ends, and its form is not predetermined by the position of the cells. Morgan (1898) showed that when a planarian is cut into fragments, regeneration to form new and complete individuals is effected by a remoulding of the original cells, and that no new cells are produced. Still more striking proof of the inadequacy of the cell as a fundamental unit is derived from Lillie's (1902) work on *Chaetopterus* larvae, where differentiation and development may occur without partition into discrete cells (fig. 1). The significance of these facts was at one time hotly debated (see Dobell, 1911) and although, to-day, our interests in cell-function are concerned with other problems, we must nevertheless be careful to avoid any tacit assumption that the cell is a natural, or even legitimate, unit of life and function.

From a mechanical point of view, cellular structure—or its equivalent—is essential, if life is to assume the forms which we actually find in Nature. Beyond a certain maximum size, a fragment of fluid protoplasm would be mechanically instable unless surrounded by a rigid envelope or endowed with a rigid internal skeleton. Those types of life which exist in the form of small drops or thin films are quite stable in the absence of cell walls—but if life is to exist in three dimensions some means must be found of affording sufficient rigidity to the mass without reducing its flexibility to a very low level. Two alternatives are available: either the protoplasmic mass can be enclosed within a rigid envelope or it must acquire an internal skeleton. The possibilities of a rigid external shell are clearly limited, whereas the existence of flexible and elastic cell walls provides the

THE CELL AS A UNIT OF LIFE

mechanical properties requisite for life in an almost infinite variety of form. A reasonable analogy is provided by a system of soap bubbles: the form of a single large soap bubble is strictly limited— whereas a large number of small bubbles adhering together can acquire, on the surface of a washtub, sufficient rigidity to endow the whole aggregate with a variety of stable forms. Cellular structure is, teleologically, a mechanical necessity for life in large and varied forms. From this point of view the cell as a unit of life is both unnecessary and unsatisfactory—it is merely the unit of mechanical stability. The real unit of life must be of a protoplasmic nature

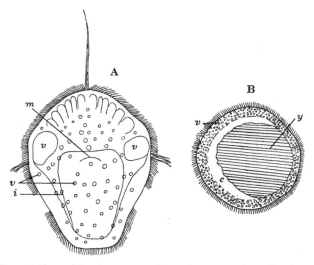

Fig. 1. Differentiation without segmentation. *A*, a normal trochophore of *Chaetopterus*, 24 hours old; *B*, a trochophore from an unfertilised egg treated with potassium chloride, 23 hours old; note the ciliation of the surface. *v*, vacuole; *y*, yolk. (Lillie, 1902.)

irrespective of whether it is subdivided to form a mechanically stable system or not: in other words, cellular structure is not in itself of primary significance. If we take a biologically heterogeneous system of growing protoplasm and proceed to a process of internal subdivision there may come a time when each phase of the system will be separated from the others by cell walls. At this stage each cell will represent a natural protoplasmic unit—but before this stage is reached the only real unit available is one which is expressed in terms independent of the process of subdivision. There can be little doubt that the most natural unit of life is the living organism, and

THE CELL AS A UNIT OF LIFE

when we find, in some cases, that its constituent cells are united by intercellular processes it is impossible to admit the validity of the cell unit without further enquiry.

On the other hand, there can be no doubt that the cell often forms a convenient physiological unit even if its individuality is not so fundamental as is sometimes supposed. Each living cell possesses, structurally, the essential machinery for independent existence;

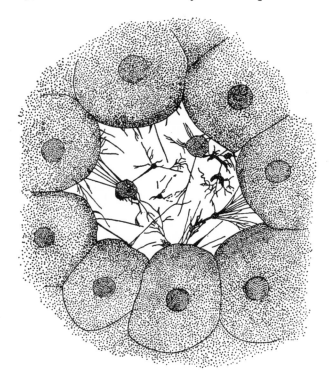

Fig. 2. Protoplasmic bridges between blastomeres of a young *Asterias* blastula (from Andrews).

each cell normally has a nucleus and is chemically and physically in equilibrium with its environment by means of its surface membranes. Much of the work on the physiology of the cell has been performed on red blood corpuscles, or egg cells, which are normally isolated units, and some caution is necessary when we attempt to apply these results to the cells which form parts of whole organisms or whole tissues.

In other ways the cell is a convenient unit, for by its use we limit the heterogeneity of the material with which we work. A single muscle fibre or a single nerve cell enables us to analyse phenomena which are masked or rendered more complex in a muscle-nerve preparation. In cell physiology we are, in fact, attempting to dissemble the machinery of an organism into its simplest component parts. When we know how each of these works in an isolated state we shall be ready to integrate the data and gain some conception of the whole organism. It is as a convenient unit of functional activity that the cell will be regarded in this book. That we cannot thereby obtain an adequate picture of all vital activities will be only too obvious, but even a cursory summary of recent physiological work will show a growing need for a precise knowledge of intracellular processes. When the organism as a whole establishes with its environment an equilibrium of profound biological significance, it does so by the machinery of its individual cells; these equilibria are seldom detectible by the statical methods of cell morphology—they are, however, detectible by experimental methods. Finally, it is from a study of the cell that we can begin to build up a conception of living matter which can be expressed to some extent in the units which have been found to be applicable to inanimate matter.

REFERENCES

ANDREWS, E. A. (1898). 'Filose activity in metazoan eggs.' *Zool. Bull.* 2, *1*.
DOBELL, C. C. (1911). 'The principles of protistology.' *Arch. f. Protistk.* 23, *270*.
DRIESCH, H. (1907). *The Science and Philosophy of the Organism.* (Aberdeen.)
LILLIE, F. R. (1902). 'Differentiation without cleavage in the egg of the annelid *Chaetopterus pergamentaceus*.' *Arch. f. Entw. Mech.* 14, *477*.
MORGAN, T. H. (1898). 'Experimental studies of the regeneration of *Planaria maculata*.' *Arch. f. Entw. Mech.* 7, *364*.
SHARP, L. W. (1921). *An Introduction to Cytology.* (New York.)

CHAPTER TWO

The Cell as a Physical Unit

A conception of the living cell as a dynamic unit can be reached from two points of view. We may either consider the relationship of cell structures to the phenomena of heredity and development, or we may regard cell structure as a molecular system, whose dynamic properties must be elucidated before we can gain any fundamental knowledge of cell function. As long as we are concerned with the mechanism of development, the units and methods which are employed are essentially biological. If, however, we are primarily interested in the non-reproductive aspects of cell activity we consciously or unconsciously attempt to use the methods and the units which are known to apply with great effectiveness to the study of particular types of inanimate systems. In other words, we attempt to define the structure and behaviour of living cells in terms of physical chemistry. The methods employed in this book frankly involve the validity of this point of view, but, before we proceed to investigate the structure and function of the cell in a mechanistic and material way, it is desirable to consider two fundamental problems which are always present, however carefully they may be disguised. Firstly, how far is it axiomatic that physical and chemical laws will apply to living organisms? Secondly, how far are populations of living units or cells really comparable to molecular systems?

When we regard the cell as a physico-chemical system we imply that the laws which govern physical and chemical systems will apply equally well to matter which is organised into the 'living' state. In one sense this may be true, but it does not follow from the nature of things, and it is a concession which can readily be abused.

Many of the fundamental laws of physical chemistry can be stated as corollaries to the kinetic theory of matter, or at any rate on this foundation they can be expressed in a concrete form. Precise and accurate as these laws appear to be in practice, they are often not statements of infallible truth (see p. 14); they are, on the other hand, expressions whose truth has such a high degree of probability that any other possibility may, for all practical purposes, be ignored. Before

THE CELL AS A PHYSICAL UNIT

we apply these laws to the living cell we ought, so far as is possible, to satisfy ourselves that there is nothing in living matter which is likely to decrease this degree of probability or which may bring other possibilities within the realm of practical observation. For the present purpose, natural laws may be divided into two categories. Firstly, there are laws which define the properties of matter in mass or in systems containing a very large number of molecules. Secondly, there are laws which define the properties of single molecules, and which are not directly concerned with the behaviour of neighbouring molecules.

The application of statistical laws to biological problems assumes that a very large number of participating units are invariably present and that we can safely ignore the behaviour of each unit as such. If these fundamental units are molecules of water or even of proteins the use of statistical laws is almost certainly justified, for even very small cells contain many millions of molecular units (see p. 9). It is, however, by no means certain that biological processes only involve reactions which depend on large numbers of individual molecules and do not involve those which depend upon comparatively small numbers of larger aggregates. Even within inanimate colloidal systems the statistical laws break down when we deal with units which are no larger than typical living cells. Using a suspension of gold particles, whose average radius was $19\,\mu\mu$, Smoluchowski (see Svedberg, 1928) found that within a volume of the suspension equal to $1064\,\mu^3$ (i.e. the approximate volume of many living cells) there was an *average* number of particles equal to 1·545. When this small volume of suspension was examined from time to time, however, the actual number of particles visible varied very considerably—sometimes no particles were visible, at other times as many as seven could be seen. This variation is, of course, due to random Brownian movement. Table I records the results of 518 successive observations. Out of a total of 518 observations, there were 168 occasions when one particle was present, 130 when two particles were present, and only one when seven were present. Instead of expressing these phenomena in terms of the relative frequency of particular states of distribution, we can express them in terms of the average time which elapses before a given state is spontaneously reformed by random movement of the particles. Thus in another case described by Svedberg, seven gold particles occurred on the average every 28 minutes when observations were made at the rate of 39 per minute. From these and other data the time interval between two corresponding states can be calculated—thus three particles are present every 2 seconds, or 17 particles once in 50,000 years. During the whole of the observations the *average* number of particles present (in $1064\,\mu^3$) over a long period of time is unchanged and is identical with that of the mass of the suspension (viz. 1·55).

A similar line of reasoning shows that if we start with a 1 per cent. difference in pressure of a gas on the two sides of a porous partition (see p. 14) and allow equilibrium to be established, we can nevertheless expect to find the original 1 per cent. difference of pressure to be spontaneously regained after an interval of $10^{10^{10}}$ years, if the volume of gas on each side of the partition is 0·5 c.c.

Table I

No. of particles present	No. of observations	Frequency of distribution	
		Obs.	Calc.
0	112	0·216	0·212
1	168	0·324	0·328
2	130	0·251	0·253
3	69	0·133	0·130
4	32	0·062	0·050
5	5	0·010	0·016
6	1	0·002	0·004
7	1	0·002	0·001

Quite clearly the degree of statistical variation which will occur in molecular systems depends largely on the magnitude of the system examined. If we examined 1 c.c. of the gold suspension the amount of variation would be relatively small: if we use a volume of $10\,\mu^3$ it will be very large indeed. As long as we are dealing with large systems, any significant variation in the distribution of matter or energy occurs for such short periods of time or on so very few occasions, that we are justified in ignoring any state other than the one which is the most probable. In other words, in large systems the statistical possibilities postulated by the kinetic theory of matter are arbitrarily eliminated by assuming the truth of the Second Law of Thermodynamics (see p. 14).

At this point the argument bears directly on biological problems. Are we justified in assuming that the essential events which occur

THE CELL AS A PHYSICAL UNIT

within a living system can be regarded as those which are the most probable from a thermodynamical point of view and which will therefore conform to statistical laws? If our living systems are large, and clearly involve a very large number of molecular or physiological units, we should expect to find that the average states of these systems conform to laws applicable to non-living systems in mass: the second law of thermodynamics would hold good, together with its corollaries the laws of mass action, diffusion, osmotic pressure and the gas laws. Accepting the absolute weight of a single atom of hydrogen as 1.63×10^{-24} gr., Cameron (1929) gives the following molecular composition for the average mammalian red blood corpuscle. (Table II.)

Table II

Cell constituent	Estimated number of molecules per cell $\times 10^6$
Water	980,000
Haemoglobin	300
Phosphatide (*lecithin*)	300
Cholesterol	230
Glucose	295
Urea	295
Glutathione	58
Creatin	26
Uric acid	7
Potassium	6,300
Magnesium	2,800
Chlorine	70

The figures suggest that even within a cell much smaller than the red blood corpuscle (e.g. *Bacillus coli*) there are many thousands of millions of molecules. As long as each molecule is an individual unit the laws of mass action can safely be applied. It is by no means impossible, however, that the essential units of living matter are each composed of a very large number of molecules and under such circumstances the number of units present may be reduced to a figure which will materially affect the validity of statistical laws. The maximum size of an ultrafilterable virus cannot be greater than $20\ \mu\mu$, and the thickness of protoplasmic films is often only a fraction of a micron. If we are to apply physico-chemical laws to the events which occur within these small systems we must be careful not to stress unduly those laws which deal only with the statistically averaged states of much larger populations. In other words, we are probably not justified in assuming the validity of the second law of thermodynamics without careful scrutiny.

The problem arises in a somewhat different form when we are concerned with larger types of living matter. A small sea urchin's egg has an approximate volume of $125,000\,\mu^3$. If the cell is dead, substances in solution distribute themselves within the egg in accordance with those laws which govern inanimate systems in mass, and we have no reason to doubt the validity of the second law of thermodynamics. As long as the cell is alive, however, the protoplasm, although apparently of a liquid nature, must be regarded as the seat of a definite structure (see Chapter V) wherein events which occur after death do not occur during life and *vice versa*. We can look upon the cell system from two points of view. Firstly, we can assume that all the essential units are present in large numbers, and that when left to themselves they will distribute themselves in accordance with the kinetic theory of matter; we may also assume that the 'living' state is definable by the average state of the whole system. If these postulates are reasonable, we should expect the cell to obey the second law of thermodynamics, and we should be justified in applying the laws of mass action to intracellular processes. One way of testing this hypothesis might be to demonstrate that the living cell is unable to absorb heat from surroundings which are at the same temperature as itself, and use this energy for maintaining vital activity. Unfortunately any direct test of this type is technically impossible owing to the intrinsic properties of living matter. As A. V. Hill (1924) remarks, 'What would a student of thermodynamics say if his machine had perforce to have food and drink and oxygen to prevent it from collapsing whilst he put it hurriedly through its Carnot cycle? How pleased would he be with an elastic body which had one set of properties at one moment, another set at another? What accuracy would he attain if his membranes began to leak as soon as he deprived them momentarily of oxygen?' Having failed to find any positive evidence to support the application of the second law to biological systems, we have to consider how far its application is useful as a working hypothesis. Here we are on firmer ground. If we assume the validity of statistical laws, the analysis of living organisms by physico-chemical methods is greatly simplified, since, under these circumstances, the conversion of energy from one form into another within the cell must conform to those limitations which are applicable to inanimate

THE CELL AS A PHYSICAL UNIT

matter. (i) The cell cannot provide a continuous supply of work by utilising the energy of bodies which are at a uniform and equal potential, and at the same time leave these bodies otherwise unchanged. For example, the living units cannot spontaneously cool themselves and give forth the resultant energy in the form of mechanical work. (ii) If the cell contains a series of machines whereby the potential energy of a molecular system is utilised for mechanical work, the maximum efficiency in every case must be the same and be independent of the nature of the machine.

$$\frac{\text{Energy transformed into work}}{\text{Energy drawn from the system}} = \frac{\theta_1 - \theta_2}{\theta_1},$$

where $\frac{\theta_1 - \theta_2}{\theta_1}$ is entirely dependent on the molecular system operating the machine. If we are prepared to make assumptions of this type we can, for some purposes, entirely ignore the physical mechanism of living processes and study the cell as a converter of energy which is strictly comparable to inanimate heat engines. Given certain measurable data concerning the cell and its environment, we can define the state of stable equilibrium without any real knowledge of the way in which this equilibrium is effected. We find in practice that many diverse phenomena can, in this way, be pieced together to form an adequate picture of biological events. For example, if we know that the surface of a red blood corpuscle is freely permeable to ions such as those of chlorine and bicarbonate, and is impermeable to potassium ions and to haemoglobin ions, then by accepting the validity of the second law we can define with accuracy the equilibrium which must exist between the corpuscle and the surrounding plasma, and these results are confirmed by experimental facts. In such cases the denial of the second law would throw the experimental facts into confusion: it seems much more satisfactory to accept the gifts which Nature provides than to remain in a rarefied atmosphere of philosophic doubt. Similarly, when we find differences of potential between one part of a cell and another, we can, by accepting the second law, proceed to analyse them by the methods which apply to inanimate systems (see Chapter XV).

The only real objection to this point of view arises from purely biological facts. The living system is thermodynamically instable, it will break down unless it is continuously supplied with free energy. In other words, no living cell is ever in strict thermodynamical equilibrium. Secondly, a fluid protoplasmic system—apparently homogeneous even under the ultramicroscope—appears to be heterogeneous from a biological point of view. The egg of a mollusc or an ascidian must have a definite structure and yet we see no obvious mechanism whereby such a structure can be maintained within the fluid cytoplasm. These and similar facts suggest some degree of caution when the cell is looked upon as a system which is always trying to reach a state of thermodynamic equilibrium. If we accept the second law, then the secret of life lies in the presence of membranes and surfaces whose mode of operation and whose efficiency are similar to those known in inanimate systems: their origin is different—for, in the cell, these machines come spontaneously into existence and set up a state of affairs which can only be produced elsewhere by the intervention of external agencies. In other words, is it thermodynamically probable that an instable system, such as a living cell, should have come spontaneously into existence? Within inanimate systems spontaneous processes invariably take place in such a manner as to reduce the free energy of the system to a minimum and to distribute the total kinetic energy at random throughout the whole population of molecules present. Spontaneous changes increase the 'entropy' of the system—they never decrease it. The only way of effecting such a decrease is by the employment of methods which involve an increase in entropy of some other external system. How far such facts are true in biological systems is unknown. For example, when the inanimate system of proteins, fat and salts found in the yolk of an egg, is converted into a living embryo, the process occurs with little or no loss of energy in the form of heat—and yet the final product has more free energy than the original system. The spontaneous origin of the living state from inanimate matter is a concept not altogether absent from the theory of biological evolution, but from a thermodynamical point of view it may be regarded as an extremely improbable phenomenon.

If we are inclined to deny the validity of the second law, we get a totally different and perhaps less satisfying picture of the living organism. Returning to Smoluchowski's suspension of gold particles, we might imagine a system which was composed of a large number of compartments each approximately $1000\,\mu^3$ in volume, and conceive of a reaction which occurred irreversibly whenever four or more particles were present within a compartment. As long as the particles were free to move from one compartment to another our imaginary reaction would always proceed to a limited extent—whereas thermodynamically it should not occur at all. We have, from this point of view, to look upon protoplasm as a series of insulated units each of which can behave independently of its neighbours but which are small enough to allow events which are intrinsically improbable in larger systems to occur fairly frequently. Such a conception of the cell may to some extent provide an analogy or suggestion concerning the 'catalytic' effect of living tissues on reactions which normally go on very slowly; at the same time it leads us into practical difficulties, since at present there is no means of verifying the postulates or employing our conclusions as working hypotheses; we have simply endowed protoplasm with the equivalent of Maxwell's demon. The means whereby chance but critical aggregations of molecules or particles are used for carrying out essential life processes would constitute the demons (or their antitheses) of the cell—and be in effect the secret of life.

The doubtful validity of the second law of thermodynamics when applied to living organisms was first pointed out by Helmholtz, and in some aspects has been discussed more recently by Guye (1925). Guye has suggested that if organisms are beyond the limitations of the second law it would be possible for a fish to 'co-ordinate to its profit' the motion due to the thermal agitation of the water in which it swims in order to transform it into the mechanical energy required for its own motion, even if the water and fish are at the same temperature. The proof that such a phenomenon does not occur is, however, no proof of the validity of the second law. If a fish were able to take energy from water at a *higher* temperature than itself and use this energy for movement, we should be almost equally surprised, although the process would be in accordance with the second law. There is overwhelming evidence to show that muscle fibres are not heat engines and therefore they are unable to draw directly on any source of thermal energy irrespective of its relative temperature. Before we can test the validity of the second law by a study of direct transference of thermal energy we must show that the organism functions as a heat engine: so far such evidence is not available.

It is doubtful how far the arguments presented in this discussion can usefully be summarised. If we tacitly accept the biological application of the statistical laws of physical chemistry we may find that we have closed the door which leads to a really fundamental knowledge of the living state, and we must be prepared to regard the development of a hen's egg as a process of increasing random distribution of energy rather than as the orderly formation of an efficient piece of machinery. On the other hand if we deny the validity of statistical laws we must regard living matter as something quite different from the molecular systems of which we have so much reliable knowledge, and we are liable to flounder in a realm of unrestricted speculation wherein our only source of comfort lies in the conviction that the average organism (so beautifully adapted to its environment and built up without any measurable dissipation of energy from gross inanimate materials) is not a convincing example of the law of progression towards a maximum probability of state. As long as speculation plays its part in human thought there will always be—as Hill (1924) aptly remarks—'some to affirm, as an act of faith and on insufficient evidence, that living cells are nothing but ordinary electrons and atoms and deny the existence of that fundamental organisation which is called life by wiser and more common-place folk: others, equally perverse, to attribute it all to "Nature" and to the inherent adaptability of the cell'.

APPENDIX

The Second Law of Thermodynamics

It is not easy to express in simple, but accurate, terms the statistical principles which underlie many of the familiar laws of physical chemistry. To those not familiar with the elementary principles of thermodynamics, the following introduction may be of use. Most, if not all, of the illustrations are taken from Hinshelwood (1926) or from Guye (1925).

Let us start with a small vessel enclosing 1 c.c. of a gas and divide it into two equal compartments by a partition which is freely permeable to the gas. If the pressures of the gas on the two sides of the partition are not equal, it follows from the kinetic theory of matter that molecules of the gas will move from one side of the partition to the other until the pressures on the two sides of the partition are equal; having reached this state, the distribution of the gas will not change. Such a phenomenon can readily be observed in practice, and the laws which express the process of equalisation of pressure can be rigidly defined for all practical purposes. It is necessary to remember, however, that when the two pressures are apparently equal there is still a movement of molecules from one side of the partition to the other: when we measure the

THE CELL AS A PHYSICAL UNIT 15

pressures and find them equal we know that, during the period of time required for any single observation, the average number of molecules passing in one direction is equal to the number passing in the other direction. It does not follow that at any instant of time there is a definite number of molecules passing the boundary in one direction and an exactly equal number passing the boundary in the opposite direction. All we can state with safety is that over a measurable period of time the average numbers of movements in the two directions are equal. At any instant of time each molecule has an equal chance of movement in either direction and, since the direction of movement is at random within the whole population, there will be times when there is an excess of molecules on one side, and other times when the excess is on the other side. Thus, if we start with exact equality of numbers on the two sides and one molecule then moves across the partition it is probable, but by no means certain, that another molecule will move in a contrary direction. On the other hand, if two molecules move in one direction the probability of there being a molecule moving in the other direction is increased. Obviously the behaviour of the gas as a whole is no indication of the behaviour of any single molecule. When we apply the gas laws to living organisms we imply that we are only concerned with very large numbers of molecular or other units and can therefore safely ignore the behaviour of each individual unit.

That statistical laws break down when the number of units involved is small can be seen from the following example which is a modification of that given by Guye (1925). If we put into a bag twenty balls, ten of which are white and ten of which are red, and select at random ten balls, there are ten different combinations possible in respect to colour. These are shown in column 1, Table III. Each combination is, however, not equally probable. It is more probable that the ten selected balls will include balls of both colours than that they should all be red or all be white. The number of ways of making any individual type of drawing can be calculated by the formula

$$\left[\frac{10!}{n!\,(10-n)!} \right]^2$$

and we can express the probability of any particular drawing by the relative frequency of the ways in which it can appear to the total number of selections possible. In this way we get the figures given in column 3, Table III. From such data it is obvious that a uniformly red sample (viz. ten red balls) will only be drawn once in 184,000 drawings—and that if we were to use one hundred balls (fifty white and fifty red) the probability of drawing out fifty red balls would be too small to play any part in an actual series of observations, although of course it is always a possibility. If, however, we were to use ten balls (five red and five white), the probability of selecting five red balls would be $\frac{1}{252}$, and if we use four balls the probability of drawing a uniformly red sample rises to one drawing in every six. It is clear that an event which is intrinsically improbable in large systems becomes more and more probable when we use smaller and smaller systems. When the number of units involved is very large, the chance of a particular grouping which is remote from the 'average' distribution of the units becomes so small that we can neglect it for practical purposes; this conception when applied to the distribution

Table III

1		2	3
Red	White	No. of possible combinations	Probability
10	0	1	0·0000054
9	1	100	0·00054
8	2	2,025	0·01097
7	3	14,400	0·0780
6	4	44,100	0·2388
5	5	63,504	0·3439
4	6	44,100	0·2388
3	7	14,400	0·0780
2	8	2,025	0·01097
1	9	100	0·00054
0	10	1	0·0000054
		184,756	1·0000

of energy or molecules of a gas is known as the Second Law of Thermodynamics. It will be noticed that the distribution of particles within a small volume of a colloidal suspension (see p. 7) is a phenomenon lying outside the scope of the Second Law.

The full significance of this law is not easily appreciated, but at the moment it is sufficient to realise that the Second Law applies only to systems containing large numbers of molecules. When we say that heat is not transferred from a cold body to one warmer than itself, or that a body does not spontaneously cool itself and perform external work, we are, in effect, saying that these events occur so seldom and to such a small extent as to play no significant rôle in natural events. We have seen, however, that events which are extremely rare or improbable in everyday life may become comparatively frequent within microscopic systems.

The conception of *entropy* may be illustrated as follows. We may imagine a bag of soot and a bag of flour. If we empty them carefully into a barrel the two powders are segregated into two heaps in contact with each other. If we begin to shake the barrel the particles of soot and flour begin to be mixed together—and the longer we shake the more the powders are mixed up: short of picking out each individual particle or using some method which discriminates between a grain of soot and a grain of flour, we cannot separate them. If we replace the flour and soot by molecules, half of which have a higher degree of thermal agitation than the other, then the longer they are left in contact the more uniformly will this energy be distributed throughout the whole mass. The degree of complete distribution of energy at random throughout the system is known as the entropy of the system. Since the most probable state is that condition where the energy is distributed completely at random, the most probable state is that of maximum entropy. Once this condition is reached in a macroscopic system of molecules it is highly improbable but not impossible that the process can reverse itself spontaneously, just as it is possible but improbable that by shaking the flour and the soot long enough the two types will again be completely segregated from each other. It will be noted that the law of progression towards maximum entropy breaks down in small systems, since the entropy of the system is a function of the probability of state.

REFERENCES

CAMERON, A. T. (1929). 'A note on the number of molecules and ions present in a single cell.' *Trans. Roy. Soc. Can.* 23, *151.*
GUYE, C. E. (1925). *Physico-chemical Evolution.* (London.)
HILL, A. V. (1924). 'Thermodynamics in physiology.' *Mem. and Proc. Manchester Phil. Soc.* 68, *13.*
HINSHELWOOD, C. N. (1926). *Thermodynamics for Students of Chemistry.* (London.)
SVEDBERG, T. (1928). *Colloid Chemistry.* (New York.)

CHAPTER THREE
Cell Dynamics

The most characteristic feature of a living cell is its power of transforming energy. In some cases this transformation is obvious; an amoeba, a ciliated cell, or a muscle fibre during their whole lives are transforming chemical potential energy into kinetic energy of movement. As soon as the cells die this transformation ceases. In most cases, however, the dynamic properties of the cell only reveal themselves in a more incomprehensible way. Every living cell, without exception, is transforming potential energy, though this is not evolved in the form of useful work but appears as heat. According to Shearer (1922) a newly fertilised egg of *Echinus miliaris* at 14·5° C. evolves $0·067 \times 10^{-6}$ gr. calories per hour. The meaning of this fundamental process is very obscure, but within certain limits we can say that the intensity of heat production over a reasonable period of time is a measure of the 'vitality' of the cell. It is strange that a process so universal should only be put to obvious practical use in the latest products of evolution; the heat production of the muscle cells of birds and mammals maintains, of course, the body temperature at a high constant level—but such an economic use of the heat produced by tissue cells is the exception rather than the rule within animate systems.

The source from which animal heat is derived is not, as a rule, difficult to locate. It is nearly always the result of oxidation of an intracellular compound; the oxygen is derived from the environment and the product of oxidation (CO_2) is liberated into the environment. This process of respiration is of profound importance for two reasons. Firstly, there is clear proof in some cases that the energy which is originally contained in the molecules of carbohydrate or fat within the cell ultimately appears in the form of useful work when the cell carries out active movements, and is also the source of the heat which is evolved when the cell is at rest. Secondly, it is only in living cells that the process of respiration occurs. If the cell is dead the carbohydrates cease to be oxidised by atmospheric oxygen in the normal manner. There are therefore two distinct problems.

CELL DYNAMICS 19

(1) How does the cell manage to oxidise carbohydrates at a low temperature when such a process does not normally occur *in vitro*? (2) When the cell has oxidised a carbohydrate to CO_2 and water, and has thereby obtained a supply of free energy, how does it use this energy before ultimately dissipating it in the form of heat? In other words, how does a cell obtain a supply of free energy in the first instance, and having got it, how does it make use of it? The machinery whereby the living cell is able to oxidise its respirable materials is bound up in the so-called oxidising enzymes (see Dixon, 1929). In the presence of these enzymes the oxidisable substances in the cell and the free oxygen of the environment cease to represent a stable inactive system and become an active system in which the oxidisable substances are rapidly converted into carbon dioxide, water and other products. Such a change might be effected in three ways. (1) The oxidisable substances, under the influence of specific enzymes, might become 'activated' and when in this state react with atmospheric oxygen. For example, when xanthine is activated by xanthine dehydrase, it is rapidly oxidised by molecular oxygen— and the process is unaffected by the presence of cyanides. (2) The oxidisable substance, although unoxidised by molecular oxygen, may be oxidised by 'activated' oxygen—although the molecules of the substrate might remain in an inactivated state. Such reactions appear to occur when the system is sensitive to the presence of cyanides, for these radicles affect the process of activation of molecular oxygen. (3) Both oxidisable substance and atmospheric oxygen might require to be activated if oxidation is to occur. Thus lactic acid when activated by dehydrases will not react with atmospheric oxygen: if however methylene blue is present, rapid oxidation occurs, since the dye acts as a carrier of activated oxygen to the activated substrate. One theory of cell respiration postulates the presence of more than one of the above types of reaction (Dixon, 1929), and implies that a significant number of different enzymes are constantly in action. Another conception has, however, been advanced by Warburg (1928). This author maintains that there is only one respiratory enzyme and that it invariably operates by activating free oxygen. The essential reaction in respiration is, from this point of view, a reaction between molecular oxygen and an organic compound of iron—which compound is, in fact, the essential 'respiratory ferment'. The reaction can be expressed by the formula

$$X.Fe + O_2 \rightleftharpoons X.FeO_2, \quad X.FeO_2 + 2A = X.Fe + 2AO,$$

where $X.Fe$ represents the respiratory ferment and A represents the oxidisable substrate. The evidence in favour of and against this view has been summarised by Dixon (1929). There can be no doubt that, in respect to inhibitory agents such as cyanides and narcotics, living systems bear a close resemblance to those oxidations *in vitro* which are known to be catalysed by iron compounds; at the same time it seems very doubtful whether all intracellular oxidations can really be attributed to reactions of this type. The respiration of many cells is not, as Warburg claimed, totally inhibited by cyanides. Even in sea urchin eggs—the material used by Warburg—the maximum inhibition is about 80 per cent. (Loeb and Wasteneys, 1911, 1913), and in other tissues may be considerably less (Gray, 1924; Dixon and Elliott, 1929). In one instance (*Chlorella*) the respiratory process is said to be completely insensitive to cyanide (Emerson, 1927).

A more adequate picture of intracellular oxidase systems is available from the work of Keilin (1925, 1926, 1929). Keilin showed that nearly all cells contain an oxidisable pigment, *cytochrome*, which, in the reduced form, exhibits a definite series of spectral bands. If a suspension of cells in an oxygenated medium is examined spectroscopically no spectral bands of cytochrome are visible; if however the oxygen be removed, four distinct bands of reduced cytochrome become visible, which again disappear on admitting oxygen. We may conclude, therefore, that oxidised cytochrome is rapidly reduced by the cell and that reduced cytochrome is rapidly oxidised by atmospheric oxygen. In other words cytochrome can act as a carrier of oxygen.

Keilin has shown that cell narcotics, e.g. urethane, can prevent the reduction of cytochrome but they do not inhibit its oxidation. For instance, if ethyl urethane is added to a yeast suspension, the cytochrome is immediately oxidised and is not reduced on the removal of oxygen. On the other hand, a trace of cyanide will immediately reduce oxidised cytochrome, and will prevent its oxidation by free oxygen. There are therefore two distinct processes at work—those concerned with the oxidation of reduced cytochrome and those concerned with the reduction of oxidised cytochrome; the cytochrome is acting normally as an intermediate carrier of oxygen between these two processes. Keilin (1929) has recently shown that the systems responsible for the reduction of oxidised cytochrome are the dehydrase system of enzymes, whereas the catalyst responsible for the oxidation of reduced cytochrome is none other than the indophenol oxidase[1] system so universal in animal tissues. All factors which influence the activity of dehydrase enzymes inhibit the reduction of cytochrome and all factors inhibiting indophenol oxidase inhibit the oxidation of cytochrome. It looks as though the 'respiratory ferment' of Warburg is really the indophenol oxidase system, and that it is only part of the whole respiratory mechanism.

Although many parts of the respiratory mechanism are but imperfectly understood, the progress of recent years has been extremely rapid and we may reasonably hope that before long we shall know the precise series of events whereby the molecules of the respirable substrates within the cell are made to yield a supply of free energy.

Bio-electric properties of the living cell

It has been known for many years that the activities of a living cell are associated with the production of electricity. If a galvanometer circuit be interposed between two regions on a cell's surface, one of which is more active than the other, a current of electricity will flow towards the more excited region. Bio-electric currents of this type are well known in muscle fibres, nerve cells, secretory cells, and even in egg cells; the detection of such currents is important, for they not only indicate that there is a difference in the electrical potential between an inactive and an active cell surface, but also that the cell is capable of generating electricity like a galvanic cell. Unlike the potential difference between the cell surface and its external medium (see p. 47) the bio-electric potential is a specific characteristic of life, for it disappears whenever the cell dies. In one way the production of an injury current or an action current is just as vital a phenomenon as respiration—and although the total energy involved is probably very small (Hill, 1921) the distribution of this energy is a fundamental factor in the organisation and behaviour of the cell. The practical significance of action currents is more clearly illustrated in muscle and nerve than in those cells with which cytologists are usually concerned. There are, however, certain points of general interest which may here be considered.

In the absence of metallic electrodes we look for the origin of the bio-electric current in localised differences in the concentration of

[1] A mixture of α-naphthol and di-methyl-para-phenylene-diamine will in the presence of this enzyme yield a blue coloration due to indophenol.

either positively or negatively charged electrolytic ions and these differences must exist within a system comparable to those galvanic cells which contain no metallic phases (see p. 393). For the moment, we may confine our attention to the so-called phase boundary systems originally investigated by Beutner (1912 c), since they lead to a definite conception of protoplasm as a generator of electricity. Beutner showed that if two unlike solutions of potassium chloride are separated by a layer of salicylic aldehyde containing salicylic acid the system is capable of acting as an electrolytic cell, the E.M.F. of which can be expressed quantitatively on the assumption that potassium ions can diffuse through the aldehyde, whereas those of chlorine are unable to do so (see p. 394). Similarly if two unequal solutions of potassium chloride are separated by a layer of orthotoluidin the mobile ions may be regarded as chlorine and the potassium can be assumed to be unable to move across the layer of

Table IV

Conc. of liquid outside the apple	Observed E.M.F.	Observed difference for $c_2/c_1 = 5$	Difference calculated by formula
$M/50$ KCl	+ 68 mv.		
$M/250$ KCl	+ 103 mv.	35	40
$M/1250$ KCl	+ 137 mv.	34	40
$M/6250$ KCl	+ 169 mv.	32	40

oil. In both systems the E.M.F.'s are equal in magnitude but opposite in sign; hence, the orientation of the potential is an indication of which type of ion is capable of moving from one phase to the other. The detailed analysis of such systems is best deferred to a later stage (see p. 393 seq.), and for the present we may confine attention to the application of such phase-boundary potentials to biological systems. Loeb and Beutner (1911) compared the E.M.F.'s given by phase boundary chains with those derived from living tissues. A cup was cut into a healthy apple and into this was placed $N/10$ KCl; the outside of the apple was in contact with a solution of potassium chloride of different concentration. The E.M.F. between the two solutions of salt was then determined (Table IV).

If the tissues of the apple were strictly comparable to a system of salicylic aldehyde then the observed E.M.F. should be given by the expression $\dfrac{RT}{F} \log_e \dfrac{c_2}{c_1}$, where c_2 and c_1 are the concentrations of potassium in contact with the outside and the inside surfaces of the apple. The observed differences (Table IV) in potential are, in fact, slightly lower than would be expected if the apple were acting as a phase or electrode which could transmit potassium ions but not chlorine ions. The sign and magnitude of the potential was found to be independent of the concentration of anions, but sensitive to changes in the concentration of cations. These results are essentially similar to those obtained by MacDonald (1900) using nerve fibres, and suggest that the cell surface acts in an electrometric chain as though it were an electrode reversible in respect to cations but not to anions. In point of fact, more than one explanation of the facts is possible (see p. 396), but Loeb and Beutner interpreted their results by postulating at the cell surface a non-aqueous layer containing an acidic radicle which enabled it to act as a carrier for cations. They also showed (1914) that artificial systems of guaiacol containing lecithin or extracts of tissue would give electromotive forces comparable to those developed by a living apple. In both cases the E.M.F. appeared to be more sensitive to potassium ions than to hydrogen ions. Loeb and Beutner naturally concluded that protoplasm is essentially a fluid which is non-miscible with water, and which is capable of acting as a reversible electrode just as is the case of salicylic aldehyde containing salicylic acid. There can be little doubt that Loeb and Beutner were wrong in attributing the electromotive phenomena of the apple to the living cells, for Michaelis and Fujita (1925) have since shown that it is more likely attributable to an inanimate cuticle. At the same time Loeb and Beutner's results remain of very considerable biological importance, for they suggest that the observed P.D. between the outside and the inside of the living cell may be due to the same cause as that characteristic of the interface between two immiscible fluids. In its simplest form, however, this conception is untenable. If a film of protoplasm is strictly equivalent to Beutner's oil phase (salicylic aldehyde and acid), there should be no observable E.M.F. when the same solution is in contact with both the inside and outside surfaces, since the only source of the potential in such a system lies in a difference in concentration of ions on the two sides. As long as the same concentration of salt is present on the two sides there can be no production of energy. This point was investigated by Osterhout, Damon and Jacques (1927), who used the large cells of the alga *Valonia*. On setting up the system

Cell sap | Protoplasm | Cell sap

a persistent E.M.F. of approximately 14·5 millivolts was observed across the protoplasm. Essentially similar results were obtained with *Nitella* cells (Osterhout and Harris, 1927 a), where the potential difference between a dead surface and a living surface was 15·9 millivolts when both were bathed in sap. Both in *Valonia* and *Nitella* the inner side of the protoplasmic surface was found to be positive to the outer surface. It is important to notice that the protoplasmic film of *Valonia* is only a few microns in thickness; if the protoplasm were comparatively thick it might well take many hours before the original outer surface came into equilibrium with the foreign solution (i.e. cell sap), and during this time the heterogeneous distribution of ions within the protoplasm itself might provide sufficient energy to give a measurable deflection on a delicate electrometer.

The full significance of these facts will be considered elsewhere, but it is quite clear that a layer of protoplasm must itself contain a supply of electrical energy and is not, as Loeb and Beutner assumed, a passive electrode; within the protoplasm itself there must be an unequal distribution of effective ions which is independent of an inequality between the solution bathing the two sides of the film. Osterhout concluded tentatively that plant protoplasm consists of three distinct layers—two outer layers X and Y of a non-aqueous nature, and a central aqueous layer W.

External medium | X | W | Y | sap.

Since a persistent E.M.F. is obtained when the external medium is replaced by cell sap, we are forced to the conclusion that there exists within the protoplasm itself such a mechanism as will maintain an intra-protoplasmic E.M.F. independent of the solutions with which it is in contact. In other words, the protoplasm is a self-charging accumulator capable of doing work; at the same time it is surprising to find such a mechanism within a system only a few microns thick.

As already mentioned, the typical bio-electric current between an injured and an uninjured surface is so orientated that the injured surface is electro-negative. Within recent years, however, Osterhout and his associates have shown that the orientation of the potential depends on the precise way in which the electrometric systems are set up. Osterhout and Harris (1927 b) showed that the injury currents of single cells of *Nitella* may be positive or negative according to the concentration of potassium with which the cell surfaces are in contact. If two drops (A and C, fig. 3) of $0.05M$ KCl (= artificial sap) are in contact with a cell and one drop (C) be saturated with

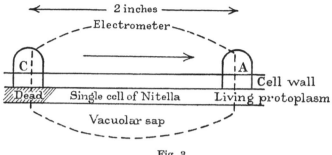

Fig. 3.

chloroform, the protoplasm at C is killed and the potential difference between A and C is entirely due to the effect of the living protoplasm at A; this yields an E.M.F. of about 20 millivolts. If chloroform is then applied to A, the potential is abolished and A becomes relatively more electro-positive to C (see fig. 4).

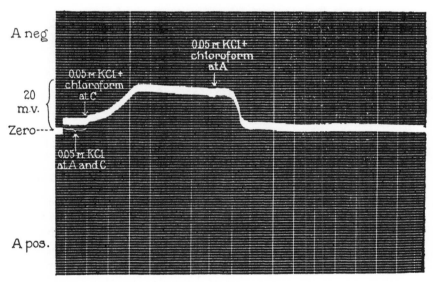

Fig. 4. Photographic record of the bio-electric changes observed in a *Nitella* cell set up as in fig. 3. When $0.05M$ KCl, saturated with chloroform, is applied at C, the point A becomes electro-negative. On applying a similar solution to A this electro-negative variation is lost. The vertical lines represent 5-second intervals.

(From Osterhout and Harris.)

This may be regarded as a typical negative current of injury. If, however, the same experiment be repeated using potassium chloride of lower concentration (0·001M KCl) the observed changes in potential are of opposite sign (see fig. 5).

Osterhout shows that these results might be expected if protoplasm contains at least two phase boundaries, each of which is capable of generating electricity. The two boundaries are probably not equally sensitive to toxic reagents, and if the outer phase boundary is affected before the inner boundary, a positive current

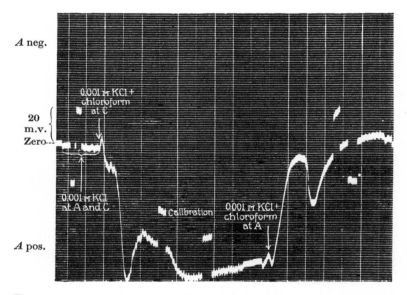

Fig. 5. Two points A and C in a *Nitella* cell are initially in contact with 0·001M KCl. Application of chloroform to C renders A electro-negative; when chloroform is subsequently applied to A the latter eventually becomes electro-neutral. Contrast this figure with fig. 4. (From Osterhout and Harris.)

of injury will result, whereas, under the reverse conditions, the injury current will be negative. This conception is illustrated in fig. 6, where the direction and length of the arrows indicates the direction and magnitude of the E.M.F. due to each phase boundary. Although hypothetical in origin, Osterhout's scheme seems applicable to a very wide range of experimental results (Osterhout and Harris, 1928 *a*, *b*).

Although it is clear that the nature of the bio-electric current can be influenced by electrolytic disturbances in the media bathing the

CELL DYNAMICS

outside and inside of the cell surface, it is by no means obvious that the normal P.D. across a protoplasmic membrane is due to such causes. As already pointed out, an E.M.F. persists when the same solution is on both sides. If we look on protoplasm as an accumulator it is possible to imagine that electricity is generated by a

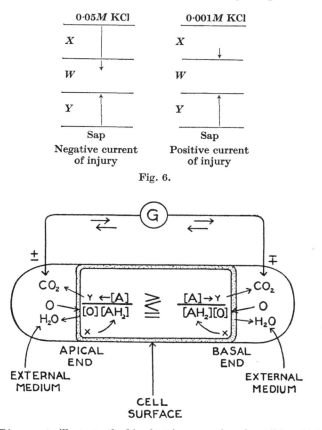

Fig. 7. Diagram to illustrate the bio-electric properties of a cell in which the intensity of respiration is not equal at two selected points (after Lund).

process of intraprotoplasmic oxidation—the energy of which is transformed into electricity. This view is supported by Straub (1929) and by Lund (1928). There is some evidence (Mond, 1927; Lund, 1928) which indicates that bio-electric currents may be sensitive to changes in the oxidative activities of the cell, but further discussion is best postponed (see p. 388).

It should clearly be understood that the correlation between bio-electric currents and phase boundary potentials is by no means certain—there are at least two alternative hypotheses (see p. 397). Whatever be the origin of these currents, however, there can be no doubt concerning their importance in cell physiology. They not only enable us to follow with accuracy the processes of protoplasmic conduction, but also enable us to realise the delicacy of the equilibria which are characteristic of life. In this respect the rhythmical waves described by Osterhout are of peculiar interest, for they illustrate the ability of the normal cell surface to change its essential properties in a reversible manner and to hand on these cycles of change to neighbouring units.

Table V (from Lund and Logan)

Current intensity in milliamps. (i)	Duration in secs. (t)	$i\sqrt{t}$
4·20	166·0 (?)	54·0 (?)
4·90	64·0	39·2
5·62	38·5	34·8
7·12	30·0	38·9
8·54	20·4	37·5
10·00	16·4	40·4
11·46	13·5	42·0
12·96	10·7	42·3
14·60	9·1	43·9

The bio-electric properties of protoplasm are intimately concerned with the response made by a cell to external stimulation. This response is often unattended by any visible change in the structure of the cell cytoplasm, but an interesting exception has been described by Lund and Logan (1924) in *Noctiluca*. The protoplasm of this protozoon consists of an emulsion in which the discontinuous phase consists of a number of vacuoles each of which is separated from its neighbours by a film of granular protoplasm. On mechanical or electrical stimulation these bounding films become instable and, since the vacuoles contain a relatively high concentration of acid, stimulation results in the liberation of free hydrogen ions into the medium. The process of electrical stimulation is curiously analogous to the stimulation of striated muscle for the product of the intensity of a stimulus and the square root of the time during which it must be applied is constant (see Table V).

Metabolism and cell structure

The dynamic activities of living cells are not restricted to processes involving the redistribution of energy; they are also detectible in other ways. Probably all cells are constantly breaking down proteins and similar bodies into simpler products, and building these products up into more complex forms. So far as can be seen, nitrogen metabolism within the cell is largely associated with the maintenance of the living machinery—and represents the removal and replacement of those parts which have suffered an irreversible process of wear and tear. At the same time, it is by no means certain that all the dynamic activities of the cell play an essential rôle in the activity of life, for in not a few cases the intensity of chemical change may vary with the concentration of particular substances in the environment without any indications that this increase is of value to the cell. According to Stephenson (1930) the greater proportion of the energy liberated by micro-organisms is not the result of essential metabolism, but is the result of unrestricted enzyme activity on a favourable medium. Stephenson puts forward the interesting suggestion that the evolution of multicellular organisms represents 'a process whereby the energy liberated in chemical activity, which in a microbe runs to waste, is so organised and disciplined that it is liberated when and where it can subserve functions such as muscular work or maintenance of temperature' (p. 21).

An intimate knowledge of cell metabolism is only of indirect interest in the study of cell structure although it raises one important fact. Every cell is endowed with the powers of effecting a well-defined series of chemical reactions and the only types of machinery which have so far been isolated from living cells are the enzymes. At the same time the microscopic structure of the cell is also known: we can recognise the nucleus, mitochondria and other morphological structures. It is a curious and significant fact that (with the outstanding exception of the chromosomes) particular functional activities of the cell have, so far, not been clearly associated with particular morphological structures. Even in those cells (e.g. muscle fibres or ciliated cells) which exhibit specific and well-defined morphological structures the function of these structures is still unknown. It looks as though the fundamental structure of the cell is of an order quite remote from that revealed by microscopic methods.

When we try to form some conception of the way in which the cell utilises the energy derived from respiratory activity or of the source from which the bio-electric current is drawn we soon find ourselves in a realm of almost unrestricted speculation. We know that the *status quo* of the living cell can only be maintained at the expense of those substances which are capable of providing free energy and that at no time is the cell in a state of static equilibrium. Apart from the production of heat, the redistribution of energy effected by respiratory oxidations may exhibit itself in other ways—notably by the production of the electricity which accompanies most if not all vital activities. We can say that in some respects the cell behaves as though it were a self-charging accumulator which is constantly maintaining a leaking condenser. Unfortunately such comparable mechanisms as occur in the inanimate world are the result of human ingenuity—they do not spontaneously come into existence nor are they composed of proteins and fats. Analogies are dangerous concepts, but it looks as though the only real conception of a living cell as a dynamic unit is provided by comparison with suitable types of inanimate machines. Consider a motor boat with a hole in the hull below the water line. The engines can be used for various purposes, but if the boat is to remain stable they must continuously pump water overboard; in addition to this the boat can, if desired, be driven through the water. Stop the engines and the boat comes into equilibrium with its external environment and sinks. So with the cell, the living mechanism is providing energy partly to maintain the *status quo* of the whole system, and partly for the performance of the useful tasks of contraction, secretion, and other processes. However kindly disposed one may be to such analogies, two difficulties arise. Firstly, why should the living cell have a hole through the hull? All one can say is that energy must be applied at certain centres or interfaces in such a way as will enable the system to remain stable and to do work. In so doing a system largely composed of proteins and water acquires the powers so far restricted to inanimate machines. The isolation of enzymes is equivalent to the isolation of parts of the working machine, and sooner or later we may be in a position to put some of these parts together. We seem driven to the conclusion that a real understanding of cell structure will rest on a knowledge of how the various parts of the cells are orientated in respect to each other, rather than on a study of the chemical constitution of these parts. One has the suspicion that the essential structure of living matter has completely eluded us and awaits, for its revelation, a new method of studying the orientation of protein molecules one to another. At the moment, however, the position is dark, and placidly to regard the cell as though its mysteries were all but revealed by the magic wand of chemistry is a sorry tribute to biological facts. These remarks are not made in any spirit of adverse criticism. If we are engaged in dissembling an aeroplane into its component parts we cannot expect to rebuild them into an effective machine until every single constituent is known to us both in form and function. It may be—to use Morgan's (1929, p. 77) phrase—that, 'When if ever the whole story can be told the problem of the adaptation of the organism to its environment and the co-ordination of its parts may appear to be a self-contained progressive elaboration of chemical compounds....'. At the same time it is unwise to minimise the difficulties. A biological problem disguised by the sparkling terminology of the chemist is too often a pathetic and rather disreputable object.

REFERENCES

BEUTNER, R. (1912 a). 'Unterscheidung kolloidaler und osmotischer Schwellung beim Muskel.' *Biochem. Zeit.* 39, *280*.
—— (1912 b). 'Die physikalische Natur bioelektrischer Potential-differenzen.' *Biochem. Zeit.* 47, *73*.
—— (1912 c). 'Potential differences at the junction of immiscible phases.' *Trans. Amer. Electro-chem. Soc.* 21, *219*.
—— (1913 a). 'Einige weitere Versuche betreffend osmotische und kolloidale Quellung des Muskels.' *Biochem. Zeit.* 48, *217*.
—— (1913 b). 'Water-immiscible organic liquids as central conductors in galvanic cells.' *Trans. Amer. Electro-chem. Soc.* 23, *401*.
—— (1913 c). 'New galvanic phenomena.' *Amer. Journ. Physiol.* 31, *343*.
—— (1913 d). 'Neue Erscheinungen der Elektrizitäterregung, welche einige bioelektrische Phänomene erklären.' *Zeit. f. Elecktrochem.* 19, *319*; *467*.
—— (1913 e). 'New electric properties of a semipermeable membrane of copper ferrocyanide.' *Journ. Phys. Chem.* 17, *344*.
—— (1920). *Die Entstehung elektrischer Ströme in lebenden Geweben.* (Stuttgart.)
DIXON, M. (1929). 'Oxidation mechanisms in animal tissues.' *Biol. Rev.* 4, *352*.
DIXON, M. and ELLIOTT, K. A. C. (1929). 'The effect of cyanide on the respiration of animal tissues.' *Biochem. Journ.* 23, *812*.
EMERSON, R. (1927). 'The effect of certain respiratory inhibitors on the respiration of *Chlorella*.' *Journ. Gen. Physiol.* 10, *469*.
GRAY, J. (1924). 'The mechanism of ciliary movement. IV. The relation of ciliary activity to oxygen consumption.' *Proc. Roy. Soc.* B, 96, *95*.
HILL, A. V. (1921). 'The energy involved in the electric change in muscle and nerve.' *Proc. Roy. Soc.* B, 92, *178*.

KEILIN, D. (1925). 'On cytochrome, a respiratory pigment, common to animals, yeast, and higher plants.' *Proc. Roy. Soc.* B, 98, *312*.
—— (1926). 'A comparative study of turacin and haematin, and its bearing on cytochrome.' *Proc. Roy. Soc.* B, 110, *129*.
—— (1929). 'Cytochrome and respiratory enzymes.' *Proc. Roy. Soc.* B, 104, *206*.
LOEB, J. and BEUTNER, R. (1911). 'On the nature and seat of the E.M.F.'s manifested by living organs.' *Science*, 34, *884*.
—— (1912). 'Über die Potential-differenzen an der unversehrten und verletzten Oberfläche pflanzlicher und tierischer Organe.' *Biochem. Zeit.* 41, *1*.
—— (1914). 'Über die Bedeutung der Lipoide für die Entstehung von Potential-unterschieden an der Oberfläche tierischer Organe.' *Biochem. Zeit.* 59, *195*.
LOEB, J. and WASTENEYS, H. (1911). 'Sind die Oxydationsvorgänge die unabhängige Variable in den Lebenserscheinungen?' *Biochem. Zeit.* 36, *345*.
—— (1913). 'Narkose und Sauerstoffverbrauch.' *Biochem. Zeit.* 56, *295*.
LUND, E. J. (1928 a). 'Relation between continuous bio-electric currents and cell-respiration. II.' *Journ. Exp. Zool.* 51, *265*.
—— (1928 b). 'The quantitative relation between E_p and cell oxidation as shewn by the effects of cyanide and oxygen.' *Journ. Exp. Zool.* 51, *327*.
LUND, E. J. and LOGAN, G. A. (1924). 'The relation of the stability of protoplasmic films in *Noctiluca* to the duration and intensity of an applied electric potential.' *Journ. Gen. Physiol.* 7, *461*.
MACDONALD, J. S. (1900). 'The demarcation current of mammalian nerve.' *Proc. Roy. Soc.* B, 67, *310*.
MICHAELIS, L. and FUJITA, A. (1925). 'Untersuchungen über elektrische Erscheinungen und Ionendurchlässigkeit von Membranen. II. Die Permeabilität der Apfelschale.' *Biochem. Zeit.* 158, *28*.
MOND, R. (1927). 'Über die elektromotorischen Kräfte der Magenschleimhaut vom Frosch.' *Pflüger's Archiv*, 215, *468*.
MORGAN, T. H. (1929). *What is Darwinism?* (New York.)
OSTERHOUT, W. J. V., DAMON, E. B. and JACQUES, A. G. (1927). 'Dissimilarity of inner and outer protoplasmic surfaces in *Valonia*.' *Journ. Gen. Physiol.* 11, *193*.
OSTERHOUT, W. J. V. and HARRIS, E. S. (1927 a). 'Protoplasmic asymmetry in *Nitella* as shown by bioelectric measurements.' *Journ. Gen. Physiol.* 11, *391*.
—— (1927 b). 'Positive and negative currents of injury in relation to protoplasmic structure.' *Journ. Gen. Physiol.* 11, *673*.
SHEARER, C. (1922). 'On the heat production and oxidation processes of the echinoderm egg during fertilisation and early development.' *Proc. Roy. Soc.* B, 93, *410*.
STEPHENSON, M. (1930.) *Bacterial Metabolism*. (London.)
STRAUB, J. (1929). 'Der Unterschied in osmotischer Konzentration zwischen Eigelb und Eiklar.' *Soc. Chim. Néerl.* 48, *49*.
WARBURG, O. (1928). *Über die katalytischen Wirkungen der lebendigen Substanz*. (Berlin.)

CHAPTER FOUR

The Cell as a Colloidal System

The conception of the cell as a colloidal system is probably one of the most important landmarks in the history of cytology. For nearly a century our knowledge of cell structure was based, not on observations of living units but upon cells which had been stabilised by fixation and artificially differentiated by staining reactions. Almost simultaneously, Fischer (1899) and Hardy (1899) demonstrated, without any element of doubt, that these standardised but arbitrary processes were of the same fundamental nature as those which effect the precipitation of inanimate protein systems and which control the uptake of dyes by the coagulated particles. This fundamental discovery showed that the only reliable guide to protoplasmic structure must lie in a study of uncoagulated colloidal systems and not in a meticulous observation of coagulated cells. For more than thirty years every student of cell structure has had, of necessity, to follow the rapid march of colloidal chemistry. To some extent the position is satisfactory: the colloidal properties of cell constituents are now realised and we are not likely to make very gross errors in the interpretation of cell structure. At the same time, we must constantly bear in mind that the living system is incomparably more complex than any which has, so far, attracted the undivided attention of the chemist. The biologist is concerned with a system in which there are many variables and whose structures are extremely instable. The chemist on the other hand is still concerned with systems which, in comparison, are extremely simple and relatively stable. Until these latter systems are more completely investigated than is at present the case, it behoves the biologist to tread warily if he is to avoid the pitfalls in his path.

The literature of colloidal chemistry is extensive, but there are many admirable sources of information: Pauli (1922), Svedberg (1928), Bogue (1924), and Freundlich (1926). For the purposes of this book it is impracticable and unnecessary to follow the numerous applications of colloidal chemistry to biological problems. We shall only attempt to indicate—as simply as possible—certain types of cytological phenomena in which colloidal phenomena quite clearly play their part.

Before proceeding further, an attempt must be made to define the colloidal state. Roughly speaking, any substance can exist in the colloidal state when it is capable of being distributed throughout a medium in the form of particles which are not smaller than $1\,\mu\mu$ and not greater than $100\,\mu\mu$ in diameter (see Table VI). Precise limitations in size cannot be defined, for they vary in different cases (see Freundlich, 1926).

Table VI (from Freundlich, 1926)

$0{\cdot}1\,\mu\mu$	$1\,\mu\mu$	$10\,\mu\mu$	$100\,\mu\mu$	$1\,\mu$	$10\,\mu$	$100\,\mu$	1 mm.
Ultramicroscopic region				Microscopic region			
Particles show Brownian movement				No visible Brownian movement			
Particles pass through an ordinary filter paper				Particles are held back by filter paper			
True solutions	Colloid solutions			Emulsions and suspensions			
$<1\,\mu\mu$	$1\,\mu\mu$–$100\,\mu\mu$			$100\,\mu\mu$–1 mm.			

A useful example of a colloidal suspension is provided by metallic particles suspended in water; if metallic gold is liberated into water in the form of particles of approximately $25\,\mu\mu$ in diameter, the colloidal solution which results is perfectly clear and red in colour; if the particles are $30\,\mu\mu$ in diameter the solution is blue.

Roughly speaking colloidal systems can be classified into two groups.

(1) *Lyophob systems*. These systems are commonly known as *suspensions* as they usually represent solid insoluble particles suspended in a fluid medium. Their properties are as follows: (i) the particles remain separate as long as there is a definite electrical potential between the surface of the particle and the medium; (ii) the sign of the surface charge depends on the hydrogen ion concentration of the medium; (iii) the magnitude of the charge depends on the nature of the other ions present; (iv) the stability of the particles is not affected by dehydrating agents or heat; (v) in concentrated suspensions the particles may adhere together to

THE CELL AS A COLLOIDAL SYSTEM

form an elastic gel. Colloidal systems of this type may possibly occur within the living cell, although so far they have not received much attention; examples are provided by particles of insoluble calcium compounds or similar substances having little or no affinity for water.

(2) *Lyophil systems*. In these cases the stability of the system is not always dependent on the presence of a surface electric charge. If the uncharged particles have a sufficiently high affinity for water the electro-neutral particles will remain separate from each other because each particle is insulated from its neighbours by an immovable film of water. The addition of any reagent which has a higher affinity for water will, if sufficiently concentrated, dehydrate the particles and enable them to adhere together—hence the term *lyophil* as applied to the hydrated particle. Such systems are upset by the presence of alcohol, concentrated neutral salts, and heat. Like lyophob systems, they can form elastic jellies if sufficiently concentrated. Most of the colloidal aggregates within a living cell belong to the lyophil type, and the proteins may be taken as the most obvious example.

The rôle which the proteins play in the economy of the cell is partly determined by the fact that they are colloids—and partly by the fact that they are amphoteric electrolytes. These two properties are, however, closely associated with each other, for it will be seen that the factors which influence the stability of the proteins as colloidal systems also influence the ability of the protein to function as an acid or as a base. In practice a study of the colloidal systems in the living cell is simplified if we first consider the proteins as ampholytes and subsequently consider certain purely colloidal features.

If the organic element of the protein molecule be denoted as R, its ampholytic powers can be represented by the formula H—R—OH. When the protein acts as a base it will ionise, as in equation (i), into negatively charged hydroxyl ions and into positively charged organic ions (HR$\cdot$)

$$\text{H—R—OH} \rightleftharpoons \text{H—R}^{\cdot} + \text{OH}' \qquad \ldots\ldots(\text{i}).$$

When the substance acts as an acid it will ionise into positively charged hydrogen ions and negatively charged organic ions.

$$\text{H—R—OH} \rightleftharpoons \text{H}^{\cdot} + \text{R—OH}' \qquad \ldots\ldots(\text{ii}).$$

Since both the acidic and basic powers of a protein are weak the presence of a strong acid or a strong base will materially affect the powers of the protein to give off hydroxyl or hydrogen ions of its own. Thus in (i) the presence of NaOH will almost entirely prevent the protein from giving off hydroxyl ions, and the reaction will proceed from right to left. Hence under these conditions the only powers possessed by the protein are exercised as in equation (ii), so that the reaction will proceed as in (iii):

$$H-R-OH \rightleftharpoons H^{\cdot} + ROH',$$

$$Na-OH \rightarrow Na^{\cdot} + OH',$$

$$H^{\cdot} + OH' \rightarrow H_2O,$$

or $\quad H-R-OH + NaOH \rightleftharpoons Na^{\cdot} + ROH' + H_2O \quad \ldots\ldots$(iii).

Similarly in the presence of a strong acid equation (ii) is suppressed and the protein only ionises as a base. It is obvious that under a particular set of conditions the tendency of the protein to give off hydrogen ions will just balance its tendency to give off hydroxyl ions and in this state the ionisation of the protein is at a minimum. This condition is known as the *isoelectric point*, for at this point the protein shows least tendency to move in an electric field.[1]

A general proof that the main bulk of the intracellular proteins act during life as acids (i.e. are on the alkaline side of their isoelectric point) can be given as follows. If an animal tissue be suspended in a suitably balanced medium, such as Ringer's solution, and the solution be acidified, the whole tissue becomes opaque when the acidity reaches an average value of about pH 3·0 and an examination under the microscope shows that intense intracellular precipitation has occurred. If instead of acidifying the Ringer solution it be made abnormally alkaline the tissue tends to absorb water and remains translucent. Precisely the same experiment can be performed with albumen or other proteins *in vitro* and is due to the fact that whenever a protein is acting as an acid it can be removed from combination with a base by adding acid but not by adding an alkali; near the isoelectric point all proteins are readily precipitated by neutral salts, hence the precipitation which occurs in the cell or in protein systems *in vitro*. An interesting variation of this experiment consists in suspending a translucent tissue, such as the gills of *Mytilus*, in isotonic sugar solutions of varying acidities, in which case the tissue does not become opaque in any of the solutions although at or below its isoelectric point—the reason being that

[1] For an alternative account of ampholytic properties see Borsook and MacFadyen (1930).

neutral salts are required to effect precipitation. A parallel experiment with albumen can be carried out at the same time.

It can readily be understood how intracellular proteins on the alkaline side of their isoelectric point must form an equilibrium with those ions within the cell which bear an opposite sign. They are, in fact, extremely sensitive to the presence of hydrogen ions which, if present in sufficient quantity, often cause visible coagulation. In both cells and protein systems *in vitro* the coagulative effect of inorganic acids is usually reversible as long as it has occurred in the presence of the salts of alkali or alkaline earth metals. If, however, the salts of heavy metals (e.g. mercury, lead, copper) are present, the resultant precipitate is only redissolved by alkali with great difficulty if at all. As will be shown later, these principles underlie the process of fixation by salts of mercury or chromium, and depend mainly on the fact that proteins are ampholytic electrolytes.

The true colloidal properties of the proteins are best illustrated by the action of heat or of alcohol. If a dilute solution of egg-white is boiled for a short time it changes from a clear transparent solution to one which is more or less opaque; the change is irreversible. An examination of the opaque system reveals the fact that it is essentially comparable to the dispersion of gold particles which has already been described. If a beam of light be passed through the solution it reveals a Tyndall effect and under suitable conditions the movement of the particles can be observed with the ultramicroscope. In the natural state these phenomena are not observed, they only occur after the albumen has been *denaturated* by boiling. At first sight the change from the natural to the denaturated state appears to be one of altered solubility. This is, however, not strictly true. If natural albumen could be dissolved in water in precisely the same manner as a typical crystalloid, it should be possible to prepare a saturated aqueous solution. This is not the case, we can go on adding albumen to water and it goes into solution with undiminished vigour; the process is more closely equivalent to mixing together two liquids which are miscible in all proportions. As will be shown in a later chapter the miscibility of protoplasm and water is of considerable theoretical interest, but at the moment we are still concerned with the effect produced on an albumen solution by heat. If we prepare two albumen solutions one of which is neutral and the other slightly alkaline or acid, and heat them, only the neutral solution is found to yield denaturated albumen, for the acid or alkaline solution remains clear and transparent. As already explained, in the presence of acid or alkali the protein is present in the ionised state, whereas at or near neutrality most of the protein is present as unionised particles. There can be little doubt that heat only denaturates unionised protein. Why the unionised state is instable at high temperatures is not clearly understood, but it is usually attributed to the fact that each molecule is normally surrounded by a film of water which insulates it from its neighbours and prevents mutual coherence—at the same time keeping the molecule in a state of 'solution'. If such a system be heated, the water film is ruptured and individual albumen molecules adhere to their neighbours in an irreversible manner—giving a denaturated suspension. Such a view is supported by the fact that undissociated sodium albuminate can be denaturated by adding to the solution any reagent with a higher affinity for water—e.g. alcohol. As agents of cytological fixation, alcohol and heat act to some degree by virtue of their power to rob the unionised protein molecules of this stabilising water film. Different proteins have slightly different properties, but there is a very large class which resemble albumen in that the stability of their unionised molecules depends on the presence of a water film—such colloidal systems are of course all of the lyophil type.

Before proceeding to consider the technique of histological fixation, it is convenient to consider another aspect of the heat coagulation of albumen. As is well known, a concentrated solution of albumen, when heated, not only goes opaque but sets to form a solid gel which is stable on cooling. So far as can be seen, all colloidal systems can form gels if the number of particles present is large enough and if these have the power of adhering to each other. In the case of albumen no true gel is formed as long as the protein is in the natural state, for the individual particles do not adhere together; when, however, their water films are irreversibly destroyed, mutual cohesion is possible, and if enough albumen is present a gel is formed. Gels may, however, be heat reversible—such as those of gelatin, and the transition from the *sol* to the *gel* state may play an important rôle in the life of the cell (see p. 68).

It will be noted that the stability of a natural protein can be upset in two ways—(i) by the use of reagents which affect the number of unionised particles present or which react with such particles to form lyophob precipitates; (ii) by such agents as heat or alcohol, which act directly on unionised particles or molecules. At this point we may consider the underlying principles of cytological fixation.

Cytological fixation

The process of fixation consists in killing a cell in such a way as to cause a minimum change in its microscopic appearance. If a cell is subjected to undue manipulation in the unfixed state, death ensues and post-mortem changes rapidly destroy all organised cell structure. By the addition of a fixative intracellular compounds are rendered inert, so that the normal consequences of death are no longer observed and at the same time the physical consistency of the cell enables it to be cut into thin sections which will absorb suitable stains. Fixation is very largely an empirical process, for different types of cell give best results with different methods of fixation; at the same time all known fixatives with two exceptions (osmic acid and formaldehyde) are powerful protein precipitants.

It has already been pointed out that most if not all intracellular proteins are on the alkaline side of their isoelectric point and there can be little doubt that, during life, they are largely combined with simple metallic bases to form partially ionised sodium or potassium proteinates. Compounds of this type are soluble in water or in salt solutions, and will form a homogeneous system as long as each unionised molecule is surrounded by a film of water molecules (see p. 35). When substances are added (e.g. alcohol) which can remove this water film the unionised proteins will be precipitated. There can be little doubt that this is the essential effect of cell fixation by alcohol. Fixation of this type is only applicable to very small or very thin cells; otherwise serious distortion takes place. Similarly, if the cell is heated beyond a certain temperature the molecules of sodium proteinate again lose their water films and are denatured; at the same time all enzymes are destroyed. The more usual fixatives are, however, either acids, salts of heavy metals, or a mixture of these two reagents. In every case they operate by precipitating intracellular proteins in a chemically inert state.

Hardy (1899) demonstrated that when a solution of a protein or a living cell is exposed to the action of a coagulative fixative, there must, from the nature of the process, ensue a separation of the protein systems into two phases, and that the final result depends more on the nature of the fixative than on the fundamental structure of the

40   THE CELL AS A COLLOIDAL SYSTEM

original protein solution or of protoplasm itself. Ordinary white of egg, like living protoplasm, exhibits no optical structure, but both systems when treated with corrosive sublimate, potassium dichromate or heat, yield a distinct network of solid structure, the interstices of which are filled with liquid. The diameter of the meshes of this precipitated network varies with the nature of the fixative: osmic acid vapour yields a close network with meshes 0.5–$0.7\,\mu$, potassium bichromate yields meshes approximately $1.3\,\mu$, whilst

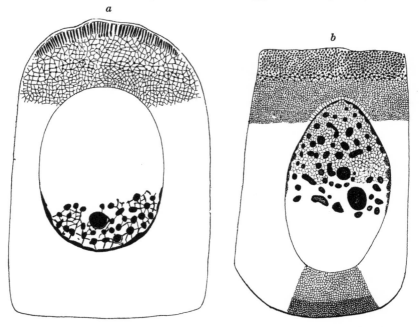

Fig. 8. Gut cells of *Oniscus* stained with iron-haematoxylin after fixation with (a) corrosive sublimate, (b) osmic acid. Note the relatively coarse reticulum in the nucleus and in the cytoplasm of (a).   (From Hardy.)

corrosive sublimate gives a still coarser structure with meshes $1.7\,\mu$ in diameter. Hardy also made the interesting observation that when coagulation occurs within a protein system which contains small suspended particles, the latter orientate themselves at the nodal points of the network, so that their final position is not a true picture of the distribution throughout the original system; also the very presence of dispersed granules may alter the diameter of the mesh of the coagulated protein, a fact which has an obvious bearing

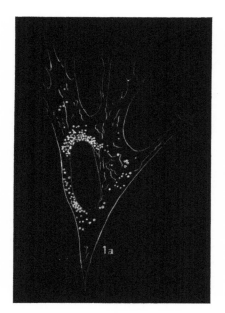 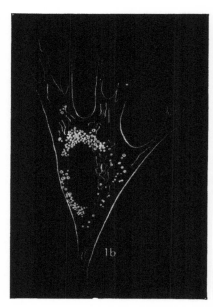

Fig. 9 A. Normal living cell (1 a) before fixation, and (1 b) after fixation by 2 per cent. osmic acid. Note almost perfect preservation of microscopic form; very slight coagulation of nucleus and cytoplasm occurs.

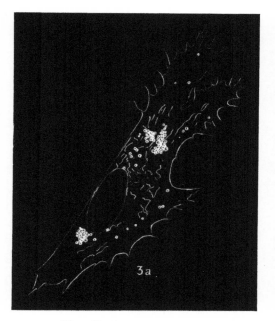 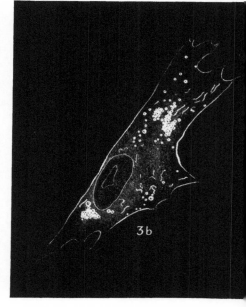

Fig. 9 B. Normal living cell (3 a) before fixation and (3 b) after fixation with strong Flemming and acetic acid. Note change in outline of cells, but the fat globules and mitachondria are normal.

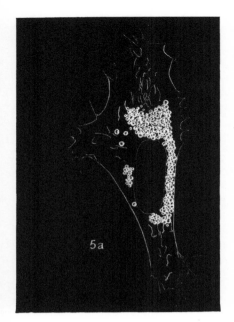
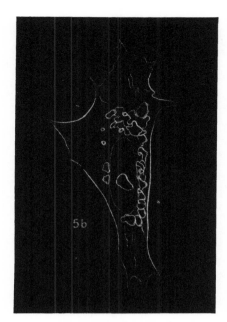

Fig. 9 C. Normal living cell (5 a) before fixation, and (5 b) after fixation by corrosive sublimate. Note coalescence of fat globules, and loss of the finer processes at the cell surface.

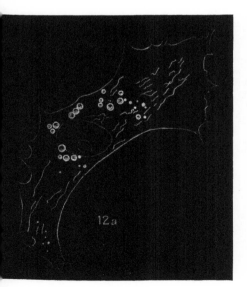

Fig. 9 D. Normal living cell (12 a) before fixation, and (12 b) after fixation by 4 per cent. neutral formalin. Note the preservation of all forms, together with fat globules and mitachondria.

on histological structure of granular cells. The coagulative effect of different fixatives is well illustrated by figs. 9 A–D, which are taken from a recent paper by Strangeways and Canti (1927). The cells shown are from a culture of the choroid of a chick and are drawn on a dark background.

Obviously the best cytological fixatives are those which stabilise the cell both physically and chemically, with a minimum alteration to its microscopic appearance. For this purpose osmic acid and neutral formalin are peculiarly useful, since except under high magnification their coagulating action is much less obvious than is that of other fixatives: unfortunately, neither osmic acid nor neutral formalin endows the cell with the physical consistency best suited for the preparation of serial sections. The precise effect of these two reagents on the physical properties of proteins is somewhat obscure (see Mann, 1902).

It is obvious that satisfactory fixation involves the coagulation of protoplasm in such a way as to form an elastic but solid gel— otherwise the visible cell structures would not remain in their natural positions. In practice the physical properties of this gel are of importance, for they determine how far it is possible to cut the cell into thin serial sections after impregnation with paraffin wax. A simple but effective demonstration of the gelation of protoplasm by fixation is provided by sea urchin eggs. A culture of *Echinus* eggs is divided into two parts and one of these is fixed by the addition of 1 per cent. neutral formalin to the sea water. After a short time both lots of eggs are transferred from sea water to $M/2$ NaCl solution. If the eggs are now gently crushed under a coverslip, the unfixed eggs will readily burst and their contents will flow out into the medium; the fixed eggs behave differently for, on crushing, they fracture much like small fragments of a gelatin jelly.

Fixation can clearly be regarded as an empirical process whereby the labile systems in the cell are converted into a stable and gelated state suitable for subsequent histological treatment. From its very nature, fixation must result in the production of artefacts which can only be recognised by careful comparison between the microscopic appearance of the living cell and the pictures which we get of cells fixed by a variety of methods. Just as fixation involves the formation of artefacts, so also dangers arise in the use of differential stains. Fischer (1899) investigated the effect of both fixation and staining on proteins *in vitro* and showed that the interpretation of the results hitherto obtained with preserved cells required careful revision. The ability of a substance to absorb and hold a given stain is not only a function of its chemical constitution, but also of the size of the particles of substance exposed to the stain. Very clear examples of this are to be found in Fischer's book *Protoplasma*; and reference to fig. 10 will show that a very slight change in technique may change the response of a cell to the same series of stains. A thorough knowledge of the chemical properties of different types of stains and of the circumstances under which they most readily combine with different types of substances would be of great value in cytology, but at present the chemistry of these processes is not sufficiently well defined to justify a departure from the empirical rules which will be found in most text books of microscopy.

Heat coagulation of protoplasm

Although fixation of cells by heat undoubtedly involves the irreversible coagulation of proteins in a denaturated state, a very simple experiment is sufficient to show that all effects of heat on a protein system are not necessarily due to this cause. If an aqueous solution of sodium mucinate be heated to boiling point no turbidity occurs. As long as the solution is cold, considerable quantities of calcium chloride can be added without affecting the transparency of the solution; on heating the mixture, however, a copious precipitate rapidly forms. In such a system the effect of heat is not due to its action on the neutral protein molecules, but is due to the fact that it enormously accelerates the rate of formation of insoluble salts of calcium (see also p. 45). Such reactions of course only occur on the alkaline side of the isoelectric point. Reactions of this type are often overlooked and the coagulative effect of heat upon living cells has often been entirely ascribed to an irreversible heat coagulation of the intracellular proteins.

From time to time authors have reported the isolation of proteins which *in vitro* coagulate at the temperature which effects coagulation *in vivo*. If this be the whole story it is curious that the general level of temperatures required for heat coagulation of proteins *in vitro* is higher than that required to kill the living cell by heat. Heilbrunn (1924) has pointed out that pure proteins do not usually coagulate below 50° C., whereas most animals die long before such a temperature is reached; this argument is far from conclusive proof that the coagulative effect of heat on protoplasm is not exercised

1 2

1 and 2. Granules of albumen precipitated by 2·5 % potassium dichromate solution and stained with Altmann's acid fuchsin and picric acid. In 1 the acid fuchsin was applied before the picric acid. Note that the small granules acquire the colour of the latter stain. In 2 the granules were stained with a mixture of the two stains. Note that the small granules are coloured by the fuchsin, although the centres of the large granules are coloured by the picric alcohol.

3 4

3 and 4. Granules of albumen (after fixation with Flemming's fluid) stained with safranin and gentian violet. In 3 the safranin was used before the gentian violet, in 4 the safranin was used after the gentian violet. Note that the small granules acquire the colour of the stain last used.

5 6

5 and 6. Nucleic acid precipitated by platinic chloride and stained with safranin and gentian violet. Inversion of staining effects as in 3 and 4.

Fig. 10. Figures illustrating the non-specific nature of differential staining.
(From Fischer.)

available for download in colour from www.cambridge.org/9781107625662

on the intracellular proteins. Heilbrunn is, however, of opinion that the heat coagulation of the eggs of *Arbacia* or *Cumingia* is due to an alteration in the state of aggregation of the fatty constituents of the cell. These eggs when exposed to 33° C. can no longer be stratified with normal ease into three layers when subjected to centrifugal force; this suggests that the high temperature has caused a marked rise in the viscosity of the protoplasm. By measuring the time of exposure to a particular temperature which is required to produce an arbitrary rise in viscosity, Heilbrunn was able to construct the curve shown in fig. 12 (see also p. 66 *seq.*).

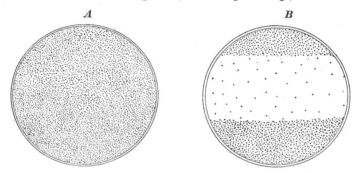

Fig. 11. Stratification of intracellular materials by centrifugal force. *A*, *Cumingia* egg before stratification. *B*, the same after stratification. Note the accumulation of fat granules at the upper pole. (From Heilbrunn.)

It is important to note that the increase of viscosity precedes any other visible death change and that, for a time, it is reversible. The results of these experiments show that the relationship between coagulation time and temperature has the same superficial form as that obtained by Buglia (1909) for serum albumen, and the temperature coefficient is of the same extremely high order (about 1·5 for each degree rise in centigrade). Heat coagulation of protoplasm is to some degree reversible, for, if the effect of the heat is not too intense, the resultant rise of viscosity is reversed on lowering the temperature. This phenomenon is parallel to the reversible heat rigor of plant protoplasm observed by Sachs (1864) and to the partially reversible heat rigor of frog's muscle observed by Gotschlich (1893). Reversibility, according to Heilbrunn, is a characteristic of protoplasmic heat coagulation as opposed to that of protein systems, since in the latter it does not occur. In contrast to the typical proteins, the fats found in living cells liquefy at comparatively low temperatures and many fat solvents are known to coagulate protoplasm (Heilbrunn). Temperature has a direct effect on the fats in a sea urchin's egg, since when heated cells are centrifuged (before coagulation has occurred) the fat, which settles to the lighter pole of the egg, is modified. Instead of being compressed into a very small cap at the pole of the cell, as in normal eggs, it is often found to be much more diffuse and, under certain circumstances, may not be visible. It looks as though, at high temperatures, the fats were

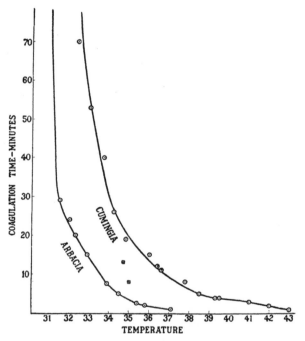

Fig. 12. Graph showing the time of exposure to various temperatures required to raise the viscosity of marine eggs to an arbitrary figure. (From Heilbrunn, 1924.)

more completely emulsified or even dissolved. Further, the coagulating effect of heat is hastened by the presence of a small concentration (1 per cent. in sea water) of ether. Heilbrunn suggests that the upper limit of temperature in which a particular type of animal can live is determined, at least in part, by the melting point of its fats.

These facts must be borne in mind when comparing the effect of heat on the coagulation of protoplasm and on simpler protein systems. At the same time, we are not justified in assuming that the reversibility of the former marks it as something distinct from the latter. On general grounds, one would expect that the coagulation of a dilute colloidal sol would involve not an increase in viscosity but a decrease: it is only when there is sufficient protein present to emesh the whole of the disperse phase and thereby form a compact gel, that one would expect heat to increase viscosity and then one would expect the process to be extremely rapid. That such a process can take place reversibly in protein systems is shown by the following experiment performed by Ringer (1890). Freshly prepared casein was treated with lime water and after twenty-four hours was filtered. To 10 c.c. of the clear solution one drop of calcium chloride solution was added without apparently producing any change in the solution. On raising the temperature to 70° C., however, the fluid rapidly became milky; on replacing the test-tube in iced water the fluid rapidly became clear again. If four drops of calcium chloride instead of one were added to 10 c.c. of cold casein solution there was still no change in the solution, but on warming, the solution quickly clotted, and on cooling, the clot dissolved and the milkiness disappeared. When the concentration of calcium chloride was increased to six drops, the fluid became slightly milky in the cold and on raising the temperature it clotted, and the clot was not completely dissolved on cooling. In some of Ringer's preparations a cloudiness appeared at low temperatures, e.g. at 20° C., and these set to a jelly at 30° C.

From the available data one is inclined to suspect that the action of heat on living protoplasm is much more complex than is the process of heat coagulation of simple protein systems. If the cell is heated rapidly to a high temperature the final result is irreversible heat coagulation of the proteins in the cell, but before this occurs there will be a period during which the normal equilibrium in the cell is upset because reactions will rapidly occur between calcium ions and protein ions which normally only take place to a very limited extent. How far such changes characterise the period of reversible heat coagulation is uncertain. A further consideration of the effects of heat on the viscosity of protoplasm will be found on pp. 66–68.

Proteins and cell activity

Leaving behind the simpler problems of heat coagulation and cytological fixation we may consider how far the behaviour of lyophil colloids *in vitro* throws real light on the activity of a cell during its natural life. Unfortunately it is difficult to make much progress. Since the cell proteins are functioning as acids we may expect to find that the cell will be extremely sensitive to the presence of free hydrogen ions as these will tend to reduce the degree of ionisation of the proteins themselves. This is in fact the case. Practically every biological activity in the cell is inhibited to some extent if an acid is allowed to penetrate into the cells. Mitosis, cell division, amoeboid and ciliary movement and respiration are all inhibited in a reversible manner by the application of acids. On the other hand, the power of the cell to 'buffer' hydrogen ions is very great (see p. 86) and it may be doubted how far significant variations in general intracellular acidity actually occur in nature. Certain well-defined examples are available, however—notably the changes in equilibrium between the tissues, plasma and blood corpuscles which are effected by the carbon dioxide of respiration; also the sensitivity of the respiratory centre to CO_2.

From the general conception of the cell proteins one would also suspect that the cell should be sensitive not only to hydrogen ions but also to all other positively charged ions, and that it might be correspondingly insensitive to relatively low concentrations of anions. Such phenomena are actually observed; many cells are peculiarly sensitive to alterations in the concentration of potassium, magnesium and calcium in their surrounding medium, and are insensitive to changes in corresponding anions such as chlorides, nitrates or even sulphates. An example of this type is provided by the ciliated epithelium on the gills of *Mytilus* (Gray, 1922) (see Chapter VI). On the other hand, when we try to obtain such data from a variety of biological sources, the whole relationship of the cell to electrolytic ions becomes extremely obscure. The confusion which exists appears to be due to two main causes: (i) there is often considerable uncertainty concerning the degree to which a cell is permeable to an ion which is present in the external medium (see Chapter XIV); (ii) when an ion has entered a cell, its activity may be seriously affected in ways quite foreign to the direct action of the ion on protein systems *in vitro*. Ions in the cell may exert their effects at organised surfaces or interfaces which are absent in protein systems *in vitro*. We meet, once again, the difficulty arising from the very complex nature of the intracellular system. The first type of difficulty sometimes disappears when we are concerned with the effect of ions on the surface of the cell only, and in such cases the cell reacts in a way remarkably parallel to inanimate protein systems. The second type of difficulty is constantly present but is likely to become less significant as one by one the intracellular variables are isolated.

A priori, one would suspect the phenomena of *antagonistic ion action* to be concerned with the equilibrium between metallic ions and protein systems, and to a significant extent this view is supported by experimental facts. There always remains, however, one outstanding difficulty. From a chemical point of view the equilibrium between proteins and potassium ions is very similar to the equilibrium between proteins and sodium ions, and yet the biological properties of the two ions are profoundly different. In some cases we may imagine that this difference is due to the differences in their rate of migration (see p. 398), but this cannot be regarded as of real significance in all cases. For example, the presence of magnesium will prevent the dispersion of the cell matrix in *Mytilus* tissue in a solution of sodium chloride, but it will not prevent dispersion if potassium has replaced the sodium (see Chapter VI). The interrelationship of the various metallic ions presents difficult problems of this type and how far they will be solved by observations *in vitro* remains uncertain. The recent work of Simms (1928) is however highly suggestive.

Electrokinetic properties of cells

Just as the protein constituents of the cell appear to be responsible for some of the cruder reactions between cells and their electrolytic environment, so there is a marked parallel between the cell system and the proteins when these are exposed to the action of an external electric field.

If a suspension of small cells is exposed to a suitable electric field, the cells all move to one or other electrode, just as is the case with the particles of denaturated albumen. We infer from this that the cells must bear an electric charge in respect to their surrounding medium. In almost every case, cells suspended in their natural media migrate to the anode and consequently must bear a negative charge in relation to these media. Red blood corpuscles, bacteria, yeast and spermatozoa, all move to the anode under normal circumstances. A number of useful references will be found in Winslow and Fleeson (1925).

It is of interest to consider the origin of this charge, how far it is dependent on the life of the cell, and how far the life of the cell is dependent on the presence of the charge. In order to do this, it is necessary to consider the properties of surface charges in inanimate systems. It has long been known that where any solid is in contact with a liquid, the interface is the seat of a definite potential. If the solid is in the form of small particles and an external E.M.F. be applied, the particles will travel to the electrode bearing the opposite sign to that of the particles; if the solid is not free to move, then the

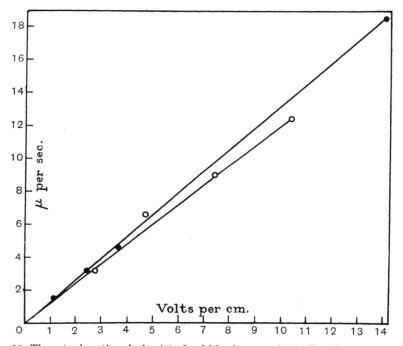

Fig. 13. The cataphoretic velocity (μ) of red blood corpuscles is directly proportional to the applied potential gradient. (From Abramson.)

fluid medium will travel to that external electrode which bears the same charge as the solid. In the first case we speak of *cataphoretic movement*: in the second case the phenomenon is known as *electrical endosmosis*.

By observing the rate of cataphoretic movement it is found that the electrostatic charge on a particle is a function of the nature of the particle and of the liquid medium

$$\text{P.D.} = \frac{4\eta v \pi}{K \cdot x} \qquad \ldots\ldots\text{(iv)},$$

where η is the viscosity of the medium, K is the dielectric constant of the surface layer, v is the velocity of migration in centimetres per second, and x is the potential gradient of the external field. Although we can, in this way, demonstrate a surface charge and measure its intensity, there is still considerable uncertainty as to its origin. Air bubbles, liquid drops, quartz particles, proteins, and living cells all exhibit a charge but it is by no means certain that the charge is strictly similar in origin in all these cases (Freundlich, 1926). In the case of the proteins, Wilson (1916) suggested that the charge is due to a Donnan equilibrium between the inside of the particle and its medium; v. Hevesy (1917), on the other hand, regards the particles as equivalent to large ions. That ions play an important rôle, there can be little doubt, since in their absence the surface charge falls to a very low value. On the other hand it is not easy to account for the charge on quartz particles or air bubbles by these hypotheses. It is probable that in some cases we are really dealing with two distinct potential differences—one being electrokinetic and the other thermodynamical in origin (see Freundlich, 1926).

Whatever be the source of the charge on the particles, there seems little doubt that each particle will repel its neighbours electrostatically. Consequently, if the charge be abolished, the particles will adhere together more readily than when a charge is present. Observations of this type have often been made. Hardy (1899) showed that denaturated proteins at their isoelectric point are very readily coagulated, and Powis (1914) showed that oil drops in aqueous emulsions adhere together as soon as the P.D. between the drops and the medium falls below 30 millivolts. In such systems the P.D. is seriously affected by the presence of neutral salts; consequent to the addition of neutral salts the P.D. falls, and if it is reduced below a critical value the particles are precipitated. Closely parallel with these phenomena are the observations of Northrop (1924) on bacteria. Northrop found that typhoid bacteria, when sensitised by immune serum, are completely agglutinated by a variety of salts when their surface charge falls below ± 13 millivolts.

As one might expect, the surface charge of particles or living cells, as measured cataphoretically, is very sensitive to the concentration of hydrogen ions present. Using red blood cells Coulter (1920–1) found that at pH 4·6 the charge is abolished—above that value the cells are positively charged and below that value the charge is negative in sign (see also Table VII).

These results are, to some extent, what one would expect to find in any system, living or otherwise, where the surface of the particles is of a protein nature. Unfortunately they are the exception rather than the rule in suspensions of living cells. In Northrop's first series of experiments the bacteria were sensitised by immune sera. If this treatment be omitted no coagulation occurs, although there may be no measurable P.D. at the surface of the cell. To explain this fact, Northrop suggests that the ability of one bacterium to adhere to another must depend upon the cohesive forces of the surface layer. After abolishing the mutual repulsion due to the electrostatic charge, two bacteria will only adhere together after making contact if a film

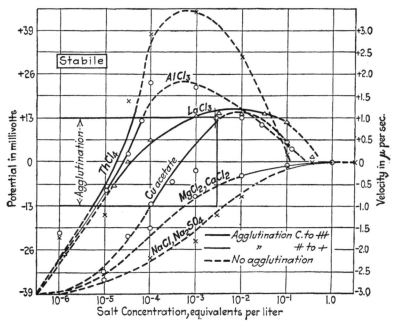

Fig. 14. Diagram showing the agglutination of *B. typhosus* when the P.D. falls below ±13 millivolts, provided that the concentration of salt required to effect this change is not too great. (From Northrop.)

with adequate cohesive properties is formed. Using non-sensitive typhoid bacteria, Northrop found that no coagulation occurred if the concentration of neutral salts required to reduce the P.D. to 13 millivolts was greater than $0.01M$, and at the same time he showed that the salts in these concentrations definitely decreased the cohesive properties of the bacteria as estimated by their power of causing two glass surfaces to adhere together. In a later chapter (Chapter VI) it will be shown that the adhesive and cohesive properties of cell surfaces are undoubtedly affected by electrolytes, and that these reagents may play an important rôle in controlling the stability of the tissue.

The effect of ions on the electrostatic charge and on the degree of dispersion of living cells has been extensively studied in the case of bacteria (see Buchanan and Fulmer, 1928) and red blood cells. It is not easy to summarise the results, but in general the controlling factors appear to be strictly comparable to those which operate on protein particles. (1) As already mentioned the sign and magnitude of the charge is very sensitive to the presence or absence of hydrogen ions. (2) Neutral salts invariably reduce the charge whether this be positive or negative (Oliver and Barnard, 1924–5 b). If the cell is negatively charged the action of neutral salts depends upon the valency and specific nature of the cations present; if the cell is positively charged it is the anions which are effective. (3) If neutral salts are present beyond a limiting concentration they will exert an action which is antagonistic to flocculation although the surface P.D. may be reduced to zero; such an affect may be due to their effect on the cohesive properties of the cell surface or be comparable to the 'lyophilic' effect observable in protein systems (see p. 35). (4) In very large concentrations neutral salts may 'salt out' the suspension —as is also the case with proteins. (5) In the case of trivalent metallic ions flocculation may be due to the cohesive effects of a deposit of metallic hydroxides on the surface of the cell (Gray, 1922).

The correlation between surface charge and precipitation is not universally admitted; Oliver and Barnard (1924–5 a) claim that there is an important difference between the surface charge of red blood corpuscles and that on particles of protein or collodion, since, when the latter are suspended in a solution of sugar at an appropriate pH, the negative charge is permanent; the charge on red blood cells, on the other hand, is liable to be transient and may fall in a few minutes to a very low value and be followed by agglutination. An abnormal environment is always liable to induce irreversible changes within the cells and some of the effects of pH on the cataphoretic velocity of red blood corpuscles may therefore be due to changes which invariably precede death rather than to a comparatively simple and reversible change in the ionic charge at the surface of a normal corpuscle (see also Abramson, 1930).

We have still to consider how far life is dependent on the presence of a surface change and *vice versa*. In 1906 Cernovodeanu and Henri observed that the spores of certain moulds in the presence of acid migrated to the cathode instead of to the anode, indicating that a reversal of charge had occurred; at the same time the spores were capable of subsequent growth. More recently Russ (1909) and Winslow, Falk and Caulfield (1923) have shown that the effect of the hydrogen ion on the cataphoresis of bacteria is unaffected by killing the cells.

Table VII

Electrophoresis of bacteria at different hydrogen ion concentrations. (From Winslow, Falk and Caulfield)

pH	1·0	1·5	2·0	2·6	3·3	3·7	5·5	6·8	7·2	7·3	7·7	9·6	10·0
Velocity of migration in μ per sec.	0	+2·8	+2·5	+1·9	−1·8	−8·9	−9·7	−7·2	−10·6	−10·3	−10·3	−8·7	−10·0

+ indicates migration to cathode. − indicates migration to anode.

It would appear therefore that the observed P.D. at the cell surface when measured cataphoretically is not to be attributed to the vital activity of the cells, but is comparable to that found on the surface of protein particles. We seem forced to conclude that the surface charge of a living cell is not dependent on the living processes of the cell. How far the reverse is true is less easy to say. If we reduce the surface charge to zero we come dangerously near to killing the cell, but the general impression given by all the facts is that the surface charge on a cell is not so much associated with the life of a cell as with the fact that the surface is of a protein nature.

Northrop has made the interesting suggestion that the abolition of electrical charges at surfaces *within* the living cell may be correlated with certain types of biological activity. In many biological reactions there is a definite latent period between the inciting cause and the response, and it is therefore of interest to note that in those cases where colloidal stability is solely dependent on surface charges there is often a long delay before coagulation is complete, although the P.D. at the surface may have been almost instantaneously reduced to zero. By assuming that the uncharged particles cannot separate from each other when once they are in contact, Smoluchowski derived a formula which will express the course of coagulation with considerable accuracy:

$$k = \frac{1}{t}\left(\frac{1}{\Sigma V} - \frac{1}{V_0}\right) 4\pi Dr \qquad \ldots\ldots(\text{v}),$$

where V_0 is the number of particles at the beginning of the coagulation, ΣV the number at time t, D is the diffusion coefficient, and r is the distance apart at which the particles attract each other (see also Freundlich, 1926). It will be noticed that equation (v) is similar to that of a bimolecular reaction. Northrop suggests that such phenomena may underlie the so-called 'latent period', characteristic of the responses of living organisms. It would be interesting to know how far coagulative processes of this type are applicable to the deposition of solid materials within the cell and in the formation of organised structures such as chromatin granules and similar bodies. At present we have no positive evidence; it would be difficult but not impossible to obtain.

It should, perhaps, be noted that the electrical potential between the surface of a cell and its surrounding medium is not the same thing as the electrical potential between the cell surface and the interior of the cell; we shall be concerned with the latter in a later chapter. The only comparable phenomenon to cataphoresis which we need consider here is the electrical endosmotic properties of living membranes. The rate of flow of fluid through a charged membrane depends fundamentally on the same factors as influence the cataphoretic movement of discrete particles. If ϕ = the volume of fluid transported in unit time across the membrane, k = dielectric constant of the solution in the membrane pores, σ = specific resistance of the solution in the membrane pores, η = viscosity coefficient of the solution, I = electric current traversing the membrane, ζ = electrokinetic P.D. between the walls of the membrane and the fluid, then

$$\phi = \frac{kI\sigma\zeta}{4\pi\eta} \qquad \ldots\ldots\text{(vi)}.$$

In other words for any given solution of constant conductivity and viscosity, the amount of water moved in unit time in a constant field will be directly proportional to the surface potential between the membrane and the external solution.

Using mammalian serous membranes Stuart Mudd (1924–5) found that liquid streams through living membranes towards the cathode of an electric field as long as the pH of the fluid is on the alkaline side of a definite critical value. If the solution is more acid than this value, the direction of flow is reversed. Usually the critical value lies between pH 4·3 and 5·3. Apparently the electro-endosmotic properties of the membrane is unaffected by the death of the tissue.

These observations obviously conform to those obtained from the cataphoresis of isolated cells.

How far the potential differences naturally set up within a living organism are sufficient to cause the movement of fluids or isolated cells is uncertain, but an interesting suggestion has recently been made by Abramson (1928). This author has shown that quartz particles move cataphoretically through soft gelatin jellies with the same velocity as through liquid sols. Since the wall of a blood capillary is very thin a comparatively small difference in potential on the two sides might be sufficient to move a leucocyte across the gelated wall just as an externally applied field will draw the cell through a soft gelatin jelly. Assuming that the side of a capillary wall in contact with an injured region of connective tissue is electropositive to the blood stream, the rate of cataphoretic movement will depend upon the magnitude of this P.D. and will be of the order of 0.5μ per sec. per volt per cm. Since a drop of potential of 1 millivolt across a membrane 0.5μ thick is equivalent to 20 volts per cm., it is quite possible that bio-electric potentials may exert within a living organism a well-defined cataphoretic influence on isolated cells.

REFERENCES

ABRAMSON, H. A. (1927). 'The mechanism of the inflammatory process. I. The electrophoresis of the blood cells of the horse and its relation to leucocyte emigration.' *Journ. Exp. Med.* 46, *987*.

—— (1928). 'The mechanism of the inflammatory process. III. Electrophoretic migration of inert particles and blood cells in gelatin sols and gels with reference to leucocyte emigration through the capillary wall.' *Journ. Gen. Physiol.* 11, *743*.

—— (1930). 'Electrokinetic phenomena. III. The "isoelectric point" of normal and sensitized mammalian erythrocytes.' *Journ. Gen. Physiol.* 14, *163*.

BOGUE, R. H. (1924). *The Theory and Application of Colloidal Behaviour.* (New York.)

BORSOOK, H. and MACFADYEN, D. A. (1930). 'The effect of isoelectric amino acids on the pH of a phosphate buffer solution. A contribution in support of the "Zwitter Ion" hypothesis.' *Journ. Gen. Physiol.* 13, *509*.

BUCHANAN, R. E. and FULMER, E. I. (1928). *Physiology and Biochemistry of Bacteria.* (London.)

BUGLIA, A. (1909). 'Ueber die Hitzgerinnung von flüssigen und festen organischen Kolloiden.' *Kolloid. Zeits.* 5, *291*.

CERNOVODEANU, P. and HENRI, V. (1906). 'Détermination du signe électrique de quelques microbes pathogènes.' *C.R. Soc. Biol.* 61, *200*.

COULTER, C. B. (1920–1). 'The isoelectric point of red blood cells and its relation to agglutination.' *Journ. Gen. Physiol.* 3, *309*.

THE CELL AS A COLLOIDAL SYSTEM 55

COULTER, C. B. (1924–5). 'Membrane equilibria and the electric charge of red blood cells.' *Journ. Gen. Physiol.* 7, *1*.

FISCHER, A. (1899). *Fixierung, Färbung und Bau des Protoplasmas.* (Jena.)

FREUNDLICH, H. (1926). *Colloid and Capillary Chemistry.* Transl. by H. Stafford Hatfield. (London.)

GOTSCHLICH, E. (1893). 'Ueber den Einfluss der Wärme auf Länge und Dehnbarkeit des elastischen Gewebes und des quergestreiften Muskels.' *Pflüger's Archiv*, 54, *109*.

GRAY, J. (1915). 'Note on the relation of spermatozoa to electrolytes.' *Quart. Journ. Micr. Sci.* 61, *119*.

—— (1920). 'The relation of spermatozoa to certain electrolytes.' *Proc. Roy. Soc.* B, 91, *147*.

—— (1922). 'A critical study of the facts of artificial parthenogenesis and normal fertilisation.' *Quart. Journ. Micr. Sci.* 66, *419*.

HARDY, W. B. (1899). 'On the structure of cell protoplasm.' *Journ. Physiol.* 24, *158*.

HEILBRUNN, L. V. (1924). 'The colloid chemistry of protoplasm. IV. The heat coagulation of protoplasm.' *Amer. Journ. Physiol.* 69, *190*.

—— (1925). 'The colloid chemistry of protoplasm.' *Colloid Symposium Monograph* 3. (New York.)

v. HEVESY, G. (1917). 'Ueber die Ladung und Grösse von Ionen und Dispersoiden.' *Kolloid. Zeits.* 21, *129*.

MANN, G. (1902). *Physiological Histology.* (Oxford.)

MUDD, S. (1924–5). 'Electro-endosmosis through mammalian serous membranes. I. The hydrogen ion reversal point with buffers containing polyvalent anions.' *Journ. Gen. Physiol.* 7, *389*.

NORTHROP, J. H. (1924). *The Flocculation and Stability of Colloidal Suspensions. Bogue's Theory and Application of Colloidal Behaviour*, 1, *70*.

NORTHROP, J. H. and DE KRUIF, P. H. (1921–2). 'The stability of bacterial suspensions. III. Agglutination in the presence of proteins, normal serum, and immune serum.' *Journ. Gen. Physiol.* 4, *639*.

NORTHROP, J. H. and FREUND, J. (1923–4). 'The agglutination of red blood cells.' *Journ. Gen. Physiol.* 6, *603*.

OLIVER, J. and BARNARD, L. (1924–5 a). 'The influence of electrolytes on the stability of red blood corpuscle suspensions.' *Journ. Gen. Physiol.* 7, *99*.

—— (1924–5 b). 'The effect of valency of cations and anions on negatively and positively charged red cells.' *Journ. Gen. Physiol.* 7, *225*.

PAULI, W. (1922). *Colloid Chemistry of the Proteins.* Eng. transl. (London.)

POWIS, F. (1914). 'Der Einfluss von Electrolyten auf die Potentialdifferenz an der Öl-Wassergrenzfläche einer Ölemulsion und an einer Glas-Wassergrenzfläche.' *Zeit. f. physik. Chem.* 89, *91, 179, 186*.

RINGER, S. (1890). 'Regarding the action of lime salts on caseine and on milk.' *Journ. Physiol.* 11, *464*.

RUSS, C. (1909). 'The electrical reactions of certain bacteria and an application in the detection of tubercle bacilli in urine by means of an electric current.' *Proc. Roy. Soc.* B, 81, *314*.

SIMMS, H. S. (1928). 'Chemical antagonism of ions I–IV.' *Journ. Gen. Physiol.* 12, *241, 259, 511, 783*.

STRANGEWAYS, T. S. P. and CANTI, R. G. (1927). 'The living cell *in vitro* as shewn by dark ground illumination and the changes induced in such cells by fixing reagents.' *Quart. Journ. Micro. Sci.* 71, *1*.

SVEDBERG, T. (1928). *Colloid Chemistry.* (New York.)

VON SZENT-GYÖRGYI, A. (1921). 'Kataphoreseversuche an kleinlebewesen Studien über Eiweissreaktionen. III.' *Biochem. Zeit.* 113, *29*.

TEAGUE, O. and BUXTON, B. H. (1906–7). 'Die Agglutination in physikalischer Hinsicht. III. Die von den suspendierten Teilchen getragene elektrische Ladung.' *Zeit. f. physik. Chem.* 57, *76*.

WILSON, J. A. (1916). 'Theory of Colloids.' *Journ. Amer. Chem. Soc.* 38, *1928*.

WINSLOW, C. E. A., FALK, I. S. and CAULFIELD, M. F. (1923). 'Electrophoresis of bacteria as influenced by hydrogen ion concentration and the presence of sodium and calcium salts.' *Journ. Gen. Physiol.* 6, *177*.

WINSLOW, C. E. A. and FLEESON, E. H. (1925). 'The influence of electrolytes upon the electrophoretic migration of bacteria and yeast cells' *Journ. Gen. Physiol.* 8, *195*.

CHAPTER FIVE

The Physical State of Protoplasm

At first sight it would appear a simple matter to determine whether cytoplasm is a solid or a liquid, and yet the data available for such a decision are far from complete. In some cases living cytoplasm clearly possesses one typical liquid property; it is capable of flowing from one place to another in the cell. Young plant cells, Amoebae, and the Myxomycetes all exhibit active flowing movements in their cytoplasm. In all these cases, however, the power to flow within the cell alters reversibly from time to time in response to changes within the cell or in the external environment. One of the most striking instances is that described by Seifriz (1923–4) in the protoplasm of the hyphae of *Rhizopus*. 'The protoplasm, in the streaming state, has a "rather low consistency". By pressure with a microdissection needle (sufficient completely to close a hypha), this streaming protoplasm of low viscosity may be caused instantly to assume the consistency of a rigid gel. Streaming ceases; but later, without further disturbance, there is a reversal of the phenomenon and the protoplasm becomes liquid again. If a hypha is torn open when the protoplasm is in the fluid condition, the contents immediately steadily flow out of the hyphal thread. If a hypha is torn open when the protoplasm is of high consistency, the gelled protoplasmic mass can be forced out of the hyphal thread by pressure with a needle, just as one would squeeze oil paint from an artist's tube. The exposed rod of protoplasmic jelly free from its supporting wall, retains its cylindrical shape in the surrounding aqueous medium for some time' (p. 437). These and similar changes are attributed by Seifriz to a reorientation of structural protoplasmic units and analogous to the mechanical breakdown and reformation of inorganic systems, such as the iron oxide gel described by Schalek and Szegvary (1923), or the cadmium gel described by Svedberg (1921, 1928). If left undisturbed these systems set to form gels which, on mechanical agitation, rapidly lose their rigidity and are transformed into typical fluid sols.

Variability in the consistency of plant protoplasm is also well illustrated by the observations of Scarth (1927). Using the mesocarp cells of the snowberry (*Symphoricarpus racemosus*) it can be shown that the protoplasmic strands which traverse the central vacuole vary in their physical properties from time to time in one and the same cell. At times the strands are inextensible, so that on slight stretching they suddenly snap across in a manner comparable to the fracture of a solid thread. Yet, almost immediately, the fractured ends begin to flow into the parietal cytoplasm. At other times these threads are the seat of active streaming movements, and in this condition they are highly extensible, and can be pulled into threads of less than $1\,\mu$ in diameter: at the same time the threads are elastic. A striking experiment, which could probably be performed on a variety of protoplasmic materials—was performed by Scarth. A needle was pushed through a rapidly moving strand in a cell of a staminal hair of *Tradescantia*. The viscid cytoplasm was carried out as a film over the point of the needle as it emerged on the other side. The stream banked round the obstruction and the motion was temporarily checked on the down-stream side but soon recovered its speed as the pent-up material started to flow round the obstacle. On moving the needle backward and forward in the axis of the strand the respective portions of the latter lengthened or shortened elastically with little or no slip of the needle. Streaming therefore is compatible not only with high viscosity but with definite elasticity.

Objections may be raised against the association of protoplasmic viscosity and the rate of intracellular streaming, since we do not know the mechanism responsible for movement, nor do we always know that the rigidity of the protoplasmic mass is not attributable to the surface layer rather than to the protoplasm as a whole. In some cases we can meet this objection by observing the Brownian movement of cell inclusions, since the magnitude of this movement is inversely proportional to the viscosity of the medium.

Brownian movement has been observed within living cells by numerous observers; of these perhaps the most recent are Gaidukov (1910), Leblond (1919), Bayliss (1920) and Heilbrunn (1928). The movement is recognisable in particles whose diameter does not exceed 4μ. As the particles decrease in size an irregular trembling motion about a fixed position is replaced by an irregular dancing movement which involves displacement in space (see Freundlich, 1926, p. 345). The movement doubtless due to the bombardment of the particle by molecules of the surrounding medium, for on this assumption the phenomenon has received peculiarly adequate quantitative analysis by Einstein, Smoluchowski, and Perrin. The nature of the movement is independent of the nature of the particle, but is influenced by the size of the particle and by the viscosity of the outside medium (see Freundlich, pp. 341–59).

Owing to the irregular nature of the motion it is impracticable to follow each displacement, but, as Perrin and Chaudesaiques have shown, the mean horizontal displacement during a period of time can be calculated from Gauss's theory of errors—so that the square of the mean displacement is proportional to the time of observation. Einstein's formula for horizontal displacement (ξ) is

$$\xi = \sqrt{\frac{RTt}{3\pi N \eta r}} \qquad \ldots\ldots(\text{vii}),$$

where $t =$ time, and $\eta =$ the viscosity of the medium surrounding a particle of radius r.

A preliminary attempt to calculate the viscosity of protoplasm from the rate of Brownian movement has been made by Heilbrunn (1928). Using *Arbacia* eggs Heilbrunn calculated the time required for a redistribution of egg granules which had been aggregated together by centrifugal force. 'If one tries to estimate the time required for the first granules to reach the pole of the egg farthest from the hemisphere to which they have been centrifuged, one finds that no exact measurement is possible.[1] The time seems to be more than half an hour and less than an hour. This would indicate a viscosity between 2·87 and 5·75 times that of water, or roughly about four times that of water.' It is obvious from equation (vii) that unless the size of the particle is known the mere observation of Brownian movement within a cell gives us no real evidence concerning the viscosity of the cell matrix. On the other hand, it would not be possible to detect movement of microscopic particles within a medium whose viscosity was comparable to that of a rigid gel. Lewis (1923) has observed Brownian movement within smooth and striated muscle cells, which indicates that their internal viscosity is not as high as might otherwise be supposed.

Baas Becking, Sande Bakhingzen and Hotelling (1928) have considered in more detail the possibility of establishing a satisfactory estimate of protoplasmic viscosity, by the use of Einstein's equation. The material used was *Spirogyra* and the method consisted in recording the displacements of a single particle having a diameter

[1] Recent observations indicate that more accurate measurements can be made with the eggs of *Sabellaria* (Cox and Harris, unpublished results).

of approximately 0.4μ. Camera lucida drawings were made at a uniform magnification of × 2100, the position of the particle was recorded on co-ordinate paper every fifteen seconds. The method is not without its difficulties, for it is hardly practicable to concentrate the eye on an individual particle for more than 180 seconds. Compensating for the difficulty in obtaining more extensive data is the fact that the standard errors of the observed data appear to be satisfactory. A more serious source of error is involved in measuring the diameter of the particles. The results obtained reveal the significant fact that the apparent viscosity (η) is not constant even when tracings were made at the same spot fifteen seconds apart: thus at 10° C. the following values for η were obtained:

0·052; 0·078; 0·052; 0·111; 0·250; 0·058; 0·062; 0·132; 0·048; 0·058.

At 10° the maximum value observed in all experiments was 0·472 and the minimum 0·022 (water = 0·01). Becking and his associates assume that these fluctuating values are due to the fact that protoplasm is essentially heterogeneous and in different regions and at different times offers different degrees of resistance to Brownian movement. This conclusion is justified as long as there has been no change in the effective radius of the moving particle. In some inanimate systems the degree of Brownian movement undergoes marked changes by variation in the layer of water immediately in contact with the particle. In suspensions of 'bentonite,' each particle may, under certain conditions, become surrounded by an immobile layer of water molecules which increases the effective radius of the particle and thereby inhibits Brownian movement (Hauser, 1929). It is not impossible that changes in hydration may occur at the surface of intraprotoplasmic granules, and if this be the case it is obvious that changes in the activity of Brownian movement will not be a reliable index of changes of protoplasmic viscosity.

Perhaps the most significant deductions from the nature of Brownian movement within living cells are those which refer to an alteration in viscosity under different physiological conditions. In such cases the same series of granules are observed, so that any decrease in the activity of their motion may be attributed to an increase in protoplasmic viscosity, as long as there is no change in the hydration of the particles. Thus van Herwerden (1925) noted that Brownian movement ceased in the leucocytes of man on exposure to saline containing 0·1 per cent. acetic acid, and that it was resumed on removing the acid. Bayliss (1920) reported a correlation between Brownian movement and pseudopod formation in *Amoeba*—but this does not appear to have been confirmed.

If the viscosity of protoplasm is liable to undergo drastic and rapid changes it is not easy to form any quantitative conception of the process, since any direct determination of viscosity occupies a considerable period of time. Whilst such changes doubtless occur in many cases, there are cells in which the average viscosity at successive moments does not appear to vary very significantly, and in these cells some indication of absolute viscosity can be obtained by means of a centrifuge. By this means visible intracellular granules can be subjected to a force (greater than gravity) which tends to move them through the protoplasmic matrix at a rate which can be measured and which is related to the protoplasmic viscosity by means of Stokes' Law.

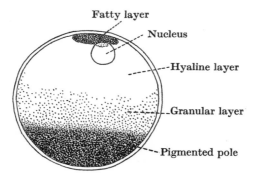

Fig. 15. Centrifuged egg of *Arbacia* showing stratification into four regions. Egg centrifuged for two minutes under 6400 gr. (After Lyon.)

The best known experiments of this type have been performed with the eggs of sea urchins. Lyon (1907) found that after exposure to a suitable centrifugal force, the eggs of *Arbacia* could be stratified into four distinct layers, one of which was perfectly clear and which alone had an affinity for cytoplasmic stains. It is in this clear hyaline zone that the mitotic figure subsequently develops, so there can be no reasonable doubt that it contains the essential elements of normal cytoplasm. It is particularly important to notice that after such drastic treatment the eggs can be fertilised and develop normally. It may therefore be concluded that the normal cytoplasm of the egg is not destroyed by experimental procedure.

More exact information concerning the fluidity of protoplasm is available from the work of Heilbrunn (1926 a–c). This author has determined the viscosity of the cytoplasm in the eggs of *Arbacia* and *Cumingia* by the application of Stokes' Law as modified by Cunningham (1910).

$$V = \frac{2cg\,(\sigma - \rho)\,a^2}{9q\eta} \qquad \ldots\ldots(\text{viii}).$$

In this formula (viii) V is the speed at which the granules travel through the cytoplasm under a centrifugal force of cg absolute units, where g is the gravity constant; σ is the specific gravity of the granules, ρ is the specific gravity of the cytoplasm itself; a is the radius of the granules, and η is the coefficient of viscosity of the cytoplasm; q is a factor which allows for the fact that in the cell there are not only a large number of granules present but that there is also a material displacement of cytoplasm as the granules travel through it. It is obvious that a measurement of V, c, a, $(\sigma - \rho)$ and q will enable η to be calculated.

As pointed out by Heilbrunn (1928) Stokes' Law in its usual form

$$V = \frac{2g\,(\sigma - \rho)\,a^2}{9\eta}$$

involves certain assumptions which are far from true in the case of centrifuged cells. The law assumes that the volume of the surrounding fluid is infinite, but when the movement of spheres occurs within a cylindrical column of fluid of limited volume the observed velocity (v) is higher than is accounted for by the true viscosity of the fluid. Applying Ladenburg's (1907) correction for this factor gives

$$V = \frac{2g\,(\sigma - \rho)\,a^2}{\left(1 + 2\cdot 4\,\dfrac{a}{R}\right)\left(1 + 3\cdot 1\,\dfrac{a}{L}\right)\eta},$$

where $R =$ radius of the cylinder and $L =$ its length. The application of this correction to Heilbronn's (1922) figures for the protoplasm of *Vicia faba* indicates that the uncorrected values are 20–30 per cent. too high. Again, if a large number of particles are present, the process of sedimentation is accompanied by a return flow of liquid, and this flow is affected by the presence of the numerous granules. Cunningham (1910) introduced a factor q, which must be applied where intracellular granules are numerous as in a sea urchin's egg,

$$V = \frac{2g\,(\sigma - \rho)\,a^2}{9q\eta},$$

where q depends on the radius of the particles (a) and on their distance apart (b). If $\sqrt{\frac{3b}{2a}}$ is not less than 2–3, q is not very significant.[1]

In all Heilbrunn's experiments c had the value 4968; the average time for stratification in *Arbacia* eggs was forty seconds and the average distance travelled was 43μ, so that $V = 0.00011$ cm. per sec. The radius of the granules was determined by a scale micrometer and found to be 0.16μ or 0.000016 cm. The specific gravity of the granules was determined by shaking the eggs to pieces; the resultant suspension was then centrifuged into sugar solutions of varying specific gravity to find a solution in which the granules just pass to the bottom of the solution. The value of σ so found was 1.10353. The value of ρ can be found from σ and the specific gravity of the whole egg, and was found to be 1.03583, so that $(\sigma - \rho) = 0.06770$. Cunningham's factor, as calculated from the size of the granules and the distance they are apart from each other, was found to be 10.54. By substituting these values in equation (viii) η is found to be 0.01830; the viscosity of water at the same temperature is 0.01, so that the protoplasm of the sea urchin's egg is, according to Heilbrunn, only 1.8 times that of water. For the protoplasm of *Cumingia* eggs Heilbrunn obtained a coefficient of viscosity four times that of water. These figures show that egg protoplasm is very much less viscous than olive oil or glycerine.

On the other hand, the application of these principles to the cytoplasm of *Paramecium* indicates that the viscosity is from 8027 to 8726 times that of water (Fetter, 1926). In these experiments the particles which were driven through the cytoplasm were either of iron or of starch, so that their specific gravity was known fairly accurately, and no correction was necessary for Cunningham's constant. Fetter suggests that the high value exhibited for *Paramecium* is due to the fact that a fibrillar neuromotor apparatus is present, which would oppose movement on the part of enclosed granules.

Heilbrunn's experiments with animal cells give results which are

[1] According to Kermack, M'Kendrick and Ponder (1929) the rate of fall of a particle under gravity in the presence of other particles is given by the equation
$$U = U_0 (1 - K\phi),$$
where U_0 is the rate of fall of an isolated particle in a large volume of fluid, the value of which is given by Stokes' Law, ϕ is the volume of suspended particles per unit volume of the suspension.

in keeping with those obtained by Heilbronn (1914) with plants. Using natural starch grains as moving particles, Heilbronn observed the velocity of fall within the cells of a section of *Vicia faba*, and concluded that the protoplasmic viscosity was less than eight times that of water. More recently the same author (1922) has estimated the viscosity of myxomycete plasmodia by introducing a small iron rod into the protoplasm and observing the number of ampères which must be passed through an electromagnet to rotate the rod. By a comparison of this current strength with that required to rotate the same rod in water, Heilbronn concluded that the protoplasm was fifteen to twenty times as viscous as water. In absolute units the viscosity of the protoplasm in the aethalia of *Reticularia lycoperdon* varies from 0·165 to 0·185, and that of the plasmodia of *Badhamia utricularis* from 0·095 to 0·105. Similar experiments have been performed by Seifriz (1923–4) on animal cells. Small nickel particles were introduced into the eggs of *Echinarachnius* and the velocity of movement in an electromagnetic field of known strength was determined. Seifriz concluded that the protoplasmic viscosity approached that of pure glycerine (800 times that of water)— a figure considerably higher than that reached by Heilbrunn. The particles used by Seifriz were 16μ in diameter and one of these was placed inside an egg of *Echinarachnius* by means of a microneedle. The pole of an electromagnet was then brought up to within 1 mm. of the egg. As soon as the magnetic field was established 'the particle immediately rushed across the central region of the egg towards the magnet and came to a standstill a short distance from the surface of the egg membrane'. On bringing the magnet to the other side of the cell a similar phenomenon at once occurred. These facts indicate that the surface of the egg (1/10 of its diameter) possesses considerable rigidity and is fairly sharply marked off from the central fluid cytoplasm. By comparing the rate at which a nickel particle travels through pure glycerine of known viscosity, Seifriz concluded that the viscosity of cytoplasm was only slightly less than that of this liquid, viz. 800 times that of water. He claims that this result harmonises with his previous estimate—based on microdissection technique—of the cytoplasm of *Tripneustes* and *Echinarachnius*.

It is of interest to note that fluid protoplasm is not elastic, and that its apparent viscosity is independent of the rate at which included particles move under the influence of centrifugal force

(Heilbrunn, 1928). If protoplasm possessed an organised structure which was destroyed by the centrifuge, the rate at which particles would move would be greatly reduced when the applied force was low. It would be interesting to apply this principle to *Paramecium* (see above).

Although some doubt may be entertained as to the exactitude with which Stokes' Law or Einstein's equation can be applied to the movement of intracellular granules, there can be no question that, in most cases examined, living protoplasm must be looked upon as essentially fluid and not solid: in other words as a colloidal sol and not as a colloidal gel.

Table VIII

Centrifugal force	Viscosity of *Cumingia* eggs in arbitrary units
310·5	3·5
552	3·7
1242	3·5
4968	3·0

The terms *sol* and *gel*, first used by Graham, are not equally easy to define. The term *sol* is usually applied to a colloidal system in which particles of a solid (or a second liquid) are suspended in a continuous phase of liquid; in other words the solid or liquid *disperse* phase is present in a *continuous* phase of liquid. In a *gel* the continuous phase is either a sol or possesses rigidity, while the dispersed phase is liquid. Under the term gel, Graham included not only jellies such as gelatin or silicic acid but also gelatinous precipitates. More recently the term has been applied to colloidal systems which possess the properties of shape and of cohesion, and which are elastic. How far the elastic 'sols' described by Hatschek (1913) should be included under gels seems uncertain. For the present purpose, however, it is possible to regard a 2 per cent. jelly of gelatin as a typical gel, and it is of interest to note that its coefficient of viscosity is 30,000 times that of water even before gelatinisation is complete.

The significance of these facts is obvious. All biological conceptions of protoplasm are essentially those of a highly complex and heterogeneous system, since cytoplasmic differentiation undoubtedly exists (see p. 88). Nevertheless, this highly organised state must exist within a system whose constituent elements are free to flow over each other with not much more difficulty than the molecules of water.

The effect of temperature on protoplasmic viscosity

The accurate determination of protoplasmic viscosity is of fundamental importance not only in respect to the physical state inside the cell, but in respect to the changes induced in the cell by changes in temperature. Chemical reactions which take place in a homogeneous medium obey with reasonable exactitude Van't Hoff's law

$$\frac{d(\log_e k)}{\delta t} = A$$

or, as it is more usually expressed, the velocity of the reaction is increased by a constant multiple (between 2 and 3) for each successive rise of 10° C. No biological reaction *in vivo* obeys this law, but each successive rise in temperature decreases the value of the temperature coefficient. At low temperatures a rise of 10° C. may increase the velocity of the reaction tenfold, whereas at higher ranges a similar rise will only produce an increase of 1·5. The reason for this discrepancy is not known, but it is possible that in dealing with the chemical reactions which occur in a heterogeneous system, such as protoplasm, the viscosity of the continuous phase should be taken into account, see Bělehrádek (1930) and Stiles (1930); before this can be adequately investigated it is necessary to know the absolute viscosity of the cytoplasm and how it varies with changes in temperature.

The effect of temperature on the viscosity of protoplasm has been investigated by Weber (1923), Heilbrunn (1925), Pantin (1924), and Becking (1928) and the results are not altogether consistent. According to Heilbrunn (1925) the eggs of *Cumingia* show a well-marked maximum viscosity about 15° C., and other maxima at 1° C. and at 31° C. (see fig. 16).

Similar results appear to have been obtained by Heilbronn (1922) using the protoplasm of Myxomycetes, although the variations observed were less marked. According to Pantin (1924), however, the viscosity of the cytoplasm of *Nereis* eggs falls steadily with rising temperature as is the case with non-living systems (see fig. 17).

The peculiar maxima observed by Heilbrunn may conceivably be due to secondary changes in physical state which in turn affect the viscosity. For example, the apparent viscosity would be materially altered by such reactions as are described on p. 45, where increasing temperature alters the equilibrium between proteins and calcium ions. Weber (1923) on the other hand suggests that the

PROTOPLASM

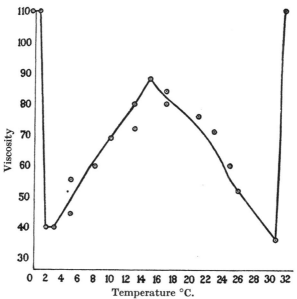

Fig. 16. The effect of temperature on the viscosity of the egg-protoplasm of *Cumingia*. Note the maxima at 1°, 15°, 31°. The viscosity was measured by the time required to statify the protoplasm under a constant centrifugal force. (From Heilbrunn, 1925.)

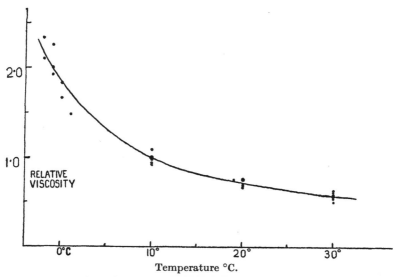

Fig. 17. Curve showing relative viscosity of egg protoplasm of *Nereis* at different temperatures. The curve is the mean of eight experiments. (From Pantin, 1924.)

viscosity of protoplasm can be 'regulated' by the cell by an undefined mechanism. It is interesting to note that if the figures given by Pantin are applicable to protoplasm in general, and are used as a correction for the value of the temperature coefficient of a cell's activity—then the three main types of protoplasmic movement (amoeboid, ciliary, and muscular) are all characterised over almost their complete temperature range by a coefficient of the same value (see Pantin, 1924, p. 33).

The effect of temperature upon protoplasmic viscosity has been studied by Becking and his associates (1928). Calculating η from observation of the displacements of a particle in Brownian movement

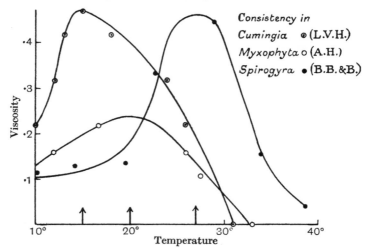

Fig. 18. Graphs showing the effect of temperature on the viscosity of protoplasm. (From Baas Becking, 1928.)

in the protoplasm of *Spirogyra*, they observed a maximum viscosity at 27° C. These results are apparently comparable to those of Heilbrunn (see fig. 18), and indicate that temperature may affect the viscosity of protoplasm in ways which are not observed in homogeneous systems where increasing temperature always reduces the viscosity and never increases it.

The gel and sol state

We have already noted that, in certain cases, the consistency of protoplasm may undergo marked changes in response to external stimuli. At one moment we may observe active streaming or

Brownian movement, whereas a moment later all visible motion ceases and the protoplasm no longer exhibits any obvious liquid properties. Such sudden and reversible changes in physical state are well-known characteristics of some colloidal systems. The best known example is, of course, gelatin. A 2 per cent. solution of gelatin below 15° C. has the properties of a fairly rigid but elastic jelly, above this temperature it is clearly a liquid capable of flow. A heat reversible 'gel' of this type is, however, not of great interest from a biological point of view, since the heat coagulation of protoplasm is of a totally different nature. On the other hand, considerable importance must be attached to those colloidal systems whose physical state depends on factors other than temperature. Of these factors the best known is mechanical agitation. Freundlich and Seifriz (1923), when investigating the viscosity of 0·85 per cent. solutions of gelatin, found that the rate of fall of a small shot through a narrow column (30 cm. in length) fell rapidly with successive trials from 22 seconds to 8·6 seconds. It looks as though the passage of the shot gradually cut a free passage through a system which under normal conditions possesses a definite and somewhat rigid structure. Parallel to this are the observations of Schalek and Szegvary (1920) who prepared elastic jellies of iron oxide which, on mechanical agitation, yielded a thin sol; these, on standing, resumed their original state of gelation. Even more marked were these phenomena when observed by Svedberg (1928) in suspensions of metallic cadmium in alcohol. It will be noted that in these cases the gel structure only exists when the system is free from agitation: when the gels are shaken the sol condition results. Comparable phenomena in biological systems are difficult to find, but may possibly occur within the pseudopodia of *Actinosphaerium* as described by Verworn (1889). The long rigid filaments of the quiescent pseudopodia immediately begin to flow on mechanical agitation. It has already been mentioned, however, that there is no evidence of an elastic or rigid structure in the protoplasm of marine eggs (see p. 65), for in these cases the apparent viscosity is independent of the rate of shear. As a rule, protoplasm appears to represent the converse state to mechanically instable gels, for stimulation results in an apparent loss of the liquid state and the acquisition of the gel condition. Clear examples of this type are provided by *Badhamia* and other Myxomycetes (see p. 57). Similar phenomena are not unknown in inanimate systems; solutions of ammonium oleate are clearly elastic when in a state of agitation—although they are inelastic when at rest (Seifriz, 1925). If several c.c.'s of oleic acid are added to 100 c.c. of 5 per cent. ammonia solution whilst rotating in a flask, an exceedingly elastic and viscous gel results. As soon as rotation ceases, this elasticity is lost within a few minutes, and the viscosity becomes so low that it will not support a nickel particle 10μ in diameter. On reshaking, the elastic state is regained.

The physical phenomena underlying the transformation of the sol to the gel state and *vice versa* are very incompletely understood. Presumably in the gel state there exist forces of cohesion between individual particles which are absent in the sol state: it is not difficult to see how gelation is effected whilst at rest, since even if cohesion is effected slowly it will progress steadily: when the system is agitated the mechanical energy applied may be sufficient to rupture the mutual surface of attachment and a sol results. On the other hand, it is not so easy to account for the converse condition where the gel state appears to depend on mechanical agitation.

It is important to notice that the transference from gel to sol and *vice versa* does not in itself effect any obvious change in the distribution of free energy and cannot in itself be responsible for active movements within a protoplasmic system. Further, apart from the obvious changes in viscosity and elasticity the transformation does not affect to any marked degree the activity of the system from a chemical point of view. Diffusion takes place through a jelly almost as rapidly as through water so long as the size of the diffusing molecules is not too great; on the other hand, the gel condition makes it possible to maintain a localised structure in a constant position, whereas this would not be the case in a sol. For example, the phenomena of Liesegang ring formation are readily observable in a gel, whereas in a fluid system the mechanical conditions are such that uniform mixing occurs except under very carefully controlled conditions. How far these facts are helpful from a biological point of view is doubtful. All the evidence suggests that when a transformation from sol to gel occurs within the normal cell it does so without any permanent or irreversible dislocation of vital activities. We must therefore conclude that the 'living' state is not dependent on such a structure as is characteristic of a colloidal sol or a colloidal gel—but is something much more fundamental. At the same time, a change from sol to gel must obviously affect the ability of protoplasm to flow from one region to another, and is probably of primary significance in the process of cell-division (see p. 206) and in amoeboid movement (see Chapter XVII).

The optical properties of protoplasm

In most cases the protoplasmic matrix of a living cell exhibits no optical structure even under the highest powers of the microscope. Exceptions to this rule are, however, not unknown. The vacuolated protoplasm of *Actinosphaerium* and other fresh-water Heliozoa are well-known examples. Whether the refractile walls together with the contents of the vacuoles are both to be regarded as essential components of the living system is not known. It is significant that the protoplasm of marine Heliozoa is much less vacuolated than that of fresh-water forms (Grüber, 1889) and is described as 'granular'. The fact that one type can be converted into another by acclimatisation to media of suitable salinity, indicates that the separation into two phases may vary considerably without violent disintegration of the essential elements. An equally regular vacuolisation of optically homogeneous cytoplasm is, on the other hand, sometimes the result of injury. Thus if *Spirostomum* is gently compressed under a cover-slip a well-defined vacuolisation precedes complete cytolysis (see also Prowazek, 1915). At first the clear transparent endoplasm develops a series of small but distinct vacuoles, these quickly enlarge so that each is separated from its neighbours by a thin wall of refractile endoplasm; eventually the vacuoles coalesce by rupture of the walls and the latter form small droplets floating in the fused contents of the vacuoles. The whole process is singularly like what occurs when a drop of olive oil containing a soap is brought in contact with water. The beautiful emulsions of water in oil which are produced in this way were regarded by Bütschli (1894) as illustrating the fundamental structure of protoplasm. This is certainly true as far as the Heliozoa are concerned, but in other cases it is open to grave doubt how far the vacuolisation is actually present during life. It is interesting to note that the process of 'vacuolisation' of protoplasm which so often heralds the death of the cell can be closely imitated in an inanimate system of globulins. If a very small drop of the concentrated globulin solution found in the yolk of *Salmo fario* be allowed to come gently into contact with the surface of water a beautiful series of vacuoles are formed prior to the final disintegration of the drop into discrete particles which in turn distribute themselves readily throughout the aqueous medium.

A definite optical structure has also been ascribed to the cytoplasm of echinoderm eggs. According to Chambers (1917) the bulk of the mature egg is to be regarded as a thick emulsion of two types

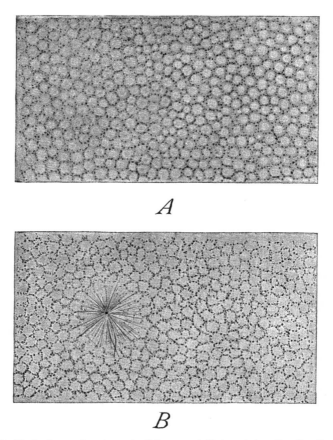

Fig. 19. Protoplasm of mature starfish egg. *A*, living state; *B*, after fixation. (From Wilson, 1928.)

of granules lying in a fluid matrix. The smaller type of granule or microsome is about $1\,\mu$ in diameter and has a high refractive index; the larger type of granule or macrosome is from $2\,\mu$ to $4\,\mu$ in diameter and has a low refractive index, and is easily destroyed. The macrosomes have however been regarded as fluid vacuoles in the walls of which lie the microsomes. If this were so, the protoplasm of the egg might reasonably be said to possess an alveolar structure. The recent observations of Wilson (1928) make this suggestion improbable. In the young immature oocytes of *Echinus esculentus* or *Asterias glacialis* both microsomes and macrosomes are relatively few in number (fig. 20) and are irregularly arranged in a clear hyaline matrix. Even in the mature egg the cytoplasm can be separated by the centrifuge into a layer rich in microsomes and a hyaline layer free from these bodies, and yet the egg is capable of development.

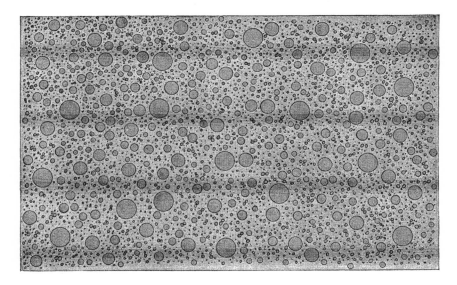

Fig. 20. Protoplasm of growing starfish egg, showing macrosomes and microsomes lying in a hyaline matrix. (From Wilson, 1928.)

The normal orderly arrangement of the microsomes in the mature egg is therefore more readily explained by the presence of a large number of macrosomes which restricts the microsomes to the space between adjacent macrosomes. We do not, therefore, need to assume a fundamental organisation for the protoplasm of the mature egg; a similar structure was observed by Bütschli in drops of oil which contained fine granules of Indian Ink.

Although protoplasm normally exhibits no optical structure, yet when submitted to a stress along a definite axis, a fibrillar appearance can sometimes be detected. The protoplasm of *Gromia* is perfectly transparent and homogeneous, but as it flows through the opening of the shell distinct fibrillae can be seen (Bütschli, 1894), and as soon as the cytoplasm ceases to be compressed the fibrillae disappear (see fig. 21). Similar phenomena have been described in *Amoeba blattae* and in the streaming protoplasm of some plant cells. These fibrillae, like those of the mitotic figure, readily disappear on applying pressure to the cell at right angles to their length. They indicate that the

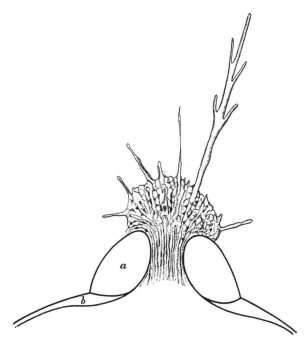

Fig. 21. Protoplasm streaming from the aperture of the shell of *Gromia dujardinii*. Note the fibrillar structure of the protoplasm in the mouth of the aperture, the alveolar extravasated protoplasm, and the hyaline pseudopodia. (From Bütschli, 1894.)

particles of cytoplasm can orientate themselves along definite axes when lying in a suitable field of force, just as do those of albumen (Hardy, 1899). The existence of such orientations is not always obvious during life; according to Lewis (1923) the visible fibrillae seen in the preserved cells of cardiac muscle are the results of fixation and are invisible in the living cells. It is, however, by no means certain that the fibrillae observed by Bütschli are present within the cell and are not restricted to the surface ectoplasm: they may, in other words, simply be wrinkles on a relatively inelastic surface.

Among those types of protoplasm which are not infrequently regarded as optically homogeneous are the ectoplasm and endoplasm of *Amoeba*. A rather different conception has, however, been advanced by Mast (1926). In *Amoeba proteus* the outside of the organism is defined by a solid elastic layer (the plasmalemma) about $0.25\,\mu$ thick. Immediately under this is a hyaline layer of variable thickness which contains granules in an active state of Brownian movement. The rest of the body consists of granular protoplasm which appears to be uniform as long as the amoeba is at rest; when the organism is in motion it can be seen that only the central core (the

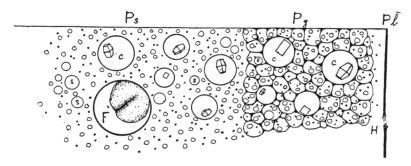

Fig. 22. Illustrating protoplasmic structure of *Amoeba* (from Mast). *Pl*, plasmalemma torn near one end (note the shreds indicating a fibrous structure); *H*, hyaline layer; *Pg*, plasmagel (the framework represented is inferred from the restricted Brownian movement of enclosed particles); *Ps*, plasmasol; *F*, food vacuoles.

plasmasol) is in motion, whereas the outer layer (the plasmagel) is stationary. Within both plasmasol and plasmagel included granules and crystals are in active Brownian movement, but those within the plasmagel are highly restricted in their movements, and do not change their relative position freely as is the case in the plasmasol. Mast concludes that the plasmagel is essentially alveolar in structure, so that an enclosed particle can only move within the confines of its own particular vacuole. This observation is of interest from two points of view. Firstly, it shows that the existence of Brownian movement is in itself no indication of a general state of protoplasmic fluidity, but only of that particular region through which the particle actually moves. Secondly, it raises the problem, discussed elsewhere (see Chapter II), of how far living matter is subdivided into small

76 PROTOPLASM

microscopic or submicroscopic phases within which different types of reaction can go on, and which may be to some extent outside the scope of statistical laws. It should, however, be noted that even if such phases exist within a sea urchin's egg they must, like the granules in the plasmasol of *Amoeba*, be capable of moving their relative position without much more difficulty than if they were suspended in water.

Miscibility of protoplasm and water

If protoplasm is to be regarded as a colloidal system whose continuous phase is largely composed of water, one would expect protoplasm to be freely miscible with water. According to Chambers

Fig. 23. Effect of injection of a large amount of distilled water into *Amoeba*. *a*, plasmalemma lifted away from granuloplasm which lies in the centre; *c, d*, note scattering of granules; *e*, normal form regained.

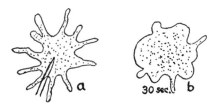

Fig. 24. Injection of small amount of $2M$ NaCl. *a*, before injection; *b*, 30 seconds after injection. Note complete mixing. (From Chambers and Reznikoff, 1926.)

(1917) this is actually the case, since when water is injected into protoplasm there is no phase boundary formed between the two. Chambers and Reznikoff (1926) have recently shown that a volume of water equal to half that of the protoplasm can be injected into an amoeba without any indication of a phase boundary; the water immediately diffuses throughout the cell.

On the other hand, Seifriz (1918) states that in no case is normal cytoplasm miscible with water. Seifriz's conclusion is based on the following facts. Isolated fragments of protoplasm from the plasmodia of Myxomycetes or from the pollen tubes of *Iris* are capable of independent existence in water, but in each case the drops quickly surround themselves by a relatively tough membrane which Seifriz regards as a precipitation membrane; such membranes, he claims, are only formed at the junction of two immiscible fluids.

Seifriz's argument appears to be based upon a misunderstanding of the conditions requisite for the formation of precipitation membranes. It is true that such membranes are readily formed at the interface of two immiscible fluids, but the essential condition for their formation is not the immiscibility of the liquids but the impermeability of the membrane to the substances which go to form it. If a drop of olive oil containing a little oleic acid is placed in contact with a mixture of alcohol and water of the same specific gravity, a membrane can be formed round the drop by adding a soluble salt of calcium to the aqueous phase: the membrane is composed of calcium oleate. On the other hand, an equally good membrane can be produced by allowing a drop of sodium oleate solution to come *gently* into contact with a drop of calcium chloride. If sodium oleate and calcium chloride solutions come into violent contact no membrane is formed, but complete mixing accompanies the precipitation of the calcium oleate. The apparent discrepancy between Chambers' results and those of Seifriz may be due to the fact that the process of micro-injection is too violent and rapid to allow of membrane formation. Such an interpretation would help to explain the fact that definite intracellular vacuoles can exist within the protoplasm of the normal cell. Kite (1913) stated that the walls of the vacuoles in *Amoeba proteus* possess a degree of cohesion much higher than the surrounding endoplasm, and this has been confirmed by Chambers (1928).

In considering the so-called miscibility of protoplasm with water, the very greatest caution must be exercised and some precise agreement must be reached concerning the real meaning of the term Many authors seem to imply that miscibility with water is synonymous with miscibility with any aqueous solution. That this is quite illegitimate can be seen in two simple experiments. If a concentrated solution of globulin comes into contact with a solution of $M/2$ NaCl complete 'miscibility' is observed: if, on the other hand, it comes into contact with water, two distinct phases are found: one which is rich in globulin and poor in water, the other rich in water and poor in globulin. Similarly, if a sea urchin's egg is crushed or ruptured in sea water the cytoplasm forms a compact and tenacious mass: if an egg is ruptured in isotonic sodium chloride solution the cytoplasm mixes freely with water (see also fig. 25). The newly formed interface between protoplasm and water is probably the seat of more than one type of reaction. If the area of the newly-formed interface increases very slowly, it is probable that there will precipitate at the surface any water insoluble protein present in the cell—these substances will perhaps take part in the formation of a definite membrane and thereby inhibit further 'mixing'. At the same time a precipitation membrane will form between any intracellular compound capable of reacting with calcium or other salts

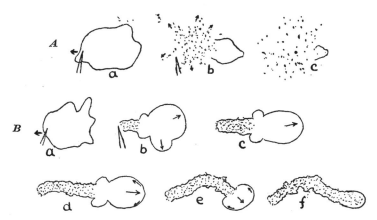

Fig. 25. Effect of tearing an amoeba. A, $M/13$ NaCl. Rapid dissipation occurs with scattering of granules; B, *Amoeba* torn in $M/13$ $CaCl_2$. Note the absence of scattered granules. The organism remains a coherent mass. (After Chambers and Reznikoff. 1926.)

in the water. If, however, the area of the interface between the protoplasm and aqueous phase increases too rapidly for membrane formation a process akin to complete miscibility will be reached.

Perhaps the nearest picture we can get of protoplasm is that of a colloidal emulsion whose continuous phase is of an aqueous nature but whose dispersed components are extremely sensitive to changes in the nature of the continuous phase. It is quite certain that the latter is not pure water, but a solution of crystalloids and electrolytes. With such a medium protoplasm is perhaps 'miscible' in all proportions, but contact with any other aqueous medium in the long run will probably lead to abnormality and death.

The chemical composition of protoplasm

The methods available for the chemical analysis of cell constituents are necessarily somewhat crude. Macroscopic analyses of cells can be carried out if the tests which are involved can be suitably modified for use with comparatively small amounts of tissue. Alternatively, if only one cell is available it is possible to apply qualitative tests for cell constituents by injecting suitable reagents into the cell. In a few cases (notably nucleic acid and certain metallic radicles), it is possible to use intracellular methods which not only give a qualitative estimate of cell constituents, but also indicate the position in the cell which these constituents normally occupy.

For macroscopic analysis, cytoplasm from the plasmodia of Myxomycetes is the most popular source of supply. This or similar material can be analysed by standard chemical methods; using *Reticularia lycoperdon*, Kiesel (1925) obtained the following composition for the water-free organic constituents of protoplasm.

Table IX

	% by weight
Oil	17·85
Lecithin	4·67
Cholesterin	0·58
Reducing carbohydrate	2·74
Non-reducing soluble carbohydrates	5·32
Glycogen	15·24
Polysaccharides	1·78
Nitrogen extractives	12·00
Proteins	29·07
Nucleic acid	3·68
Lecithoproteins	1·20
Unknown substances	5·87
	100·00

Such analyses are necessarily very rough approximations to the composition of homogeneous protoplasm, but when we compare the analyses of a number of different types of cell we get fairly reliable evidence to indicate that protoplasm invariably contains proteins, and that with the proteins there are combined or associated certain derivatives of the fats—all other organic constituents being present in varying quantities or even absent altogether.

Macrochemical analyses of large cell masses have been made on many occasions, but from a cytological point of view do not yield results of any great significance. It is interesting to note, however, that the composition of some types of cell seems to depend on the composition of the external medium or with the physiological condition of the cells. For example, the protein present in some bacteria (e.g. *Azotobacter*) may vary from 12 per cent. dry weight if a culture is grown on a solid medium to 30·5 per cent. if grown in a liquid medium (Hunter, 1923); on the other hand, the protein content of *Aspergillus niger* is independent of the medium (Terroine, Würmser, and Montané, 1922). The influence of age or physiological condition on cell composition may be illustrated by the figures given by Kruse (1910) for the nitrogen content of *Mucor* (Table X).

Table X

Days of growth	Dry weight	% nitrogen
4	0·46	16·2
6	0·51	7·1
8	1·12	5·6
10	1·67	6·2
12	1·72	5·2
14	1·83	3·7

Qualitative analyses of protoplasm in mass can be confirmed by microanalysis. Cells treated with Millon's reagent (a mixture of mercurous and mercuric nitrites) yield a positive reaction for proteins; similarly fats can be detected by treatment with osmic acid (see Bolles Lee, 1928, p. 465 seq.). Starch and certain other carbohydrates can be detected by means of the iodine reaction. Further examples of this type of qualitative analysis will readily occur to the reader, and as a rough guide to the composition of cell constituents such reactions are useful, although they are in few cases quantitative and in no case do they throw much light on the real molecular structure of the cytoplasmic ground work of the cell. Before leaving reactions of this type it is worth mentioning an instance of intracellular analysis made possible by microinjection of reagents into the interior of a cell. By this technique Pollack (1928) was able to detect the presence of intracellular calcium by the injection of sodium alizarin sulphate into an *Amoeba*. This alizarin compound reacts with calcium to form purplish-red crystals of calcium alizarinate. Shortly after injection the interior of the cell shows the presence of fine red granules, while the whole cytoplasm is itself coloured pale red. According to Pollack an *Amoeba* possesses a considerable reserve of calcium salts, some of which are set free on pseudopod formation. It seems probable that the method of microinjection of reagents could be extended, with success, to other intracellular reactions.

To some extent, organic cell constituents can be identified by taking advantage of their differential staining properties. This method was elaborated by Unna and Tielemann (1917) in their analyses of various tissue-cells and of amoebae. The cells are first fixed to a slide with osmic acid vapour and are then subjected to suitable stains after treatment with various solvents. Unna and Tielemann divide the main organic cell constituents into basic proteins, acid proteins, and lipoids. Similar analyses have been made of bacterial and yeast cells by Gutstein (1925).

Table XI

Compound	Properties	Location in *Amoeba*
(a) Basic proteins	Not easily soluble in water Stain in haematoxylin Digested by 1 % trypsin	Nucleus Endoplasm Ectoplasm
(b) Basic proteins	Soluble in water	Endoplasm
Protamines	Stained red in Giemsa stain Not digested by pepsin	Surface of nucleus only
Globulin	Stained in methylene blue Soluble in 2 % NaCl	Interior of nucleus Endoplasm, ectoplasm
Lipoids	Soluble in acetone, alcohol or benzine	Endoplasm, ectoplasm

The colorimetric tests applied by Unna and others are of interest because they indicate with some precision the locality occupied in the cell by the various constituents, and in this respect resemble the better known intracellular tests for inorganic radicles.

The inorganic constituents of the cell were examined microchemically by Macallum (1905). Roughly speaking, the methods employed are those used for larger analyses, but suitably modified for the small quantities involved in a single cell. In every case, the success of the method depends on the formation of a specific precipitate within the cell, or in a specific differentiation in staining properties.

Cobalti-nitrite test for intracellular potassium. The reagent employed is a solution of sodium cobalti-nitrite which is prepared by dissolving 20 gr. of cobalt nitrite and 35 gr. of sodium nitrite in 75 c.c. of water containing 10 c.c. of glacial acetic acid. As soon as the evolution of nitrogen peroxide has ceased, the solution is filtered and diluted to 100 c.c. with water; it is then ready for use. If a small quantity of this solution is added to a solution containing a potassium salt, an orange-yellow crystalline precipitate of the triple salt is formed. This can subsequently be converted into cobalt sulphide by the addition of ammonium sulphide and the original presence of potassium thereby made more obvious. In practice the sodium cobalti-nitrite is added to the freshly-teased tissue or frozen section and left for 20 minutes: the tissue is then washed several times with ice cold water until the washings are colourless. The preparation is then mounted on a slide in a mixture of equal parts of 50 per cent. glycerine and of concentrated ammonium sulphide solution. The location of the potassium can then be determined microscopically from the position of the opaque cobalt sulphide. The preparations are relatively stable. According to Macallum potassium may or may not be present in the cytoplasm, but it is invariably absent from the nuclei. The cytoplasm of many protozoa (e.g. *Vorticella*) appears to be devoid of potassium, but in other cases potassium can be detected quite readily within the cytoplasm—often in highly localised areas (see fig. 26). Thus in *Spirogyra* the potassium is strictly limited to the margin of the chromatophore; in striated muscle-fibres the dim bands are rich in potassium, the light bands giving a negative test. In plant cells, also, potassium is accumulated at regions of active growth, e.g. at the root tips of the spores of *Equisetum arvense* or in the early stages of growth of pollen tubes. Nerve cells contain no potassium.

Microchemistry of iron. Free iron salts can be detected within the cell by a variety of methods of which perhaps the best is that developed by Macallum (1897). The test depends on the fact that haematoxylin forms with the salts of heavy metals deeply coloured insoluble higher oxides; no such reaction occurs if the iron is incorporated into an organic molecule. Organic compounds of iron can be detected only if the iron is first converted into a metallic salt by acid hydrolysis. By means of this test Macallum (1897) demonstrated the presence of iron within the nucleus: the position of the iron being marked by the deposition of the dye as a blue-black or

PROTOPLASM 83

blue colour. An alternative test for free iron salts is provided by the use of ferrocyanide solutions or ammonium sulphide.

Microchemistry of phosphorus. Macallum's test for intracellular phosphorus is described in detail by Mann (1902, p. 295). The cells are exposed for some time to a solution of ammonium molybdate. The phosphates then form a phospho-molybdate which may render the cells yellow to the eye. The phospho-molybdate is then reduced

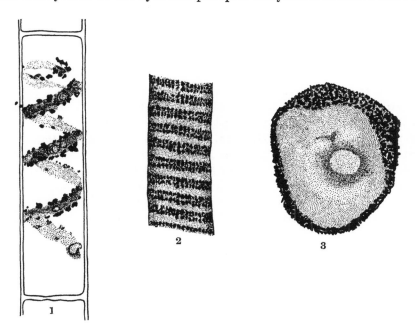

Fig. 26. *Intracellular distribution of potassium.* The regions rich in potassium are shown black after treatment with sodium cobalti-nitrite and ammonium sulphide. (From Macallum.)
  1. Cell of *Spirogyra* showing condensation of potassium on the surface of the chromatophore.
  2. Wing muscle of *Aeschna*.
  3. Nerve cell from the Gasserian ganglion of a dog.

by washing the cells in a 2 per cent. solution of phenyl-hydrazin hydrochloride. Originally pyrogallol was used as a reducing agent —but this is to be avoided as it also reduces the free ammonium molybdate. The presence of the phospho-molybdate is revealed after reduction by the formation *in situ* of the dark green oxide of molybdenum. By suitable preliminary treatment of the cell it is possible to determine whether the phosphate detected is attributable to free phosphates, phospho-proteins, or phospho-lipins. In every case the nucleus is found to be rich in phosphorus, and in muscle the dark bands are much richer than the light—as is also the case for iron.

The Feulgen reaction for nucleic acid. One of the most valuable microchemical tests is the reaction elaborated by Feulgen (1923) for the detection of thymo-nucleic acid *in situ* in the cell. The principle of the test depends on the fact that partial acid hydrolysis of thymo-nucleic acid yields an aldehyde grouping which reacts colorimetrically with fuchsin containing free sulphurous acid; the latter reaction is Schiff's test for aldehydes in which a positive reaction is indicated by the development of a red or violet colour. The test for nucleic acid *in vitro* can be performed as follows (Feulgen, 1923, p. 1055). A solution of 0·3 per cent. sodium salt of thymo-nucleic acid in water is prepared: to 1 c.c. of the solution is added 1 c.c. $N/10$ H_2SO_4 and the mixture incubated for 10 minutes in a water bath. The solution is then cooled and neutralised by $N/10$ NaOH. A few drops are then added to the test solution of fuchsin (see Feulgen, p. 1063). After some minutes a brilliant coloration is observed. The test is sensitive to 0·01 mg. of nucleic acid. When used for the detection of thymo-nucleic acid within the nucleus the test is highly specific if proper precautions are observed. The nucleic acid from yeast yields negative results. The test provides a very valuable check on the relative distribution of nucleic acid and of 'chromatin' with living cells.

Except in the particular case of the Feulgen reaction it is not very easy to assess the cytological value of microchemistry. When we state that protoplasm invariably contains proteins and derivatives of the fats, the contribution to cytology is comparable, in aeronautics, to a statement that an aeroplane invariably contains iron and copper. Either the essential constituents of the living machine are composed of protein and lipoids or they are orientated in a matrix of these compounds. Both for a study of the cell and of an aeroplane we require to know not only the shape and function of the various chemical constituents, but we want to know how they are orientated in respect to one another to form a working and useful unit. By purely chemical methods we can at times detach, from the whole living system, parts of the machine which will continue their normal function *in vitro*, and a study of such parts—enzymes in particular—may eventually enable us to form some adequate picture of the whole engine. We get the impression, however, that during this process the true structure of living protoplasm is eluding us, and that no real advance will be made until some new method of analysis enables us to see how the various protein molecules and aggregates are orientated in respect to each other. When that happy state has been reached it will be interesting to see how far the chemical constitution of the protoplasmic framework is only of secondary significance.

The hydrogen ion concentration of protoplasm

The hydrogen ion concentration of intracellular protoplasm is highly significant but is not easy to determine with accuracy. Within certain limits its value can be assessed by injecting, into the cell, indicators whose colour changes are sufficiently well defined in the region of absolute neutrality. Observations of this type were first made by Needham and Needham (1925, 1926) and later by Chambers and Pollack (1927). From the results obtained by these authors we may conclude that the hydrogen ion concentration of living protoplasm is almost the same (6·9 ± 0·1) for all the cells examined—*Amoeba* (Needham and Needham, 1925; Chambers, Pollack and Hiller, 1927), echinoderm eggs (Needham and Needham, 1926; Chambers and Pollack, 1927), and cells of the frog and mammals (Chambers, Pollack and Hiller, 1927).

The recent work of Chambers and others shows that considerable care must be exercised in the interpretation of colorimetric observations of intracellular pH. If basic dyes such as neutral red are injected in the cell, the dye quickly accumulates into discrete granules whilst the hyaline cytoplasm remains more or less colourless. The colour of these granules is not a reliable guide to the cytoplasmic pH. If, on the other hand, acid dyes such as brom-cresol-purple, phenol red, and cresol red are used, they do not accumulate into granules but give a permanent and diffuse colour to the cytoplasm. Under normal conditions the colour of the cytoplasm is equivalent to a pH of 6·9 ± 0·1. According to Chambers (1928) this value cannot be altered by the presence of alkali or acid as long as the cell is alive; the only constituents of the cell which alter in pH under such circumstances are the granules which have a high affinity for basic dyes. A typical observation was as follows. Echinoderm eggs, stained with neutral red, were immersed in a hanging drop of sea water coloured with cresol red, and were then injected with phenol red. The cytoplasm stained yellow and, embedded in the cytoplasm, were bright red granules containing neutral red; the external sea water containing cresol red was yellow. The whole system was then exposed to ammonia gas until the alkalinity of the sea water was sufficient to give an alkaline (red) reaction with cresol red. As soon as this occurred the cytoplasmic granules were found to be yellow (alkaline reaction of neutral red), while the hyaline matrix remained yellow (acid) in the phenol red. The picture was then reversed by admission of carbon dioxide—both the sea water and the neutral red granules became acid, but the cytoplasm retained its constant hydrogen ion concentration of pH 6·9. Similarly Chambers found that when using brom-cresol-purple, it was impossible to make the cytoplasm acid by the admission of free carbon dioxide as long as the cell remained alive. From these experiments and from those of Pollack (1928) it may be concluded that protoplasm has a very high buffering power—and that if its pH value changes significantly the cell rapidly dies. At the same time it is not easy to harmonise these conclusions with evidence from other sources. An acid such as CO_2 undoubtedly enters the living cell—so that if (as appears to be the case) the hydrogen ions accumulate in discrete granules, it is these granules which must be responsible for the apparent buffering power of the cytoplasm. However efficient the granules may be as absorbents of hydrogen ions there must come a time when no more acid can be absorbed—and at all times there must be an equilibrium between the amount of acid in the granules and the amount in the cytoplasm. Consequently there must, theoretically, be a change in the cytoplasm whenever the total amount of acid in the cell is varied. This change may not be detectible by indicators but it is possibly of profound physiological importance. Practically every physiological activity is sensitive to the presence of acids which penetrate the cell. Ciliary and amoeboid movement, mitosis, cell-division, and respiration are all profoundly depressed by the presence of intracellular carbon dioxide or any other rapidly penetrating acid. It is difficult to believe that these activities are really dependent on the reaction of neutral red granules—it is much more probable that our methods of observing the hydrogen ion concentration of hyaline protoplasm are too crude for physiological purposes.

The failure to observe changes in protoplasmic pH in the presence of free CO_2 is surprising in view of the fact that a marked change can

PROTOPLASM 87

be effected by the injection of calcium salts (Reznikoff and Pollack, 1928). This latter change is what one would expect in a protein system—but if hydrogen ions liberated by the addition of calcium can effect a pH change colorimetrically and if the 'acid of injury' can effect a similar change, it is difficult to see why the injection of free hydrogen ions does not act in the same way.

Electrical conductivity of protoplasm

The internal electrical conductivity of protoplasm has been determined by Brooks (1924) and by Gelfan (1928): Brooks used the extravasated protoplasm of the Myxomycete *Brefeldia maxima*. Individual plasmodia yielded 5–20 c.c. of protoplasm whose electrical conductivity was measured in the usual way. The conductivity of the protoplasm was found to be equivalent to that of a $0{\cdot}00145N$ NaCl solution and about 2·8 times that of the medium with which the plasmodium is normally in contact. Brooks concluded that the conductivity of protoplasm always bears a definite relationship to that of the medium surrounding the tissue.

Gelfan's (1927) method was preferable to that of Brooks in that it did not involve extensive injury to the cells; two specially prepared microelectrodes were inserted into the protoplasmic layer of the cells of *Nitella*, and the conductivity was found to be equivalent to that of a $0{\cdot}04N$ KCl solution, whereas that of the cell sap was slightly higher. The rate of streaming of the protoplasm appears to have no appreciable effect on the conductivity, although it is interesting to note that manipulation of the electrodes usually caused a temporary cessation of streaming. The following table from Gelfan (1928) shows the conductivity of protoplasm in different types of cell.

Table XII

Species	Normality	Specific conductance in reciprocal ohms per cm.
Amoeba proteus	0·01	$1{\cdot}225 \times 10^{-3}$
Paramoecium	0·06	$6{\cdot}9 \times 10^{-3}$
Nitella protoplasm	0·04	$4{\cdot}7 \times 10^{-3}$
Starfish oocytes	0·25	$26{\cdot}2 \times 10^{-3}$

It will be noted that the internal conductivity of protoplasm is relatively high—much higher than that of the whole cell; this fact is of great theoretical significance and will be discussed in a later chapter (Chapter XIV).

Cytoplasmic differentiation

If we restrict our view of protoplasmic structure to such facts as have already been described, we picture to ourselves an optically homogeneous material in which the included phases or particles are free to move with comparative ease. Within such a system it is not easy to see how one region of cytoplasm can remain essentially different to another unless these adjacent regions represent separate submicroscopic phases, each of which represents a physico-chemical entity distinct from but in equilibrium with its neighbours. That some such heterogeneity must exist is shown by the variety of chemical reactions which go on in an orderly manner as long as the cell structure remains intact, and which cease to proceed in this way when the cell dies. As Bayliss remarked, 'Protoplasm is an extraordinary complex heterogeneous system of numerous phases and components'. To this there can be no denial, but the position is unsatisfactory; thirty years ago biological heterogeneity was associated with microscopic fibrils, granules or alveoli, to-day we have their parallel in phase boundaries, monomolecular films, and membranes; all of these are possible, none of them is certain. The outstanding facts show that from a biological or even biochemical point of view, protoplasm must be heterogeneous, but the essential nature of this heterogeneity has not yet proved susceptible to direct physico-chemical attack. Until the chemist can contribute something more than submicroscopic models of protoplasmic heterogeneity there will be a danger of underestimating the complexity of the whole problem; it is important that the biologist should continue to regard the situation from his own peculiar angle.

The physical homogeneity of a sea urchin's egg is, to some extent, in harmony with Driesch's conception of a homogeneous and totipotent substance capable of arbitrary subdivision into discrete cells; it is equally in harmony with Wilson's (1903) demonstration that fragments of cytoplasm can be removed from the eggs of *Cerebratulus* prior to maturation without affecting the ability of the remainder to develop into normal larvae. Since it is immaterial in this case which region of cytoplasm is removed from the egg, the whole of the cytoplasm must be biologically homogeneous or the powers of self-regulation and of cytoplasmic regeneration must be extremely well developed. In both cases the system may be regarded as 'totipotent'. It must be remembered, however, that the immature *Cerebratulus* egg is not typical of egg-cells in general, for in many cases the cytoplasm is essentially heterogeneous in that distinct phases are distributed throughout the cell in an orderly manner in respect to each other. Even the eggs of *Cerebratulus* during their later stages of maturation exhibit functional heterogeneity, since larvae which develop from fragments of the cell cut from the polar region of the egg are usually deficient in the digestive tract and lateral lappets, whereas larvae from the antipolar region possess these organs. This particular example and others have been admirably reviewed by Wilson (1925), Conklin (1924), and Morgan (1927).

As pointed out by Conklin (1924), many egg-cells exhibit true cytoplasmic heterogeneity even before fertilisation and this heterogeneity can be recognised in the so-called structural polarity of the cells. Nearly all eggs are polarised in the sense that the germinal vesicle is acentric, being nearer to the animal pole than to the vegetative pole. In the frog's egg the polarity is visible to the eye, since the animal pole is pigmented, whereas the vegetative pole is not. In nearly all animals the subsequent ectoderm of the embryo is derived from the cytoplasm lying at the animal pole of the egg, whereas the endoderm is derived from the cytoplasm at the opposite or vegetative pole. Even in eggs where there is no macroscopic evidence of polarity, there is, in a large number of cases, definite evidence of the existence of two fundamental poles which are marked by the position of the polar bodies or by the accumulation of specific pigments. As examples one may quote the stratified pigment in the eggs of *Strongylocentrotus* (Boveri, 1901), *Dentalium* (Wilson, 1904), *Myzostomum* (Driesch, 1896), *Styela* (Conklin, 1905). The position of the polar bodies or of specific pigments is not in itself evidence of a fundamental heterogeneity of protoplasm; such structures simply show that the egg is polarised in the sense that one series of events is happening or has happened at one pole, whereas another series of events has happened at the other. As Conklin (1924) has, however, clearly pointed out, it is from a study of this visibly polarised structure that we may hope to gain some conception of the underlying mechanism of protoplasmic heterogeneity.

That the heterogeneous distribution of pigment granules is not necessarily evidence of their association with a specific type of cytoplasm was shown by Morgan and Spooner (1909), who showed that when the eggs of *Strongylocentrotus* are centrifuged, the position of the granules can be altered and may subsequently find themselves in a region of the embryo in which they do not normally occur.[1] At the same time there is good reason to believe that in a normal egg such pigments actually do mark out regions of cytoplasm which are fundamentally different from each other. This evidence is divided into two types. Firstly, it is possible to operate on the cells and remove regions of cytoplasm which are specific in the sense that they are characterised by the presence of particular pigments or inclusions. Secondly, we can remove regions of cytoplasm which are specific in that they lie in a definite position in respect to the polar axis of the egg. The second type of evidence has priority in point of time. Driesch and Morgan (1895) and Fischel (1897) showed that the eight rows of ciliated plates in the ctenophore *Beroë* are derived from the peripheral cytoplasm at the animal pole of the unsegmented egg. If part of this region be removed, then one or more ciliated bands are missing in the subsequent embryo. Similarly, Crampton (1897) found that if the so-called polar lobe (which is entirely cytoplasmic) be removed from the egg of *Illyanassa*, the subsequent larva failed to develop mesoblast bands.

It may be argued that the process of removal of cytoplasmic fragments is relatively crude and that the results may, to some extent, be due to a generalised disturbance of the whole system. This objection can hardly be urged against the data obtained by Wilson (1904 a) from the eggs of the scaphopod *Dentalium*. When the eggs are laid three distinct regions of cytoplasm are visible (fig. 27, 3), owing to the presence of an equatorial band of brown pigment which separates two colourless areas, one at each end of the polar axis. If the egg is subsequently fertilised, the process of maturation is completed and the egg begins to segment. The first cleavage-plane does not, however, develop simultaneously at the two poles, but appears first at the apical pole which is marked by the second polar body; meanwhile a well-marked protrusion appears at the opposite pole. This protrusion consists of the colourless polar material already mentioned and as cleavage proceeds, so this 'polar' lobe becomes more and more definite (see fig. 27); finally the polar lobe remains attached to one blastomere (CD) by a narrow bridge of cytoplasm, although it is finally absorbed into this cell before the second cleavage. Apart from the existence of the polar lobe, cleavage is

[1] In other respects the larvae are normal.

equal so that after it has been absorbed into the cell *CD*, this is distinctly larger than its neighbour *AB*. The second cleavage is essentially similar to the first, and in this case a polar lobe protruded from *CD* is incorporated into the daughter blastomere *D*; the blastomere *AB* forms no polar lobe. Wilson carried out a number of experiments which clearly show that the presence of the cytoplasmic polar lobe is essential if certain well-defined organs are to be present

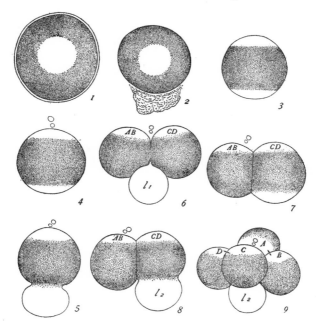

Fig. 27. Cleavage of eggs of *Dentalium*. (From Wilson, 1904.)

1, Outline of egg soon after release from the ovary; 2, the same egg 20 mins. later, after throwing off surface membrane; 3, similar egg from the side; 4, egg 1 hour after fertilisation, with polar bodies; 5, beginning of cleavage, showing polar lobe; 6, trefoil stage, $1\frac{1}{4}$ hours after fertilisation; 7, two-celled stage; 8, formation of second polar lobe; 9, second cleavage at its height. l_1, first polar lobe; l_2, second polar lobe.

in the subsequent larva. If the unsegmented fertilised egg is divided by an equatorial cut into polar and antipolar portions, and the nucleus be left in the former, then segmentation proceeds without the formation of antipolar lobes. Many such eggs, however, failed to give embryos, whereas others developed into larvae which were deficient in apical organ and post-trochal regions (see fig. 28 (*b*)). If, on the other hand, the portion containing the nucleus was derived from the antipolar region of the egg, subsequent segmentation occurred with the formation of normal polar lobes and the resultant larvae were small but normal. If the original cut were parallel to the polar axis, both halves developed normally as long as each region contained part of the white antipolar material. These experiments suggest that the presence of the polar lobe material is essential for the formation of the apical organ and the post-trochal region of the larva; and this conclusion is strengthened by the result of removing whole blastomeres instead of simply polar lobes. Wilson showed that if the two first blastomeres (AB and CD) are separated, they

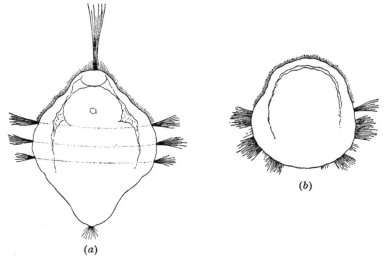

Fig. 28. Trochophores of *Dentalium* 24 hours old. (a) From normal egg. (b) From egg whose first polar lobe had been removed. Note absence of apical tuft of cilia, and of the post-trochal region. (From Wilson.)

each continue to segment, but the larva from AB lacks both apical organ and post-trochal regions, whereas both these regions are present in the larva from CD by which the polar lobe had previously been absorbed. Similarly if each of the first four blastomeres (A, B, C, and D) are separated, three of them (A, B, and C) develop without the formation of polar lobes and without the formation of apical organs or post-trochal regions: the fourth blastomere (D) produces, however, polar lobes, apical organ, and post-trochal region. Finally, Wilson showed that if the first polar lobe be removed, the remainder of the egg produces larvae without

PROTOPLASM 93

apical organs and post-trochal regions; whereas if the removal of the polar lobe is deferred until its second appearance, the resultant larva possesses an apical organ but has no post-trochal region. It is extremely difficult to avoid the conclusion that the cytoplasmic material located in the polar lobe is specific in respect to the formation of well-defined organs.

Perhaps the most spectacular description of cytoplasmic differentiation is that given of the egg of the ascidian *Styela partita* by Conklin (1905, 1912). The unripe egg is radially symmetrical, having a central mass of grey yolk, a large transparent germinal vesicle near the animal pole and a superficial layer of yellow cytoplasm. After maturation the maturation spindle lies in a region of

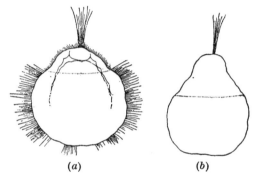

Fig. 29. Larva of *Dentalium* 24 hours old. (a) From egg whose second polar lobe had been removed. Note the absence of the post-trochal region, but that the apical organ is present. (b) From blastomere *CD* whose second polar lobe had been removed. Note reduction of post-trochal region. (From Wilson.)

transparent material which is itself derived from the germinal vesicle. At this stage the egg can be fertilised by a spermatozoon which enters the egg from the opposite or vegetative pole. Immediately this occurs the superficial layer of yellow cytoplasm flows to the point of entry and there forms a yellow cap, while at the same time the clear material derived from the germinal vesicle forms a definite zone above the yellow cap. The egg is still radially symmetrical, but prior to the formation of the first cleavage spindle the yellow cap and clear zone move towards the posterior pole and the former takes the form of a crescent round the posterior side of the egg; whilst on the anterior side of the egg there arises a light grey crescent. In this way the original radial symmetry of the egg is replaced by a bilateral symmetry and the first cleavage furrow passes through the plane of symmetry and divides the egg into two symmetrical halves. In a normal egg Conklin identified five distinct cytoplasmic regions and was able to trace the distribution of each region among the tissues of the differentiated larva. (i) The yellow crescentic *myoplasm* gives rise to the muscles of the tail, (ii) a light yellow *chymoplasm*, derived from the horns of the crescent, produces the mesenchyme, (iii) the light grey crescent produces the notochord and neural tube, (iv) the *endoplasm* yields the endoderm, and (v) the transparent *ectoplasm* yields the definitive ectoderm. Conklin showed that if the fertilised eggs are centrifuged in such a way that they cannot rotate within their membranes, the yellow substance, clear plasma, and grey substance can be displaced from their normal positions, and if cleavage occurs whilst they are so displaced, then the distribution of the substances among subsequent blastomeres is abnormal; subsequently a parallel dislocation arises in respect to the various organs of the larva. Thus if the eggs are centrifuged during the telophase of the first cleavage the whole of the yellow crescent can be forced into one of the two blastomeres, with the result that, at a later stage, the larval muscle cells are formed entirely on one side of the larvae. By a similar process, 'larvae may be turned inside out, the endoderm, muscles, and chorda being on the outside; ectoderm, neural plate cells and sense organs on the inside' (Conklin, 1925, p. 579). Since it is known that displacement of nuclei does not influence cell differentiation, we have no alternative than to believe that the cytoplasm of an egg cell is a highly differentiated system. It must, however, be remembered that this differentiation is something more fundamental than the heterogeneous distribution of pigment granules or obvious cell inclusions. It will be recalled that Morgan and Spooner (1909) showed that pigment granules can be moved by the centrifuge without effecting any permanent change in the distribution of cytoplasmic areas; optical differences in cytoplasm are probably the result and not the cause of an underlying specificity. Recently Morgan (1927) has criticised Conklin's results and suggested that his experimental results are due to pressure rather than to displacement of cytoplasmic phases under centrifugal force. That cytoplasmic differentiation exists there can be no doubt, the point at issue is whether particular regions can be displaced by centrifugal force. It is certainly surprising that differences in structure which must be of a highly complex and subtle nature should exhibit sufficient differences in specific gravity to be affected by comparatively weak centrifugal forces.

The ontogenetic origin of cytoplasmic differentiation has been investigated in comparatively few cases, but as far as the evidence goes, it looks as though it occurred relatively early in the history of the egg cell. Yatsu (1904) found that if the egg of the nemertine *Cerebratulus* be cut into fragments before maturation, all the fragments will develop into complete but small larvae. If, however, the egg is first fertilised and then cut into fragments, the number of complete larvae formed is very much reduced. The critical period in the cycle appears to coincide with the completion of the maturation divisions (which follow fertilisation) and not with the breakdown of the germinal vesicle. Yatsu also made the interesting observation that although practically no normal larvae are obtained from eggs from which a portion of the cytoplasm is removed between the fusion of the male and female pronuclei and the onset of the first cleavage, yet when the two-celled stage is reached each blastomere is capable of complete development.

The results of Brachet's (1905) experiments with the frog's egg indicate that in this case differentiation of the cytoplasm occurs about two hours after fertilisation, since prior to this period portions of cytoplasm can be withdrawn without influencing the normal course of development.

The eggs of *Cerebratulus* and of the frog show that when cytoplasmic differentiation takes place it does so independently of the total mass of undifferentiated cytoplasm. This is also true of *Crepidula*; Conklin (1917) centrifuged the eggs of this mollusc and thereby induced the formation of very large polar bodies. Although this reduced the egg to one-half of its normal size, yet the egg developed normally.

Although the available facts indicate fairly clearly that the cytoplasm of an egg is to be regarded as a mosaic structure in which the constituent parts are gradually segregated from each other by cell division, it is by no means easy to form a satisfactory picture which will cover all the known facts. The data derived from regeneration experiments show clearly that specific types of cells sometimes contain the potentiality of forming other types if necessary (see also p. 297). In the regeneration of crustacean limbs ectodermal cells produce muscle fibres. In other words, although the segregation of mesoderm from ectoderm occurs at a very early stage in the development of the egg, we cannot assume that this segregation is complete and irreversible. Again, differentiation is not wholly dependent on cell-division, for in *Chaetopterus* differentiation may occur without any cell boundaries (Lillie, 1902). It is not easy to piece together these data, and although the biological complexity of cytoplasm is best pictured as a type of mosaic, we must be careful not to eliminate potentialities of differentiation which may only reveal themselves at a later stage in development or under abnormal circumstances. If we start with the simplest cytoplasmic state, we can imagine that there is no differentiation—as indeed appears to be the case in *Cerebratulus* eggs prior to maturation, or in the early segmentation stages of the Coelenterata, where Hargitt (1904, 1906) has shown that cleavage is normally irregular and yet where normal larvae always result. Sooner or later, however, differentiation of parts is effected, although the time at which this process occurs in different types may vary. Apparently it is reached earlier in the so-called cases of determinate cleavage than in indeterminate cleavage—it is earlier in *Dentalium* than in the typical sea urchin. Concerning the mechanism of this differentiation we know nothing, but the work of Spemann does, at least, give us a reasonable although purely biological point of view. We can imagine that within the cell there are centres of organisation (e.g. the grey crescent of the frog's egg) which do, in fact, control the fate and position of other undifferentiated cytoplasmic regions: once these regions reach a definite phase in their development their fate is irretrievably fixed, whereas in the earlier stages this is by no means the case. Such a view is clearly in harmony with the facts of amphibian development; if we can imagine the dorsal lip of a blastopore marshalling the indeterminate cells of a urodele blastula into their appropriate positions and endowing them in some way with their appropriate characters, we can equally well believe that the streaming movements, whereby the cytoplasmic pattern of *Styela* eggs (Conklin) is brought into being, is due to an essentially similar cause. On this view there is no fundamental difference between determinate and indeterminate cleavage, there is simply a difference in respect to the phase of development at which the parts of the cells and the cells themselves cease to depend on an organising centre for their subsequent mode of development.

When we attempt to picture the facts of cytoplasmic differentiation in physical terms, the difficulties are overwhelming. How can we provide a mosaic of three-dimensional states within a medium which is only ten times as viscous as water? Having postulated a series of ultra-microscopic membranes, surfaces and liquid crystals and having endowed them with all possible degrees of complexity, it may be doubted whether such conceptions are really useful. It seems more rational to regard living material as a state of matter where the constituent molecules are organised in a way quite unknown in the inanimate world, and which will only be elucidated by methods of analysis which have yet to be discovered. The study of protoplasmic structure clearly illustrates the limitations of a purely physical conception of biological problems. Without doubt the underlying mechanism of cytoplasmic differentiation is of an atomic or electronic nature, but until we have the means to explore the situation in much greater detail, a knowledge of our ignorance is perhaps the most valuable asset we can hope to possess.

REFERENCES

BAAS BECKING, L. G. M., BAKHINGZEN, H. V. D. S. and HOTELLING, H. (1928). 'The physical state of protoplasm.' *Verh. d. kon. Akad. v. Wet. t. Amst.* 25, 5.

BAYLISS, W. M. (1920). 'The properties of colloidal systems. IV. Reversible gelation in living protoplasm.' *Proc. Roy. Soc.* B, 91, *196*.

BĚLEHRÁDEK, J. (1930). 'Temperature coefficients in biology.' *Biol. Rev.* 5, *30*.

BOLLES LEE, A. (1928). *Microtomists' Vade-mecum.* 9th edition. (London.)

BOVERI, T. (1901). 'Die Polarität von Ovocyte, Ei und Larve des *Strongylocentrotus lividus*.' *Zool. Jahrb.* 4, *630*.

BRACHET, A. (1905). 'Recherches expérimentales sur l'œuf de *Rana fusca*.' *Arch. de Biol.* 21, *103*.

BROOKS, S. C. (1924). 'The electrical conductivity of pure protoplasm.' *Journ. Gen. Physiol.* 7, *327*.

BÜTSCHLI, O. (1894). *Investigations on Microscopic Foams and on Protoplasm.* Eng. transl. (London.)

CHAMBERS, R. (1917). 'Microdissection studies. I. The visible structure of cell protoplasm and death changes.' *Amer. Journ. Physiol.* 43, *1*.

—— (1928). 'Intracellular hydrion concentration studies. I. The relation of the environment to the pH of protoplasm and of its inclusion bodies.' *Biol. Bull.* 55, *369*.

CHAMBERS, R. and POLLACK, H. (1927). 'Micrurgical studies in cell physiology. IV. Colorimetric determination of the nuclear and cytoplasmic pH in the starfish egg.' *Journ. Gen. Physiol.* 10, *739*.

CHAMBERS, R., POLLACK, H. and HILLER, S. (1927). 'The protoplasmic pH of living cells.' *Proc. Soc. Exp. Biol. and Med.* 24, *760*.

CHAMBERS, R. and REZNIKOFF, P. (1926). 'Micrurgical studies in cell-physiology. I. The action of the chlorides of Na, K, Ca, and Mg on the protoplasm of *Amoeba proteus.*' *Journ. Gen. Physiol.* 8, *369.*
CONKLIN, E. G. (1905). 'Mosaic development in ascidian eggs.' *Journ. Exp. Zool.* 2, *145.*
—— (1912). 'Experimental studies on nuclear and cell-division in the eggs of *Crepidula.*' *Journ. Acad. Nat. Sci. Phil.* 15, *503.*
—— (1917). 'Effects of centrifugal force on the structure and development of the eggs of *Crepidula.*' *Journ. Exp. Zool.* 22, *311.*
—— (1925). 'Cellular differentiation.' *General Cytology.* (Chicago.)
CRAMPTON, H. E. (1897). 'Experimental studies on gasteropod development.' *Arch. f. Entw. Mech.* 3, *1.*
CUNNINGHAM, E. (1910). 'On the velocity of steady fall of spherical particles through fluid medium.' *Proc. Roy. Soc.* B, 83, *357.*
DRIESCH, H. (1896). 'Betrachtungen über die Organisation des Eies und ihre Genese.' *Arch. f. Entw. Mech.* 4, *75.*
DRIESCH, H. and MORGAN, T. H. (1895). 'Zur Analysis der ersten Entwicklungsstadien des Ctenophoreneies.' *Arch. f. Entw. Mech.* 2, *204.*
FETTER, D. (1926). 'The determination of the protoplasmic viscosity of *Paramoecium* by the centrifuge method.' *Journ. Exp. Zool.* 44, *279.*
FEULGEN, R. (1923). 'Die Nuclealfärbung.' *Abderh. Handb. biol. Arbeitsm.* 5, ii, *1055.*
FISCHEL, A. (1897). 'Experimentelle Untersuchungen am Ctenophorenei. I.' *Arch. f. Entw. Mech.* 6, *109.*
FREUNDLICH, H. (1926). *Colloid and Capillary Chemistry.* (London.)
FREUNDLICH, H. and SEIFRIZ, W. (1923). 'Über die Elastizität von Solen und Gelen.' *Zeit. f. physik. Chem.* 104, *233.*
GAIDUKOV, N. (1910). *Dunkelfeldbeleuchtung und Ultramikroskopie in der Biologie und in der Medizin.* (Jena.)
GELFAN, S. (1927). 'The electrical conductivity of Protoplasm and a new method of its determination.' *Univ. Cal. Publ. Zool.* 29, *453.*
—— (1928). 'The electrical conductivity of protoplasm.' *Protoplasma,* 4, *192.*
GOTSCHLICH, F. (1912). 'Allgemeine Morphologie und Biologie der pathogenen Mikroorganismen.' *Handb. der path. Mikroorganismen,* 1, *30.*
GRÜBER, A. (1889). 'Biologische Studien an Protozoen.' *Biol. Centrb.* 9, *14.*
GUTSTEIN, M. (1925). 'Ueber den Kern und den allgemeinen Bau der Bakterien.' *Centralb. f. Bakteriol.* Abt. 1, 95, *357.*
HARDY, W. B. (1899). 'On the structure of cell protoplasm.' *Journ. of Physiol.* 24, *158.*
HARGITT, C. W. (1904). 'The early development of *Eudendrium.*' *Zool. Jahrb.* 20, *257.*
—— (1906). 'The organisation and early development of *Clava leptostyla.*' *Biol. Bull.* 10, *207.*
HATSCHEK, E. (1913). 'Die Viskosität der Emulsoidsolen und ihre Abhängigkeit von der Schergeschwindigkeit.' *Koll. Zeit.* 13, *88.*
HAUSER, E. A. (1929). 'Ueber die Thixotropie von Dispersionen geringe Konzentration.' *Koll. Zeits.* 48, *57.*
HEILBRONN, A. (1914). 'Zustand des Plasmas und Reizbarkeit. Ein Beitrag zur Physiologie der lebenden Substanz.' *Jahrb. f. wiss. Bot.* 54, *357.*

HEILBRONN, A. (1922). 'Eine neue Methode zur Bestimmung der Viskosität lebender Protoplasten.' *Jahrb. f. wiss. Bot.* 61, *284*.

HEILBRUNN, L. V. (1925). 'The colloid chemistry of protoplasm.' *Colloid Symposium Monograph*, 3. (New York.)

—— (1926 a). 'The centrifuge method of determining protoplasmic viscosity.' *Journ. Exp. Zool.* 43, *313*.

—— (1926 b). 'The absolute viscosity of protoplasm.' *Journ. Exp. Zool.* 44, *255*.

—— (1926 c). 'The physical structure of the protoplasm of sea-urchin eggs.' *Amer. Nat.* 60, *143*.

—— (1928). *The Colloid Chemistry of Protoplasm*. (Berlin.)

HERTWIG, O. (1893). 'Über den Werth der ersten Furchungszellen für die Organbildung des Embryo.' *Arch. mikr. Anat.* 42, *622*.

VAN HERWERDEN, M. A. (1925). 'Reversible Gelbildung in Epithelzellen der Froschlarve und ihre Anwendung zur Prüfung auf Permeabilitätsunterschiede in der lebenden Zellen.' *Arch. f. exper. Zellforsch.* 1, *145*.

—— (1927). 'Umkehrbare Gelatierung und Temperaturerhöhung bei einer Süsswasseramoebe.' *Protoplasma*, 2, *271*.

HOWLAND, R. B. (1924). 'Dissection of the pellicle of *Amoeba verrucosa*.' *Journ. Exp. Zool.* 19, *263*.

HUNTER, O. W. (1923). 'Protein synthesis by *Azotobacter*.' *Journ. Agric. Research.* 24, *263*.

JACOBS, M. H. (1920). 'The production of intracellular acidity by neutral and alkaline solutions containing carbon dioxide.' *Amer. Journ. Physiol.* 53, *457*.

—— 1922. 'The influence of ammonium salts on cell reaction.' *Journ. Gen. Physiol.* 5, *181*.

JOST, L. (1904). *Vorlesungen über Pflanzenphysiologie*. (Jena.)

KAYSER, E. (1890). 'Études sur la fermentation du cidre.' *Ann. de l'Inst. Pasteur*, 4, *321*.

KAYSER, M. E. (1894). 'Études sur la fermentation lactique.' *Ann. de l'Inst. Pasteur*, 8, *737*.

KERMACK, W. O., M'KENDRICK, A. G. and PONDER, E. (1929). 'The stability of suspensions. III. The velocities of sedimentation and of cataphoresis of suspensions in a viscous fluid.' *Proc. Roy. Soc. Edin.* 49, *170*.

KIESEL, A. (1925). 'Beitrag zur Kenntnis der chemischen Bestandteile der Myxomycetenfruchtwandung.' *Zeit. f. physiol. Chem.* 150, *102*.

KITE, G. L. (1913). 'Studies on the physical properties of protoplasm. I. The physical properties of the protoplasm of certain animal and plant cells.' *Amer. Journ. Physiol.* 32, *146*.

KRUSE, W. (1910). *Allgemeine Mikrobiologie*. (Berlin.)

LADENBURG, R. (1907). 'Ueber die innere Reibung zäher Flüssigkeiten und ihre Abhängigkeit vom Druck.' *Ann. der Physik*, 22, *287*.

LEBLOND, M. (1919). 'Le passage de l'état de gel à l'état de sol dans le protoplasma vivant.' *C.R. Soc. Biol.* 82, *1150*.

LEWIS, W. H. (1923). 'Observations on cells in tissue-cultures with darkfield illumination.' *Anat. Record*, 23, *387*.

LILLIE, F. R. (1902). 'Differentiation without cleavage in the egg of the annelid *Chaetopterus pergamentaceus*.' *Arch. f. Entw. Mech.* 14, *477*.

LYON, E. P. (1907). 'Results of centrifugalizing eggs.' *Arch. f. Entw. Mech.* 23, *151*.

MACALLUM, A. B. (1895). 'On the distribution of assimilated iron compounds, other than haemoglobin and haematin in animal and vegetable cells.' *Quart. Journ. Micr. Sci.* 38, *175*.
—— (1897). 'On a new method of distinguishing between organic and inorganic compounds of iron.' *Journ. Physiol.* 22, *92*.
—— (1898). 'On the detection and localisation of phosphorus in animal and vegetable tissues.' *Proc. Roy. Soc.* B, 63, *467*.
—— (1905). 'On the distribution of potassium in animal and vegetable cells.' *Journ. Physiol.* 32, *95*.
—— (1908). 'Die Methoden und Ergebnisse der Mikrochemie in der biologischen Forschung.' *Ergeb. der Physiol.* 7, *552*.
MACALLUM, A. B. and MENTEN, M. L. (1906). 'On the distribution of chlorides in nerve cells and fibres.' *Proc. Roy. Soc.* B, 77, *165*.
MANN, G. (1902). *Physiological Histology.* (Oxford.)
MAST, S. O. (1926). 'The structure of protoplasm in *Amoeba*.' *Amer. Nat.* 60, *133*.
MORGAN, T. H. (1927). *Experimental Embryology.* (New York.)
MORGAN, T. H. and SPOONER, G. B. (1909). 'The polarity of the centrifuged egg.' *Arch. f. Entw. Mech.* 28, *104*.
NEEDHAM, J. and NEEDHAM, D. M. (1925). 'The hydrogen-ion concentration and the oxidation reduction potential of the cell-interior. A microinjection study.' *Proc. Roy. Soc.* B, 98, *259*.
—— (1926). 'The hydrogen-ion concentration and oxidation reduction potential of the cell-interior before and after cleavage. A micro-injection study on marine eggs.' *Proc. Roy. Soc.* B, 99, *173*.
PANTIN, C. F. A. (1924). 'Temperature and the viscosity of protoplasm.' *Journ. Mar. Biol. Assoc.* 13, *331*.
POLLACK, H. (1928). 'Intracellular hydrion concentration studies. III. The buffer action of the cytoplasm of *Amoeba dubia* and its use in measuring the pH.' *Biol. Bull.* 55, *383*.
PROWAZEK, S. v. (1915). 'Zur Morphologie und Biologie von *Colpidium colpoda*.' *Arch. f. Protistk.* 36, *72*.
REZNIKOFF, P. and POLLACK, H. (1928). 'Intracellular hydrion concentration studies. II. The effect of injection of acids and salts on the cytoplasmic pH of *Amoeba dubia*.' *Biol. Bull.* 45, *377*.
SCARTH, G. W. (1924). 'Colloidal changes associated with protoplasmic contraction.' *Quart. Journ. Exp. Physiol.* 14, *99*.
—— (1927). 'The structural organisation of plant protoplasm in the light of micrurgy.' *Protoplasma*, 2, *189*.
SCHALEK, E. and SZEGVARY, A. (1920). 'Über Eisenoxydgallerten.' *Kolloid. Zeits.* 32, *318*.
—— (1923). 'Die langsame Koagulation konzentrierter Eisenoxydsolen zu reversibele Gallerten.' *Kolloid. Zeits.* 33, *326*.
SCOTT, F. H. (1900). 'The structure, micro-chemistry and development of nerve cells, with special reference to their nuclein compounds.' *Univ. of Toronto Studies (Physiological Series)*, 1; *Trans. of the Canadian Institute*, 1898–9.
SEIFRIZ, W. (1918). 'Observations on the structure of protoplasm by the aid of microdissection.' *Biol. Bull.* 34, *307*.
—— (1921). 'Observations on some physical properties of protoplasm by aid of microdissection.' *Ann. of Bot.* 138, *269*.

SEIFRIZ, W. (1923–4). 'The structure of protoplasm and of inorganic gels: an analogy.' *Brit. Journ. Exp. Biol.* 1, *431*.
—— (1925). 'Elasticity and some structural features of soap solution.' *Colloid Symposium Monograph*, 3, *285*. (New York.)
—— (1926 a). 'Elasticity as an indicator of protoplasmic structure.' *Amer. Nat.* 60, *124*.
—— (1926 b). 'The physical properties of erythrocytes.' *Protoplasma*, 1, *345*.
—— (1928). 'Die physikalischen Eigenschaften des Protoplasmas.' *Arch. exp. Zellforsch.* 6, *341*.
—— (1929 a). 'The alveolar structure of cytoplasm and nuclei.' *Protoplasma*, 9, *177*.
—— (1929 b). 'The contractility of protoplasm.' *Amer. Nat.* 63, *410*.
STILES, W. (1930) 'Viscosity of protoplasm as determining the rate of biological reactions.' *Biol. Rev.* 5, *171*.
SVEDBERG, T. (1921). 'Discussion on the physical properties of elastic gels.' *Rep. Faraday Soc. and Phys. Soc. of London*, 1.
—— (1928). *Colloid Chemistry*. (New York.)
TERROINE, E. F., WÜRMSER, R. and MONTANÉ, J. (1922). 'Influence de la constitution des milieux nutritifs sur la composition de l'*Aspergillus niger*.' *C.R. Acad. Sci. Paris*, 175, *541*.
UNNA, P. G. and TIELEMANN, E. T. (1917). 'Zur Chemie der Amöben.' *Centralbl. f. Bakteriol.* Abt. 1, 80, *66*.
VERWORN, M. (1889). *Psycho-physiologische Protisten-studien*. (Jena.)
WEBER, F. (1923). 'Methoden der Viskositätsbestimmung des lebenden Protoplasmas.' *Abderh. Handb. biol. Arbeitsm.* Abt. 11, Teil 2, *655*.
WILSON, E. B. (1903). 'Experiments on cleavage and localisation in the Nemertine egg.' *Arch. f. Entw. Mech.* 16, *411*.
—— (1904 a). 'Experimental studies on germinal localisation. I. The germ regions in the egg of *Dentalium*.' *Journ. Exp. Zool.* 1, *1*.
—— (1904 b). 'Experimental studies on germinal localisation. II. Experiments on the cleavage mosaic in *Patella* and *Dentalium*.' *Journ. Exp. Zool.* 1, *197*.
—— (1923). 'The physical basis of life.' *Science*, 57, *277*.
—— (1925). *The Cell in Development and heredity*. (New York.)
—— (1926). 'Newer aspects of the alveolar structure of protoplasm.' *Amer. Nat.* 60, *105*.
—— (1928). *The physical basis of life*. (New Haven.)
YATSU, N. (1904). 'Experiments on the development of Egg-fragments in *Cerebratulus*.' *Biol. Bull.* 6, *123*.
—— (1912 a). 'Observations and Experiments on the Ctenophore egg. I.' *Journ. Coll. Sci. Imp. Univ. Tokio*, 32, *1*.
—— (1912 b). 'Observations and experiments on the Ctenophore egg. III.' *Annot. Zool. Japon.* 8, *5*.

CHAPTER SIX

Cell Membranes and Intercellular Matrices

In many cases, although not in all, it can be shown that the physical state of the superficial protoplasm is not the same as that in the interior of the cell, and a steadily increasing body of evidence points to the conclusion that whereas the mass of the intracellular cytoplasm may be of a fluid nature, the peripheral layers of the cell are essentially solid. Before presenting this evidence it is desirable to define, as far as possible, the terms which will be employed. In the present discussion three terms will be used—elasticity, plasticity and rigidity. An *elastic* substance is one which is capable of changing its shape when subjected to an external strain in such a way that the degree of distortion is limited and is proportional to the strain applied: as soon as the strain is removed, a perfectly elastic body completely regains its original shape. A *plastic* body, on the other hand, also changes its shape when subjected to an external strain, but in this case no deformation occurs unless the force applied reaches a critical minimum value; as long as a supra-minimal but constant force is applied, the substance continues to be deformed and no equilibrium condition is reached. When the deforming force is removed there is no return of a perfectly plastic body towards its original form. A *rigid* body is one which tends to resist a change in shape when exposed to an external force. It will be noted that the term *rigid* is relative—and that elastic bodies can also exhibit plastic flow; further, it has already been noted that liquid suspensions may be elastic. A typically elastic substance is rubber, a typically plastic substance is clay.

It is evident that we can test the mechanical properties of the periphery of the cell by applying a suitable distorting force. A convenient method of applying such a force to a localised area of the cell is provided by a microdissecting needle. This was first done by Kite (1913), who showed that the surface ectoplasm of *Amoeba* can be drawn into strands which on release return more or less completely to their normal position in the surface. Very striking results of this type were obtained by Howland (1924) with *Amoeba verrucosa* (see fig. 30). We conclude, therefore, that the surface ectoplasm is elastic; but if stretched too much, it will begin to exhibit plastic flow. This result, in conjunction with the fact that the surface ectoplasm offers considerable resistance to the movement of enclosed particles (see p. 105), points to the fact that the surface layer is of a solid nature. This conclusion is in obvious harmony with the earlier observations of Pfeffer (1906). 'It is easy to see how the firmer ectoplasm of the plasmodia of *Chondrioderma, Aethalium* and other Myxomycetes is produced from fluid endoplasm and may be reconverted into the latter. The ectoplasm may be 0·01 mm. thick

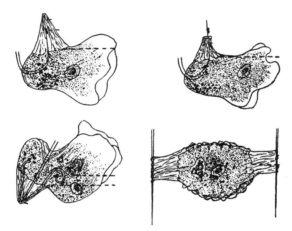

Fig. 30. Dissection of the surface pellicle of *Amoeba verrucosa* by means of microneedles. Note the surface corrugations on the stretched hyaline layer. (From Howland, 1924.)

and is, therefore, more than a mere surface-tension film, and is much thicker than the ectoplasmic membrane.... By using plasmodial threads of about 0·3 mm. in thickness, in which the surface-tension effect is small, Pfeffer was able to determine that the consistency was about that of a jelly and the same is shewn by the way in which moving particles are repelled from the surface layers without producing any perceptible deformation or inducing any streaming movement. Similarly, oil drops and vacuoles passing through a tube of ectoplasm are compressed and distorted without producing any bulging in the tube. The appearance closely resembles that shown when fluid gelatin containing suspended particles is passed through a fine glass tube kept lined with a layer of solidified gelatine' (Pfeffer, vol. 3, pp. 279, 280).

Similar evidence of a solid peripheral membrane is derived from sea urchin eggs. In *Echinus esculentus* the peripheral region of the

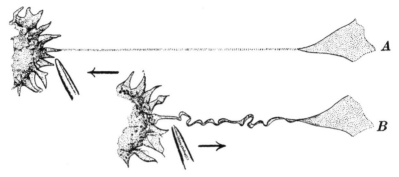

Fig. 31. Mechanical extension of a surface film of a leucocyte of *Asterias* by means of a needle. On removing the needle (*B*) note the crinkled surface of the imperfectly elastic film. (From Fauré-Fremiet.)

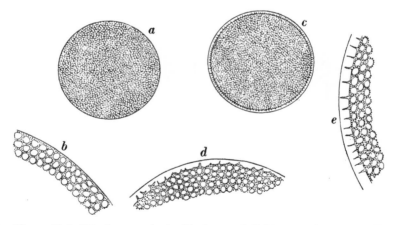

Fig. 32. *a*, Unfertilised or newly fertilised egg of *Echinus esculentus* (fertilisation membrane omitted). *b*, Surface of unfertilised or newly fertilised egg enlarged. Note the small (black) microsomes lying near the surface. *c*, Fertilised egg 30 mins. after fertilisation. Note the well-defined hyaline layer at the surface. *d* and *e*, Surface of fertilised egg (enlarged).

newly fertilised egg exhibits an appearance identical with that of the deeper regions of the cell; after some minutes, however, the surface layers are differentiated by an absence of granules, and within half an hour a clearly defined hyaline layer can be identified without any difficulty (Gray, 1924). This peripheral layer can be drawn out into elastic strands, which, like the ectoplasm of *Amoeba*, sooner or later begin to exhibit plastic flow.

An attempt to measure the elasticity of a surface layer of protoplasm was made by Seifriz (1924–5). A particle of nickel 16 μ in diameter was embedded in the cortical protoplasm of a mature unfertilised egg of *Echinarachnius* and the egg exposed momentarily to a magnetic field of force: the electromagnet being 0·08 mm. from the particle. Under the influence of the magnet the nickel particle moved towards the magnet and in doing so stretched the superficial layers of the cell for a distance of 9 μ. When the tip of the magnet was 1·5 mm. from the particle the distance travelled by the particle was only 5 μ. For a perfectly elastic medium the distance through which the particle moved should be inversely proportional to the square of the distance between the distance of the particle from the

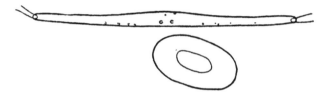

Fig. 33. Erythrocyte of *Cryptobranchus*, normal, and stretched between two needles. (From Seifriz.)

magnet. The actual results are obviously inexact in this respect, but they indicate a possible means of measuring the absolute value of the elasticity of cell membranes. It may be noted that Seifriz's experiments show very clearly the difference in consistency between the cell interior and the peripheral layers, since nickel particles which travel rapidly through the interior under a magnetic force meet with much greater resistance at the peripheral surface. Another instance of surface elasticity is provided by red blood corpuscles. Seifriz (1929) has shown that the erythrocytes of vertebrates have a highly elastic surface—a fact which is also apparent from the observations of Krogh (1922). Fig. 33 shows an erythrocyte of *Cryptobranchus* stretched between two needles until its long axis is three times as great as that of the normal cell.

Seifriz (1921) showed that if the surface ectoplasm be removed from an actively moving plasmodium of a Myxomycete a new film rapidly forms. Similarly, the hyaline surface of a fertilised echinoderm egg is readily reformed at a cut surface of the egg. In other words, wherever protoplasm comes into contact with a suitable aqueous environment, a coherent gel is formed at the surface. Since surface films of this type are well known in inanimate protein systems we may enquire how far the existence of a hyaline layer at the cell surface is simply the mechanical effect of bringing two liquid surfaces into contact, or on the other hand, how far this characteristic layer is an integral part of the living cell rather than a mechanical product or surface 'secretion'. If we bring an alkaline solution of casein into contact with a drop of calcium chloride a surface membrane of calcium caseinate is formed. Similarly we can form a membrane by using sodium oleate instead of sodium caseinate. In each case membrane formation depends on two factors. (i) The formation of an insoluble phase at the surface of the drop, (ii) the particles of this precipitated phase when in intimate contact with each other must form a coherent mass with sufficient rigidity to maintain its stability as part of a membrane. Since we know that the cell is rich in proteins, it is reasonable to enquire how far formation of ectoplasm is due to a deposition of protein molecules in such a form as to render them capable of forming a coherent mass.

The cohesive properties of protein aggregates is largely influenced by their electrolytic environment (Hardy and Wood, 1908), and therefore it is of some interest to enquire how far the surface layers of living cells is also dependent on the same series of factors. So far such an enquiry has not been carried out with isolated cells to any extensive degree, but more complete evidence is available from another source. When cells are united to form definite aggregates or tissues, the cohesion between the cells is entirely due to their hyaloplasmic surfaces. This is very easily shown in the case of echinoderm eggs (Gray, 1924); and Galtsoff (1925) has shown that, when isolated cells of the sponge *Microciona* reunite to form new aggregates by fusion of their hyaloplasmic layers, the deeper masses of granuloplasm remain separate from each other. It is therefore clear that the intercellular matrix or cement which binds cells together is homologous to the hyaline surface layer of isolated cells. From a study of the effect of ions on the cohesive properties of an intercellular matrix we can, to some extent, determine how far its normal stability is dependent on those factors which are known to affect the cohesive properties of protein gels (Gray, 1926). A study of this phenomenon has been found to provide a convenient intro-

INTERCELLULAR MATRICES

duction to the effect of ions on biological processes, and for this reason the facts may be considered in some detail.

The factors which influence the cohesive properties of cell surfaces

It has long been known that the hyaline surface of a cell is instable in the absence of divalent cations (e.g. calcium, magnesium) from the external medium. In the absence of calcium no hyaline surface

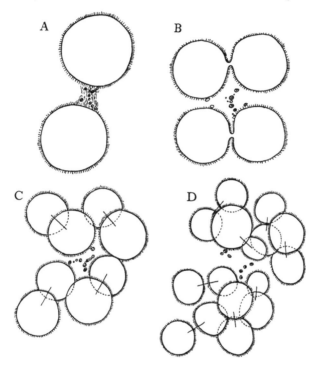

Fig. 34. Separation of blastomeres of *Echinus microtuberculatus* in calcium free seawater. Note the disorganisation of the hyaline layer and the spherical form of the cells. (From Herbst.)

forms on an echinoderm egg, and in 1900 Herbst showed that in calcium free sea-water the intercellular matrix of an echinoderm larva is dissolved and the cells separate freely from each other. In the case of *Mytilus* tissues the matrix is disorganised if magnesium is absent (Gray, 1926), whilst Galtsoff (1925) found that calcium and magnesium are both required for the reunion of artificially separated sponge cells. We thus reach the conclusion that the normal elastic surface of the cell is a gel whose cohesive properties depend upon the presence of divalent metallic ions. The only compounds which appear to have the property of forming cohesive membranes in the presence of calcium appear to be the proteins and possibly some of the polysaccharides. If substances such as gluten, casein, or mucin are powdered and rubbed with a little water they readily yield a tough coherent mass which is practically insoluble in

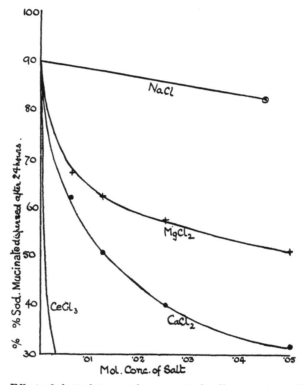

Fig. 35. Effect of electrolytes on the amount of sodium mucinate dispersed after 24 hours. (From Gray, 1926.)

water, and which can be pressed out into a more or less transparent membrane. The factors influencing the stability of such protein gels are strikingly similar to those which influence the stability of the hyaline gel which binds together the cells on the gills of *Mytilus edulis*.

In general, the cohesion of a protein gel depends on the degree of ionisation of the protein and on its ability to combine with water to form a hydrated electro-neutral particle. The facts, so far as they are known to apply to a system on the alkaline side of its isoelectric point, are as follows.

1. Membranes of gluten when exposed to low concentrations of an alkali gradually lose their coherence and pass into the dispersed condition. The dispersion by alkali is, however, markedly reduced by the presence of neutral salts (Hardy and Wood, 1908). It is

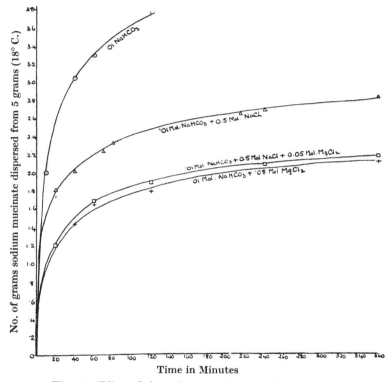

Fig. 36. Effect of electrolytes on the rate of peptisation of sodium mucinate. (From Gray, 1926.)

primarily the cations of the salts which affect the system, and the efficiency of any cation largely depends on its valency, the divalent ions being much more effective than the monovalent. This stabilising action of cations against the dispersive power of alkali can properly be regarded as an *electrostatic effect* comparable to the effect produced on the cataphoretic potential of dispersed particles (see p. 50). It is important to note that the anions present in the system play little or no part in this respect, but that all cations on the other hand antagonise the action of hydroxyl ions. Robertson and Miyake (1916) showed that salts affect the dispersal of casein by alkali in essentially the same way, and figs. 35 and 36 show the inhibiting effect of cations on the dispersion of sodium mucinate.

Precisely comparable data have been derived from the study of an intercellular matrix. If portions of the gill of *Mytilus* are placed in isotonic urea of the same pH as normal sea water, the inter-

Table XIII

pH of mol. urea solution	Degree of dispersion of hyaline matrix after			Meaning of symbols
	15 min.	40 min.	120 min.	
8·0	+++	+++	+++	Complete dispersion = +++
7·2	++	+++	+++	Approx. 50 % of
6·6	++	+++	+++	cells set free = ++
6·0	+	+++	+++	Approx. 25 % of
5·4	Very slight	++	++	cells set free = +
4·8	0	+	++	No cells set free = 0
4·0	0	0	0	

Table XIV

pH of $M/2$ NaCl	Time for dispersion in minutes	Average velocity $1000/T$	Log $1000/T$
4·0	∞	—	—
5·0	250	4·0	0·60
6·2	120	8·3	0·92
7·2	80	12·5	1·10
8·5	45	22·2	1·35
9·0	35	28·6	1·45
10·0	23	43·5	1·64
10·3	19	52·6	1·72

cellular matrix is entirely dissolved within fifteen minutes at a room temperature of about 15° C. If, however, acid is added to the medium the dissolution of the matrix is very strongly inhibited (see Table XIII).

If the same experiment is performed when urea is replaced by isotonic sodium chloride solution, we see very clearly that the presence of the salt decreases the dispersive power of the hydroxyl ions just as it does in systems of mucin or casein.

INTERCELLULAR MATRICES

From Tables XIII and XIV and fig. 37 it is clear that the presence of sodium chloride delays the dispersal of the matrix by hydroxyl ions from 15 to 60 minutes, although the rate of dispersion continues to be proportional to the concentration of hydroxyl ions. Divalent cations act in precisely the same manner as sodium ions, but are very much more powerful (see Tables XV and XVI).

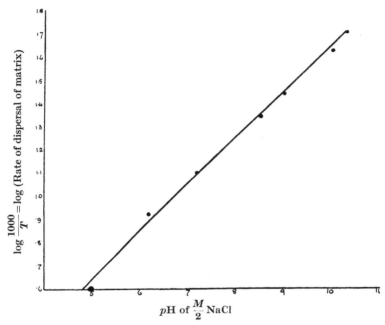

Fig. 37. Graph showing relation between concentration of hydroxyl ions and rate of dispersion of the intercellular matrix of *Mytilus*. (From Gray, 1926.)

Table XV

Concentration of Mg·· in mol. urea solution pH 7·8	Degree of dispersion of hyaline matrix after			Meaning of symbols	
	20 min.	80 min.	120 min.		
0·0	+++	+++	+++	All cells separated	= +++
0·01	0	⊕	+	A few cells separated	= +
0·02	0	⊕	+		
0·03	0	0	⊕	Very few cells separated	= ⊕
0·04	0	0	⊕		
0·05	0	0	0		

Table XVI

Concentration of Ca·· in mol. urea solution pH 7·8	Degree of dispersion of hyaline matrix after			Meaning of symbols
	30 min.	45 min.	60 min.	
0·0	+++	+++	+++	Complete dispersal = +++
0·02	⊕	⊕	⊕	Gradual dispersal
0·04	⊕	⊕	⊕	in granular form = ⊕
0·06	0	⊕	⊕	
0·08	0	⊕	⊕	

We may conclude that, other things being equal, the mechanical stability of the cell membrane depends on just those electrostatic factors which operate on a casein membrane. The real dispersive agent in both cases is the hydroxyl ion since this ionises the membrane or matrix: all salts inhibit this process, divalent cations being much more powerful than monovalent cations.

2. Although the dispersal of a protein membrane most readily takes place when it is ionised, it usually only does so if the constituent particles have an affinity for water, and if the attraction between the water and the particles is sufficient to overcome the cohesion between adjacent particles. Thus the sodium salts of the proteins are readily dispersed because their molecules have a high affinity for water—they are in fact soluble. Under certain conditions, however, the divalent metals form compounds with proteins which have little or no affinity for water (van Slyke and Winter, 1914; van Slyke and Bosworth, 1913), so that membranes of such substances do not disperse unless very strong forces are used, and particles of such compounds can only remain dispersed if electrically charged.

The formation of these basic salts does not occur unless protein *ions* are present. If the ionisation of a protein system is reduced by the presence of an excess of monovalent cations, the formation of basic insoluble salts is inhibited. The capacity of such metals as calcium to form insoluble compounds will be referred to as the *chemical* effect of ions.

It is the distinctive property of magnesium that it is the only metal which will, at the pH of normal sea water, prevent the dispersal of the *Mytilus* cell matrix and at the same time maintain the translucency and the healthiness of this matrix. Although all other divalent cations resemble magnesium, in exerting a strongly antagonistic action to the dispersive power of the hydroxyl ions, they all involve a change in the matrix which leads to its precipitation in a granular form. This invariably occurs with all trivalent cations. There is thus a marked biological difference between the action of magnesium ions and those of calcium. It should further be noted that the presence of magnesium inhibits the precipitating effect of calcium (Gray, 1926). Precisely comparable phenomena occur in casein solutions. If calcium chloride is added to sodium caseinate at pH 7·3 a marked opacity is observed followed by a definite precipitate. If, however, the ionisation of the sodium caseinate is depressed by the presence of magnesium no opacity occurs on adding calcium. The distinctive effect of magnesium on both protein and living systems appears to be its capacity of exerting an electrostatic effect on negatively charged surfaces, without seriously affecting the capacity of the particles to exist in the condition requisite for forming a cohesive membrane. We can now progress a stage further in the analysis of the factors determining the stability of the cell surface—for we can say that unless the surface is unionised it must react with calcium in the normal environment; this calcium compound may or may not have sufficient cohesion to form a membrane.

Although the electrostatic and chemical effects of cations are the dominating factors which influence the stability of a protein on the alkaline side of its isoelectric point, there are other factors which are less clearly understood. For example, in the absence of di- or tri-valent cations a specific action may be exerted by different anions which has been described as *lyophilic* in nature. Thus the effect of hydroxyl ions is inhibited by the anions in the following order:

$$I' < Br', NO_3' < Cl' < CH_3COO'.$$

A lyophilic effect can, however, also be exerted by undissociated molecules or by non-electrolytes, e.g. the 'salting out' of proteins and the inhibition of glycerine on the dispersal of casein by hydroxyl ions (Robertson and Miyake, 1916). Whereas, therefore, the electrostatic and chemical action of salts is restricted to the cations and only operates on ionised particles, lyophilic effects on the other hand operate on unionised particles and can be exerted by molecules or by ions of either sign. Since lyophilic effects are only operative if the colloid is otherwise free to disperse, it is not surprising to find that these effects are all but completely masked when the powers of dispersion are strongly reduced by electrostatic action. Thus in physiologically balanced solutions containing calcium and magnesium the lyophilic effect of anions almost entirely disappears (Gray, 1922).

That the different sodium salts exert a well-marked lyophilic action on the cell matrix is shown by the following table:

Table XVII

$M/2$ salt, pH 8·0	Time of dispersion of *Mytilus* matrix in minutes
NaCl	30
NaNO$_3$	30
NaBr	22
NaI	19
Na acet.	80

The anions fall into well-marked lyophil series as was pointed out by R. S. Lillie (1906). The series is practically the same as that found for protein dispersals. The fact that the lyophil series is all but entirely unobservable in the presence of divalent cations (see Gray, 1922) is entirely in keeping with the hypothesis that the lyophil action can only apply to those parts of the matrix which are electrostatically free to disperse. By increasing the concentration of NaCl to twice the isotonic value, the inhibition of dispersion is greatly increased, and is approximately equal to that of $M/2\,\text{Na}_2\text{SO}_4$.

The stability of the hyaline intercellular matrix of *Mytilus* tissue is apparently determined, therefore, by just those factors which control the degree of cohesion of a protein gel. Briefly summarised, the facts show that in normal sea water the stability of the matrix is maintained because the dispersive effect of the hydroxyl ions is antagonised by the combined effect of all the metallic cations present: of these cations magnesium is the most important, since it has a high antagonistic effect against hydroxyl ions and also prevents the calcium in the sea water from giving a brittle granular precipitation within the matrix. All these facts find their parallel with casein membranes.

The facts as analysed in this way must not disguise the existence of others less easily interpreted. Firstly, although magnesium is a much more powerful antagonist to hydroxyl ions than sodium, it cannot completely stabilise the cell membrane: sooner or later in a solution of urea and magnesium the cell matrix becomes instable: the latter is only completely stable when sodium, potassium, magnesium and calcium are present in the proportions in which they occur in sea water—and one type of ion cannot completely compensate for the absence of another (cf. also Galtsoff, 1925). Secondly, there is a marked biological difference between potassium ions and sodium ions which is not apparent in inanimate systems. Whereas the presence of sodium and magnesium ions will render the matrix comparatively stable, the substitution of $M/2$ KCl for $M/2$ NaCl renders the matrix instable even in the presence of relatively high concentrations of magnesium ions. Thirdly, the cell matrix becomes instable in the absence of oxygen (as in *Ctenolabrus* eggs (Loeb, 1905), or in the presence of CO_2 (Loeb) or of certain drugs

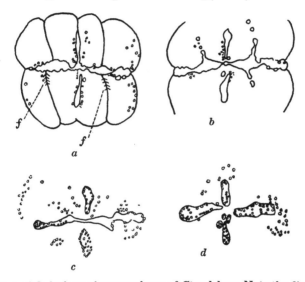

Fig. 38. Effect of O_2 lack on cleavage planes of *Ctenolabrus*. Note the disorganisation of the hyaline phase into discrete globules. (From Loeb.)

(Carter, 1926), and these reagents are apparently without action on inanimate systems. Although these facts may eventually be found to have a comparatively simple physical explanation, for the moment it is necessary to be cautious before accepting too readily the conception that the stability of the periphery of the cell is solely determined by factors which also operate on inanimate membranes. We have considered the special case of *Mytilus* cells in some detail, since it appears to form a striking case in which the action of ions on a biological system runs parallel to their action on other colloidal systems. The analysis is, however, open to criticism in that it is a special case. In fact, we know that however efficient magnesium may be as a stabiliser of cell membranes in *Mytilus* it is not efficient in the case of other tissues. In *Fundulus*, echinoderm larvae, and in *Spirogyra*, calcium alone can stabilise the cell surface. These exceptions do not, however, invalidate the argument. We have seen that the union of calcium with the cell surface is inhibited by magnesium in the case of *Mytilus*, because the latter metal reduces the ionisation of the surface to a very low level. If, however, the

Fig. 39. Dissolution of hyaline membrane and cleavage planes in eggs of *Ctenolabrus* owing to excess of CO_2; *a*, normal cells; *b*, cells after 14 mins. exposure to CO_2; *c*, later stage. (From Loeb.)

isoelectric point were different, it might well be that even in the presence of magnesium the surface would bear an electro-negative charge. In this case calcium ions in the sea water would react with the matrix to form a stable compound. In other words a matrix of a calcium compound, similar to those which apparently exist in echinoderms and in *Fundulus*, would be formed. Further, this type of membrane might well occur in the case of tissues normally in equilibrium with waters containing little or no magnesium. The calcium type of matrix can thus be very simply derived from the more generalised colloidal type such as exists in *Mytilus*. It might be suspected that the calcium type of matrix would occur in the case of fresh-water organisms and this is supported by Benecke's (1898) experiments with *Spirogyra*. Such membranes in the case of marine animals must, however, resemble those of casein, in that when in combination with calcium they still retain a certain degree of coherence. In the case of animals living in more acid environments calcium would not form basic compounds so readily, but would tend to do so if the waters became more alkaline.

The evolution of intercellular secretions may perhaps be depicted as follows:

Table XVIII

Hyaloplasmic layer of mucoid protein

```
                Hyaloplasmic layer of mucoid protein
                              |
              ┌───────────────┴───────────────┐
         Mg·· present                    Mg·· absent
        ┌──────┴──────────┐                   │
   Electro-neutral   Electro-negative    Reacts with Ca··
        ↓                  ╲                  │
   Matrix as in         Reacts with Ca··      │
   gills of Mytilus          ╲                │
                              ↘               ↓
                          E.g.         Basic cal-      E.g. matrix
                       echinoderm ←── cium salt ──→ of freshwater
                          larvae       of protein    algae, etc.
                                           │
                                     Reacts with CO₂
                                           ↓
        Reorganisation               Deposition of
        of protein molecule          inorganic salts
              │                      of calcium
              ↓                            ↓
        Conchiolin, chitin           E.g. calcareous
           cartilage                   exoskeletons
```

Diagram illustrating the possible relationships between different types of intercellular matrices.

It should perhaps be mentioned that the above analysis differs somewhat from the currently accepted explanation of the effect of salts on the stability of cell surfaces. According to R. S. Lillie (1906), the surface layers of the cell consist of a calcium compound of an organic radicle; if the cell is exposed to pure sodium chloride solution, this calcium compound undergoes double decomposition with the formation of a soluble sodium salt. The inadequacy of this suggestion is seen from the fact that the cell surface is often instable in the absence of electrolytes, and it ignores the fact that the prime factor in dispersing the matrix is the hydroxyl ion.

Two further points may be noted. Firstly, the effect of ions on the stability of the cell surface offers no real evidence for antagonistic action between metallic ions. In their electrostatic effect all ions act in essentially the same manner—viz. all cations antagonise the hydroxyl ions—but, except in the case of the hydrogen ion, all the monovalent ions are much less efficient than the divalent ions. Secondly, we reach an important problem which will be discussed in a later chapter, but which must here be mentioned. The surface of living cells is not freely permeable to electrolytes and the work of Osterhout (1906) has shown that the factors which affect this impermeability are closely parallel to those which are here shown to be concerned with the mechanical stability of a protein gel. Since, however, protein gels, however coherent, are freely permeable to electrolytes, one must conclude that the mechanism responsible for cell impermeability is not simply the gel condition of the hyaline surface, but that probably in this gel there is embedded another mechanism which is responsible for the slow rate of penetration of electrolytes into the living cell.

REFERENCES

BENECKE, W. (1898). 'Mechanismus und Biologie des Zerfalles der Conjugatenfäden in einzellen Zellen.' *Jahr. f. wiss. Bot.* 32, *452*.

CARTER, G. S. (1926). 'On the nervous control of the velar cilia of the nudibranch veliger.' *Brit. Journ. Exp. Biol.* 4, *1*.

CHAMBERS, R. (1924). *General Cytology.* (Chicago.)

FAURÉ-FREMIET, E. (1928). 'Constitution et propriétés physico-chimiques des éléocytes d'*Amphitrite johnstoni* (Malmgren).' *Protoplasma*, 5, *321*.

GALTSOFF, P. S. (1925). 'Regeneration after dissociation. I.' *Journ. Exp. Zool.* 42, *183*.

GRAY, J. (1922). 'The mechanism of Ciliary Movement. II. The effect of ions on the cell-membrane.' *Proc. Roy. Soc.* B, 93, *122*.

GRAY, J. (1924). 'The mechanism of cell-division. I. The forces which control the form and cleavage of the eggs of *Echinus esculentus*.' *Proc. Camb. Philos. Soc. Biol. Series*, 1, *164*.
—— (1926). 'The properties of an intercellular matrix and its relation to electrolytes.' *Brit. Journ. Exp. Biol.* 3, *167*.
HARDY, W. B. and WOOD, T. B. (1908). 'Electrolytes and colloids. The physical state of gluten.' *Proc. Roy. Soc.* B, 81, *38*.
HERBST, C. (1900). 'Über das Auseinandergehen von Furchungs- und Gewebezellen in kalkfreiem Medium.' *Arch. f. Entw. Mech.* 9, *424*.
HOWLAND, R. B. (1924). 'Dissection of the pellicle of *Amoeba verrucosa*.' *Journ. Exp. Zool.* 40, *263*.
KITE, G. L. (1913). 'Studies on the physical properties of protoplasm. I. The physical properties of the protoplasm of certain animal and plant cells.' *Amer. Journ. Physiol.* 32, *146*.
KROGH, A. (1922). *The Anatomy and Physiology of Capillaries.* (New Haven.)
LILLIE, R. S. (1906). 'The relation of ions to contractile processes. I. The action of salt solutions on the ciliated epitheliums of *Mytilus edulis*.' *Amer. Journ. Physiol.* 17, *89*.
LOEB, J. (1905). *Studies in General Physiology*. Pt. I. (Chicago.)
OSTERHOUT, W. J. V. (1906). *Injury, Recovery and Death.* (Philadelphia.)
PFEFFER, W. (1906). *The Physiology of plants*. Eng. transl. Vol. 3. (Oxford.)
ROBERTSON, T. B. and MIYAKE, K. (1916). 'The influence of alkali and alkaline earth salts upon the rate of solution of casein by sodium hydroxide.' *Journ. Biol. Chem.* 25, *351*.
SEIFRIZ, W. (1918). 'Observations on the structure of protoplasm by aid of microdissection.' *Biol. Bull.* 34, *307*.
—— (1921). 'Observations on some physical properties of protoplasm by aid of microdissection.' *Ann. of Bot.* 138, *269*.
—— (1924–5). 'An elastic value of protoplasm, with further observations on the viscosity of protoplasm.' *Brit. Journ. Exp. Biol.* 2, *1*.
—— (1929). 'The contractility of protoplasm.' *Amer. Nat.* 58, *410*.
VAN SLYKE, L. L. and BOSWORTH, A. W. (1913). 'Reparation and composition of unsaturated or acid caseinates or paracaseinates.' *Journ. Chem.* 14, *211*.
VAN SLYKE, L. L. and WINTER, O. B. (1914). 'Preparation, composition and properties of the caseinates of magnesium.' *Journ. Biol. Chem.* 17, *287*.

CHAPTER SEVEN

The Nucleus

The nucleus is one of the few visible structures common to the great majority of living cells, and the changes which it undergoes during the life of the cell are more obvious and more precise than those exhibited by any other cell constituent. In the great majority of cases the nuclei of living animal cells are optically homogeneous except for the presence of definitive nucleoli. The typical nucleus, when viewed by transmitted light, is a clear spherical or ovoid body, whose position in the cell is relatively constant. When viewed by reflected light, or even on a dark field with intense lateral illumination, a healthy nucleus as a rule exhibits no internal structure. Obvious examples of homogeneous nuclei are provided by the eggs of sea urchins (Albrecht, 1898); the skin cells of *Triton* (Gross, 1917), tissue-culture cells (Lewis and Lewis, 1924), and the numerous nuclei examined by Chambers and Rényi (1925). A few exceptions to this rule should perhaps be mentioned. In 1917 Gross described a granular structure in the nuclei of the living cells of the salivary glands of *Limnea* and *Unio*, and a similar appearance was identified in the nuclei of the red blood corpuscles of *Triton* by Commandon and Jolly (1913, 1918). According to Shiwago (1926) the nuclei of

---

Fig. 40. Showing structure of nucleus and course of mitosis in living plant cells. (From Martens, 1927.)
1. Early prophase: living nucleus.
2. The same nucleus as in 1 after fixation.
3. Living nucleus from cell of *Listera ovata*.
4. Spireme in living nucleus of *Arrhenatherium*.
5. Two chromosomes from 4 after fixation.
6. End of prophase in *Arrhenatherium*: living nucleus.
7. Same nucleus as 6, three minutes later.
8. Same nucleus as 6, 7, five minutes after 6.
9. Same nucleus as 6–8, ten minutes after 6.
10. Anaphase after fixation and staining.
11. Anaphase from living cell.
12. End of anaphase, from living cell.
13. Beginning of telophase, from living cell.
14. Late telophase, from living cell.

NUCLEUS

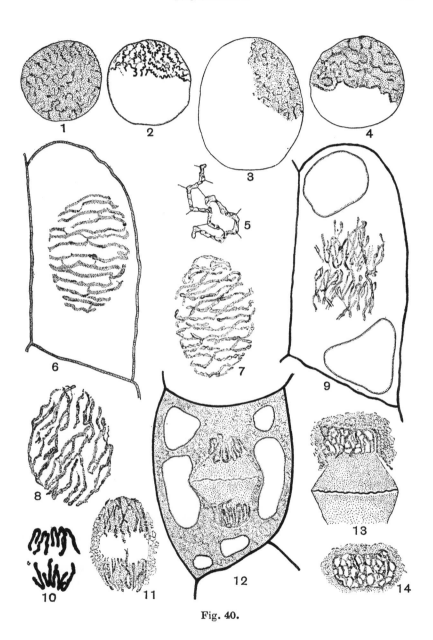

Fig. 40.

the leucocytes of the frog contain a series of fine but distinct interlacing filaments not unlike that described by Fauré-Fremiet (1910) in certain infusorian nuclei.

Since all nuclei exhibit a visible granular or fibrillar structure after coagulative fixation, it is generally supposed that the structures seen in preserved preparations or in moribund nuclei are to be regarded as purely artificial products of coagulation, which cannot be correlated with the fundamental structure of a living nucleus. This view, developed many years ago by Hardy (1899), is now accepted by the majority of animal cytologists (Champy, 1913; Policard, 1922; Lewis and Lewis, 1924; Chambers and Rényi, 1925; Schitz, 1925; Burrows, 1927). See, however, Bělăr (1928).

Among plant cytologists, however, there appears to be some difference of opinion and the essential homogeneity of the interkinetic nucleus has been questioned by Martens (1927). According to Martens, the normal uninjured nuclei of the cells of the stigma of *Arrhenatherium elatius* show a distinct intranuclear structure; numerous granules are visible which are attached to each other by extremely fine threads, thereby forming a loose network of interlacing fibrils throughout the nucleus. When the cell is fixed by Bouin's fixative, no change occurs in the distribution of either the chromatin granules or of the interlacing linin threads; both structures simply become more obvious to the eye (fig. 40). From careful observations of this type, Martens concludes that linin threads and chromatin granules probably have a real existence in all interkinetic nuclei, although in many cases their refractive index is so close to that of the nuclear sap as to render them invisible in life. It hardly seems reasonable, however, to believe that there should be three nuclear components all with indices of refraction almost identical; further, in those cases where the nuclear matrix is composed of a fluid with low viscosity (e.g. the nuclei of invertebrate oocytes—see below) the intrinsic probabilities are against the existence of a solid network. If Martens' observations are correct—and there is no reason to suggest the contrary—then in the case of the nuclei of *Arrhenatherium* the whole of the coagulable part of the living nucleus is already in a relatively high state of aggregation, otherwise the addition of fixatives would increase the number or size of the solid granules seen in the nucleus, and this does not appear to be the case. In most animal cells, however, we may suppose that the normal degree of aggregation of the colloidal constituents is very much less, and that these elements are so highly dispersed as to render them beyond the limits of optical resolution.

When an apparently homogeneous nucleus spontaneously gives rise to a series of discrete chromosomes, some form of molecular aggregation must occur, and in all cells there comes a time during the prophase of mitosis when these aggregations become visible in the living cell and are the obvious forerunners of the chromosomes. In a very crude, yet justifiable way, we can look upon chromosome formation as an increase in the degree of aggregation of certain nuclear constituents, so that sooner or later more than one nuclear phase becomes visible to the eye. The same thing occurs in fixation, so that this process will always give a relatively faithful picture of a living structure, if that structure is already in a state of considerable aggregation; this is probably the case in all cells during the closing stages of the prophase, and is always the case for the nuclei examined by Martens. The only real ground of divergent opinion covers the differential effects produced by different types of fixative. If the visible effect of fixation were always the same it might be possible to agree with Martens that chromatin granules and linin threads are present in a very finely dispersed state in all nuclei, but this is by no means certain (see p. 40).

Before proceeding further it is useful to bear in mind that an optically homogeneous nucleus is periodically resolved into a system of concrete units (the chromosomes), whose number and form are specific for each organism; further, the chromosomes themselves can be further resolved into a series of biological units whose rôle in the transmission of hereditary characters can be accurately deliminated. Whatever be the true nature of genetical units, it is difficult to avoid the conclusion that they are represented in a chromosome by a physical organisation of high complexity. Since nearly all interkinetic nuclei are derived from and give rise to specific aggregates of chromosomes, it seems unreasonable to deny to the interkinetic nucleus a fundamental structure of very considerable specificity and complexity. Presumably, biological structures of this type must be based on physical or chemical heterogeneity between different regions of the chromosome or nucleus: Unfortunately, the physical units which underlie the fundamental biological structure are not aggregated together in sufficient masses to enable us to resolve them even with the most efficient optical appliances. In short, in both nucleus and cytoplasm we encounter the same paradoxical divergence between our ideas of biological complexity and our conceptions of optical or physical homogeneity. The problem of the nucleus is even more striking than is the case of cytoplasm, since the biological constitution of the nucleus is known with much greater accuracy than is that of the cytoplasm, whilst at the same time its optical homogeneity is often very clearly defined.

It is obvious that an intranuclear architecture of considerable complexity might exist within an optically homogeneous system, if the system had the properties of a solid gel; if a living nucleus possessed a continuous solid phase, one might imagine that included phases even of molecular dimensions were prevented from undergoing a random distribution by their inability to move through the meshes of the continuous solid phase. The earliest attempts to determine the physical state of living nuclei (Kite, 1913) led to the belief that nuclei were, in fact, solids which could be cut into discrete fragments. Peebles in 1912 reported that the macronucleus of *Paramecium* resembled a stiff jelly, but at certain periods of the life history the gel was liquefied leaving, however, always a solid membrane at the surface. More recent observations, however, have cast considerable doubt on the solid nature of the nucleus. Chambers (1924) has shown that fluids can be injected into the living nucleus without the formation of distinct vacuoles, and that microdissecting needles move through the nucleus without leaving any evidence of shearing stresses. That the large nuclei of the oocytes of *Asterias* or of *Echinus* are undoubtedly composed of a fluid is seen very clearly from the behaviour of the included nucleoli (Gray, 1927). The nucleoli move freely through the nucleus and always orientate themselves in respect to gravity by coming to rest on the floor of the nucleus. The velocity of movement of the nucleoli is readily observable although in absolute units it is very slow: viz. 1·5 mm. per hour, which probably indicates a viscosity about twice that of water.

If we can judge the concentration of solid matter within a living nucleus by the intensity with which it absorbs nuclear stains, it would appear that large nuclei, which are quite clearly largely composed of a fluid material, contain nucleo-proteins in a more dilute or more highly dispersed condition than do the smaller but more compact nuclei characteristic of somatic cells. It would seem premature to assume that all nuclei are essentially fluid masses in the sense that their continuous phase is a liquid wherein the solid

NUCLEUS

phase is dispersed; it is more likely that in some cases the solid phase or phases are aggregated to give a continuous phase, just as in the case of a relatively strong solution of gelatin.

An adequate analysis of the physical states within a nucleus is hampered by the fact that any severe disturbance of the cytoplasm, or even of the surface of the cell, rapidly involves profound changes in the nucleus. This is particularly clear from the experiments of Chambers (1924). Chambers found that if a nucleus be removed from an echinoderm egg it either swells up until it bursts or it becomes coagulated and can then be cut into separate fragments. Rough handling will also produce coagulation (see also p. 130). It seems

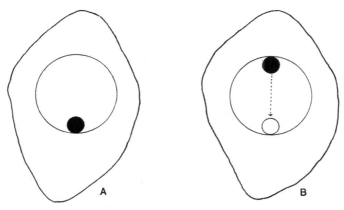

Fig. 41. Lateral view of oocyte of *Echinus* showing in *A* the position of equilibrium of the nucleolus; *B* shows the path followed by the nucleolus after rotating the oocyte through 180° on a horizontal axis. (From Gray, 1927.)

clear that whether the nuclear matrix be liquid or solid it represents a highly instable system whose physical state is changed much more readily than is that of inanimate colloidal systems, with the possible exception of the gels described elsewhere (see p. 57).

The ease with which a nucleus will pass into a solid state or yield solid derivatives (chromosomes) has sometimes been associated with the properties of nucleic acid. Mathews (1915) put forward the ingenious suggestion that the gelatinous condition of typical chromosomes is due to the power possessed by certain salts of nucleic acid to exist in the gel state; Chambers also associates the rapid postmortem gelation of injured nuclei with similar substances. It is quite true that a 1 per cent. solution of the sodium salt of thymus nucleic acid forms, in the cold, an elastic gel. This gel is, however, heat reversible, whereas when nuclei and chromosomes are exposed to high temperatures they coagulate irreversibly. Further, the chromosomes of plants are similar to those of animals in their physical consistency (Chambers and Sands, 1923); yet the nucleic acid of plants does not yield a reversible gel with sodium hydroxide; plant chromosomes must owe their consistency to some other cause than the gelatinising properties of the salts of nucleic acid. As with protoplasm, we can only ascribe to nuclei the properties of a highly instable and delicately poised system, whose instability is probably an essential feature of the living state and which finds no adequate parallel in inanimate systems.

Chemical properties of the nucleus

The chemical composition of the nucleus was first investigated by Miescher in 1876. It had long been known that living tissues contained considerable quantities of phosphorus, and that this was present in at least three forms, (*a*) as inorganic phosphates, (*b*) in organic union with fatty substances, such as lecithin, (*c*) in union with protein substances. Miescher found that this third fraction could be extracted in large quantities by means of dilute alkali from the cells found in pus. He concluded that it was derived from the cell nuclei, since pus cells have large nuclei and but little cytoplasm; he called it *nuclein*. This body, which contained from 0·9 to 4 per cent. of phosphorus, was quickly found to be present in all cells containing large nuclei; e.g. spermatozoa, spleen cells, and yeast. In 1887 Altmann found that by treating nuclein with pepsin, an organic acid containing 8·9 per cent. of phosphorus could be obtained, and this he called *nucleic acid*. By suitable technique an almost pure suspension of nuclei can be obtained from the heads of ripe spermatozoa and these have been analysed. In every case they have been found to consist of nucleic acid in association with a basic protein or protamine. As far as can be judged by direct analysis the nucleic acid of all animal nuclei has the same composition and properties irrespective of its origin.

Animal nucleic acid consists of four hexose molecules, associated with two pyramidine, and four phosphoric acid radicles. The nucleic acid of plants, however, differs in its carbohydrate radicles from that of animals. So far, no conclusive proof has been given that nucleic acid occurs in the cell outside the nucleus; the yield of nucleic acid from any given tissue is roughly proportional to the volume of the nuclei present—see Table XX (Whiteside, 1923).

Although it looks as though all the nucleic acid derived from animals has the same general composition and properties, there is a certain degree of variation in the basic substances with which the acid is associated. Sometimes the basic constituents of the nucleus

Table XIX

Source of nucleic acid	% composition			
	Carbon	Hydrogen	Nitrogen	Phosphorus
Sperm of *Alosa* ...	36·3	5·0	16·0	8·1
Sp·¹ of *Maranoesox*	37·5	4·4	16·0	9·7
Human placenta ...	37·4	4·3	15·0	9·7

Table XX

Tissue	Average vol. of cell in μ^3	Average vol. of nucleus	$\dfrac{\text{Vol. of nucleus}}{\text{Total vol. of cell}} \times 100$	% yield of nucleic acid
Liver	2803	121·0	4·32	0·7
Pancreas	1521	67·9	4·46	0·8
Thyroid	1242	144·0	11·5	1·1
Thymus	216	48·0	22·0	3·1

Table XXI

% amino acids in protamines	*Arbacia*	Salmon
Alanin	6·8	—
Serin	—	3·2
Valin	—	1·6
Arginin	88·9	88·9
Prolin	3·8	4·3

are proteins, but in the case of spermatozoa they are always of the protamine type. If the constituent amino-acids which form the protamines (of the nucleus in the spermatozoa) in the sea urchin *Arbacia* be compared with those of the salmon, it will be seen that the chemical composition of the nucleus, at present, gives no adequate picture of the biological complexity of the structures so closely associated with heredity (Table XXI).

It is conceivable that the marked biological differences between the nuclei of different animals are to be associated with isomeric differences in the composition of individual nucleic acids or of the bases with which they are associated. As was originally pointed out by Miescher, there are almost unlimited possibilities for isomerism with molecules of these dimensions. At present, however, the chemical evidence does little more than sound a note of warning to those who are inclined to regard the nucleus as a heterogeneous collection of units, each of which has a specific chemical constitution. The specificity of a nucleus is more likely based on specificity of structure than of composition; a Ford motor car and a Rolls-Royce motor car have approximately the same composition—their essential differences depend on the form and orientation of parts whose chemical nature is only of secondary significance.

The universal presence of nucleic acid in animal nuclei leaves little doubt that this compound is essentially the same as the 'chromatin' of histologists. Free nucleic acid has just those affinities for basic stains as are characteristic of chromatin, and the variation in the staining properties of the kinetic nucleus are paralleled by the staining properties of nucleic acid when saturated to varying degrees by combination with basic proteins. The evidence supporting an identity of chromatin and nucleic acid is greatly strengthened by the recent work of Feulgen and others. If a tissue be exposed for a suitable time and at a suitable temperature to a solution of mineral acids the distribution of chromatin can be detected by subsequent staining with acid fuchsin. As already mentioned, Voss (1925) interprets this technique as a preliminary acid hydrolysis of nucleic acid whereby aldehyde groups are set free in the localised regions originally occupied by nucleic acid, whilst the subsequent staining with fuchsin containing SO_2 is identical with Schiff's colour reaction for aldehyde groupings (see p. 84). If we accept Feulgen's technique as a specific indication of the presence or absence of nucleic acid, it seems certain that neither nucleic acid nor nucleo-proteins occur in the cytoplasm of normal cells, except in those cases where 'chromatin' is extruded from the nucleus at particular cycles of cell-activity (e.g. oocytes of molluscs, arthropods).

The cytoplasm of the anucleate Cyanophyceae presents an interesting problem. In a recent study of *Nostoc*, Mockeridge (1926) has shown that although nucleic acid cannot be detected as such within the cells, yet all its typical derivatives are undoubtedly present. All tests for nucleic acid proved negative, but the radicles of phosphate, pentose, adenine, guanine, cytosine, and uracil were found in quantities sufficient to allow of definite identification. How far the condition in *Nostoc* is to be regarded as a primitive state which phylogenetically preceded the formation of a definite nucleus containing synthesised nucleic acid, or how far the Cyanophyceae are to be looked upon as a degenerate group where 'primitive' nucleic acid is no longer formed, is unknown.

Biological properties of the nucleus

For many years considerable interest has been associated with the problem of how far the presence of the nucleus is necessarily associated with the normal functions of the cell. The experimental difficulties associated with this work are considerable, since it is usually impossible to remove the nucleus from a cell without causing extensive damage to the whole system. Certain cell functions can undoubtedly continue more or less unchanged in the absence of the nucleus: an enucleated nerve will continue to conduct a stimulus, and both ciliary (Peter, 1899) and amoeboid movements continue for a considerable period. There is, in fact, no evidence that the nucleus plays any essential rôle in the processes of stimulation, conduction, or contraction. Nor is the nucleus necessary for the normal osmotic properties of the cell surface; enucleated fragments of echinoderm eggs exhibit the same osmotic relationships as complete eggs.

To what extent the nucleus plays an essential rôle in the metabolic processes of the cell is a matter of some doubt. That it is not entirely inert is certain, but in estimating the value of experimental results, it must be borne in mind that the injury inflicted to a cell by removal of the nucleus may in itself have profound metabolic results. That injury to the cell rapidly induces changes in the nucleus was observed by Chambers (1924) in the male germ cells of insects (*Dissosteira carolina*). The normal nucleus lies more or less in the centre of the cell and is uniformly hyaline in appearance. Any form of mechanical injury to the cell, however, soon induces the formation of intranuclear granules and filaments apparently similar to the chromatin granules and linin threads seen in preserved material. Again, if many tissues are stained *intra-vitam* with methylene blue, the stain never enters a living nucleus but remains in the cytoplasm: if the cell is injured the stain very readily enters the nucleus which stains much more intensely than the cytoplasm. A similar phenomenon was observed by Osterhout (1917). The cells of *Monotropa uniflora* contain a colourless chromogen which darkens in the presence of oxygen. In the living cell the nucleus is quite colourless, but if the cell be injured the darkening of the chromogen is first seen in the nucleus. These facts suggest the need for caution in the use of evidence derived from injured cells.

One of the clearest proofs of the part played by the nucleus in the cell is that given by the well-known experiments of Townsend (1897), in which he showed that enucleated fragments of the protoplasm of

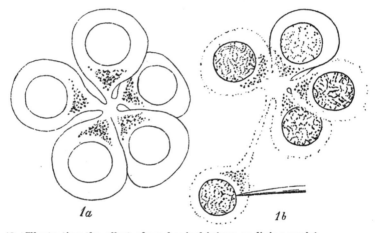

Fig. 42. Illustrating the effect of mechanical injury on living nuclei.
1 a. Group of normal spermatogonia of *Dissosteira*.
1 b. One spermatogonium has been seized by a needle. Note the coagulation of all the nuclei, and the disintegration of four cells. (From Chambers, 1924.)

roothairs in *Marchantia* fail to form a new cellulose wall unless in organic connection with another fragment containing a nucleus. Verworn (1889) states that an enucleated *Amoeba* loses the power of digesting fragments of food, although the powers of ingestion are not lost. According to Lynch (1919) an enucleated fragment of *Amoeba* is able to make use of glucose as a source of food, but unlike the whole organism is unable to synthesize nitrogenous compounds from glucose and urea. All these results seem to indicate that the nucleus is possibly associated with the anabolic powers of the cell. This is in keeping with the fact that a fragment of macronucleus is necessary for the regeneration of fragments of *Stentor* (Lillie, 1896).

Numerous workers have from time to time suggested that the nucleus is closely associated with the respiratory activity of the cell. This view was first put forward by Spitzer (1897) and was supported by R. S. Lillie (1902, 1913), Loeb (1899), and Mathews (1907). Loeb (1899) showed that during mitotic division the mass of the nuclei in a tissue may increase considerably and that this did not occur in the absence of oxygen. It therefore looked as though oxygen were necessary for nuclear synthesis. R. S. Lillie (1913) based his conclusions on the fact that when cells are placed in a mixture of para-phenylene-diamine and α-naphthol, the indophenol formed is aggregated at the surface of the nucleus. In 1910 Warburg showed that whereas the nucleated red blood corpuscles of birds possess a fairly high level of respiration, those of the mammals which have no nucleus have a very low respiratory level. On treating bird corpuscles in such a way as to isolate the nuclei, the latter were found to retain a moderate power of respiration, whereas the remainder of the cell absorbed little or no oxygen. It may be doubted how far any of these arguments are convincing. If the nucleus is essential for respiration, enucleated fragments of *Amoeba* should be quite independent of the presence of free oxygen, and binucleate cells might be expected to have an abnormally high oxygen consumption; neither of these conclusions are supported by facts. Warburg's experiments are not quite conclusive, since they do not show that the oxygen absorbed by the extracted nuclei is not similar to the rise in oxidation produced by injury; in other words, it is by no means clear that the oxygen absorbed by isolated nuclei is really part of the normal respiration of the cell and is not simply a pre-mortem phenomenon. Further, if the nucleus is intimately associated with oxidation, it is curious to find that the latter is independent of the presence of a nuclear membrane, since cells in which active mitosis is taking place have the same order of respiratory level as cells in which the nucleus is bounded by a membrane (Gray, 1925).

There is, however, one other type of evidence which indicates that the nucleus is associated with cell-metabolism. In certain glands, notably the spinning glands of insects, the cells have much branched nuclei whose form is said to change with changes in the secretory activities of the cell. Similarly, Conklin (1924) found that the size of the nuclei in the liver cells of *Crepidula* varied with the degree of secretory activity. A cell filled with secretory products has a nucleus only one-quarter the size of a cell containing little or no secretion.

The relationship between the nucleus and the cytoplasm

The experiments of Lillie (1896) and of Townsend (1897) show that the presence of the nucleus exerts a definite effect upon the activity of the cytoplasm; it is equally true that the cytoplasm exerts an influence on the nucleus. In some eggs one of the polar bodies is not formed until the sperm enters the cytoplasm: as soon as this occurs, the polar body is thrown off, although the male pronucleus lies dormant during the process. In the hybrid *Echinus* ♀ × *Mytilus* ♂ the female pronucleus is thrown into a state of activity, although the sperm nucleus degenerates. These facts suggest that the act of fertilisation causes a change in the cytoplasm which then induces changes in the nucleus. It has already been noted that substances may pass out of the nucleus into the cytoplasm, and it appears that the converse may be true. Danchakoff (1916) describes the passage of a chromatin-like substance from the cytoplasm into the nucleus of *Asterias* eggs.

In 1892 Sachs stated that the size of a meristematic cell in plants bore a constant relationship to the size of the nucleus; and in the following year Strasburger (1893) arrived at the same conclusion. A few years later, this relationship was investigated in animal cells by Hertwig (1903) and by Minot (1908), both of whom drew different conclusions as to its significance. Both agreed that the 'Kernplasmarelation' was of fundamental importance, but whereas Hertwig believed that a relative increase in the size of the nucleus was associated with the onset of senescence, Minot believed that the ageing of the cell was associated with a relative increase in the size of the cytoplasm. That there is usually a rough correlation between the size of the nucleus and the size of animal cells was shown by Boveri (1905); enucleated fragments of echinoderm eggs, when fertilised, yielded blastulae with a normal number of cells, but each was half the normal size, and of course each nucleus contained only half the normal number of chromosomes. Similar results were obtained for the cells of the frog by Brachet (1910) and Herlant (1911).

In spite of a good deal of research, no general conclusions are available as to the theoretical significance of the kernplasmarelation. Popoff (1908) found that during the growth of Protozoa the nucleus grows less rapidly than the cytoplasm, so that the ratio N/P becomes less. This state of affairs leads to a condition wherein the volume of the cytoplasm begins to exert a 'cytoplasmic strain' on the small nucleus. At this stage the nucleus begins to grow rapidly and cell division occurs. When the nucleus has undergone mitotic division the combined volume of the two newly constituted daughter nuclei is considerably larger than that of the original nucleus before cleavage, so that the value of N/P for newly divided organisms is larger than that of the parent individual and the 'strain' is removed (see also p. 272). That the combined volume of two newly formed daughter nuclei of segmenting eggs of echinoderms may be considerably larger than that of the original parent nucleus was also observed by Loeb (1910), who regarded the mitotic phase as one of nuclear synthesis. Loeb suggested that the nucleus was built up from a compound of glycero-phosphoric acid and lecithin in the cytoplasm, which became incorporated into the

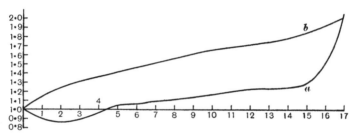

Fig. 43. Graph showing the relative growth of nucleus (*a*) and cytoplasm (*b*) between two successive divisions of *Frontonia*. The volume of both nucleus and cytoplasm is doubled between each division, but until immediately before division the relative rate of growth of the nucleus is much lower than that of the cytoplasm. (From Popoff.)

nucleus whenever the nuclear membrane broke down. Fauré-Fremiet (1913) failed, however, to find any change in the lecithin content of *Ascaris* eggs during segmentation.

A more general study of this problem, however, soon showed that nuclear growth during mitosis is a very variable phenomenon. In the eggs of *Cynthia* it is very slight (Conklin, 1924), in *Vespertilio murinus* it is entirely absent during the earlier cleavages. Even in echinoderm eggs (Table XXII) the work of Godlewski (1908) shows that the volume of the nucleus does not increase in geometrical progression with each successive division as Loeb supposed.

The value of such determinations is, however, limited by the fact that observed changes in the volume of the nucleus do not distinguish between a real synthesis of intranuclear material and an absorption of water. Both Masing (1910) and Schackell (1911) failed to find any appreciable change in the nucleic acid content of echinoderm eggs between fertilisation and the development of the blastula. Again, Erdmann (1908) found that the rate of increase in the total volume of the chromosomes was considerably less than the increase in the total volume of nuclei (see also p. 271 *seq.*). A valuable review of the facts which bear on the problem of the nucleo-cytoplasmic ratio will be found in Fauré-Fremiet's (1925) text book, p. 72 *seq.*

An attractive hypothesis of nuclear-cytoplasmic relationship has been put forward by Robertson (1923). The process of growth is held to depend upon the concentration of an autocatalyst whose distribution is effected by the breakdown of the nuclear membrane. "During the periods between nuclear division, each nucleus retains the charge of autocatalyst with which it was originally provided, and adds to it in the course of nuclear synthesis which is rendered

Table XXII

State of cleavage	Nuclear volume
Zygote	1,300 μ^3
1st cleavage	1,582
2nd ,,	2,084
5th ,,	19,938
6th ,,	30,262

possible by its presence. At the next division the autocatalyst is shared between the nuclear materials and the surrounding medium in a proportion determined, in part by its relative solubility in the two media, and in part by its affinity for chemical substances within the nucleus. At the end of this redistribution the autocatalyst stands in equilibrium between the external solvent or pericellular medium on the one hand, and on the other hand the nuclear substances with which it combines or in which it is dissolved. The nuclear membrane is then reformed, and the autocatalyst within the nucleus is again shut off from dispersal into the surrounding medium until the occurrence of the next succeeding division." It has been pointed out (Gray, 1925) that such a cycle of nuclear activity is not in harmony with observed data, unless the process of growth is entirely independent of all oxidative changes in the cell. For Robertson's views on the relation between the kernplasma-relation and cell-differentiation the original monograph should be consulted (1923).

The rôle of the nucleus during development

The inheritance of paternal characters has long suggested that the process of cell differentiation in a developing organism is to some extent controlled by the constitution of the two parent nuclei, since the possession of a nucleus and the ability to transmit hereditary characters are almost the only factors common to the ovum and spermatozoon. Further, there is every reason to believe that specific differences can exist in the homologous chromosomes of gametes at maturation and that these differences are the cause of specific differences in cell differentiation at an early or late stage in the development of the organism. The establishment of these facts should, however, not be allowed to mask the significance of other data.

In 1885 Weismann put forward his well-known hypothesis as an attempt to correlate the facts of heredity with the process of cell differentiation. He suggested that the nuclei of the germ cells contained all the factors necessary for complete differentiation. By successive nuclear divisions these factors were gradually segregated from each other, so that each type of somatic cell eventually contained nuclear characters different from any other type, and these characters controlled the ultimate differentiation of somatic cells. Definite experimental work, however, soon threw considerable doubt on this conclusion. Driesch (1900) proved that a complete larva can arise from isolated blastomeres of a sea urchin egg. Hertwig (1893) showed that when a frog's egg develops under pressure and is subsequently released, a normal embryo develops, although certain of the nuclei have been forced into different regions of the larva. Driesch performed similar experiments with sea urchin eggs, and both he and subsequent workers have confirmed Hertwig's results. It is, therefore, impossible to assume that successive nuclear cleavages involve the segregation of specific characters during development. It might, however, be argued that at least the first three divisions of the sea urchin egg are morphologically equivalent, since the blastomeres are all of approximately the same size, and that nuclear differentiation in these forms is deferred to a later stage. This criticism cannot be applied to the case of *Nereis* which was investigated by Wilson (1896). In this case the third division of the normal egg separates off four granular micromeres from four larger macromeres and only the latter contain oil drops. These four basal cells can be recognised in the larva for a considerable period and they alone eventually form the archenteron. If any nuclear differentiation occurs during development then, at the third cleavage, there must be four macromeric nuclei containing endodermic characters

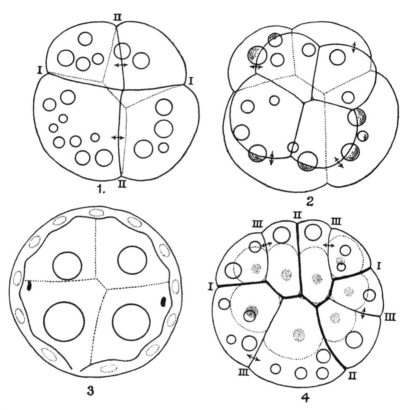

Fig. 44. Effect of pressure on the segmentation of *Nereis* eggs. 1, Normal 4-cell stage; 2, normal 8-cell stage with the first quartet of micromeres lying above the basal quadrant; 3, young trochophore showing four entomeres surrounded by the twelve cells of the prototroch; 4, 8-cell stage produced under pressure, the dotted lines show the outlines of the 8 micromeres which form after the release of the pressure. (From Wilson, 1896.)

(in addition to other characters segregated in later quartets of micromeres), and four micromeric nuclei containing no endodermic characters. Wilson showed, however, that if subjected to pressure the third cleavage can give rise to a flat plate of eight cells, instead of two tiers each of four cells. On releasing the pressure some of these eggs divided so as to give two tiers of eight cells each. The larger tier consisted of typical endodermic cells containing oil drops, and the smaller tier of typical granular micromeric cells. Some of these larvae develop, and are normal except for the fact that the archenteron is formed from eight instead of from four macromeres. The mosaic theory of nuclear cleavage demands that in the normal egg the third division produces four cells free from endodermal characters, yet under experimental conditions it is obvious that this is not the case, since eight nuclei are capable of differentiation into endoderm cells. These experiments were repeated by Morgan (1910), who pointed out that the larvae obtained were not absolutely normal, but the fact remains that eight endodermic cells may be produced by differential cleavage, whereas the mosaic theory of normal development only provides for four.

The attempt to correlate the biological conception of nuclear or chromosomal differentiation with the limited physico-chemical data at our disposal is obviously a depressing task. The theory of inheritance leads to a picture of chromosomal structure which is as clearly defined and as strongly established as any biological conception can hope to be; any physical conception of the nucleus must cover the biological facts if it is to be of any use. Try as we will, it seems impossible to point to any inanimate system (endowed with the known chemical or physical properties of the nucleus) which in any real way possesses the requisite complexity whereby it might reflect even feebly the biological facts. Much that has been written concerning the structure of protoplasm applies with equal force to the structure of the nucleus; and as in other cytological problems the gulf between biology and physical chemistry is certainly no narrower than is usually supposed.

REFERENCES

ALBRECHT, E. (1898). 'Untersuchungen zur Struktur des Seeigeleis.' *Sitz. Ber. Ges. Morph. Phys. München*, 14, *133*.

BĕLĂŘ, K. (1928). 'Über die Naturtreue des fixierten Präparats.' *Zeit. f. Indukt. Abstammungs- und Vererbungslehre*, Suppl. Band 1, *402*.

BOVERI, T. (1905). *Zellen-Studien. V. Über die Abhängigkeit der Kerngrösse und Zellezahl der Seeigel-larven von der Chromosomenzahl der Ausgangszellen.* (Jena.)

BRACHET, A. (1910). 'Recherches sur l'influence de la polyspermie expérimentale dans le développement de l'œuf de *Rana fusca*.' *Arch. de Zool. expér. et gén.* 6, *1*.

BURROWS, M. T. (1927). 'The mechanism of cell division.' *Amer. Journ. Anat.* 39, *83*.
CHAMBERS, R. (1914). 'Some physical properties of the cell nucleus.' *Science*, 40, *824*.
—— (1924). 'Physical structure of protoplasm.' *General Cytology*. (Chicago.)
—— (1924 a). 'Les structures mitochondriales et nucléaires dans les cellules germinales mâles chez la Sauterelle.' *La Cellule*, 35, *107*.
CHAMBERS, R. and RÉNYI, G. S. (1925). 'The structure of the cells in tissues as revealed by microdissection. I. The physical relationships of the cells in epithelia.' *Amer. Journ. Anat.* 35, *385*.
CHAMBERS, R. and SANDS, H. C. (1923). 'A dissection of the chromosomes in the pollen mother cells of *Tradescantia virginica*.' *Journ. Gen. Physiol.* 5, *815*.
CHAMPY, C. (1913). 'Recherches sur la spermatogénèse des Batraciens et les Éléments accessoires du testicule.' *Arch. Zool. Exp.* 52, *13*.
COMMANDON, J. and JOLLY, J. (1913). 'Démonstration cinématographique des phénomènes nucléaires de la division cellulaire.' *C.R. Soc. Biol.* 75, *457*.
—— (1918). 'Étude cinématographique de la division cellulaire.' *Journ. Physiol. et Path. gén.* 17, *573*.
CONKLIN, E. G. (1912). 'Cell size and nuclear size.' *Journ. Exp. Zool.* 12, *1*.
—— (1924). 'Cellular differentiation.' *General Cytology*. (Chicago.)
DANCHAKOFF, V. (1916). 'Studies on cell-division and cell-differentiation. I.' *Journ. Morph.* 27, *559*.
DRIESCH, H. (1900). 'Die isolierten Blastomeren des Echinidenkeimes.' *Arch. f. Entw. Mech.* 10, *361*.
—— (1908). *The Science and Philosophy of the Organism*. (London.)
ERDMANN, R. (1908). 'Experimentelle Untersuchungen der Massenverhältnisse von Plasma, Kern, und Chromosomen in dem sich entwickelnden Seeigelei.' *Arch. f. Zellforsch.* 2, *76*.
FAURÉ-FREMIET, E. (1910). 'Étude physiologique sur la structure de noyaux de type granuleux.' *C.R. Acad. Sci. Paris*, 150, *1355*.
—— (1913). 'Le cycle germinatif chez l'*Ascaris megalocephala*.' *Arch. Anat. Micro.* 15, *435*.
—— (1925). *La Cinétique du Développement*. (Paris.)
GODLEWSKI, E. (1908). 'Plasma und Kernsubstanz in der normalen und der durch äussere Faktoren veränderten Entwicklung der Echiniden.' *Arch. f. Entw. Mech.* 26, *278*.
GRAY, J. (1925). 'The mechanism of cell-division. II. Oxygen consumption during cleavage.' *Proc. Camb. Philos. Soc. Biol. Series*, 1, *225*.
—— (1927). 'The mechanism of cell-division. IV. The effect of gravity on the eggs of *Echinus*.' *Brit. Journ. Exp. Biol.* 5, *102*.
GROSS, R. (1917). 'Beobachtungen und Versuche an lebenden Zellkernen.' *Arch. f. Zellforsch.* 14, *279*.
HARDY, W. B. (1899). 'On the structure of cell protoplasm.' *Journ. Physiol.* 24, *158*.
HERLANT, M. (1911). 'Recherches sur les œufs di- et tri-spermiques de grenouille.' *Arch. de Biol.* 26, *173*.
HERTWIG, O. (1893). 'Ueber den Werth der ersten Furchungszellen für die Organbildung des Embryo.' *Arch. mikr. Anat.* 42, *662*.

HERTWIG, R. (1903). 'Ueber Korrelation von Zell- und Kerngrösse und ihre Bedeutung für die Differenzierung und die Theilung der Zelle.' *Biol. Zentralb.* 23, *49, 108.*
KITE, G. L. (1913). 'Studies on the physical properties of protoplasm.' *Amer. Journ. Physiol.* 32, *146.*
LEWIS, W. H. and LEWIS, M. R. (1924). 'Behaviour of cells in tissue cultures.' *General Cytology.* (Chicago.)
LILLIE, F. R. (1896). 'On the smallest parts of the *Stentor* capable of regeneration.' *Journ. Morph.* 12, *239.*
LILLIE, R. S. (1902). 'On the oxidative properties of the cell nucleus.' *Amer. Journ. Physiol.* 7, *412.*
—— (1913). 'The formation of indophenol at the nuclear and plasma membranes of frogs' blood corpuscles and its acceleration by induction shocks.' *Journ. Biol. Chem.* 15, *237.*
LOEB, J. (1899). 'Warum ist die Regeneration kernloser Protoplasmastücken unmöglich, usw.?' *Arch. f. Entw. Mech.* 8, *689.*
—— (1910). 'Über den autokatalytischen Charakter der Kernsynthese bei der Entwicklung.' *Biol. Zentralb.* 30, *347.*
LYNCH, V. (1919). 'The function of the nucleus of the living cell.' *Amer. Journ. Physiol.* 48, *258.*
MARTENS, P. (1927). 'Recherches expérimentales sur la cinèse dans la cellule vivante.' *La Cellule,* 38, *69.*
MASING, E. (1910). 'Über das Verhalten der Nucleinsäure bei der Furchung des Seeigeleis.' *Zeit. f. physiol. Chem.* 67, *161.*
MATHEWS, A. P. (1907). 'A contribution to the chemistry of cell-division, maturation and fertilization.' *Amer. Journ. Physiol.* 8, *89.*
—— (1915). *Physiological Chemistry.* (London.)
MIESCHER, H. (1897). *Histochemische und Physiologische Arbeiten.* (Leipzig.)
MINOT, C. S. (1901). 'Senescence and Rejuvenation.' *Journ. Physiol.* 12, *97.*
—— (1908). *The problems of age, growth, and death.* (New York.)
MOCKERIDGE, F. A. (1926). 'An examination of *Nostoc* for nuclear materials.' *Brit. Journ. Exp. Biol.* 4, *301.*
MORGAN, T. H. (1910). 'The effects of altering the position of the cleavage planes in eggs with precocious specification.' *Arch. f. Entw. Mech.* 29, *205.*
OSTERHOUT, W. J. V. (1917). 'The rôle of the nucleus in oxidation.' *Science,* 46, *367.*
PEEBLES, F. (1912). 'Regeneration and regulation in *Paramoecium caudatum.*' *Biol. Bull.* 23, *154.*
PETER, K. (1899). 'Das Centrum für die Flimmer- und Geisselbewegung.' *Anat. Anz.* 15, *271.*
POLICARD, A. (1922). *Précis d'histologie physiologique.* (Paris.)
POPOFF, M. (1908). 'Experimentelle Zellstudien.' *Arch. f. Zellforsch.* 1, *245.*
ROBERTSON, T. B. (1913). 'Further explanatory remarks concerning the chemical mechanics of cell division.' *Arch. f. Entw. Mech.* 35, *692.*
—— (1923). *The Chemical Basis of Growth and Senescence.* (Philadelphia.)
SACHS, J. (1892). 'Beiträge zur Zellentheorie.' *Flora,* 75, *57.*
—— (1893). 'Ueber einige Beziehungen der specifischen Grösse der Pflanzen zu ihrer Organisation.' *Flora,* 77, *49.*
—— (1895). 'Weitere Betrachtungen über Energiden und Zellen.' *Flora,* 81, *405.*

SCHITZ, V. (1925). 'Étude sur l'évolution des éléments génitaux chez les Mollusques Ptéropodes. I. La Spermatogénèse.' *Biolog. Generalis*, 1, *537*.

SCHACKELL, L. F. (1911). 'Phosphorus metabolism during early cleavage of the echinoderm egg.' *Science*, 34, *573*.

SHIWAGO, P. (1926). 'Über die Beweglichkeit der Fadenstrukturen im lebenden "Ruhekerne" der Froschleukozyten.' *Biol. Zentralb.* 46, *679*.

SPITZER, W. (1897). 'Die Bedeutung gewisser Nukleoproteide für die oxidative Leistung der Zelle.' *Arch. f. ges. Physiol.* 67, *615*.

STRASBURGER, E. (1893). 'Ueber die Wirkungssphäre in den Kerne und die Zellgrösse.' *Histol. Beitr.* 5, *97*.

—— (1895). 'Karyokinetische Probleme.' *Jahrb. f. wiss. Bot.* 28, *151*.

—— (1897). 'Ueber Cytoplasmastrukturen, Kern- und Zelltheilung.' *Jahrb. f. wiss. Bot.* 30, *375*.

TOWNSEND, C. O. (1897). 'Der Einfluss des Zellkerns auf die Bildung der Zellhaut.' *Jahrb. f. wiss. Bot.* 30, *484*.

VERWORN, M. (1889). *Psycho-physiologische Protisten-Studien.* (Jena.)

VOSS, H. (1925). 'Untersuchungen mit Nuklealreaktion.' *Verh. anat. Gen.* 34, *118*.

WARBURG, O. (1910). 'Ueber die Oxydationen in lebenden Zellen nach Versuchen am Seeigelei.' *Zeit. f. physiol. Chem.* 66, *305*.

WHITESIDE, B. (1923). 'A comparison of the nucleic acid content of various tissues.' *Anat. Record*, 26, *91*.

WILSON, E. B. (1896). 'On cleavage and mosaic work.' *Arch. f. Entw. Mech.* 3, *19*.

CHAPTER EIGHT
Mitosis

The first observations of the living nucleus in a dividing cell were made, about 1850, on the oocytes of invertebrates. In the immature condition, these cells are characterised by very large conspicuous nuclei, which can easily be seen under low magnifications. As soon as maturation begins, the outline of the nucleus disappears and is not again visible until segmentation into two daughter cells has occurred, when a small definitive nucleus can be seen in each cell. Most of the earlier observers were agreed that, prior to each cell division, the nucleus disappeared *in toto* and a new nucleus was reformed *de novo* after each segmentation. When, however, it became possible to preserve the eggs with osmic acid or other fixative, and to apply a differential method of staining, the apparent disappearance of the nucleus, prior to cell division, was traced to a profound reorganisation of the nucleus rather than to its complete incorporation into the cytoplasm. As the methods of fixation and staining became more and more effective, the whole series of changes undergone by a dividing nucleus was found to be essentially the same in all plants and animals.

Within more recent times, however, the validity of the results obtained from fixed preparations have been questioned, so that it becomes important to know how far the phenomena of mitosis, as usually described, conform to what is known to occur within the living cell. It has already been mentioned that a 'resting' or interkinetic nucleus in the living condition is usually optically homogeneous with the exception of definitive nucleoli, whereas in the preserved state the nucleus appears to be divided into at least two phases, only one of which has an affinity for basic stains and is distributed throughout the nucleus as a series of granules or filaments. It would thus appear that such preparations are a very doubtful guide to intranuclear structure. Nevertheless, it will be shown later that as far as the chromatic constituents of the kinetic nucleus are concerned, fixed preparations give a remarkably faithful picture of the events which occur during life. It is only in respect to the achromatic portions of the dividing nucleus that real misinterpretation may be caused by fixation.

As studied in permanent preparations, the main outlines of typical mitosis are as follows. First, there appear on the surface of the nucleus two small areas or granules which pass to opposite poles of the nucleus, and which may or may not be surrounded by a series of diverging rays. These bodies are known as the centrosomes and the rays, if present, are called astral rays. Soon after the centrosomes have reached the poles of the nucleus, the latter elongates along the polar axis and its chromatic constituents become incorporated into a definite number of chromosomes; this is the end of the *prophase*. Quite suddenly the outline of the nucleus disappears and the chromosomes are seen lying as an irregular group midway between the centrosomes, and the area formerly outlined by the nucleus is occupied by a series of fibrils which converge to each centrosome. The fibrillar area is the *spindle*. At the close of this *metaphase* condition the chromosomes cease to be scattered over the centre of the spindle, but are organised on a flat plate across its equator. The *anaphase* follows; each chromosome segregates into two longitudinal halves, one of which passes towards each centrosome. Finally during the *telophase*, each group of daughter chromosomes forms an irregular mass of interfolding chromosomes or (in other cases) forms a mass of swollen vesicles which, losing their affinity for stains, are gradually built up into a new nucleus.

The forces involved in mitosis have long been the subject of speculation, and many unfruitful theories have been based on the assumption that every detail seen in a coagulated cell is a true picture of the living nucleus. Within recent years, however, it has proved possible to see every stage of division in a living nucleus, and the chances of misinterpreting preserved material is thereby greatly diminished. The following description is taken from that given by Strangeways (1922) who observed the choroidal cells of the chick *in vitro*. When about to divide, the cells withdraw their characteristic pseudopodia and assume an oval or rounded shape (fig. 77). The first observable change in the nucleus is the conversion of the organised nucleoli into hazy granules; these are then transformed into definite chromosomes which can be seen lying within the nucleus as a number of fine threads, often writhing about 'like eels in a box'. At this stage, the outline of the nucleus suddenly

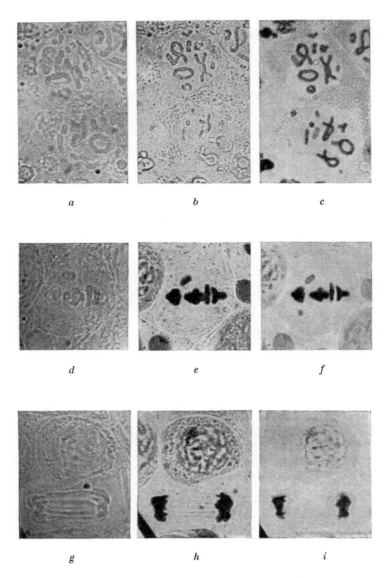

Fig. 45. Effect of fixation and staining on the dividing nuclei of spermatocytes (*Chorthippus*). a, d, g, living cells; b, e, h, fixed with osmic acid vapour and Flemming's solution; c, f, i, stained preparations. (From Bĕlăr.)

disappears and the chromosomes rapidly arrange themselves at right angles to the spindle which, with the centrosomes, can now be seen. It is interesting to note that no trace of the spindle can be

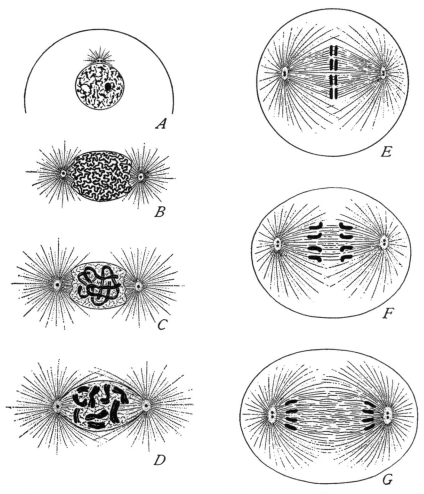

Fig. 46. Diagrammatic figures of typical mitosis. (From Wilson, 1928.)

recognised until the outline of the nucleus has disappeared. About five to ten minutes after the first appearance of the spindle, the chromosomes begin their anaphasic movement toward the centrosomes and are clearly seen as finger-like processes; at this stage the

MITOSIS

cell is oval in form. Having reached the poles, the chromosomes seem to fold in upon themselves and form a single faint irregular body, which is later surrounded by a clear zone unmistakably differentiated off from the cytoplasm. Finally a definitive nucleus with a nucleolus becomes visible in each daughter cell.

From this account and others (Bĕlǎr, 1929) it can be realised that the classical accounts of mitosis are undoubtedly correct in their main outlines except in so far as the achromatic parts of the nucleus are concerned. Particularly convincing in this respect are the ob-

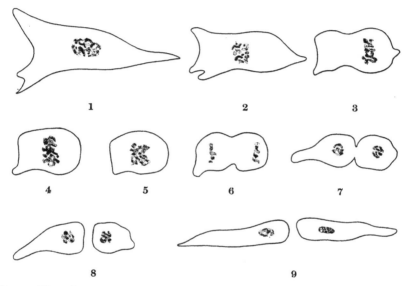

Fig. 47. Nine observations at 10-minute intervals of connective tissue cell of a rat (34° C.). Note withdrawal of pseudopodia during cleavage. (From Lambert and Hanes.)

servations of Martens (1922) and Bĕlǎr (1929), who have carefully compared the effect of fixation on plant and animal cells respectively at different stages of mitosis. Even the finest structures seen in the living nucleus appear unchanged after an adequate process of fixation (figs. 40, 45).

Velocity of mitosis

In order that the events of mitosis should be seen in true perspective, it is of value to consider, at this early stage, the length of time occupied by each phase of the process. It is only by observing the living nucleus that the velocity of mitosis can be observed with any accuracy. The velocity of mitosis means the period of time which elapses between the onset of the prophase and the completion of the telophase. This must be carefully distinguished from the frequency of mitosis which means the number of nuclear divisions per unit of time. The distinction is clearly seen in the mitotic divisions of cells grown *in vitro*. Strangeways (1922), Fischer (1925 b) and others have shown that at about 38° C. fibroblasts and other cells divide once in every forty-eight hours, although only a small fraction of this time is occupied by actual mitosis; on the other hand, the mitoses seen in the early stages of the segmentation of an egg follow each other without any period of interkaryokinetic rest. At 17° C. a nuclear division of an egg of *Echinus miliaris* occupies about thirty-four minutes and is immediately followed by the next division (Gray, 1927).

Table XXIII
Mitotic cycles in *Echinus miliaris* (17° C.). (Gray, 1927)

1st to 2nd cleavage	33 min.
2nd ,, 3rd ,,	32 ,,
3rd ,, 4th ,,	36 ,,
4th ,, 5th ,,	35 ,,
5th ,, 6th ,,	35 ,,
6th ,, 7th ,,	33 ,,

It is curious to find that the mitotic cycle of a large echinoderm egg is completed in practically the same time as that of the much smaller nuclei of the rat (Lambert, 1913) and of the chick (Wright, 1925), although its temperature is much lower. According to Lambert (1913) the connective tissue cells of the rat, when grown *in vitro*, require for each mitotic cycle at 38° C. about 21 to 29 minutes and at 35° C. from 35 to 50 minutes.

The period of time occupied by the different phases of division is not easy to determine and the observed figures show some variation from each other. Direct observations of living vertebrate cells have been made by Levi (1916) and by Lewis and Lewis (1917). With one exception the time tables of these authors are in agreement and may be illustrated by the time table given by Lewis and Lewis, who observed the mesenchyme cells of chick embryos; the cells were observed in Locke's solution at 39° C.

In a fairly comprehensive series of observations, Lewis and Lewis found that very considerable variation may occur in the velocity of each individual phase (see Table XXIV). The only significant difference between the observations of Lewis and Lewis and those of Levi is that the former allow from 30 to 60 minutes as the normal period of the prophase, whereas Levi restricts the same phase to a period of 5 to 20 minutes. It is obviously difficult to determine in the living cell the precise moment at which the prophase begins, but as pointed out by Wright (1925) the period can be accurately determined from

Table XXIV

	Average time in minutes
Prophase	30–60
Metaphase	2–10
Anaphase...	2–3
Telophase...	3–12
Reconstruction of nucleus	30–120
	2–3 hours

Table XXV (from Wright)

	Early prophase	Spireme	Metaphase	Anaphase	Telophase	Reconstruction of nucleus
No. of mitoses found	140	115	91	66	91	86
% of total found, equalling % of total time required for whole cycle	24	19	15	12	15	15
Time in minutes for each phase Total time: 34 min.	8	6½	5	4	5	5

fixed preparations if the duration of any other phase of the division is accurately known. Thus, assuming that the period of the telophase is 5 minutes, then by observing the relative numbers of prophases and telophases seen in a culture, the period of the prophase can be calculated, since the number of each phase present must be proportional to the time occupied for the completion of that particular phase.

The above table shows the results obtained by Wright from cells of the chick's heart incubated at 37° to 38° C. It would appear that the whole cycle occupies about 34 minutes and that the time allowed by Lewis and Lewis for the prophase is too long.

In echinoderm eggs Ephrussi (1926) gives the following time table for *Paracentrotus lividus*.

Table XXVI

	Time in minutes after fertilisation	
	At 18·5°	At 26°
Formation of the amphiaster	18	16
Disappearance of nuclear membrane	48	28
Metaphase plate	52	32
Division of chromosomes	56	36
Beginning of anaphase	59	39
End of anaphase	65	43
Reconstruction of nucleus	74	50

In any investigation of the velocity of mitosis it is important to remember that the velocity of mitosis is very much reduced if the conditions of incubation are at all unfavourable. The presence of even slight traces of CO_2 are sufficient to reduce the velocity in *Echinus* eggs (Gray, 1927).

We may now, perhaps, consider how far experimental methods have thrown light on the various phases of mitosis in the hope that it may be possible to form some conception of the types of mechanisms involved. It is convenient to start by considering each phase as a separate entity.

Prophase

The first sign of approaching division in a living nucleus consists in a loss of the normal homogeneity of the nuclear contents owing to the formation of threads or granules which have a high affinity for basic stains. As far as we can tell, this process must represent the segregation of the nucleic acid in the nucleus from the more basic constituents. As long as nucleic acid is associated with strong basic radicles or is in the form of a nucleo-protein it has no affinity for basic stains, for this property is peculiar to the free acid. Since chromatin is, in all probability, nucleic acid, we may infer that when chromosomes are formed inside the nucleus they contain free nucleic acid, whereas the more basic portions of the nucleo-proteins remain elsewhere or are in some way destroyed. Apart from this one

MITOSIS

conclusion, the chemistry of the dividing nucleus is entirely unknown, and when we remember that the nucleic acid derived from all animal cells, irrespective of group, genus, or species, is identically the same in composition, it becomes obvious how a purely chemical conception of living matter fails at present to give even a reasonable picture of biological facts (see p. 128). Genetically a chromosome must possess a highly complex constitution which is specific to each species of organism; all we know, chemically, is that one invariable constituent is nucleic acid. Perhaps the nearest picture we can get is that of Mathews (1921, p. 175), wherein the nucleic acid is regarded as a gelatinised matrix in which are concealed those highly specific entities known as genes.

Fig. 48. Metaphase chromosomes of *Dissosteira* being torn by needles. (From Chambers, 1924.)

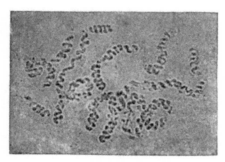

Fig. 49. Metaphase chromosomes in pollen mother cell of *Tradescantia* showing spiral structure. (From Sakamura.)

The chromosomes of animal cells are undoubtedly solid bodies in that they possess individuality of form; the same is true of plant chromosomes, whose consistency can be demonstrated by microdissection (Chambers and Sands, 1923). For the direct investigation of plant chromosomes Chambers and Sands (1923) used the pollen mother cells of *Tradescantia*, and by means of fine hooked needles were able to seize a single chromosome by each end and subject it to a longitudinal pull (see also fig. 48). By such methods they concluded that these particular chromosomes are elastic 'jelly-like' cylinders whose cortex differs from its central less refractive core.

150 MITOSIS

The actual segregation of organised chromosomes within a homogeneous nucleus has been observed in at least three different types of cell; in the chick cells grown *in vitro* by Strangeways (1922), in the oocytes of *Asterias*, and the spermatocytes of the grasshopper *Dissosteira* (Chambers, 1924). Chambers' observations on *Dissosteira* are of peculiar interest for they show that the opening phases of

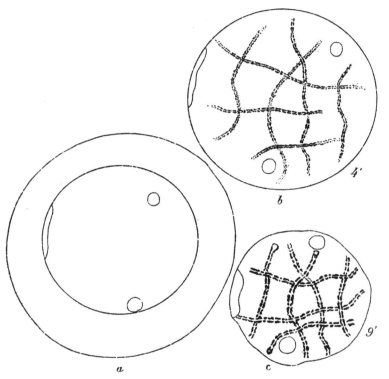

Fig. 50. Showing effect of puncture on a living spermatocyte of *Dissosteira*. *a*, Normal cell; *b*, four minutes after puncture: the cytoplasm has degenerated and fine filaments have appeared in the nucleus; *c*, nine minutes after puncture: note formation of 'chromosomes'. (From Chambers, 1924.)

normal nuclear activity can be induced by localised injury. Within a minute of being punctured, fine granular streaks can be seen within the spermatocyte nucleus and these soon grow into distinct filaments on whose surface is arranged a series of refringent granules. The granules grow in size and each thread thickens. Ten minutes later, the whole nucleus contracts somewhat, and the granules on the

MITOSIS

filaments fuse together, thus converting the latter into a number of homogeneous hyaline bodies. There can be little doubt that these bodies represent true chromosomes, and as Chambers points out, their origin from hyaline axial threads with surface granules had previously been described by Martens in 1922 in *Paris quadrifolia*.

In some spermatocytes, on the other hand, nuclear puncture results in the formation, not of typical prophasic chromosomes, but in the direct formation of bodies which strangely resemble chromosomes of the metaphase (fig. 51). Apparently the result of puncture depends on the stage to which the nucleus has normally progressed towards mitosis before the operation; in each case puncture simply accelerates the normal process of development. An interesting observation was

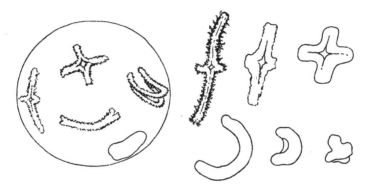

Fig. 51. Chromosome formation within a hyaline nucleus of *Dissosteira*, following puncture. (From Chambers, 1924.)

also made by puncturing a nucleus in which the chromosomes could already be detected. In this case the nuclear membrane disappeared, but the network of chromosomes persisted. On seizing a loop of the network, it was possible to draw it out as an elastic thread. On releasing it, it regained its original form (fig. 52).

The opening stages of the normal mitotic prophase in living plant cells has been carefully studied by Martens (1927). In these cells the chromatic elements can be seen prior to the prophase, and the chromosomes (during the prophase) are delimated by a rupture of the interchromosomal junctions which are characteristic of the interkinetic nucleus; no spireme is formed, but each condensing chromosome becomes orientated parallel to its mates on the short axis of the nucleus: the whole prophase occupies from 36 to 45 minutes (fig. 40).

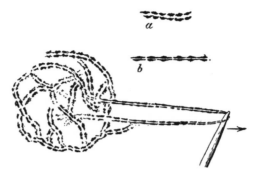

Fig. 52. Intranuclear filament seized by needle. a, Fragment of filament before stretching; b, after stretching. (From Chambers, 1924.)

The metaphase

Soon after the differentiation of the intranuclear chromosomes, the outline of the nucleus disappears. How far this is due to the dissolution of a definite nuclear membrane is uncertain, although a membrane appears to be present at the surface of the nucleus of an *Asterias* oocyte (Chambers, 1921). At this stage the prophase ends and the metaphase begins. The chromosomes now become arranged in a flat equatorial plate across the centre of the spindle. It seems clear that the force which compels the chromosomes to move to the equator of the spindle is located at the poles of the spindle or in the polar asters. This conclusion follows from the observation of F. R. Lillie (1912), who showed that in monastral *Nereis* eggs the chromosomes remain scattered at the conclusion of the prophase and showed no tendency to form a metaphase plate. Similarly, if the eggs of echinoderms are subjected to ether before fertilisation, the male pronucleus often fails to fuse with the female pronucleus; in such cases a bipolar aster may develop in association with the male nucleus, but only a single aster develops near the female nucleus. In these circumstances (fig. 53) the male chromosomes form a metaphase plate, whereas those of the female remain scattered (Wilson, 1902). That the arrangement of the chromosomes on a metaphase plate is not haphazard was shown by R. S. Lillie (1905) and later by Cannon (1923). Both of these authors have shown that the chromosomes on the metaphase plate arrange themselves in the same pattern as a

MITOSIS

series of magnets, all orientated in such a way as to repel one another in a strong external field (fig. 54). Such a system can be arranged as follows (Cannon). A number of corks, through each of which is fixed a similar small rod magnet, are floated in a vessel containing water in such a way that all the magnets are pointing vertically upwards and all with the same pole uppermost; the north pole for instance. Since similar magnetic poles repel each other, these floating magnets will also repel each other and so collect at the sides of the vessel. If now a strong south pole is brought over the water on which the magnets are floating the latter, while still exerting a

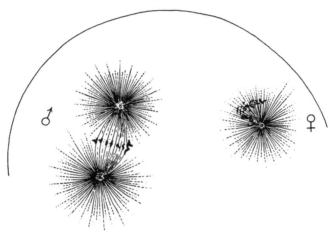

Fig. 53. Fertilised egg of *Toxopneustes* after treatment with ether. The male chromosomes are arranged normally at the equator of a bipolar spindle. The female nucleus has only one aster; there is no spindle, the chromosomes remain scattered and do not form a metaphase plate. (From Wilson.)

repellent action on each other, are attracted towards the south pole and they now arrange themselves in equilibrium in a definite order. It is found that any number up to five will arrange themselves at the corners of a regular figure. If, however, there are six magnets, five are arranged at the corners of a pentagon, whilst the sixth passes to the centre of the figure. With ten magnets on the other hand, two magnets lie within an octagon formed by the remainder; with fifteen magnets there are five within the outer ring. The actual arrangements of chromosomes on the equatorial plate being identical with that occupied by the magnetic model (see fig. 54), it seems reasonable to assume that the chromosomes are orientated by two

154   MITOSIS

forces, one of which is exerted by the poles of the spindle. It should, however, not be forgotten that both Strangeways (1922) and Chambers (1924) have observed spontaneous movements by the chromosomes themselves. Strangeways describes the motion as 'eel-like', whereas

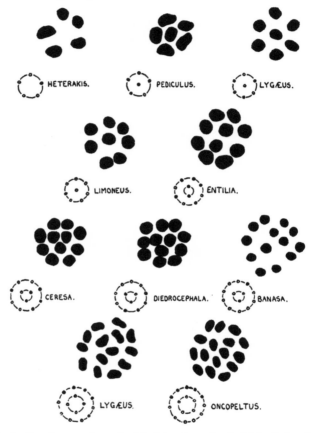

Fig. 54. Illustrating figures of equatorial plates, in each of which all the chromosomes are practically of the same size and shape. Under each plate is the name of the genus from which it is taken and in front of this name is figured the predicted arrangement of the chromosomes. (From Cannon.)

Chambers' description is that of an amoeboid type of movement, in which one part of a chromosome expands at the expense of the rest. The close of the metaphase of mitosis is usually marked by the longitudinal cleavage of each chromosome. Concerning this process almost nothing is known beyond the fact that it may occur at a much earlier

MITOSIS

stage, so that on arrival at the equator of the spindle each chromosome is made up of two longitudinal halves. The process by which these halves are formed as separate entities is independent of a bipolar spindle, since it occurs in monastral systems (*Nereis*).

The achromatic mechanism

From a kinetic point of view, the anaphase is the most interesting phase of mitosis, for it is at this period that the daughter chromosomes migrate relatively rapidly to the poles of the spindle. Before attempting to evaluate the numerous theories which have been put forward to elucidate this phenomenon, it is essential to consider in greater detail the nature of the spindle in which the chromosomes move.

In every mitotic division of a nucleus, there exists a clear spindle-shaped body at each pole of which there can usually be detected a granule or area known as the *centrosome*. In the cells of most of the higher plants, these structures compose the whole of the achromatic apparatus of the dividing nucleus, but in many animal cells there exist round each centrosome a series of diverging rays known as the *aster*. There are therefore two types of achromatic figure, (i) an *anastral* type, and (ii) an *astral* type. Since the kinetic phenomena of mitosis are alike in the two cases, one may conclude that however necessary the asters are during the process of cell division (see Chapter IX) they play no essential rôle in the division of the nucleus; the essential structure for nuclear division appears to be the spindle. On the other hand, much of the evidence suggests that the formation of a spindle only occurs when the centrosomes are present, and that these bodies give rise to both spindle and asters. Unfortunately, the centrosomes are always extremely small and may be visible in the living cell. Whenever two asters come into contact with each other a spindle is formed between them and the resultant figure is known as an *amphiaster*. In view of the diversity of opinion concerning the origin of centrosomes and other parts of the achromatic figure it is difficult to give a logical account which will cover all the known facts.

In preserved material the mitotic spindle is almost universally represented by a series of fibres converging at the two poles (spindle fibres); at the metaphase some of these fibres are closely applied to the chromosomes. Spindle fibres have never been seen in the living cell, for during life the spindle appears as a homogeneous hyaline body enclosing the chromosomes. That spindle fibres are the artificial products of fixation is confirmed by the work of M. R. Lewis (1923), who showed that they only appear in living cells if the medium is sufficiently acid in reaction. In a medium of pH 4·6 fibres become visible in the spindles of cells dividing *in vitro*; on removing the acid the fibres disappear, and this phenomenon can be repeated several times without killing the cells. Since nearly all cytological fixatives contain considerable quantities of acid, there can be little doubt that visible spindle fibres must be regarded as artefacts. At the same time the fibres seen in fixed or coagulated cells are probably indications of a field of force between the poles of the spindle. Hardy (1899) showed that if a homogeneous film of albumen is coagulated whilst in a state of tension, definite fibrillae are formed in a plane parallel to the direction of stretching.

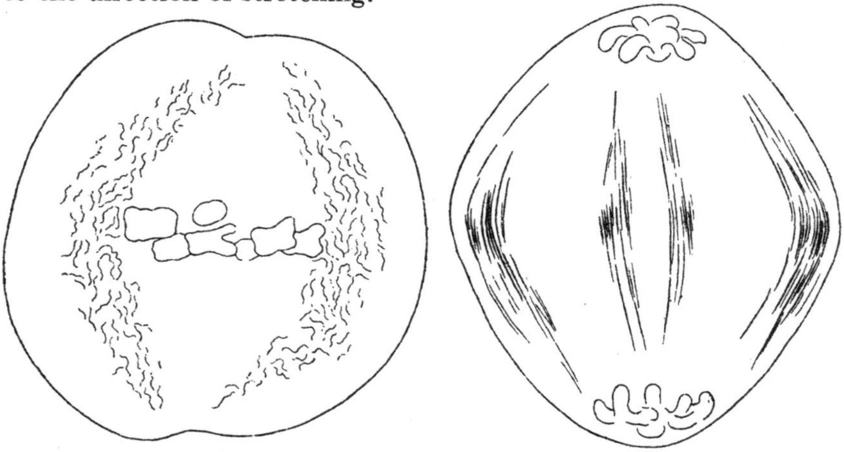

Fig. 55. Living primary spermatocyte of *Dissosteira*. 1, Note the chromosomes on metaphase plate and the clear spindle surrounded by mitachondria; 2, late anaphase showing the two polar graphs of chromosomes and the linear arrangement of mitachondria. (From Chambers, 1924.)

All recent evidence supports the view that the spindle is a comparatively rigid structure. This was first shown by Morgan (1910), who found that it was possible to move the spindle bodily by centrifugal force through the matrix of an echinoderm egg; Shearer (1910) found that the spindle of *Histriobdella* was also a rigid structure capable of isolation from the cell. The later work of Chambers (1917) (see fig. 56) shows quite clearly that the spindle is capable of considerable mechanical distortion and manipulation; it is an elastic gel and not in any way comparable to a fluid with low viscosity. Similarly in plant cells (Chambers and Sands, 1923), the spindles in the pollen mother cells of *Tradescantia* can be dissected out as a whole and are highly elastic.

MITOSIS

It is of importance to note that the length of the major axis of the spindle determines the distance moved by the chromosomes during the anaphase of mitosis. If an echinoderm egg is allowed to segment in sea water containing a small trace of ether (0·05 per cent.), the spindle is abnormally small and it is found that the two daughter nuclei, although perfectly normal, are abnormally close together.

The origin of the spindle appears to differ in different types of cell. In some plants and in some protozoa it is formed entirely from the nucleus, since it is well developed inside that body before the nuclear membrane disappears at the end of the prophase. In the higher plants, on the other hand, the spindle is formed from polar caps lying outside the nucleus; in most animal cells it arises outside the nucleus although in close association with it.

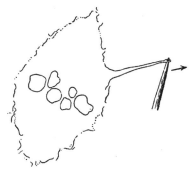

Fig. 56. Metaphase spindle of *Dissosteira* dissected out of the cell: note the extensile nature of the spindle when drawn out by a needle. (From Chambers, 1924.)

The formation of a spindle is closely associated with the presence of the two polar centrosomes, but it is useful to remember that the facts can be considered from two distinct points of view. From the first of these, the spindle is regarded as a definite cell organ formed from definite material known as *archiplasm*, which may be located inside or outside the nucleus. From the other point of view, the spindle is an entirely transient structure which is the result of the mutual effect of two centrosomes on material of either nuclear or cytoplasmic origin. If the centrosomes lie within the nucleus the spindle will be formed of nuclear material; if on the other hand the centrosomes lie outside the nucleus, the spindle may be formed from the cytoplasm. In cases (e.g. the echinoderm egg) where the centrosomes lie very close to the surface of the nucleus, it seems almost certain that the spindle is composed of the achromatic parts of the nucleus. Immediately before the dissolution of the nuclear membrane, the nucleus is elongated between the two centrosomes, the nuclear boundary suddenly disappears and the area which it enclosed is then clearly defined as the spindle.

The polar asters

In many embryonic animal cells the poles of the mitotic spindle are marked by well-defined *asters*, whose rays pass outward from the pole and extend into the cytoplasm of the cell. In the case of an echinoderm egg the rays of the aster appear to be marked out by the radial arrangement of granules which, elsewhere in the cell, are more or less evenly distributed throughout the cytoplasm (see fig. 57). During the metaphase of nuclear division these asters are very conspicuous objects and are seen to consist of a large number of rays each composed of a hyaline homogeneous substance radiating out from a central clear area of protoplasm (centrosphere). According to Chambers (1917), the rays are broadest at their base and gradually taper along their length until they are indistinguishable from the general hyaline matrix of the protoplasm. It is interesting to note that the smaller the aster the more clearly defined are the rays; as the aster grows in size, so the rays become longer but less distinct. According to Chambers, the tongues of granular cytoplasm lying between the rays are rigid, whereas the rays and the centrosphere consist of a liquid of relatively low viscosity. The rigidity of the

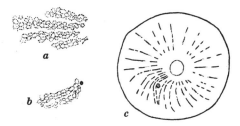

Fig. 57. *a*, Astral rays greatly enlarged; *b*, tip of ray bent by needle; *c*, astral rays bent by insertion of needle. (From Chambers, 1917.)

astral cytoplasm renders the whole aster rigid, so that it can be moved about mechanically in the cell, and can be distorted in form by means of needles (see fig. 57). An aster can be twisted into a spiral and on being released from distortion it may or may not resume its normal shape. If the above description applies to asters in general, it would seem that an apparent absence of asters in some cells may be due to an absence of cytoplasmic granules, since it is these structures which make it possible to differentiate optically between the asters and the surrounding cytoplasm.

It seems probable that we ought to regard the asters not as transient cell organs but as the expression of a specific dynamic activity. Thus, if a dividing cell is subjected to a low temperature, or to such an anaesthetic as ether, the whole of the rays instantly disappear leaving the centrosphere as a clearly defined homogeneous mass. This phenomenon occurs whenever a dividing cell is subjected

MITOSIS

to any form of inhibition, e.g. tearing the asters by means of needles, or mechanical pressure.

The origin of the asters has been sought by a variety of methods and is still uncertain. In some cases it seems quite clear that they are formed at or near the surface of the nucleus. In echinoderm eggs

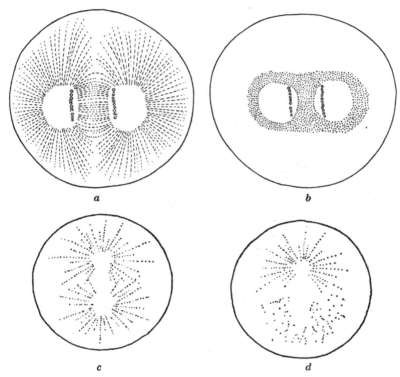

Fig. 58. Destruction of asters by ether or by mechanical disturbance. *a*, Normal egg 45 minutes after fertilisation. The course and extension of the astral rays were drawn with all possible accuracy; *b*, same egg exposed to ether, note disappearance of astral rays leaving clearly defined astrospheres; *c*, normal egg; *d*, one aster destroyed by mechanical disturbance with a needle. (*a* and *b* from Wilson; *c* and *d* from Chambers.)

the first aster to appear arises from a centre near the middle piece of the spermatozoon soon after the latter has entered the egg. This aster rapidly enlarges in size as the male pronucleus moves towards the egg nucleus, so that when fusion occurs the whole egg is filled by its radiations. The rigidity of this aster is shown by the fact that it is able to distort the spherical form of the cell (Gray, 1924), and by the fact that it can be mechanically pushed and rolled about in the cytoplasm without losing its form (Chambers, 1917). As the astral rays increase in length, the centrosphere grows in size, so that there appears to be a centripetal flow of fluid towards the centre. According to Chambers, the male nucleus is only in equilibrium when it is lying near the centre of the aster, since if the nucleus be displaced it always tends to resume its central position. Similarly, as long as the female pronucleus lies outside the confines of the astral rays it remains stationary, but as soon as the rays reach it, the female nucleus moves with increasing velocity towards the centre of the aster, and is thus brought into contact with the nucleus of the spermatozoon. Not only the nuclei, but even oil drops move towards

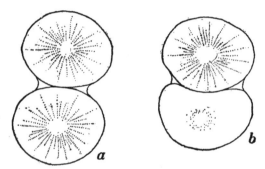

Fig. 59. *a*, Normal egg; *b*, the aster has been destroyed in the lower cell: note the compression of the latter cell against the more rigid cell containing the aster. (From Chambers, 1924.)

the centre of the aster, so that there appears to be a force directed towards the centre of the aster which acts throughout that region of the cell traversed by the rays. As soon as the male and female pronuclei have fused together, the large male aster gradually fades away and is replaced by the small amphiaster which will be described below. According to Chambers (1917) the disappearance of the monastral rays is attended by a loss of rigidity of the cytoplasm, since a needle can now be drawn through the egg without disturbing the structures lying on each side of the needle.

Additional evidence to show the rigidity of an aster is illustrated in fig. 59 *b* (after Chambers), which shows the compression of a blastomere whose aster has been mechanically destroyed; the second blastomere which contains an aster remains spherical, although

MITOSIS

both blastomeres are subjected to the same pressure from the hyaline membrane of the cells (see Chapter IX).

The existence of a single aster is of very considerable theoretical importance; it shows that the amphiaster is not the expression of a bipolar force of the type of electricity or magnetism. Large monasters can be produced in egg cells by a variety of methods; drugs, such as strychnine or chloral hydrate (Hertwig, 1887), hypertonic sea water (Morgan, 1896; Wilson, 1901; Herlant, 1918), and by mechanical agitation (Boveri, 1903; Painter, 1915). Wilson found that when the eggs of *Toxopneustes* were exposed to hypertonic sea water for a short period, large monasters developed subsequently in normal water. Under these conditions the whole of the normal cycle of nuclear changes occurred (including the division of the chromosomes) but the daughter chromosomes failed to move apart, and the cell did not divide. Such monastral cycles of nuclear activity occurred rhythmically in the same egg for as many as six times, after which normal bipolar cleavages might occur (Wilson, 1901).

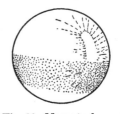

Fig. 60. Monastral egg of *Strongylocentrotus*. (From Painter.)

More than one observer (Boveri, 1903; Painter, 1915) has observed that towards the end of the monastral cycle the large monaster moves towards the cell periphery, thereby flattening the centrosphere into a curved disc parallel to the surface of the egg; meanwhile the surface of the egg furthest from the centrosphere exhibits extensive cytoplasmic activity. This centrifugal movement of the aster is of importance when we come to consider the rôle of the asters during cleavage of the cell.

As mentioned above, the large male aster of echinoderms appears to have its origin at the originally posterior end of the sperm nucleus, and in certain cases there is strong evidence to support the view that the aster actually arises from the 'centrosome' lying in the middle piece of the spermatozoon. Boveri believed that the middle piece of the sperm alone was capable of producing an aster, and that the nucleus itself was not directly involved. The origin of the sperm aster from the middle piece of the spermatozoon is, however, contradicted by the work of Lillie (1912) on *Nereis*, where the middle piece of the spermatozoon does not enter the egg and where the sperm aster arises at the proximal pole of the sperm nucleus. Lillie further showed that the aster can arise at any fractured surface of the male pronucleus. This he demonstrated by centrifuging the fertilised egg in such a way as to break the male pronucleus into two portions, only one of which entered into the egg. In every case a male aster developed at the proximal surface of the nuclear fragment. The size of the male aster appeared to be proportional to the size of the male pronucleus. Lillie's observations suggest that the aster arises from a constituent of the nucleus, and that the so-called 'centrosome' is essentially of nuclear origin and is not derived from a permanently extra-nuclear source. It looks as though both centrosome and aster are the result of changes in or near the surface of a nucleus which has reached a definite phase of nuclear activity.

The amphiaster

In all typical cases of astral mitosis, two asters arise near the surface of the nucleus. At first they are small, although the rays are very distinct. As the intervening spindle increases in size, the centrospheres of the asters grow and the rays project further and further into the cytoplasm of the cell. No rays are visible inside the spindle. When the cell divides into two halves, each daughter cell contains one aster which fades away soon after division is complete; at the next mitosis two small asters again appear at the surface of each daughter nucleus.

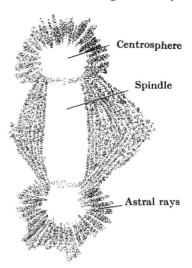

Fig. 61. Amphiaster of *Echinarachnius* egg. (From Chambers.)

If we regard an aster as the product of a definite cell organ, then it would appear that the latter is capable of regular subdivision just as is the case of a chromosome. This view was accepted by Boveri who gave the name *archiplasm* to the specific material giving rise to both asters and spindle. Similarly where the centre of each aster is occupied by a definite centrosome it seems reasonable to believe that each daughter centrosome is derived as a cleavage product from a pre-existing centrosome.

There is, unfortunately, a considerable amount of confusion involved by the word centrosome. As Wilson (1925, p. 675) points out, it is used in at least four different senses. The most convenient usage is that the term *centrosome* denotes that structure (at the centre of the astral rays) which does not disappear when the rays disappear. This permanent structure may or may not contain a small highly staining granule, the *centriole*.

In a few cases the centriole appears to be present during the nondividing phases of the cell's existence, although this is by no means clearly established. It must be remembered that the centriole or definitive centrosome is very minute and can only be recognised during mitosis from the fact that it lies at the focus of the astral rays where there are no other cell granules. During the interkinetic period it is not easy to differentiate clearly between a centriole and other cell inclusions having similar staining properties.

The main support of morphological cytology is afforded to the view that the asters are formed from permanent cell organs which are usually located on the outer surface of the nucleus. At the same time it must be admitted that during the interkinetic phases no trace can often be found of centrosomes or centrioles. The evidence from experimental enquiry gives, however, more support to the view that the centrosomes and their subsequent asters are essentially transitory organs which can arise *de novo* and independently of any pre-existing centrosome. In 1896 Morgan described the origin of asters in the cytoplasm of echinoderm eggs after treatment with hypertonic sea water. These asters are commonly known as cytasters to distinguish them from those which normally arise near the surface of the nucleus. These cytasters were carefully examined by Wilson (1901) who showed, by examination of the living eggs of *Toxopneustes*, that they were capable of division and capable of deforming the surface of the cell just as is the case with normal asters (fig. 85). Since cytasters appeared to develop in different regions of the cytoplasm simultaneously, it seemed extremely probable that they arose *de novo*, although at the centre of each was a centriole comparable to that of a normal aster. On the other hand, the exposure of the egg to hypertonic sea water undoubtedly leads to a precocious activity of the normal nuclear asters, leading to the formation of multipolar spindles and it is conceivable that the cytasters were derived by division from normal asters. Wilson attempted to analyse the position further by shaking eggs into fragments and exposing the enucleate fragments to hypertonic sea water. In such fragments cytasters always developed. Unfortunately mechanical agitation undoubtedly upsets the normal asters (see above), and although this objection appeared to have been met by the work of McClendon (1908) on

Asterias eggs, Yatsu (1908) later showed that cytasters will only develop in enucleated fragments of *Cerebratulus* eggs if the original germinal vesicle had broken down prior to enucleation.

An attempt to form an adequate theory of the origin of asters was made by Conklin (1902), who suggested that asters can develop in the cytoplasm wherever there is present an essential derivative of the nucleus, which need not necessarily take the form of a definitive granule. This view seems to cover most of the facts: we may imagine that the nucleus can so affect the neighbouring cytoplasm that an aster is formed. In a normal cell this action is greatest at the surface of the nucleus; on the other hand, if we so treat the egg that this nucleo-cytoplasmic reaction occurs much more readily, it seems reasonable to suppose that the products of the germinal vesicle having probably been distributed all over the cell, asters will develop here and there in the peripheral cytoplasm, and at the same time there will be an abnormal number of asters formed at the surface of the nucleus itself.

The real interest afforded by asters lies in the fact that they are almost certainly the active agents of cell division, although they are not directly concerned with nuclear division. The rôle of the asters in cell division is discussed in the following chapter, but it is interesting to note that derivatives of centrosomes are undoubtedly concerned in other types of cell activity. The Henneguy-Lenhossek theory (see Gray, 1928) postulates that the basal granule of a cilium is homologous with the nuclear centrosome, and there can be no possible doubt that the two bodies are derived from a common source. Jordan and Helvestine (1922) claim that the ciliated cells in the epididymis of the rat divide amitotically because the division centres of the cells are functioning as basal granules. The recent work of Kindred (1926) indicates that in frog's epithelia there is still left near or in the nuclei the potentiality of forming centrosomes after the original centrosomes have become basal granules of cilia. It would seem that, in this case, centrosomes can be regenerated by a nucleus just as occurs in the sperm head of *Nereis*. As long as it was reasonable to regard spindle fibres as real and contractile elements in the cell, the change in function from a centrosome to a basal granule was of very great theoretical importance, for it suggested that the forces exerted by the asters or in the spindle were comparable to those which move a cilium. This position is, however, no longer tenable, and the centrosomal force remains obscure.

The anaphase

Having considered the nature of the achromatic spindle in some detail, we may return to a consideration of the movements seen during normal mitosis. It is obvious that during the anaphase there is a force which moves the daughter chromosomes towards the poles of the achromatic spindle, and it is generally assumed that the energy for movement is derived from the spindle or its poles, and not from the chromosomes themselves. The nature of this force remains unknown, although it has been the subject of much ingenious speculation. Among the earliest theories of mitotic movement were those of Klein (1878) and van Beneden (1883), who looked upon the fibres seen in the preserved spindle as contractile fibrillae comparable to muscle fibres. Some of these fibrillae were assumed to be attached at one end to the pole of the spindle and at the other end to the chromosomes. Since the pole of the spindle was rigidly fixed in position, the daughter chromosomes moved to the poles when the fibrillae attached to them underwent the process of contraction. This theory was accepted and elaborated by Boveri (1887) who showed that, during the fertilisation of *Ascaris*, the chromosomes appeared to be drawn on to the spindle by the activity of their attached fibrils; these fibrils became shorter and thicker during the process of contraction, just as is the case in a muscle fibre. To some extent Boveri differed from van Beneden, since he believed that the astral rays and spindle fibres were formed anew during each mitotic cycle from the archiplasm of the cell, whereas van Beneden and his contemporaries looked upon the rays and fibrils as permanent cell organs which, although orientated to form asters and spindle-fibres during mitosis, were present in an unorganised manner during the interkinetic periods. At a later date, Heidenhain (1894, 1896) constructed his well-known models of anaphasic movement in which the fibrillae were represented by elastic strands of rubber. Heidenhain's belief in the contractile properties of the mitotic fibrillae was strengthened by his discovery of a large permanent aster in the motile leucocytes of *Salamandra*; similar fibrillae were observed in the pigment cells of fish by Solger (1891) and Zimmermann (1893). These facts, together with the homology of the mitotic centrosome with the basal granule of a contractile cilium, suggested that in each case the fibrillae arising from the basal bodies are contractile in nature.

From time to time objections were urged against the contractile theory of mitosis, and numerous alternatives were suggested. The fact that spindle fibres are not visible in the living cell means, however, that unless it is very extensively modified the whole theory is of very little value. The fibres described by Boveri and Heidenhain are undoubtedly due to a field of force between the centrosomes and the chromosomes, but they throw no light on its nature, and cannot possibly be regarded as permanent structures comparable to muscle fibres. At the same time it is conceivable that the force which moves the chromosomes is of the same nature as that which causes a change in the shape of a muscle. It is well known that contraction is not dependent on the presence of fibrillae (amoeboid movement, contraction of cardiac muscle cells), so that some form of active contraction along definite lines may be possible within the mitotic spindle. If such be the case, one would expect that any active movement through a viscous spindle during the anaphase would be marked by an increase in the energy expended by the whole cell, and since mitosis ceases in the absence of oxygen, it might be possible to associate the movement of the chromosomes with an increase in the oxygen consumption of the cells. The experimental difficulties attending such investigations are considerable, but are not insuperable, and it has been shown (Gray, 1925) that there is no detectible change in the oxidative activity of the cell during mitosis. These observations were made on the segmenting eggs of *Echinus* which form peculiarly suitable material for investigation since, with proper precautions, it is not difficult to ensure that all the eggs are in the same mitotic phase at any particular time. At the same time it must be remembered that the anaphase only lasts for about five minutes, and unless the amount of oxygen involved was significantly large compared to that required for other cell purposes it would not be easy to measure. We must at present conclude, however, that there is little hope of estimating the energy changes involved during anaphasic movement, since if such changes exist they are too small to be estimated by the methods at present available; this means that one hopeful line of attack remains closed.

The conception of the asters and spindle as a field of force dates from 1873, when Fol pointed out their similarity to the field between two unlike magnetic poles. If the achromatic structures of a dividing nucleus are regarded as transient structures and not as a rearrangement of pre-existing material, a reasonable explanation is forthcoming for the ease with which the astral rays can be made to disappear in the presence of anaesthetics and other substances. Whenever the mitotic forces cease to exist, the field of force disappears just as the field of an electromagnet disappears when the current ceases to flow. Unfortunately the dynamic conception of mitosis does not carry us very far. Some of the earlier hypotheses are obviously untenable, and have been summarised by Meves (1896, 1898) and by Prénant (1910). The more modern theories are as yet incapable of experimental analysis. Before considering any of these in detail it is important to stress the fact that, so far, there is no direct evidence to show that the existence or activity of an achromatic system involves an expenditure of energy by the cell, and for this reason it is impossible to form any idea of the magnitude of the forces involved. Were it possible to detect an increased evolution of heat, or an increased consumption of oxygen whenever a nucleus is dividing mitotically, it would be possible to form a reasonable conception of the whole process. This is, however, not the case: there are no appreciable changes in heat production (Meyerhof, 1911) or of oxygen consumption (Gray, 1925) during mitosis.

Most of the earlier dynamical theories of mitosis involved the conception of an electrical or magnetic field of force and a full description of these theories will be found in the works of Prénant (1910), Meek (1913), and Wilson (1925). Apart from the difficulty of explaining the existence of monasters, all the electrical theories of mitosis are unacceptable in that they are purely speculative. The available experimental evidence is, in fact, definitely against the view that the chromosomes move to the poles of the spindle because there exists an electrostatic field of a particular type. It is true that the behaviour of a nucleus in an electric field shows that the chromatin bears a negative charge relative to the rest of the nucleus, but McClendon (1910) has shown that to upset the normal orientation of chromosomes within a spindle it is necessary to apply currents far more powerful than those which are likely to exist in living cells. McClendon exposed the growing root tip of hyacinth or onion bulbs to a current of 110 volts for varying periods by application of copper sulphate electrodes to the surface of the tissue. In most of his experiments the current was passed for 30 minutes and varied from 0·00001 to 0·01 ampères. As soon as the current reached 0·00005 to 0·001 ampères the basophil constituents of the nucleus and of the cytoplasm showed a marked tendency to move to the anode, as was also the case of the cytoplasm. With currents of 0·01 ampères the pressure of the migrating chromatin was sufficient to distort the nucleus into a pear-shaped structure.

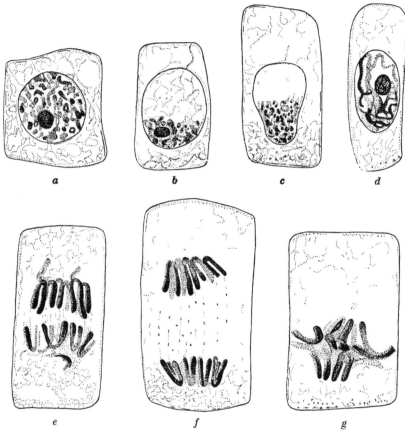

Fig. 62. Effect of a direct current on mitosis in the root tip of onion bulb. (From McClendon.) *a*, Control: note uniform distribution of chromatic elements in nucleus and cytoplasm; *b*, 0·0001 ampères: note basophil material carried to anode; *c*, 0·01 ampères: note anodic distortion of nucleus; *d*, 0·0015 ampères: the spireme of the prophase has moved; *e*, 0·0002 ampères: no movement; *f*, 0·0006 ampères: the whole of the spindle has moved bodily; *g*, 0·003 ampères: the whole of the spindle has moved.

Pentimalli (1909) stated that during the progress of mitosis the chromosomes became more and more sensitive to an external electric field, but this was not confirmed by McClendon, who showed that some at least of Pentimalli's results were due to displacement of chromosomes not by the electric field but by the microtome knife. McClendon states that as the process of mitosis advances the chromatin becomes less and less sensitive to the current; whereas 0·0015 ampères are sufficient to move the prophase spireme, at least twice that current is required to move the spindle. Further, during the anaphase the whole spindle with the contained chromosomes move as one unit, and there is no displacement of chromosomes relative to the spindle.

It is obvious that the general configuration of astral rays and 'spindle fibres' in a normal cell is no evidence of the nature of the force which is their underlying cause. Models of amphiastral figures can be made not only from magnetic or electrical fields but also from those of thermal, osmotic, and other types of energy. Bütschli's (1876) figure, here reproduced (fig. 63 b), shows that particles of gelatin will orientate themselves suitably when they are subjected to the

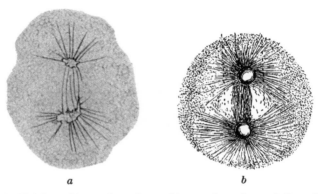

a   *b*

Fig. 63. *a*, Artificial amphiaster formed round two centres of coagulation. The matrix is a solution of globulin in alkali, and the coagulation is effected by platinic chloride. (From Fischer.) *b*, Artificial amphiaster formed round two air bubbles enclosed in a warm matrix of gelatin. Fixed in formol and chromic acid and stained by acid fuchsin. (From Bütschli.)

stresses caused by two bubbles of air contracting within a gelatin matrix. Similarly, the figures (figs. 64 and 65) obtained by Leduc (1914) are equally good mitotic models obtained in osmotic fields. Until it can be shown that the mitotic field is sensitive to forces of one particular type it is quite arbitrary which theory to adopt; presumably one gives preference to whichever theory provides the most picturesque model.

When an attempt is made to analyse the nature of the force operating in the spindle by applying an external field of a particular type, some support is forthcoming for the conceptions of an osmotic field of diffusion, for at least we know that the stability of the spindle and the behaviour of the chromosomes are upset by disturbances in osmotic pressure (see fig. 69).

As already mentioned there is, in all echinoderm eggs, an unmistakable accumulation of fluid at the poles of the spindle during

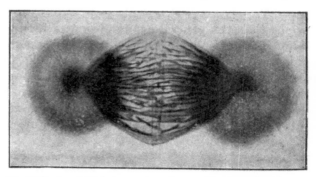

Fig. 64. Diffusion field between two drops of hypertonic salt solution with a hypotonic drop between them. The lines of diffusion are marked by particles of Indian Ink. (From Leduc.)

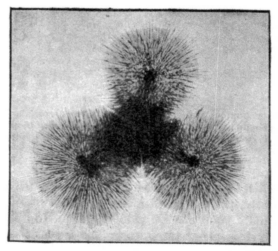

Fig. 65. A triastral field of diffusion between three contiguous drops. (From Leduc.)

the whole period of mitosis; Bütschli (1876) suggested that the anaphasic movement of the chromosomes might be due to similar protoplasmic currents set up in the spindle itself; these currents he assumed to be caused by localised changes in surface tension.

MITOSIS

Elaboration of these conceptions have been made from time to time by Rhumbler (1896–1903) and others, but it is difficult to picture such phenomena when we remember that the spindle is an elastic solid. Perhaps one of the most interesting of the hydrodynamic theories of mitosis is that put forward by Lamb (1908). This author suggested an application of Bjerknes' (1900) observations on the field of force surrounding adjacent bodies which are oscillating or pulsating in a fluid medium. If, within a fluid, two bodies are pulsating synchronously and in opposite phase they repel one another; or if they are pulsating synchronously in the same phase they attract one another. Between the aphasic pulsating or oscillating bodies there is a field of force which is morphologically identical in form with that between *unlike* magnetic poles (see fig. 66). Lamb

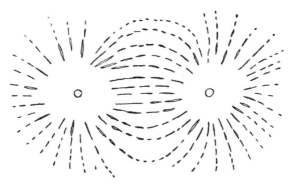

Fig. 66. Diagram of the field of force between two bodies pulsating out of phase with each other.

suggested that at one pole of the mitotic spindle there is a body (the centrosome) which pulsates or oscillates synchronously in opposite phase to its mate at the other pole. This hypothesis would account for the fact that the two centrosomes move apart as mitosis proceeds and yet are united by a field of force similar to that between the unlike poles of a magnet. Since bodies heavier than the surrounding medium are attracted by the poles of such a hydrodynamic field whereas lighter bodies are repelled, Lamb suggested that a change in the specific gravity of the chromosomes might account for the fact that they are at first (metaphase) located as far as possible from the centrosomes, whereas afterwards they are attracted to the poles. Lamb put forward his suggestion as an *ad hoc* hypothesis of only 'hypothetical value' and no direct evidence in its favour has, as yet, become available. An extension of the suggestion has, however, been made by Cannon (1923) by applying the hypothesis to 'an hypothetical, isolated, ideal cell' and then extending these deductions to the results of actual experiment. That such a procedure is fascinating there can be no doubt, but it is at the same time a negation of the experimental method. There is no evidence that the chromosomes are suspended in a fluid medium; all the experimental facts support the view that the spindle is a structure possessing considerable rigidity. If the chromosomes lie within a gelated matrix of this type it is quite improbable that they would move in the way outlined in Bjerknes' experiments. The second objection to Lamb's hypothesis lies in the fact that there is no evidence which suggests that at each end of the spindle is a pulsating or oscillating body. Until these two objections have been met, it hardly seems profitable to attempt to apply the theory itself.

Summary of mitotic movement

It is obvious that no theory of mitotic movement has yet established itself as a working hypothesis for the obvious reason that none of the proposed theories indicates a conceivable line of experimental analysis. From a purely mechanical point of view it would be of interest to know the kind and magnitude of force which can move a particle of nucleic acid through a gelated matrix, bearing in mind that the speed of movement may be extremely slow (e.g. 1 mm. per day). Until something of this sort is attempted, it seems futile to put forward theories for which there is no real foundation in fact. That there is a force which operates between the chromosomes and the poles of the spindle seems almost certain and until it is identified with any of the known physical forces there is no harm done by giving it a name. In this respect there is something to be said for Hartog's (1905, 1914) mitokinetism.

From an experimental point of view it seems useful to stress the following facts:

(i) The only structure which is necessary for the anaphasic movement of chromosomes is a bipolar or multipolar spindle. Neither asters nor definitive centrosomes are universally present. (ii) The longitudinal division of the chromosomes is independent of the presence of a spindle. (iii) In a few cases chromosomes are capable of autonomous movement, but daughter chromosomes do not move apart except when enclosed in a spindle. The distance which these

MITOSIS

chromosomes move is proportional to the length of the spindle. (iv) Since fibrils are found in the spindle when the latter is coagulated, it is reasonable to suppose that there is normally a state of tension or other localised distribution of energy along the axis of the spindle. (v) The orientation of the chromosomes at the equatorial metaphase suggests that the spindle itself generates the forces which are responsible for the orientation of the chromosomes.

The whole phenomenon of mitotic movement is baffling in the extreme. We are dealing with the movement of particles of very small size, in an elastic medium at a very low velocity, and all attempts to detect an attendant output of energy by the cell have failed.

Modification of mitosis by external reagents

Like other properties of the cell, mitotic activity can be readily and reversibly inhibited by such reagents as cold, acids, or anaesthetics. The effect of variations in temperature on the successive phases was investigated by Jolly (1904) and more recently by Ephrussi (1926), who concluded that each phase has its own temperature coefficient—that of the prophase being high (2·5–1·66) whereas that of the anaphase movement is low (1·0–1·22).

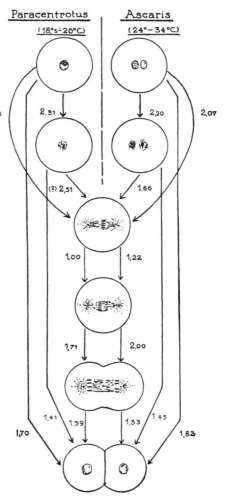

Fig. 67. Diagram illustrating the temperature coefficients of the various phases of mitosis. (From Ephrussi.)

The application of cold, quinine, anaesthetics, or KCN (Mathews, 1907), during the period of mitosis leads to a disappearance of the asters and spindle, although the chromosomes themselves do not appear to be affected. It is perhaps curious to find that an exposure to X-rays does not affect mitosis (Strangeways and Hopwood, 1926), whereas the rays can prevent a cell undergoing further mitoses when it is irradiated during the interkinetic period. The effect of acid on mitosis has not as yet been investigated in detail, but Smith and Clowes (1924) have shown that, when the CO_2 tension of sea water is increased, the velocity of cell division in *Arbacia* and in *Asterias* is decreased, and that, at a definite tension of the gas, mitosis ceases altogether.

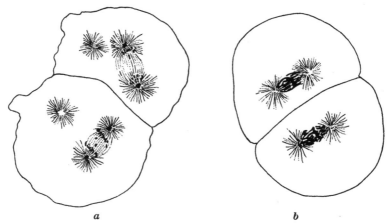

Fig. 68. *a*, Accessory asters formed in fertilised egg of *E. esculentus* after exposure to hypertonic sea water; *b*, egg of *E. esculentus* exposed to hypertonic sea water after fertilisation. Note the normal asters and the very abnormal chromosomes. (From Gray, 1913.)

In the above cases, the inhibiting agent simply reduces the velocity of mitosis, and if its effect is not too intense mitosis is resumed at its normal pace when the inhibitor is removed. On the other hand, mitosis is extremely sensitive to many external changes which seriously affect the process in an irreversible manner. As a type of such action we may take exposure to an abnormally high osmotic pressure (Konopacki, 1911; Gray, 1913). If the eggs of echinoderms are exposed to concentrated sea water it is frequently found that one or more chromosomes fail to move to the poles during the anaphase. If the osmotic pressure reaches a critical figure the whole mitotic system may be affected, the asters disappear, and the whole of the chromosomes form an irregular mass drawn out between the poles of the spindle (see figs. 68 and 69). It is interesting to note that many of these experimentally induced irregularities occur naturally when an egg is fertilised by the sperm of a foreign species. When the eggs of the sea urchin *Sphaerechinus granularis* are fertilised by the sperm of *Strongylocentrotus lividus* the mitoses of the segmenting eggs are usually quite normal, although in a few cases abnormalities occur which recall the effects of hypertonic sea water on other types of egg (fig. 69). In the reverse hybrid *Strongylocentrotus* ♀ × *Sphaerechinus* ♂, many of the paternal chromosomes fail to pass to the

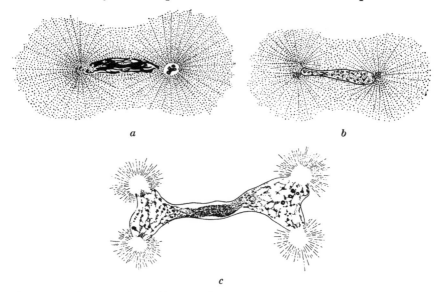

Fig. 69. *a* and *b*, The effect of hypertonic sea water (50 c.c. sea water + 8·8 c.c. 2½M NaCl) on the nucleus of a fertilised egg of *Strongylocentrotus*. (From Konopacki.) *c*, Abnormal zygote nucleus of hybrid egg, *Sphaerechinus* ♀ × *Strongylocentrotus* ♂. Note tetraster and compare the form of the nucleus with that in *b*.

poles of the first mitotic spindle (fig. 70) and are omitted from the daughter nuclei (Baltzer, 1910). Similarly the abnormal mitoses found in the cross fertilised eggs of *Echinus esculentus* × *E. acutus* (Doncaster and Gray, 1913), can be artificially induced in normally fertilised eggs of *E. acutus* by treatment with hypertonic sea water (Gray, 1913). There can be little doubt that abnormal mitoses can be the result of a variety of causes, and that the final effect on the nucleus is largely independent of the particular agent employed—heat, high or low osmotic pressure, electrolytic ions

Fig. 70. Cleavage spindles of hybrid *Strongylocentrotus* ♀ × *Sphaerechinus* ♂. Note that some of the chromosomes have failed to pass to the poles of the spindle. (From Baltzer.)

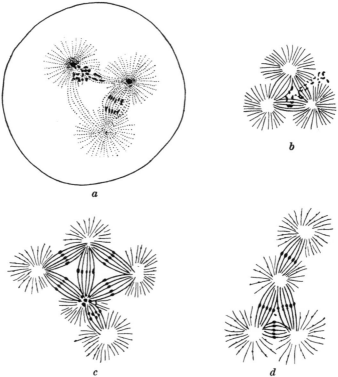

Fig. 71. Polyastral figures in echinoderm eggs: *a*, induced by treatment with hypertonic sea water (Konopacki); *b–d*, by treatment with strychnine (Hertwig).

MITOSIS

and some drugs all induce the same type of irregularity in the asters or chromosomes; it is therefore difficult to gain from such experiments any clear conception of the mitotic mechanism, although the results may have an important application to the study of malignant growths.

Mitotic stimulation

The reasons which cause a cell nucleus to undergo the process of division are difficult to analyse, since mitosis is usually only part of a long and complicated series of cell changes which go on hand in hand with growth (see Chapter XI). Growing cells often exhibit a regular rhythm of mitosis and newly divided cells usually grow; hence, peculiar interest is attached to those cases where mitosis can be made to occur in tissues whose rate of growth is reduced to a minimum. Dustin (1921, 1922) found that if 1–3 c.c. of a foreign blood serum was injected into the peritoneal cavity of a mouse, a remarkable outburst of mitosis occurred in the cells of various organs after a latent period of 2–3 days. The organs involved were the thymus gland, skin, lymphatic ganglia, and intestinal epithelium. The induced mitoses, having followed their normal course, were not followed by others although other injections might be given. Dustin correlates the onset of mitosis with the liberation of a substance which is released from other cells in the process of degeneration and in this respect his views are supported by Gutherz (1925), who believes that degenerating nuclei produce substances (necrohormones) which are the cause of the premature and incomplete maturation divisions observed in the oocytes of sexually immature mammals and in the formation of oligopyrene spermatozoa in molluscs.

Somewhat parallel to Dustin's observations are those of Isawaki (1925), who found that the injection of 1/60th c.c. of a bacillus culture into the caterpillar of *Galleria melorella* induced active mitoses in certain types of leucocytes. In one case the number of mitoses rose from three per thousand cells to 136 (fig. 72); and the intensity of the effect produced by the injection varied markedly with the temperature of incubation. A somewhat different type of mitotic stimulant is that isolated by Chambers and Scott (1925), who found that, during autolysis, malignant tumour cells produce a substance which increases the rate of growth of tumour tissues *in vitro*, and this stimulant they believe to be derived from the nuclei of the autolysed cells. Perhaps the most active supporter of specific mitotic stimulants is Haberlandt (1922, 1923), who has elaborated the theory of mitotic hormones in plants. Haberlandt cut thin sections of potato tuber and after five or six days found numerous mitoses whenever a section had included a portion of the vascular bundles. Sections without vascular bundles showed no mitoses. If, on the other hand, such sections were closely attached by a film of agar to a vascularised section, both fragments of tuber showed mitoses. Similarly mitosis could be induced by contact with freshly triturated vascular tissues. From these facts Haberlandt

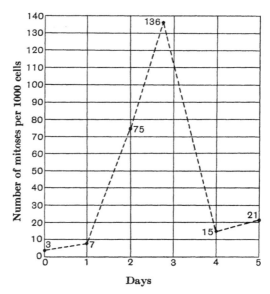

Fig. 72. Graph showing induction of mitosis in blood cells of *Galleria* by injection of bacilli. The abscissa shows the time in days after injection. (From Isawaki, 1925.)

concludes that there must be associated with the vascular bundles a substance capable of inducing nuclear division.

Somewhat analogous to Haberlandt's wound hormones are the 'desmones' described by Fischer (1925 b). This author states that isolated animal cells when grown *in vitro* fail to divide because they lack an essential constituent which can only be supplied by way of the intercellular bridges which he claims unite all the cells in a normal culture. These conclusions have, however, been denied by Wright (1925), who believes that isolated cells in a tissue-culture are capable of division, although they usually divide synchronously.

MITOSIS

Wright interprets this simultaneous type of mitosis as the result of a uniform environment on a homogeneous population of cells. It may be doubted, however, how far this explanation is complete. An isolated bacterium in a satisfactory medium soon gives rise to a population which is heterogeneous in respect to the moment of cleavage, whereas when organic continuity between nuclei is clearly

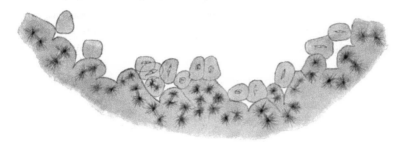

Fig. 73. Periblast of embryo of *Belone*. Note synchronisation of mitosis.
(From Gurwitsch.)

Fig. 74. Section of testis of Salamander showing synchronisation of mitosis.
(From Gurwitsch.)

established there is very clear evidence of synchronisation of cleavage. In syncytia all the nuclei divide together (fig. 73), and simultaneous mitosis is very common in the spermatocytes within a testicular tubule (fig. 74). How far these phenomena are due to the fact that all the nuclei concerned are or may be in protoplasmic connection with each other is doubtful; in this respect the observations of Polowzow (1924) are of interest. This author found that if fertilised eggs of sea urchins are exposed to dilute concentrations of alcohol, the mitotic figures are abnormally small and that nuclear division proceeds without cell cleavage: in such cases all the nuclei within any single cell divide simultaneously (fig. 75). If an egg is allowed to develop normally in sea water the first two blastomeres usually cleave simultaneously, whereas if one blastomere is separated from its mate nuclear division often falls out of step. Between the early blastomeres of many echinoderm eggs fine protoplasmic bridges undoubtedly exist (Andrews, 1899; Gray, 1925), and it is possible that the presence of these bridges enables adjacent cells to undergo simultaneous nuclear division. It

Fig. 75. Syncytial mitosis in sea urchin eggs after treatment with alcohol. Note that all the nuclei in one blastomere are dividing simultaneously: in the other cell no nuclei are dividing. (From Polowzow, 1924.)

would be of interest to know whether all the cells of a developing larva divide together as long as they are in organic connection with each other and whether, when such cells as 4 D in a trochophore larva remain quiescent, organic connection with their neighbours has been lost.

Balfour was inclined to ascribe the variation in cleavage rhythm noticeable during the segmentation of eggs to a variation in the amount of yolk present, e.g. the yolk-laden blastomeres of the lower hemisphere divide slower than those at the animal pole. There can, however, be little doubt that this explanation is insufficient to account for the marked variation in cleavage rhythm noticeable during invertebrate development. As pointed out by Wilson (1925, p. 997), the two upper sister cells in each quadrant of the 16-celled stage of annelids and molluscs differ markedly from each other in their rhythm. One divides many times in quick succession, the other divides only twice, although there is no visible difference in yolk content between them. The cleavage rhythm is much more closely related to the functional requirements of the embryo than to simple mechanical causes (F. R. Lillie).

By far the most original conception of mitotic stimulation is, however, that of Gurwitsch (1923 *seq.*). This author, together with his pupils, claims that certain types of living tissues emit a specific type of ray (mitogenic rays) which induces the onset of mitosis in

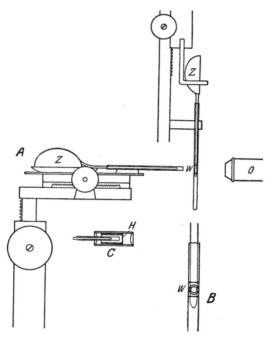

Fig. 76. Gurwitsch's apparatus for investigation of mitogenic rays. A rapidly growing root-tip of an onion bulb is arranged horizontally in a glass tube and is focussed on a slowly growing root arranged vertically. (See Gurwitsch, 1926.)

neighbouring cells. Gurwitsch's original experiment consisted in exposing one side of a root-tip of an onion bulb to contact with the tip of another actively growing root (fig. 76). After a short period of inductance, the number of mitoses found on the exposed side of the induced root was significantly greater than that on the opposite side. That this result is not due to the diffusion of materials from one root to the other is shown by the fact that actual contact is unnecessary (1924). From this experiment it is inferred that induced mitosis is caused by rays of very short wave length, since the 'mitogenic' rays can readily pass through glass and quartz, but are absorbed by gelatin from which it looks as though their wave length was of the order of 2000 Å. Not only actively growing root-tips are able to induce mitosis in other roots, for inductance occurs when the 'inducting' root is replaced by yeast (Baron, 1926), tadpoles (Gurwitsch, A. and L., 1925), or oxygenated blood (Sorin, 1926). In nearly all his experiments Gurwitsch has used the root-tip of plant bulbs as 'detectors' of the mitogenic rays, although Baron (1926) found that yeast cells can be induced to divide more rapidly by exposure to growing plant roots. In one experiment the number of budding yeast cells rose from 8 to 22 per 100 cells when exposed to root-tip emanations, whereas in the root-tip the number of mitoses rose by 22–26 per cent. on exposure to actively sprouting yeast. So far no animal cells have been used as detectors.

Table XXVII

Number of mitoses in sections of root-tip of *Vicia faba*

	Spireme	Monaster	Diaster	Telophase
Normal	37	30	19	33
After ½ hour in 0·125 % ether	160	61	40	25

It is not altogether easy to assess the significance of Gurwitsch's experiments, but if the results are substantiated by further work they must represent an entirely new line of biological research. It is unfortunate, perhaps, that it has been too often assumed that an arbitrary difference of numbers of mitoses between the control section and the induced section of the root-tip is truly significant, since a statistical justification of this point is fundamental to all the experimental conclusions. That the root-tips of plants are peculiarly sensitive to mitotic stimulation is confirmed by Mainx (1924), who found that after exposure to low dilutions of ether the number of mitoses was greatly increased.

From time to time the determining cause of mitosis in animal cells has been sought by variation in environmental factors. In this connection it seems fairly clear that mitosis tends to occur after a period of starvation. For example, many amphibia are capable of withstanding prolonged starvation during which the cytoplasmic portion of their cells is markedly reduced in volume. As soon as food is provided, rapid mitosis frequently occurs (Kornfeld, 1922); whether in plants the rhythmical formation of food-stuffs during daylight is the factor responsible for the prevalence of mitoses during the night (Karsten, 1915, 1918) is not certain. On the other hand, active mitosis has been described in the tissues of starving animals [pigeons (Morpurgo, 1888); cats (Hofmeister, 1890); salamanders (Morgulis, 1911)]. There can be no doubt that, at times, the phenomena of mitosis and of growth are independent of each other, for in dividing eggs the rate of increase in the amount of respiring protoplasm changes quite independently of the mitotic rhythm (Gray, 1927). Even in Protozoa, where the cells maintain the same average size over a number of successive cleavages, the absolute size of the cell depends on the nutrient level of the surrounding medium and mitosis does not necessarily occur when the cell has reached a critical size.

REFERENCES

ANDREWS, E. A. (1899). 'Filose activity in metazoan eggs.' Zool. Bull. 2, 1.
BALTZER, F. (1910). 'Ueber die Beziehung zwischen dem Chromatin und der Entwicklung und Vererbungsrichtung bei Echinodermenbastarden.' Arch. f. Zellforsch. 5, 497.
BARON, M. A. (1926). 'Über mitogenetische Strahlung bei Protisten.' Zeit. f. wissen. Biol. Abt. D, 108, 617.
BĚLAŘ, K. (1929). 'Beiträge zur Kausanalyse der Mitose. II. Untersuchungen an den Spermatocyten von Chorthippus (Stenobothrus) lineatus Pauz.' Arch. f. Entw. Mech. 118, 359.
VAN BENEDEN, E. (1883). 'Recherches sur la maturation de l'œuf, la fécondation et la division cellulaire.' Arch. de Biol. 4, 265.
VAN BENEDEN, E. and NEYT, A. (1887). 'Nouvelles recherches sur la fécondation et la division mitosique chez l'Ascaride mégalocéphale.' Bull. Acad. Roy. Belg. III, 14, 215.
BJERKNES, V. (1900). Vorlesungen über hydrodynamische Fernkräfte. (Leipzig.)
BOVERI, TH. (1887). 'Ueber die Befruchtung der Eier von Ascaris megalocephala.' Sitz. Ber. Ges. Morph. Phys. München, 3, 71.
—— (1887). 'Zellenstudien. II.' Jen. Zeit. 22, 685.
—— (1895). 'Ueber das Verhalten der Centrosomen bei der Befruchtung des Seeigeleies nebst allgemeinen Bemerkungen über Centrosomen und Verwandtes.' Verh. phys.-med. Ges. Würzburg, N.F. 29, 1.
BOVERI, T. (1903). 'Über das Verhalten des Protoplasmas bei monocentrischen Mitosen.' Sitz. Ber. d. phys.-med. Ges. Würzburg, 12.
BÜTSCHLI, O. (1876). 'Studien über die ersten Entwicklungsvorgänge der Eizelle, die Zelltheilung, und die Konjugation der Infusorien.' Senckenb. Naturforsch. Ges. 10, 213.
CANNON, H. G. (1923). 'On the nature of the centrosomal force.' Journ. Genetics, 13, 47.
CHAMBERS, H. and SCOTT, G. (1925). 'On a growth promoting factor in tumour tissue.' Brit. Journ. Exp. Path. 7, 33.

CHAMBERS, R. (1917). 'Microdissection studies. II. The cell aster. A reversible gelation phenomenon.' *Journ. Exp. Zool.* 23, *483*.
—— (1921). 'Microdissection Studies. III. Some problems in the maturation and fertilization of the echinoderm egg.' *Biol. Bull.* 41, *318*.
—— (1922). 'New apparatus and methods for the dissection and injection of living cells.' *Anat. Record*, 24, *1*.
—— (1924). 'Les structures mitochondriales et nucléaires dans les cellules germinales mâles chez la Sauterelle.' *La Cellule*, 35, *107*.
CHAMBERS, R. and SANDS, H. C. (1923). 'A dissection of the chromosomes in the pollen mother cells of *Tradescantia virginica* L.' *Journ. Gen. Physiol.* 5, *815*.
CONKLIN, E. G. (1902). 'Karyokinesis and cytokinesis in the maturation, fertilisation, and cleavage of *Crepidula* and other gasteropods.' *Journ. Acad. Nat. Sci. (Phil.)*, 12, *5*.
—— (1917). 'Mitosis and amitosis.' *Biol. Bull.* 33, *396*.
DANCHAKOFF, V. (1916). 'Studies on cell-division and cell-differentiation. I.' *Journ. Morph.* 27, *559*.
DONCASTER, L. and GRAY, J. (1913). 'Cytological observations on the early stages of segmentation of *Echinus* hybrids.' *Quart. Journ. Micr. Sci.* 58, *483*.
DUSTIN, A. P. (1921 *a*). 'Déclenchement expérimental d'une onde cinétique par injection intrapéritonéale de sérum.' *C.R. Soc. Biol.* 85, *23*.
—— (1921 *b*). 'L'onde de cinèses et l'onde de pycnoses dans le thymus de la souris après injection intrapéritonéale de sérum étranger.' *C.R. Soc. Biol.* 85, *260*.
—— (1922). 'Influence d'injections intrapéritonéales répétées de peptone sur l'allure de la courbe des cinèses.' *C.R. Soc. Biol.* 87, *371*.
DUSTIN, A. P. and CHAPEAUVILLE, J. (1922). 'Les caractères de l'onde cinétique déclenchée par une injection intrapéritonéale de peptone.' *C.R. Soc. Biol.* 86, *509*.
EPHRUSSI, B. (1926). 'Sur les coefficients de température des différentes phases de la mitose des œufs d'oursin (*Paracentrotus lividus*) et de l'*Ascaris megalocephala*.' *Protoplasma*, 1, *105*.
FISCHER, A. (1925 *a*). 'Beitrag zur Biologie der Gewebezellen. Eine vergleichendbiologische Studie der normalen und malignen Gewebezellen *in vitro*.' *Arch. f. Entw. Mech.* 104, *220*.
—— (1925 *b*). *Tissue Culture*. (London.)
GALLARDO, A. (1909). 'La division de la cellule phénomène bipolaire de caractère électro-colloïdal.' *Arch. f. Entw. Mech.* 25, *125*.
GRAY, J. (1913). 'The effects of hypertonic solutions upon the fertilised eggs of *Echinus*.' *Quart. Journ. Micr. Sci.* 58, *447*.
—— (1924). 'The mechanism of cell-division. I. The forces which control the forms and cleavage of *Echinus esculentus*.' *Proc. Camb. Philos. Soc. Biol. Series*, 1, *164*.
—— (1925). 'The mechanism of cell-division. II. Oxygen consumption during cleavage.' *Proc. Camb. Philos. Soc. Biol. Series*, 1, *225*.
—— (1927). 'The mechanism of cell-division. III. The relationship between cell-division and growth in segmenting eggs.' *Brit. Journ. Exp. Biol.* 4, *313*.
—— (1928). *Ciliary Movement*. (Cambridge.)

GURWITSCH, A. (1923). 'Die Natur des spezifischen Erregers der Zellteilung.' *Arch. f. mikr. Anat. u. Entw. Mech.* 100, *11*.
—— (1924). 'Physikalisches über mitogenetische Strahlen.' *Arch. f. mikr. Anat. u. Entw. Mech.* 103, *490*.
—— (1926). *Das Problem der Zellteilung physiologisch betrachtet.* (Berlin.)
GURWITSCH, A. and GURWITSCH, L. (1925). 'Weitere Untersuchungen über mitogenetische Strahlungen.' *Arch. f. mikr. Anat. u. Entw. Mech.* 104, *109*.
—— —— (1926). 'Die Produktion mitogener Stoffe im erwachsenen tierischen Organismus.' *Zeit. f. wissen. Biol.* Abt. D, 107, *829*.
GURWITSCH, A. and GURWITSCH, N. (1924). 'Fortgesetzte Untersuchungen über mitogenetische Strahlung und Induktion.' *Arch. f. mikr. Anat. u. Entw. Mech.* 103, *68*.
GURWITSCH, LYDIA (1924). 'Untersuchungen über mitogenetische Strahlen.' *Arch. f. mikr. Anat. u. Entw. Mech.* 103, *483*.
GUTHERZ, S. (1925). 'Über vorzeitige Chromatinreifung an physiologisch degenerierenden Säugeoozyten des frühen Wachstumsperiode.' *Zeit. f. mikr. Anat. Forsch.* 2, *1*.
HABERLANDT, G. (1922). 'Über Zellteilungshormone und ihre Beziehungen zur Wandteilung, Befruchtung, Parthenogenesis, und Adventivembryonie.' *Biol. Zentralb.* 42, *145*.
—— (1923). 'Wundhormone als Erreger von Zellteilungen.' *Beit. z. allgem. Bot.* 2, *1*.
HARDY, W. B. (1899). 'On the structure of cell protoplasm.' *Journ. Physiol.* 24, *158*.
HARTOG, M. (1905). 'The dual force of the dividing cell. I. The achromatic spindle figure illustrated by magnetic chains of force.' *Proc. Roy. Soc.* B, 76, *548*.
—— (1914). 'The true mechanism of mitosis.' *Arch. f. Entw. Mech.* 40, *33*.
HEIDENHAIN, M. (1894). 'Neue Untersuchungen über die Centralkörper und ihre Beziehungen zum Kern und Zellprotoplasma.' *Arch. mikr. Anat.* 43, *423*.
—— (1896). 'Ein neues Modell zum Spannungsgesetz der centrierten Systeme.' *Anat. Anz.* 12, *67*.
HERLANT, M. (1918). 'Comment agit la solution hypertonique dans la parthénogénèse expérimentale (Méthode de Loeb). I. Origine et signification des asters accessoires.' *Arch. Zool. exp.* 57.
HERTWIG, R. (1903). 'Über Korrelation von Zell und Kerngrösse und ihre Bedeutung für die geschlechtliche Differenzierung und die Teilung der Zelle.' *Biol. Zentralb.* 23, *49*.
HERTWIG, O. and HERTWIG, R. (1887). 'Ueber den Befruchtungs- und Teilungsvorgang des tierischen Eies unter dem Einfluss äusserer Agentien.' *Jen. Zeit.* 20, *120, 477*.
HOFMEISTER, F. (1890). 'Über den Hungerdiabetes.' *Arch. f. exp. Path. u. Pharm.* 26, *355*.
ISAWAKI, Y. (1925). 'Sur le déclanchement expérimental des ondes de cinèse dans le sang de quelques insectes.' *Ann. de Physiol. et de Physiochem. biol.* 1, *580*.
JOLLY, J. (1904). 'Recherches expérimentales sur la division indirecte des globules rouges.' *Arch. d'Anat. micr.* 6, *455*.

JORDAN, A. E. and HELVESTINE, F. (Jnr.) (1922). 'Ciliogenesis in the epididymis of the white rat.' *Anat. Record,* 25, *7.*
KARSTEN, G. (1915). 'Über embryonales Wachstum und seine Tagesperiode.' *Zeit. f. Bot.* 7, *1.*
—— (1918). 'Über die Tagesperiode der Kern und Zellteilungen.' *Zeit. f. Bot.* 10, *1.*
KINDRED, J. E. (1926). 'Cell division and ciliogenesis in the ciliated epithelium of the pharynx and oesophagus of the tadpole of the green frog, *Rana clamitans*.' *Journ. Morph. and Physiol.* 43, *267.*
KLEIN, E. (1878). 'Observations on the structure of cells and nuclei.' *Quart. Journ. Micr. Sci.* 18, *315.*
KONOPACKI, M. (1911). 'Ueber den Einfluss hypertonischer Lösungen auf befruchtete Echinideneier.' *Arch. f. Zellforsch.* 7, *139.*
KORNFELD, W. (1922). 'Über den Zellteilungsrhythmus und seine Regelung.' *Arch. f. Entw. Mech.* 50, *526.*
LAMB, A. B. (1908). 'A new explanation of the mechanics of mitosis.' *Journ. Exp. Zool.* 5, *27.*
LAMBERT, R. A. and HANES, F. N. (1913). 'Beobachtungen an Gewebskulturen *in vitro*.' *Virchow's Arch.* 211, *89.*
LEDUC, S. (1914). *The Mechanism of Life.* (London.)
LEVI, G. (1916). 'Il ritmo e le modalità della mitosi nelle cellule viventi coltivate *in vitro*.' *Arch. ital. d. anat. e di embriol.* 15, *243.*
LEWIS, M. R. (1923). 'Reversible gelation in living cells.' *Johns Hopkins Hosp. Bull.* 34, *373.*
LEWIS, W. H. and LEWIS, M. R. (1917). 'The duration of the various phases of mitosis in the mesenchyme cells of tissue cultures.' *Anat. Record,* 13, *359.*
LILLIE, F. R. (1912). 'Studies of fertilisation in *Nereis*. III. The morphology of the normal fertilisation of *Nereis*. IV. The fertilising power of portions of the spermatozoön.' *Journ. Exp. Zool.* 12, *413.*
LILLIE, R. S. (1905). 'On the conditions determining the disposition of the chromatic filaments and chromosomes in mitosis.' *Biol. Bull.* 8, *3.*
MCCLENDON, J. F. (1908). 'Segmentation of eggs of *Asterias forbesii* deprived of chromatin.' *Arch. f. Entw. Mech.* 26, *662.*
—— (1910). 'On the dynamics of cell-division. I. The electric charge on colloids in living cells in the root tips of plants.' *Arch. f. Entw. Mech.* 31, *80.*
MAINX, F. (1924). 'Versuche über die Beeinflussung der Mitose durch Giftstoffe.' *Zool. Jahrb.* 41, *553.*
MARTENS, P. (1922). 'Le cycle du chromosome somatique dans les Phanérogames. I. *Paris quadrifolia*.' *La Cellule,* 32, *331.*
—— (1927). 'Recherches expérimentales sur la cinèse dans la cellule vivante.' *La Cellule,* 38, *69.*
MATHEWS, A. P. (1907). 'A contribution to the chemistry of cell-division, maturation and fertilization.' *Amer. Journ. Physiol.* 18, *89.*
—— (1921). *Physiological Chemistry.* (London.)
MEEK, C. F. U. (1913). 'The problem of mitosis.' *Quart. Journ. Micr. Sci.* 58, *567.*
MEVES, F. (1896). 'Zelltheilung.' *Ergeb. d. Anat. u. Entw.* 6, *285.*
—— (1898). 'Zelltheilung.' *Ergeb. d. Anat. u. Entw.* 8, *430.*
MEYERHOF, O. (1911). 'Untersuchungen über die Wärmetönung der vitalen Oxydationsvorgänge im Seeigelei (I–III).' *Biochem. Zeit.* 35, *246.*

MORGAN, T. H. (1896). 'The production of artificial astrospheres.' *Arch. f. Entw. Mech.* 3, *339*.
—— (1899). 'The action of salt solutions on the unfertilised and fertilised eggs of *Arbacia*, and of other animals.' *Arch. f. Entw. Mech.* 8, *448*.
—— (1910). 'Cytological studies of centrifuged eggs.' *Journ. Exp. Zool.* 9, *593*.
MORGULIS, S. (1911). 'Studies of inanition in its bearing upon the problem of growth.' *Arch. f. Entw. Mech.* 32, *169*.
MORPURGO, B. (1888). 'Sul processo fisiologico di neoformazione cellulare durante la inanizione acuta dell' organismo.' *Arch. Sci. med.* 12, *395*.
NASSONOV, D. (1918). 'Recherches cytologiques sur les cellules végétales.' *Archives Russes d'Anatomie, d'Histologie, et d'Embryologie*, 2, *95*.
PAINTER, T. S. (1915). 'An experimental study in cleavage.' *Journ. Exp. Zool.* 18, *299*.
PENTIMALLI, F. (1909). 'Influenza della corrente elettrica sulla dinamica del processo cariocinetico.' *Arch. f. Entw. Mech.* 28, *260*.
POLOWZOW, W. (1924). 'Über die Wirkung der Alkoholnarkose auf die Entwicklung der Seeigeleier. II. Polysyncytien.' *Arch. f. mikr. Anat. u. Entw. Mech.* 103, *1*.
PRÉNANT, A. (1910). 'Théories et interprétations physiques de la mitose.' *Journ. de l'Anat. et Phys.* 46, *511*.
RAWIN, W. (1923). 'Weitere Beiträge zur Kenntnis der mitotischen Ausstrahlung und Induktion.' *Arch. mikr. Anat. u. Entw. Mech.* 101, *53*.
RHUMBLER, L. (1896). 'Versuch einer mechanischen Erklärung der indirekten Zell und Kerntheilung. I. Cytokinese.' *Arch. f. Entw. Mech.* 3, *527*.
—— (1897). 'Stemmen die Strahlen der Astrosphäre oder ziehen sie?' *Arch. f. Entw. Mech.* 4, *659*.
—— (1903). 'Mechanische Erklärung der Aehnlichkeit zwischen magnetischen Kraftliniensystem und Zellteilungsfiguren.' *Arch. f. Entw. Mech.* 16, *476*.
SAKAMURA, T. (1926). 'Chromosomenforschung an frischem Material.' *Protoplasma*, 1, *537*.
SHEARER, C. (1910). 'On the anatomy of *Histriobdella homari*.' *Quart. Journ. Micr. Sci.* 55, *287*.
SMITH, H. W. and CLOWES, G. H. A. (1924). 'The influence of carbon dioxide on the velocity of division of marine eggs.' *Amer. Journ. Physiol.* 68, *183*.
SOLGER, B. (1891). 'Über Pigmenteneinschlüsse in der Attraktionsphäre ruhender Chromatophoren.' *Anat. Anz.* 6, *282*.
SORIN, A. N. (1926). 'Zur Analyse der mitogenetischen Induktion des Blutes.' *Zeit. f. wissen. Biol.* Abt. D, 108, *634*.
STRANGEWAYS, T. S. P. (1922). 'Observations on the changes seen in living cells during growth and division.' *Proc. Roy. Soc.* B, 94, *137*.
STRANGEWAYS, T. S. P. and HOPWOOD, F. L. (1926). 'The effects of X-rays upon mitotic cell division in tissue culture *in vitro*.' *Proc. Roy. Soc.* B, 100, *283*.
STRASBURGER, E. (1884). 'Die Controversen der indirekten Kernteilung.' *Arch. mikr. Anat.* 23, *246*.
—— (1897). 'Über Cytoplasmastrukturen, Kern- und Zelltheilung.' *Jahrb. f. wiss. Bot.* 30, *375*.

WILSON, E. B. (1901). 'Experimental studies in cytology. I. A cytological study of parthenogenesis in sea-urchin eggs.' *Arch. f. Entw. Mech.* **12**, *529*.
—— (1902). 'Experimental studies in cytology. II. Some phenomena of fertilisation and cell division in etherized eggs. III. The effect on cleavage of artificial obliteration of the first cleavage furrow.' *Arch. Entw. Mech.* **13**, *354*.
—— (1925). *The Cell in Development and Heredity.* (New York.)
—— (1928). *The Physical Basis of Life.* (New Haven.)
WRIGHT, G. P. (1925). 'The relative duration of the various phases of mitosis in chick fibroblasts cultivated *in vitro*.' *Journ. Roy. Micr. Soc.* *414*.
YATSU, N. (1908). 'Some experiments on cell-division in the egg of *Cerebratulus lacteus*.' *Annot. Zool. japon.* **6**, *267*.
ZIMMERMANN, K. W. (1893). 'Studien über Pigmentzellen.' *Arch. mikr. Anat.* **41**, *367*.

CHAPTER NINE
Cell Division

A cleavage of the whole cell is the natural sequence to the division of the nucleus and the two phenomena are usually part of one complete cycle of cellular activity. During this cycle, the cell undergoes more drastic changes in its internal architecture and in its external form than at any other stage of its existence; nevertheless, the mechanism of cleavage is still the subject of acute controversy.

Since a dividing cell undergoes a spontaneous change in form, there are two possible ways whereby its activities might be analysed. Firstly, the cell may be regarded as a small elastic sphere which can only be divided into two parts by an expenditure of sufficient energy to overcome the natural tendency of the cell to resist deformation of form. On the analogy of other types of cell movement, such a mobilisation of energy might be expected to show itself by an alteration in the rate at which the cell evolves heat or absorbs oxygen. Observations of this nature are of peculiar importance, since they should give a measure of the forces involved during cleavage even when the actual mechanical causes of the phenomenon are invisible under the microscope. Failure to obtain an insight into the energy changes involved would, on the other hand, restrict the analysis of cleavage to the visual observation of cell structures and to the response made by these to a variable environment. Our knowledge of the energetics of cell division is, unfortunately, very slight and most of the data are of a negative character. For this reason it is convenient first to consider the visual phenomena of cleavage.

The visual phenomena of cell division

In selecting material for a study of the visual phenomena of cleavage it is important to remember that the form of any given tissue cell is almost always determined by or affected by the presence of its neighbours, and that this disturbing effect will operate during the actual process of division. Such a factor is absent in the case of cells which, after cleavage, live isolated from each other. It is very unfortunate that cells of this latter type (e.g. leucocytes or fibroblasts) exhibit during division a highly irregular form which is exceedingly difficult if not impossible to resolve into simpler components. Division of this type is brought about by a slow and disjointed process of separation into two daughter cells, and whilst this is going on little or no change is visible in the interior of the cell. The cleavage of a living fibroblast has been observed on many occasions and by many authors: the following description is largely based on the account given by Strangeways (1922).

During the early stages of mitotic activity all isolated cells, however irregular their original outline, withdraw their pseudopodial projections and assume a spherical or ellipsoidal form (fig. 77, 1–5). Soon after the anaphasic movement of the chromosomes, the cell begins to elongate along the mitotic axis. This elongation is accompanied by two sets of visual phenomena at the surface of the cell. Firstly, the elongation of the mitotic axis involves a diminution in length of the equatorial axis of the cell, except in the regions of the poles, where the diameter of the cell remains practically unchanged. Secondly, there appears at the poles a series of blunted projections or 'blisters' which slowly develop and slowly disappear by reincorporation into the body of the cell (fig. 77, 7–10). This curious activity at the poles of the cell lasts, as a rule, for about three or four minutes, and during the whole of this time the equatorial constriction gradually becomes more obvious. Eventually the two poles of the cell appear to attach themselves to the substratum by the formation of fine pseudopodia very similar to those characteristic of non-dividing cells. From this point onwards the two halves of the cell separate from each other by a disjointed but active process of mutual separation—until, with increasing velocity, the two daughter cells move away from each other leaving until the last moment fine strands of hyaline material which connect one daughter cell to the other. The whole process of division occupies at 37° C. from 30 to 45 minutes. Owing to its low velocity, disjunctive cleavage is not readily followed as a series of successive events which are clearly marked out as such to the eye of the observer. For this reason the artificial acceleration made possible by the process of cinematography affords considerable assistance in establishing an adequate time relationship between the different phases of the whole process. Films of

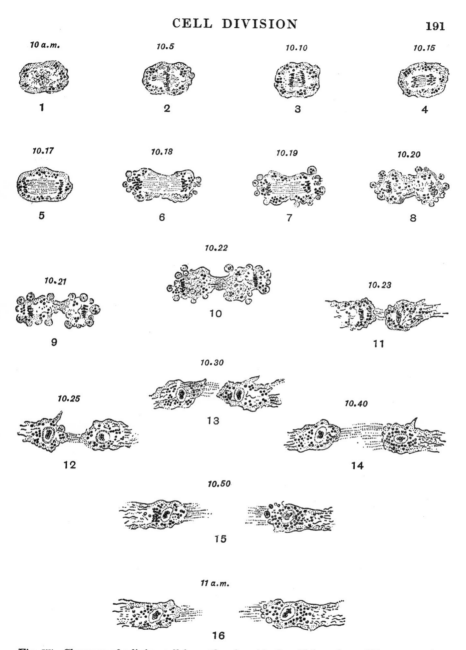

Fig. 77. Cleavage of a living cell from the choroid of a chick embryo (Strangeways). The whole cycle occupied about fifty minutes. Note the 'bubbling' at the poles of the cell which accompanies the formation of the polar furrow.

this nature (e.g. those of Dr Carrel, and those of Dr Canti) show very clearly the polar activities of the dividing cell, and the curious pulling effect of one daughter cell against the other. At the same time, the interpretation of accelerated films demands some degree of caution if they are used as a basis for understanding the underlying processes of division.

No comprehensive theory of disjunctive cell division has yet been put forward. There are, however, two features which seem worthy of comment. Firstly, the so-called bubbling activity at the poles of the cell. This process is not restricted to dividing cells, it occurs over the whole surface of the cell when the latter becomes moribund and is about to die; it can also be induced to occur in other types of cell

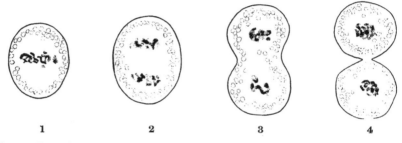

Fig. 78. Four observations at 5-minute intervals of the cleavage of a connective tissue cell of a rat. Note the elliptical form of the cell immediately before the development of the cleavage furrow. During the later stages the spherical form of the daughter cells is well marked. (Lambert and Hanes.)

(e.g. eggs of molluscs and echinoderms) by a weakening of the hyaline layer with resultant protrusion of temporary exovate processes. When illustrated by an accelerated cinematograph film both eggs and moribund fibroblasts give the same impression of active 'bubbling' as the poles of a normally dividing cell. One gets the impression that bubbling is invariably associated with a weakening of the hyaline layer of the cell. The second point of interest involves the drawing apart of the two daughter cells. It is more than probable that the active organs of movement are located in the pseudopodial processes (as in normal cell movement), and it is significant that the pseudopodia make their appearance in the regions where the 'blisters' have recently been forming; it is conceivable that the pseudopodia, like the blisters, possess a thinner and less well-defined surface layer than does the rest of the cell.

CELL DIVISION

Cell cleavage of the disjunctive type clearly involves three factors: (i) the organisation of the cell into an ellipsoid mass followed by the development of an equatorial furrow, (ii) a weakening of the hyaline surface at the poles of the long axis, and (iii) an active drawing apart of the two halves by pseudopodial movement. The possible significance of this interpretation must be deferred to a later paragraph.

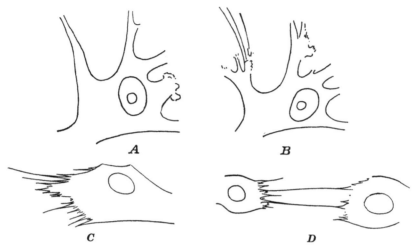

Fig. 79. Cells form a culture of the heart of a chick. *A*, Vegetative cell before the withdrawal of the large pseudopodium seen at the top left corner; *B*, same cell after withdrawal of the pseudopodium, note the long protoplasmic filaments; *C*, a vegetative cell withdrawing pseudopodia; *D*, two daughter cells moving apart after division. Note the protoplasmic junctions: these are eventually ruptured. (From Seifriz.)

Astral cleavage

In contrast with the disjunctive cleavage, so typical of leucocytes and other isolated cells, are the orderly and geometrical changes in form which characterise the division of a spherical egg cell into two contiguous blastomeres. Not only are the changes in form of a comparatively simple nature, but during cleavage the internal architecture of the cell is of a type which exists at no other phase of activity. It is not surprising, therefore, that marine eggs provide the favourite material for the investigation of cell division. At the same time they introduce a complication. When an echinoderm egg divides into two blastomeres, each of the latter remains adherent to its neighbour, and neither of them (in nature) ever regains the spherical form of the undivided egg. A considerable body of evidence shows that the precise form of the cleavage furrow is the result, not only of the cleavage mechanism, but also of the mechanism which keeps the two daughter cells in contact with each other when the cleavage itself has been completed (Gray, 1924).

Before proceeding to aspects of astral cell division, which are either speculative or controversial, it is useful to consider those facts which are generally accepted as true. As pointed out elsewhere (Chapter VI), the fertilised egg of a sea urchin (*Echinus esculentus* in particular) possesses a well-defined hyaline layer at the surface of the cell (fig. 32). Until the process of cleavage begins, this peripheral or hyaloplasmic layer is uniformly distributed over the egg surface and fine protoplasmic strands pass across it between the cytoplasm of the cell and the external boundary of the layer itself (Gray, 1924). This external boundary undoubtedly consists of a solid membrane whose properties are quite well defined. It is imperfectly elastic; it can be drawn into fine threads which, on release, recover to some extent their original form, leaving distinct although sometimes transitory traces of disturbance. The chemical nature of the membrane is unknown, but it has two important reactions: firstly, it is soluble in sea water if the latter contains no calcium; secondly, in the presence of dilute acid it contracts and becomes tightly compressed to the surface of the cytoplasm. The existence of this hyaline membrane has been clearly demonstrated in a considerable variety of material, but it is much more clearly defined in the eggs of *E. esculentus* than in *E. miliaris*, *Arbacia*, or *Echinarachius*. Just as in the case of disjunctive cleavage, so the onset of astral cleavage is marked by an elongation of the egg axis which is at right angles to the plane of the future cleavage furrow. Until this polar elongation begins, the hyaloplasmic layer is uniformly distributed over the surface of the egg; as soon as the egg elongates, the hyaline material begins to flow from the poles of the cell to the equatorial furrow. As the furrow deepens, so the hyaline material becomes more and more aggregated into the equatorial region of the egg until, with completion of cleavage, a well defined intercellular plate of hyaloplasm separates each cell from its mate, whilst the intercellular plate is continuous with a very thin layer of similar hyaline substance which still covers the polar regions of each cell (fig. 80). The protoplasmic processes, which, prior to cleavage, traverse the hyaloplasm, are withdrawn during division, only to be reformed after this process is complete.

The more pertinent of these facts have not been questioned, but

CELL DIVISION

are accepted with one important proviso by three recent investigators whose interpretations differ widely from each other (Just, 1922; Chambers, 1919, 1924; Gray, 1924). The main point at issue involves the precise moment at which the hyaloplasm begins to flow from the poles of the cell to the equator. According to Just (1922), this redistribution proceeds to an appreciable extent before the cytoplasm

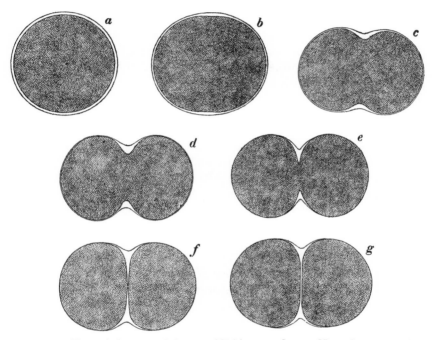

Fig. 80. *a–g*, Normal cleavage of the egg of *Echinus esculentus*. Note the gradual flow of hyaloplasm (white) from the poles of cell to the equator. In *f* note that the cytoplasm is almost completely divided. In *g* the hyaloplasm has joined in the centre and the two masses of cytoplasm are completely divided off from each other but are completely surrounded by hyaloplasm.

of the egg begins to lose its spherical form, whereas other observations (Gray, 1924) lead to the belief that these two processes begin simultaneously. As will be obvious later, this point is of fundamental importance.

In discussing the process of mitotic division, it was pointed out that the period between the fusion of the two pronuclei and the anaphase of the first division is marked by a gradual increase in the size of the two mitotic asters. As this increase in size takes place, it can be seen that each aster is an almost perfect sphere, but as the spheres increase in volume the individual rays tend to become less distinct. This phase of increase in size of the two asters is illustrated by fig. 81. At the anaphase (see fig. 81, 3) the form of the asters begins to change. They continue to enlarge in volume but they are no longer spherical, for they flatten against each other in the median line (figs. 58 a and 81, 4–7). As soon as the two daughter nuclei are formed, practically the whole of the cytoplasm of the egg is occupied by the two asters whose rays extend almost to the periphery of the cytoplasm. So far, the cell has retained its spherical outline, but at this moment there is a visible change in the shape of the whole egg. The polar elongation begins.

The changes in the size and in the form of the asters are admittedly not so clear in the living egg as they are in preserved material (see, however, fig. 58), and some difference of opinion exists concerning the precise moment at which the asters reach their maximal development. Just (1922) concludes that the astral rays, in *Arbacia* eggs, fade away before cleavage of the cytoplasm begins: this is not the case in *E. esculentus*, for a well-defined aster can be seen after the completion of cleavage, both in preserved and in living material; during the actual process of division almost the entire volume of the egg is occupied by the astral rays, only the region of the equatorial furrow is exempt. This uncertainty concerning the time relationship of cleavage to astral development is of importance, although not decisive from a theoretical standpoint.

In some cells the development of the cleavage furrow is accompanied by a visible streaming of peripheral cytoplasm towards the equatorial furrow. This has been described by Erlanger (1897), Loeb (1895), and more recently by Spek (1918).

So much for the facts. All recent theories of cell cleavage hinge on three points. Firstly, what rôle, if any, is played by the hyaloplasm? Secondly, what rôle, if any, is played by the mitotic asters? Thirdly, what part is played by the peripheral streaming of the cytoplasm?

It is convenient to consider the hyaloplasm first. Just believes that this layer is the active mechanism of cleavage. In *Arbacia* eggs he has observed in this layer some degree of amoeboid movement which he regards as an indication of sufficient power to move to the equator of the cell and, by a process of active constriction, divide the cell into two halves. Obviously, on this interpretation, there is no

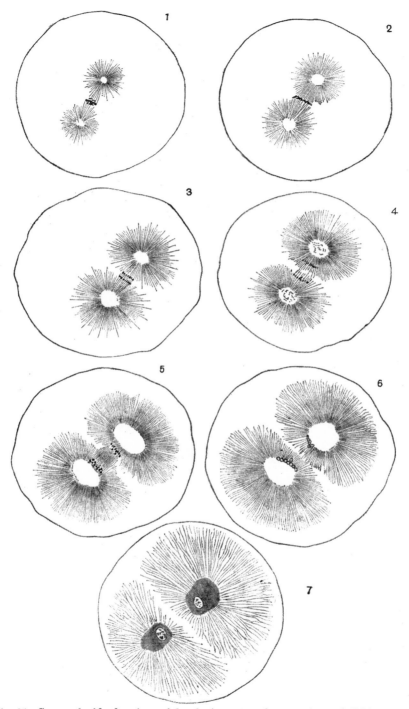

Fig. 81. Camera lucida drawings of developing asters from sections of *Echinus* eggs fixed in corrosive sublimate. Note change in shape of the asters after the anaphase stage is reached: also note loss of definition of astral rays with increase in the size of the asters.

reason why an equatorial accumulation of hyaloplasm should not occur before the actual process of cytoplasmic cleavage begins, since the polar elongation of the cell would be the natural result of equatorial compression. A direct test of Just's interpretation is available by removing the hyaloplasmic layer from the egg either before or during cleavage. As already mentioned, the hyaloplasmic membrane is soluble in calcium-free sea water, and it is for this reason that segmented eggs disintegrate into their constituent cells when in such a medium (see Chapter VI). If eggs of *Echinus esculentus* are reared in normal sea water until 10–15 minutes before cleavage, and are then placed in calcium-free sea water, they soon reveal two significant facts. Firstly, the hyaline layer has been dissolved. Secondly, the

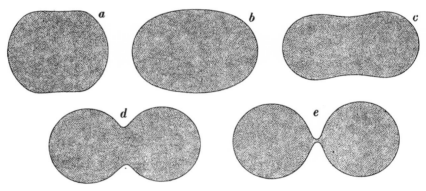

Fig. 82. Cleavage of the egg of *Echinus* in calcium-free sea water. Note that the absence of the hyaline layer causes a marked increase of the elongation of the polar axis of the cell and that the latter is eventually resolved into two spherical blastomeres.

polar elongation of the cells, far from being decreased, is actually increased (see fig. 82). It is extremely difficult to harmonise these facts with Just's hypothesis. That the presence of the hyaline layer tends to reduce rather than increase the polar elongation of the dividing cell is strongly supported not only by the effect of its removal during (or after) cleavage by calcium-free sea water but also by the action of two entirely independent pieces of evidence.

By a fortunate coincidence there is a marked difference in the osmotic properties of the hyaline membrane and of the egg cytoplasm. The former is freely permeable to electrolytes, whereas the latter is not: consequently when the eggs of *Echinus esculentus* are placed in hypertonic sea water the volume of the cytoplasm is

CELL DIVISION 199

decreased, whilst the volume contained within the hyaline membrane is, if anything, increased. Under such conditions the compression normally exerted by the hyaloplasm on the cytoplasm is relieved and the relative length of the polar axis of the cell is definitely increased (fig. 83). The form of the dividing cell under such circumstances is similar to that of eggs exposed to calcium-free sea water, although of course the cell volume is less. Just as the compression exerted by the hyaloplasmic layer on the cytoplasm can be relieved

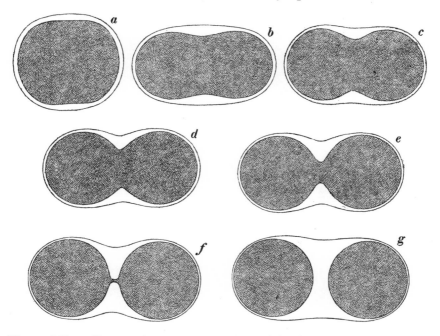

Fig. 83. Effect of hypertonic sea water on the form of the cleaving eggs. *a–g* show the effect of transferring the normal eggs shown in fig. 80 *a–g* to hypertonic sea water. Note change in form of cytoplasm, with increased elongation of the polar axis.

by hypertonic sea water, so it can be reinstituted by returning the egg to normal sea water, or better still by exposing the egg to acid sea water. Under such circumstances the dividing cell invariably responds by shortening its polar axis. Since the hyaloplasm is bounded by a solid extensible membrane which will only change in form if we exert on it a definite force, and since it is undoubtedly the agent whereby the fully divided cells adhere to each other in their naturally compressed form, we are driven to conclude from the above evidence that the rôle of the hyaloplasm during cleavage is the same as after cleavage; in both cases it tends to reduce the polar axis of the cell when the latter tends to elongate under the influence of other forces. We cannot regard the hyaloplasm as the active cause of polar elongation or of cell division: it affects the form of the cleaving cell, but is not part of the active cleavage mechanism.

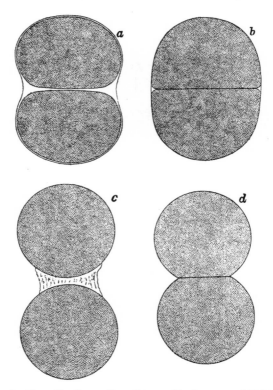

Fig. 84. Effect of acid on the form of contiguous blastomeres of *Echinus*. *a*, Normal egg; *b*, same egg in acid sea water; *c*, egg with hyaloplasm partially removed by calcium-free sea water; *d*, same egg in calcium-free sea water + acid. Note that the compression of blastomeres is restricted to the area over which hyaloplasm is still present.

When a dividing cell is denuded of hyaloplasm (see fig. 82), its form suggests quite clearly that division into two daughter cells is being effected by the resolution of the cytoplasm into two spherical regions, each of which when surrounded by its normal hyaloplasm responds as though it were an elastic shell. The suggestion at once

CELL DIVISION

arises—can each of these spherical regions be correlated with the regions occupied by each of the two mitotic asters?

The rôle of the asters during cleavage

The most direct proof that the asters form an active part of the cleavage mechanism is provided by the fact that any irregularity in the size or position of these structures is invariably accompanied by an irregularity in the form and position of the cleavage furrow.

In order that cleavage should occur in echinoderm eggs it is essential that there should be two asters each of which must be located within a short distance of the cytoplasmic periphery.

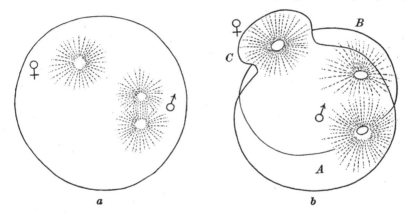

Fig. 85. Fertilised egg of *Toxopneustes* after exposure to ether. *a*, Before cleavage; *b*, after cleavage. Note that the male amphiaster has divided the cell into two blastomeres *A* and *B*, whilst the female monaster has deformed the surface of the cell at *C*. (From Wilson.)

A single aster near the periphery will deform the surface of the cell but it will not produce a cleavage furrow; this can be observed in the case of the large monaster typical of the period prior to fusion of the two pronuclei (Gray, 1924): it is also illustrated by the observations of Wilson (1901) (fig. 85). Whenever there are two asters present which are equal in size and whose rays extend to the periphery of the cell, a cleavage furrow will form between them; if there are three asters, there will be three cleavage furrows; if there are four asters, there will be four cleavage furrows. Similarly if two asters (of adequate size) are present—but one is larger than its mate —then unequal cleavage results. If the line joining the centre of the two asters does not pass through a diameter of the egg by the time the astral rays reach one side of the egg, then the cleavage furrow develops first on that side, and only later (as the rays reach the opposite side of the equator) does the furrow develop on the other side. All these phenomena can be observed in the natural cleavages of various types of cell. That the irregularity of form and position of the asters is the cause and not the result of irregular cleavage is suggested by the fact that similarly irregular cleavages can be induced to occur in *Echinus* eggs by experimental means (Gray, 1924).

E. B. Wilson (1901 *b*) showed that the normal astral radiations disappear if the eggs of *Sphaerechinus* are exposed to sea water containing ether (see p. 159). On replacing the eggs in normal sea water the radiations reappear: they do not, however, reach their normal size before again fading away. The result is that the egg may subsequently form a cleavage furrow which fails to cleave the egg. As soon as the asters fade away all development of a cleavage furrow ceases. This experiment has been repeated, and Wilson's results have been confirmed.

If eggs of *E. esculentus* are allowed to develop in normal sea water until the anaphase of the first division and are then transferred to 3 per cent. solution of ether in sea water, the astral rays very rapidly (2–3 minutes) disappear, and a clear irregular space appears in the centre of each of the original asters (fig. 58 *b*). If these eggs are now returned to normal sea water, the astral rays reappear within about 15 minutes. In some of these eggs the rays rapidly extend to the periphery of the cell, and the latter cleaves normally into two cells. In other eggs, however, the reformed rays do not reach the periphery of the cytoplasm before fading away. In such eggs no cleavage occurs until the second nuclear division.

If eggs are allowed to develop in normal sea water until the telophase stage is reached, and are then etherised and returned to normal sea water again, it is found that whereas two large asters existed at the beginning of the treatment, yet when the asters reform in sea water, they appear not as two large asters, but as four asters much smaller in size. The two original daughter nuclei divide in conjunction with the four new asters and the cell divides into four normal blastomeres. In other words, the first cleavage has been entirely omitted (see also fig. 94).

Finally, if eggs are etherised in 2·5 per cent. ether solution, and are then transferred to sea water containing a very small concentra-

CELL DIVISION

tion of ether, e.g. 0·05 per cent., the asters which reform in the sea water always remain small. Nuclear division occurs normally but, owing to the small size of the spindle and of the asters, the nuclei remain close together and no cleavage occurs. The cell thus becomes a well-marked syncytium (fig. 86 a). Eventually, however, the small asters become sufficiently numerous to extend throughout the whole egg, and at this moment multiple cleavage occurs. In some eggs there appears to be a tendency for the nuclei with their asters to collect at the surface of the egg (see fig. 86 b), and when the astral rays of these nuclei extend to the egg surface there is a distinct tendency for segmentation furrows to appear between them, although these furrows are never complete.

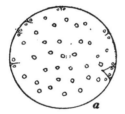 

Fig. 86. *Echinus* eggs segmenting in the presence of 0·05 per cent. ether. a, Note numerous nuclei forming a syncytium; b, note peripheral arrangement of nuclei and incomplete cleavage planes.

There can be very little doubt, therefore, that the nature and extent of the cleavage furrow are very closely associated with the position and the size of the asters. If, then, the appearance of a cleavage furrow is the direct mechanical effect of asters whose rays extend to within a critical distance of the periphery of the cytoplasm, and these asters are made to disappear before the cleavage furrow is actually completed, further cleavage should cease and the egg should tend to resume its spherical form. This is the case. If eggs are allowed to develop in normal sea water until the cleavage furrow is just beginning and are then etherised in 2·5 per cent. ether, the condition shown in fig. 87 can be obtained. In these eggs a well-defined cleavage furrow exists but there are no asters. On transferring such eggs to sea water the cleavage plane is gradually lost. If the original cleavage furrow was shallow, then on transference to normal sea water after etherisation, the egg gradually becomes completely spherical, at the same time the surface of the hyaloplasm is thrown into distinct folds. If, however, the original furrow was

204 CELL DIVISION

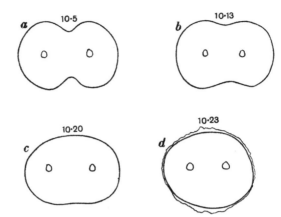

Fig. 87. Egg of *Echinus* with normal cleavage furrow placed in 2·5 per cent. ether solution for 20 minutes, then transferred to normal sea water. Note absence of asters and gradual loss of cleavage furrow; also the crinkled hyaline membrane in *d*.

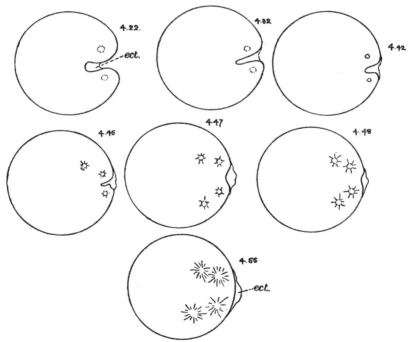

Fig. 88. Egg of *Echinus* transferred from 2 per cent. ether to normal sea water. Note asymmetrical astrospheres without astral rays, also cleavage furrow, gradual obliteration of cleavage furrow, and displacement of hyaloplasm. Note subsequent division of astrospheres and development of astral rays.

CELL DIVISION

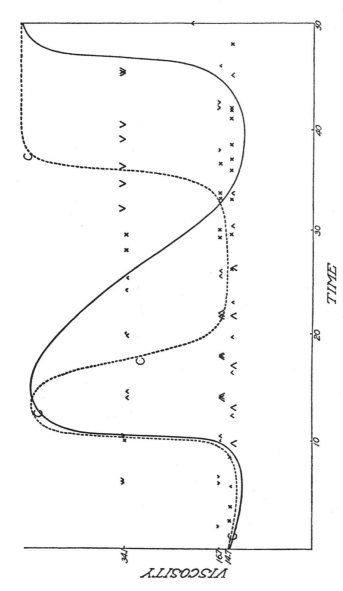

Fig. 89. Viscosity changes during mitosis in the sea urchin egg. The unbroken line is plotted from the centrifuge data of Heilbrunn, and the dotted line is a rough estimate from the microdissection experiments of Chambers. (From Heilbrunn.) The inverted V's represent observations of high viscosity; the V's represent low viscosity. The first rise in viscosity may be associated with the development of the sperm aster and the second with the formation of the amphiaster (see however Heilbrunn, 1921).

well developed, then on return to sea water the egg tends to retain an elongated form, although the furrow itself disappears. At the same time the wrinkling of the hyaloplasm is extremely obvious in the equatorial region. The elongated form of these eggs is, however, rapidly lost as soon as the astral rays of the next nuclear division begin to approach the periphery of the cytoplasm.

Besides ether, many other reagents tend to inhibit the normal development of the asters, and yet allow the nucleus to divide normally. Among such reagents is slightly hyperalkaline sea water. The asters are small and asymmetrically situated and produce a cleavage furrow on one side of the egg only. The same thing occurs in hypertonic sea water. A deficiency of calcium or potassium has the same effect.

The form of the cleavage furrows in many of these reagents is extremely irregular. It is difficult to see how such furrows could be the result of a differential interfacial tension at the poles and at the equator of the cell; they are, however, explicable on the assumption that the furrow is being brought about by a redistribution of the different phases of the egg.

The close relationship which exists between the position and form of the asters with the position and form of the cleavage furrow is readily understood if we assume that the asters are an essential part of the active cleavage mechanism, and this view provides a reasonable working hypothesis for further analysis.

Before discussing the nature of the force which may be exerted on the cytoplasm by means of the asters, it is of value to recollect that the evidence from microdissection (Chambers, 1917) and from the use of the centrifuge (Heilbrunn, 1921), indicates fairly clearly that the region of the cytoplasm occupied by the astral rays is of a more rigid or viscous nature than the non-radiate regions. The initial increase in the viscosity of the egg which takes place soon after fertilisation is directly associated with the existence of the fully formed sperm aster which pervades the whole egg. As soon as this aster fades away the cytoplasm again resumes the fluid state. Similarly, Chambers (1917) has shown that the asters during cleavage are areas of considerable rigidity when compared with the peripheral regions of the cell.

Theory of astral cleavage

If we are prepared to regard the asters as elastic spheres possessing a definite degree of elastic rigidity (see p. 158) and if we are prepared to admit that they grow in size at the expense of the fluid cytoplasm, it seems possible to formulate a tentative theory of astral cleavage. Consider two solid and perfectly rigid spheres separated from each other and each surrounded by a film of liquid material (fig. 90 *A*). If these two spheres are allowed to come into contact with each other they will adhere together by their liquid films, and if the latter are sufficiently voluminous there will be an accumulation of fluid at the region of contact, since in this way the total free surface of liquid is reduced to a minimum. Similarly if we start with two clean contiguous spheres and add to their surface a liquid, the latter will distribute itself as in fig. 90 *B*, and as the amount of liquid is increased, so the external surface of the liquid will more and more approximate to that of a sphere. The two contiguous spheres are, of course, held together by the surface tension of the fluid phase and can only be separated by applying a force sufficient to overcome this tension and

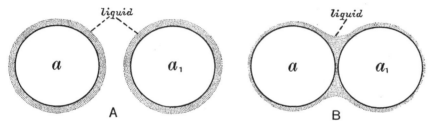

Fig. 90. *A*, Two solid spheres with a surface layer of liquid; *B*, the same spheres in contact. Note the aggregation of liquid at the equator of the system.

the viscous resistance set up by the fluid when in a state of flow. If instead of using two rigid spheres in the above experiment, we use two elastic spheres whose degree of elasticity is such that the tension exerted by the common liquid surface is sufficient to distort the spheres in an obvious way, then the region of contact between the spheres will be marked by a flat interface and each sphere will be compressed along its polar axis. An extreme case of such a system is provided by soap bubbles where the liquid phase is extremely thin, so that the equatorial accumulation of fluid is very small. Another example is provided by drops of water immersed in a drop of oil of the same specific gravity (Gray, 1924). If a fairly large drop of olive oil be immersed in a mixture of alcohol and water of the same specific gravity, it is possible to inject into the oil two drops of the external medium. If these drops are gradually enlarged, the external surface of the oil remains spherical until the diameter of each of the enclosed and equal sized water drops is nearly half the diameter of the oil. If, now, the volume of the water drops be increased, or if the volume of the oil be decreased, a marked change in the form of the system takes place. The whole system elongates along its polar axis, the enclosed water drops flatten equatorially and are separated by a film of oil; simultaneously the oil flows away from the poles and collects at the equator as shown in fig. 91.

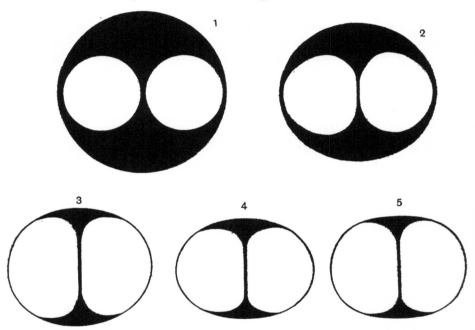

Fig. 91. Two water drops enclosed in a drop of olive oil. Note changes in the distribution of the oil and in the form of the water drops which occur when the relative volumes of water and oil are altered. The oil is black, the water white.

Starting from the assumption that the asters represent two spherical elastic spheres which increase in volume at the expense of the peripheral fluid cytoplasm, we can apply the above principles to the cleavage of an egg. One would expect the egg to maintain its spherical outline until the combined diameters of the two asters was equal to that of the spherical cytoplasm. At this point three things must happen: (i) the asters will be pressed against each other in the equatorial plane, (ii) the polar axis of the egg will increase in length as soon as the elastic force exerted by the asters

CELL DIVISION 209

is sufficient to overcome the tendency of the cytoplasmic and hyaline surfaces to resist a change in form, (iii) as the polar axis increases with the growth of the asters, so the fluid material round the asters (i.e. the peripheral cytoplasm and the fluid between the hyaline membrane and the cytoplasm) will begin to flow to the equator of the egg and this flow will continue until (a) the asters cease to grow, (b) the force required to keep the fluid material in motion is equal to that of the elastic forces exerted by the compressed asters. It will be recalled that a peripheral flow of cytoplasm from the poles of the cell to the equator has actually been observed during normal cleavage (Erlanger, 1897).

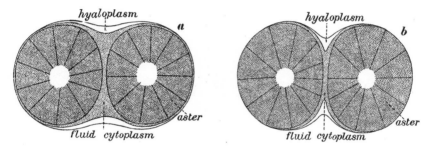

Fig. 92. Diagram to illustrate the conversion of fluid cytoplasm into two elastic spheres (marked by astral rays). During this process the fluid phases will distribute themselves largely in the equatorial furrow.

The theory of astral division here outlined stands or falls by the possibility of regarding the asters as elastic spheres capable of generating sufficient elastic energy to overcome the resistance offered by the peripheral cytoplasm and by the hyaline layer. An adequate proof of this fundamental point is not easy to obtain: it is, however, a reasonable interpretation of the observations of Chambers (1917) and of Heilbrunn (1921). It is also supported by the fact that when the astral rays fade away after cleavage the two blastomeres are more readily compressible than is the case when the asters are still *in situ* (see also fig. 59).

There is one series of facts which are at once a support and a criticism of these conceptions. Most non-spherical cells divide at right angles to their longest axis. If the two asters are to be looked upon as two spherical masses free to move within the viscous cytoplasm, they will naturally orientate themselves along the long axis of the cell, as is the case in Nematode eggs (see fig. 93). On the other

210   CELL DIVISION

hand, such an orientation is absent during the incomplete cleavages of crustacean eggs and in the unequal cleavages characteristic of ovarian maturation divisions. In such cases, the asters do not appear to be free to move within the cytoplasm but are apparently anchored in an eccentric position. It is just conceivable that in these cases the viscous resistance of the cytoplasm is greater than the elastic or plastic strength of the cytoplasmic surface, so that as the asters grow they distort the surface of the cell instead of moving bodily through the cytoplasm and long before their combined diameters are equal to that of the whole egg.

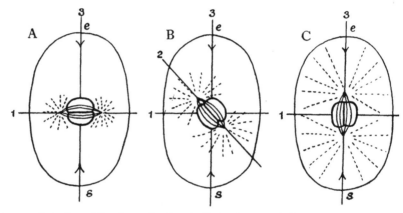

Fig. 93. Rotation of the first mitotic spindle in the egg of *Ascaris*; the arrows *e* and *s* show the path of approach of the male and female pronuclei. Note final position of spindle in long axis of the egg. (From Korschelt and Heider.)

To some extent, the utility of any biological hypothesis depends on its width of application, and an obvious objection to be urged against any theory of cleavage which ascribes a specific rôle to the mitotic asters is the fact that some cells exhibit no asters during cleavage. As far as is known, however, when astral rays are absent from a dividing animal cell, the orderly formation of a geometrical cleavage furrow is never clearly marked, although it is said to occur in certain plant cells (Farr, 1918). It will be noted, however, that the essential character of the aster is not the presence of astral rays, but the possession of a definite degree of elastic rigidity. It is quite possible that such a physical state might exist locally in the cell without a visible radiate structure. The presence of two such regions would account for the ellipsoidal form of dividing leucocytes. In other words, the cleavage mechanisms of an egg and of a fibroblast may be essentially the same, except that, in the latter case, there is no visible sign of the elastic regions at the poles of the cell.

Burrows (1927) has recently ascribed the cleavage of fibroblasts to the 'centrospheres' (= asters) and although this conception harmonises with the hypothesis sketched above, it must be admitted that no very obvious proof exists to show that centrospheres are actually present in such cells as rigid structures.

When all has been said in its favour, any theory of cell division is at present open to the fundamental objection that we are unable, by direct experiment, to measure the magnitude, nature and distribution of the forces generated within the cell. All we can do is to rely on visual phenomena and interpret them as best we may. Unfortunately, visual phenomena are seldom identical in any two types of cell. An observer is likely to attribute greater importance to a particular phenomenon if this is more strikingly obvious than another in the particular material he is observing: in another type of cell the clarity of the two phenomena may be reversed and a different perspective is obtained. For this reason, it is not hard to reach a position more adapted to dialectic skill than scientific enquiry.

Alternative theories of astral cleavage

Just's (1922) theory has already been mentioned. This author regards the hyaline layer as the effective mechanism of cleavage and claims that an equatorial accumulation of hyaloplasm occurs *before* the polar axis begins to elongate and before any equatorial furrowing begins. If this be true, then the theory outlined above must be revised. A very careful observation of the normal eggs of *Echinus esculentus* leads to the belief that Just's conclusions are not applicable to this material; only in abnormal eggs, which fail to divide, is there any local aggregation of hyaloplasm whilst the egg is in the spherical condition. There can be no doubt whatever, that in *Arbacia* or *Echinarachnius* eggs the hyaline layer is much thinner than in *E. esculentus*, and for this reason close observation is more difficult. Just's point is, however, of fundamental importance and should be determined without delay: there is no reason why such a difference of opinion should exist, since the truth could be established by cinematography with reasonable ease. At the same time, the behaviour of eggs denuded of the hyaline layer seems to preclude Just's main conception of cleavage.

In marked contrast to Just's theory is the hypothesis put forward by Chambers (1919). This author, far from regarding the hyaline layer as the active cause of division, denies that it plays any part in determining the form of the dividing egg. Chambers claims that the hyaline layer can be removed (by microdissection) from the egg

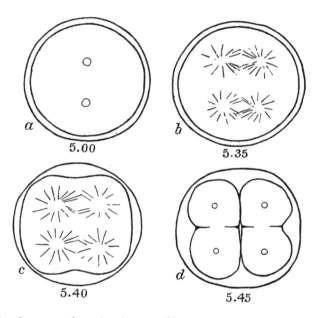

Fig. 94. Development of an *Asterias* egg after manipulation with a needle so as to suppress the first cleavage furrow. The egg ultimately yields a normal larva. (From Chambers.)

prior to division, and that, after such treatment, normal cleavage ensues. This view overlooks the fact that the hyaline layer is very rapidly regenerated in normal sea water, and that it unquestionably controls the form of the cell during the intercleavage periods. On the other hand, Chambers was the first author to suggest that the asters are to be regarded as rigid bodies which are the active agents of cleavage. 'The segmentation process may be explained as consisting essentially in a growth within the egg of two bodies of material through a gradual transformation of the cytoplasm. This transformation is associated with a change in the physical state of the

CELL DIVISION

protoplasm, two semisolid masses growing at the expense of the more fluid portions of the cytoplasm' (Chambers, 1919, p. 52).

By means of specially prepared needles Chambers was able to operate on the dividing eggs of *Arbacia* and *Asterias*. In one experiment the amphiaster was destroyed by mechanical agitation, and consequently no cleavage furrow formed until new asters had developed preparatory to the second cleavage into four cells. That practically the whole cytoplasm

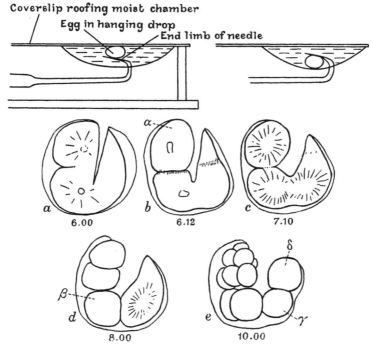

Fig. 95. By means of the needles shown at the top of the figure, an egg of *Asterias* was cut as shown in *a* when the asters were well developed. Note that the form of the cut surface is retained for some minutes. (From Chambers.)

of the egg possesses considerable rigidity when the asters are fully developed is shown by fig. **95**. If a cut is made into an egg at this stage, the form of the egg is retained, and the edges of the cut are well defined: this is not the case if a cut is made before the asters are well developed. The results of these experiments lead to the conclusion that the changes in shape during cleavage are due to the formation of the two rigid asters, and it is this process which leads to the elongation of the egg axis prior to cleavage.

The three hypotheses now considered all attribute cleavage to the mechanical pressure exerted either by the hyaline layer, or by the growing asters, or both. In this they differ from the type of hypothesis put forward some years ago by Robertson (1909–13), McClendon (1912, 1913), and more recently by Spek (1918). These authors attribute cell cleavage to localised alterations in the 'surface tension' of the cytoplasm. It has already been mentioned that during the development of a cleavage furrow there is a peripheral flow of liquid cytoplasm with its granules from the poles of the cell to the equatorial furrow. Whereas earlier observers were inclined to regard these moving currents as the active cause of cleavage, Spek suggests that they are caused by localised changes in surface tension on the cell surface and that it is these changes which are the direct cause of both cleavage and current formation. As Chambers (1924) points out, the currents seen in the slowly cleaving echinoderm eggs are much less obvious than those seen in the more rapidly dividing nematode eggs observed by Spek.

The theories of Robertson, McClendon, and Spek are all based upon analogies drawn from the behaviour of oil drops floating or immersed in water. McClendon infers that cell cleavage is due to a reduction in the interfacial tension at the poles of the cell; Robertson on the other hand infers that the reduction takes place at the equator. For the details of these arguments reference must be made to the original papers, but it is necessary to draw attention to the fact that the models put forward will only work successfully under certain conditions. In practice, an oil drop can only be made to divide by alterations of interfacial tension when (a) the drop is of considerable size, (b) when the rate at which the differential surface tension develops is very rapid, and in order that this may be the case very powerful reagents must be used. If the drop of oil be very small, any difference set up in the interfacial tension at one point is rapidly transmitted over the whole surface, and only a momentary disturbance in the form of the drop is observed. In the case of a larger drop, cleavage only occurs when the alkali employed for the local change in interfacial tension is sufficiently strong to act with great rapidity; otherwise the whole surface comes into equilibrium before cleavage can occur. The living cell is extremely small in comparison to the oil drops used for such experiments, and the application of such reagents as were used by McClendon or by Robertson would immediately cause the death of the cell. Again, the process of normal cleavage is relatively slow; it may take at least half an hour.

The velocity at which a cleavage furrow cuts through the equator of a cell does not appear to have been recorded, although it is of some interest. A few observations on the eggs of *Echinus* indicate that the velocity at which a furrow cuts through the cell is very slow; if the cell is 100μ in diameter, cleavage may occupy about 30 minutes at $11 \cdot 0°$ C.—this is equivalent to a velocity of about 1 cm. in two days. At lower temperatures the process is very much slower. It is interesting to note that the velocity of cleavage is more or less independent of the size of the cell, so that large cells take longer to divide than do smaller cells—a fact which is in marked contrast to the mitotic division of the nucleus (see p. 146).

There is no reason to suppose that the energy required to cleave an echinoderm egg is of a different order to that required to cleave a *Paramecium*, and this (according to Mast and Root, 1916) is of the order of at least 383 dynes per square centimetre. When we compare this with the differences in interfacial tension which can be set up at an oil/water surface, it is difficult to accept the view that there is any real comparison to be drawn between the cleavage of a single-phase oil drop and that of a living cell.

Both Robertson and McClendon make one very important assumption. They assume that the surface of the living cell is of a liquid nature. Further, both authors leave their analogy at the point where the cleavage is just complete. One of the most striking features of the fully cleaved cell is that the two resulting blastomeres show no tendency to fuse with each other. Newly cleaved oil drops, however, fuse together readily as soon as they are again in contact; they can only be prevented from fusing if a third phase be present which forms a protecting layer on the surface of the oil sufficiently strong to oppose the operation of surface forces (see p. 250). There is no evidence that either McClendon's or Robertson's experiments would succeed under such conditions.

One significant fact is often overlooked. If, after a cleavage furrow has begun to form, it be brought to a standstill by mechanical destruction of the aster—by cold, or by chemical means—the furrow itself is relatively stable, and is only slowly obliterated (see figs. 87, 88). All incomplete furrows would be highly unstable if the cell surface were liquid or were under mechanical tension. Since the egg surface is solid, such forms are readily explicable. The hyaloplasmic membrane is extensible but is not perfectly elastic, so that when once it has been elongated by the growth of the asters, it tends to prevent the cell regaining its original spherical form when the asters are removed from the partially cleaved cell.

The energetics of cleavage

Although the elastic properties of the cell surface can be demonstrated with ease, it is by no means a simple matter to obtain a quantitative estimate of the force which must be applied to effect a change in surface area equal to that produced by normal cleavage.

An attempt to obtain such data has been made by Vlès (1926), whose estimate of the surface energy of a spherical cell is based on the fact that when in equilibrium with a distorting force the degree of departure from the spherical form is a function of the surface energy. When a spherical egg cell is resting on a horizontal plane and is in equilibrium with gravity, the surface energy (a) of the cell surface can be expressed by the empirical formula

$$a = 10\sqrt{\frac{b}{a-b}(d_0 - d_e)},$$

where a is the horizontal diameter of the egg and b is the vertical diameter; d_0 is the density of the egg and d_e the density of the medium. At pH 8·0—which is the appropriate value for sea water—the surface forces have a value of approximately 20 dynes per square centimetre, or roughly that characteristic of an olive oil/water interface. Prior to and during the process of cleavage the eccentricity of the egg in equilibrium with gravity fluctuates considerably and immediately prior to the normal elongation of the mitotic axis there is a marked increase in the power of the egg to resist deformation (see fig. 96).

It will be noted that the distorting force of gravity is opposed by more than the so-called surface tension of the egg surface. This surface is composed of an elastic solid, so that Vlès' observations are probably a measure of the elasticity of the surface rather than of the intensity of forces strictly comparable to surface tension. It is also rather doubtful how far it is legitimate to assume that the only force opposed to gravity is located at the egg surface for this would only be true if the interior of the cell is entirely fluid; if it has, at any time, a finite degree of rigidity, the resistance offered to gravity would be temporarily increased. Since the mitotic asters possess rigidity, the marked increase in the value of a immediately prior to cleavage may be due to changes inside the cell rather than at its surface. It would be interesting to consider how far Vlès' data would enable us to predict the amount of energy required to increase the surface area

CELL DIVISION

of an egg by 25 per cent., which is approximately that produced by equal cleavage.

Apart from this isolated observation all our data concerning the energy disturbances during cell division are based on indirect methods of attack. In 1904 Lyon attempted to measure the intensity of carbon dioxide production during the cleavage of the eggs of *Arbacia*. His results, admittedly based on a somewhat indifferent technique, indicated that the actual process of cell division was accompanied by an increased production of CO_2. In apparent contrast

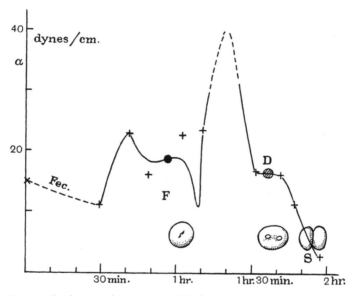

Fig. 96. Surface tension, expressed in dynes per sq. cm., of an egg of *Paracentrotus* prior to and during cleavage. (From Vlès, 1926.)

to this observation, Lyon found that this period of increased carbon dioxide production did not coincide with that period of the division cycle during which the egg is most susceptible to lack of oxygen or to the presence of potassium cyanide. 'When oxygen is most necessary and presumably is being used in largest amount, CO_2 is produced in least amount.' Lyon made it abundantly clear that he did not regard his experiments as of sufficient accuracy to warrant far-reaching conclusions, but he inferred tentatively that the energy for cell division is derived from a non-oxidative reaction. Some years later Warburg (1908) investigated the rate of oxygen consumption of echinoderm eggs at varying stages of segmentation, and showed quite conclusively that as development proceeds so the demand for oxygen increases. There is some tendency to accept these facts as an indication that the process of cleavage is intimately associated with increased respiration and, at first sight, this point of view is supported by the more recent observations of Vlès (1922). This author observed the changes in the hydrogen-ion concentration

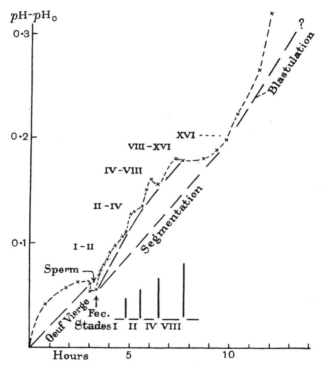

Fig. 97. Correlation between evolution of CO_2 and cell division (after Vlès). Each cleavage appears to initiate an outburst of CO_2.

of the sea water in immediate contact with the dividing eggs of *Paracentrotus*, and reported marked cyclical changes in respect to each cleavage cycle. Vlès' data are embodied in fig. 97. The respiratory changes observed by Vlès are obviously different to those described by Lyon, since in the former case the period of most intense CO_2 production follows cleavage, whereas in the latter case it marked the actual period of cleavage itself.

CELL DIVISION 219

In view of the theoretical significance of a change in respiratory activity during cell cleavage, it may perhaps be permissible to point out one or two peculiarities in Vlès' curve. It would appear that just prior to the third cleavage there is an actual absorption of CO_2, unless this point is due to experimental error. If this be so, however, many of the points employed to show that there is a periodic formation of CO_2 also lie within the experimental error. Again, after the third cleavage, the eggs do not appear to have given off any CO_2 for more than one hour. Before attaching implicit faith to these facts one would like to know more precisely the conditions under which the experiment was carried out. In experiments dealing with small variations in the respiration of cells it is essential that the conditions should be accurately defined. For this reason, the results of Vlès cannot strictly be compared to those described below. In Vlès' experiments the CO_2 was allowed to accumulate and the eggs were not agitated.

In order that an accurate estimation of the rate of carbon dioxide production or of oxygen absorption may be made at different stages during the whole mitotic cycle, it is necessary that the conditions of the experiment should be such as will allow the eggs to develop normally and at the same rate, so that practically all the cells cleave at the same time. Again, in order that several determinations can be made during the comparatively short time occupied by the cleavage process, it is advisable to prolong this period by carrying out the whole experiment at a fairly low temperature. An attempt to fulfil these conditions was made by Gray (1925). The eggs of *Echinus esculentus* were used, their oxygen consumption being measured by means of a differential manometer. In these eggs the first cleavage furrow cuts through the egg in about 30 minutes (at $11 \cdot 0°$ C.) after its first appearance; the time required for subsequent cleavages is shorter. The results of such experiments (see figs. 98, 99) indicate that there is no measurable change in the rate of oxygen consumption associated with the act of cell division. At the same time it is important to remember that a cell when deprived of oxygen will not divide. This was first shown by Loeb (1895).

Mathews (1907) found that agents such as cold, quinine, and anaesthetics which are known to reduce oxidations also prevent cell division, and cause a disappearance of the astral rays. Mathews concluded that the whole process of cell division is intimately associated with the oxidative processes in the egg, and that the periodicity of the former is due to a periodicity in the capacity of the cell to carry out oxidations involving the use of atmospheric oxygen, this periodicity in oxidative power being due to the periodic liberation from the nucleus of an oxidase whenever

220 CELL DIVISION

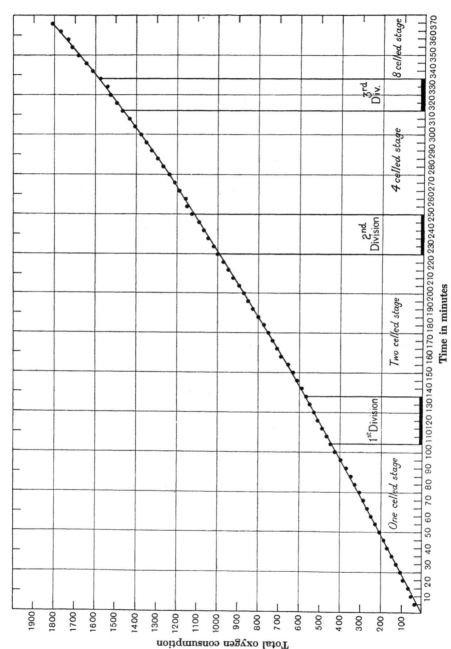

Fig. 98. Graph showing the oxygen consumption of fertilised eggs of *Echinus* prior to, during, and after cleavage. This curve is comparable to that of Vlès (see fig. 97). Note however the absence of any change in respiratory activity which can be associated with the act of cleavage. (Gray.)

CELL DIVISION

the nuclear membrane breaks down. In opposition to this conclusion is the fact established by Warburg that the level of oxygen consumption of the eggs can be maintained almost unchanged when the periodic changes of the nucleus are entirely inhibited.

Table XXVIII. Mm. pressure O_2 used per half hour by fertilised eggs of *Echinus*

No. of experiment	30 minutes immediately previous to division	30 minutes during division	30 minutes immediately after division	No. of division
A	13·0	12·3	14·0	1st
A	14·8	15·0	13·7	2nd
A	16·4	15·9	18·3	3rd
B	14·1	13·6	13·0	1st
B	13·3	13·4	13·7	2nd
C	6·5	6·1	7·1	1st
D	21·0	19·0	21·0	1st
Totals	99·1	95·3	100·8	

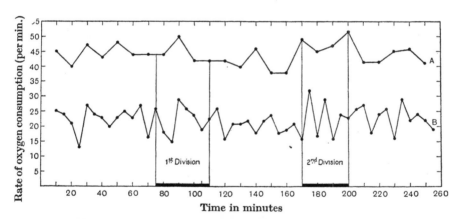

Fig. 99. Graph showing the rate of oxygen absorption during the process of cleavage. Note the absence of any measurable change in oxygen consumption prior to or after cleavage. The upper graph shows the rate of oxygen consumption during successive ten minute intervals; the lower graph shows the rate during successive five minute intervals. (Gray.)

It seems necessary to distinguish between a general disturbance of the cell's activity which accompanies a lack of oxygen and those specific reactions within the cell which require atmospheric oxygen.

From the evidence available we may draw one of two conclusions. Either the process of cleavage does not involve redistributions of energy which involve oxidative reactions; or, the oxidative changes which provide the energy for cleavage (being less than 2 per cent. of the total oxidations of the whole egg) cannot be detected by the methods so far employed. The same unfortunate position is reached from a consideration of the heat production during cleavage. Meyerhof's (1911) curve was not based on experiments having the particular objective now involved, but its close resemblance to the curve of oxygen consumption suggests that the two processes are closely associated with each other. There is, in other words, no positive evidence to indicate a disturbance in the rate of heat production during cleavage. The more recent work of Rogers and Cole (1925) is less easily interpreted, but it does not seem to elucidate the nature of the cleavage process.

It is clear that a spherical egg with its plastic surface layer cannot be divided into two parts without the expenditure of energy; during this process there must be a thermal disturbance. It seems reasonable to suppose that this disturbance could be measured if the technique employed were sufficiently accurate. From the data at present available, we can only conclude that the energy which the cell expends during cleavage forms only a very small fraction of the total energy which the cell requires to maintain its normal life. From the point of view of energetics, cleavage, like growth, is a comparatively insignificant process in the life of the cell.

The effect of cell division on the form of epithelial cells

Just as the form of the first cleavage furrow of a sea urchin's egg is in part determined by the mechanism which afterwards binds one cell to another, so to a much greater extent would one expect the same factor to operate when a dividing tissue cell is completely surrounded by its neighbours. The changes in form of an epithelial cell undergoing division have not been observed in life, and would form an interesting object of study. In some cases it would appear as though a prismatic epithelial cell becomes spherical prior to division (Seeliger, 1893; Korschelt, 1888) and that its cleavage is not unlike that of an unsegmented spherical egg. Whilst this may sometimes be the case, it can hardly be true of all epithelial divisions. By subjecting the segmenting eggs of *Echinus esculentus* to gentle pressure, it is possible to observe the cleavage of cells which are in

CELL DIVISION

intimate contact with their neighbours, and in no case has a cell been seen to become spherical before cleavage. The first sign of approaching cleavage is an elongation of the polar axis which is accompanied by a bending of the equatorial interfaces towards the centre of the cleaving cell (fig. 100, 2). This distortion of the interfaces involves marked changes in the form of the adjacent cells which accommodate themselves accordingly. The actual cleavage furrow cuts through the dividing cell in a plane at right angles to the long

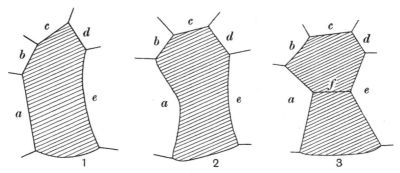

Fig. 100. Camera lucida drawings of cleaving cell in *E. esculentus* blastula. Note that the neighbouring cells a and e have accommodated their form to that of the dividing cell. The furrow f cuts vertically through the cell.

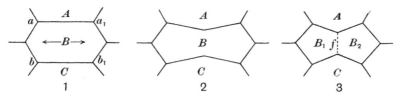

Fig. 101. Diagram of division of a hexagonal epithelium cell. The equatorial interfaces aa_1 and bb_1 are distorted so that—after the vertical furrow f has cut through the dividing cell B—each daughter cell is a pentagon and the two adjacent cells A and C have each acquired a new interface, making a gross increase of six interfaces.

axis and at right angles to the plane of the epithelium, so that when cleavage is complete the two daughter cells have acquired between them four new interfaces and the adjacent cells have each acquired one extra interface (see figs. 100, 101).

The cells in the wall of a blastula, like those in a typical epithelium, have a variable number of interfaces, although the average number is no doubt six (see p. 257). Taking a six-sided cell as a typical unit, Lewis (1926, 1928) has considered in some detail the

224 CELL DIVISION

theoretical effect of cell division. If a regular hexagonal cell (fig. 102) divides with the mitotic spindle orientated along one of the three major axes, the cleavage plane will lie along one of the axes a, b, or c. Cleavage will result in the production of two very irregular pentagons. If the cell interfaces are plastic and tend to reduce the transverse section of each cell to a regular pentagon, the net result of division is the production of two pentagons and two heptagons (see fig. 102). Each product of division (a and a_1) is a pentagon, whereas two neighbouring cells (b and b_1) acquire an additional surface. Similarly, if in a sheet of hexagonal cells any one cell and its six neighbours divide simultaneously, there will result eight new hexagonal cells, and six new pentagonal cells, while two peripheral non-dividing cells will become heptagonal, and two octagonal; the average number of sides possessed

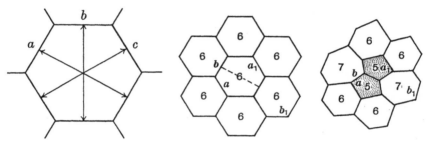

Fig. 102. Diagrammatic division of a hexagonal cell undergoing division and giving rise to two pentagonal and two heptagonal cells. (Lewis.)

by all the new or reorganised cells being exactly six. The existence of pentagonal, heptagonal, or octagonal cells in a normal epithelium of hexagonal cells may therefore be the direct result of cell division. Lewis accepts this point of view and has put forward possible hypotheses which account for the existence of cells with less than five or more than eight surfaces in transverse section. It may be mentioned, however, that an alternative interpretation of the various types of polygon was put forward by Wetzel (1926), who stresses the disturbing effect of unequal growth on the symmetry of a hexagonal system (see Chapter X).

There can be little doubt that in the wall of a blastula the irregularity in the number of facets possessed by different cells is largely due to initial irregularities of size, rather than to the effect of cell division or to variation in the growing powers of the cell. The ideal system considered by Lewis is none the less of interest and of considerable importance when applied to plant tissues. In animal cells the junction between two interfaces is not fixed, but can move in response to altered stresses throughout the system, and consequently the length of any given interface (as seen in section) may vary considerably when a neighbouring cell divides. Whether this is true in plant cells is doubtful, and if Lewis is correct in assuming that it is not the case in *Cucumis*, interesting problems at once arise. By actual measurement Lewis found that in section every cell interface is approximately of the same length irrespective of the number of sides possessed by the cell, and this implies that the act of cell division involves a marked change in the area covered, not only by the dividing cell, but also by its neighbours. According to Lewis (1928) the area covered by a cell is roughly $(n-2)\,a$, where a is a constant and n the number of sides in transverse section, so that the area of the original hexagonal cell A is $4a$, and that of each daughter pentagon is $3a$. The two daughter cells therefore together cover an area 50 per cent. larger than the cell from which they were derived. Similarly each of the cells b and b_1, see fig. 102, by acquiring a seventh side increase from $4a$ to $5a$—or together add an area equal to 50 per cent. of a hexagonal cell. Thus the total and immediate effect of cell division is materially to increase the area covered by the whole series of cells affected. Lewis associates the increase in surface area postulated at cell division with the growth of the cells involved, but it is difficult to understand how this could occur with such rapidity. Until the volume of the cells in the neighbourhood of dividing plant cells has been assessed, and until the process of division has been seen in life, it seems doubtful how far further speculation is advantageous.

If a hexagonal cell divides so as to give two daughter pentagons whose total area is equal to the original hexagonal cell, and if the two daughter pentagons are regular, then the length of one of the sides of the pentagon must be 0·869 times the length of the side of the original hexagonal cell. This could only occur if there were a corresponding reduction in the length of all the sides of the adjacent cells. Some such adjustment probably occurs in animal tissues, but in plants Lewis suggests that all the non-dividing cells can retain their normal length of side by growth on the part of the pentagons and of two of the neighbouring cells as above described. By similar reasoning, a non-dividing pentagon in contact with two dividing cells would increase its surface area by more than 60 per cent.

It is interesting to note that epithelial cells do not appear to divide when they attain a critical size, for the main factor associated with the cleavage of *Cucumis* cells appears to be the number of interfaces present. The higher the number of interfaces the greater is the percentage of cleavages observed. Since the volume of the cell appears to increase with the number of interfaces, it follows that the larger the cell the more likely it is to divide.

The factors which determine the direction of the planes of cleavage

From a biological standpoint the cleavage of a cell involves more fundamental changes than a quantitative division of the cell mass. Since the entire animal with its differentiated parts owes its ultimate shape to the size, position, and form of its constituent cells, the factors which determine the direction of each cell cleavage are of very real significance in an attempt to understand the mechanism which underlies the processes of embryology. Is an animal's form determined by those mechanical conditions which control the form, size and position of the constituent cells? Alternatively, do particular cells cleave in a particular manner and orientate themselves in a particular position because of an inherent property of the living material of which they are formed? A discussion of the problem in this form would lead far away from the scope of this work: all that concerns us here is the evidence which throws light on those forces which can influence, if not determine, the plane in which a given cell will divide. Roughly speaking, the evidence can be divided into two sections: an analysis of the relation of a cleavage plane to an organised or predetermined axis of the organism on the one hand, and to the external environment of the cell on the other.

One of the most obvious features of all cleavage planes is the fact that they are at right angles to the long axis of the existing mitotic spindle; it therefore follows that the direction of the resultant cleavage plane in respect to the whole cell is determined by those forces which are responsible for the orientation of the whole mitotic figure. If we consider an inter-kinetic nucleus lying towards the centre of a spherical cell, it is obvious that the long axis of the subsequent spindle is determined (so far as the nucleus is concerned) as soon as the centrosomes have orientated themselves at the poles of the nucleus. In the case of some eggs, the poles of the nucleus (as defined by the position of the centrosomes) bear a definite relationship to the cytoplasmic elements of the cell. This is clearly the case

CELL DIVISION

in the frog's egg where the first cleavage plane passes through the animal and vegetable poles of the cell. We can, however, go back a step further. In frogs' eggs Roux (1885) and afterwards Morgan and Boring (1903) showed that the first cleavage plane, in the majority of cases at least, marks out the median line of the future embryo. At the same time the first cleavage furrow passes through or near the point of entry of the spermatozoon into the egg, e.g. in frogs' eggs (Roux, 1887), in the sea urchin *Toxopneustes* (Wilson and Mathews, 1895) and in *Nereis* (Just, 1912).

The series of events which relate the entry of the spermatozoon with the cleavage furrows were clearly described by Wilson and

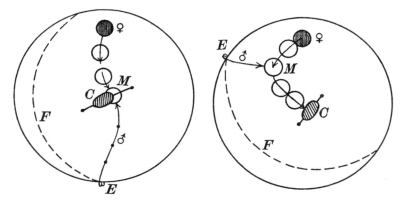

Fig. 103. Diagrams from successive camera lucida drawings of the living eggs of *Toxopneustes*. E, Point of entry of sperm. M, Position of fusion between male and female pronuclei. C, Axis of first mitotic spindle. F, First cleavage plane. (From Wilson and Mathews.)

Mathews (1895). The fertilising spermatozoon may enter at any point on the surface of the egg of *Toxopneustes* and its point of entry is marked by the so-called 'entrance cone' (see p. 424). After inclusion into the cell, the sperm nucleus is marked by a sperm aster and both structures move into the interior of the cell on a path closely approximating to, but not coinciding with, a radius of the egg (fig. 103). During the time occupied by the approach of the female pronucleus, the sperm nucleus may be slightly deflected from its original penetration path, but the angle of deflection is slight; after fusion, the zygote nucleus passes to the centre of the cell. Within certain limits, therefore, the sperm nucleus travels straight to the centre of the egg, so that the central point of the male astrosphere lies on or near a line joining the centre of the zygote nucleus to the point of entry of the spermatozoon. The asters of the first cleavage are apparently formed by a polar migration of 'archiplasm' from the centre of the male aster as described elsewhere (see p. 162), so that the axis of the first mitotic spindle is at right angles to the path of entry of the spermatozoon. Since the cleavage plane is at right angles to this mitotic axis, it follows that the cleavage furrow must pass through or near the point of entry of the spermatozoon. This close relationship between the point of entry of the spermatozoon and the direction of the first cleavage plane is found also in frogs' eggs (Roux, 1887) and in the polychaet *Nereis* (Just, 1912). Curiously enough, the rule is not absolute. In Just's experiments (in which the point of entry was determined by a trail of Indian ink

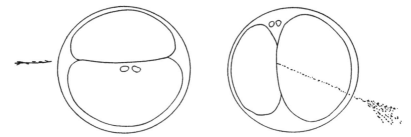

Fig. 104. Cleavage of the egg of *Nereis* showing cleavage plane passing through the point of entry of the spermatozoon. This point is marked by the trail of Indian ink left in the track of the spermatozoon through the gelatinous egg membrane. (From Just.)

left in the cortical jelly of the egg by the fertilising spermatozoon (see fig. 104)), out of a total of fifty-six eggs the first cleavage furrow passed through the point of entry of the spermatozoon in forty-six cases, whereas in ten cases this rule was not obeyed. How far these exceptions to the rule are only apparent is uncertain; it is possible that in some cases the egg rotates within the cortical jelly before the cleavage furrow develops, so that the trail of the spermatozoon through the jelly no longer indicates the point of entry into the egg. This remarkable relationship between the first cleavage furrow and the point of entry of the spermatozoon is of considerable theoretical importance. As already mentioned, the first cleavage furrow is often either coincident with, or closely approximates to, one of the main axes of symmetry of the resulting organism, so that if the spermatozoon can enter at any point on the egg's surface, it follows that the axes of symmetry cannot be predetermined before the egg is fertilised, but are determined by the point of entry of the spermatozoon. In this respect the recent work of Morgan and Tyler (1930) is highly significant. These authors found that the degree of correlation between the first cleavage plane and the point of entry of the spermatozoon varies in different types of egg: in *Cumingia* it is as high as 78 per cent., in *Chaetopterus* it is only 41 per cent. Morgan and Tyler's results are summarised in Table XXIX.

In all these cases the first cleavage is unequal and passes to the right or to the left of the polar axis (as defined by the position of the polar bodies) of the egg. In *Cumingia* either the first or the second cleavage furrow may become the median plane of the body. These two facts must be taken into account in any attempt to define the factors which control the direction of the early cleavages (see p. 235).

Table XXIX

Species	Total number of eggs observed	Percentage of eggs with first cleavage through point of entry of sperm	Percentage of eggs with first cleavage within 45° of point of entry of sperm	Percentage of eggs with first cleavage between 45° and 90° of point of entry of sperm
Cumingia	98	78	14	8
Chaetopterus	116	41	30	29
Nereis	64	51	27	22

In egg cells, which are not spherical or in which there is a marked physical heterogeneity in different parts of the cell, the direction of the first cleavage furrow is clearly not solely dependent on the point of penetration of the spermatozoon. In nematode eggs (Erlanger, 1897) the polar axis of the zygote nucleus lies at first along one of the shorter axes of the cell, but gradually rotates so as to lie in the plane of the long axis; eventually the first cleavage furrow is formed at the equator of the cell (see fig. 93). A comparable phenomenon can sometimes be seen during the second cleavage of echinoderm eggs (Gray, 1927).

From this point onwards two fairly well-defined lines of enquiry appear to be open. We may follow the relationship between particular cleavage planes in early ontogeny and the major axes of the subsequent organism; or we may seek by experiment to alter the

230 CELL DIVISION

natural course of cleavage and, by changing the normal plane of cell division, gain some insight into the nature of those forces which are the controlling factors in natural development. Since one of the main purposes of this book consists in delving into the mechanical characters of cell activity, we shall first follow the latter line and then, retracing our steps, attempt to see how far its course diverges or approaches the path of experimental embryology.

The effect of mechanical pressure upon cleavage planes

Roughly speaking, the laws associated with the names of O. Hertwig (1893) and of Pflüger (1884) can be summed up in one sentence,

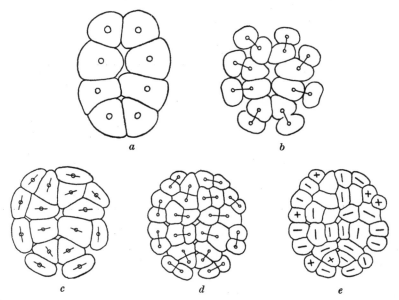

Fig. 105. Egg of *Echinus microtuberculotus* segmenting under pressure; *a*, the third cleavage yields a flat plate of eight cells instead of two tiers of four each; *b*, 16-celled stage showing tangential fourth division; *c*, the lines show the mitotic axes of the fifth division, in every case this is in the long axis of the cell; *d*, shows the resultant 32-celled stage; *e*, 64-cells: + signifies a vertical division; a line indicates a horizontal division. (From Ziegler.)

'The cleavage plane is at right angles to the longest axis of the protoplasmic mass'. The real significance of this fact is, however, more clearly expressed in Pflüger's dictum, 'The mitotic figure elongates in the line of least resistance'; consequently the cleavage plane is at

CELL DIVISION

right angles to this line. Pflüger's conclusion was based on a series of experiments in which the eggs of the frog were induced to cleave under pressure; in such circumstances the cleavage plane was always in the direction of or in the same plane as the applied pressure. Pflüger's results were confirmed and amplified by Roux, and were extended to echinoderm eggs by Driesch (1893 a), Ziegler (1894), and Yatsu (1910). Ziegler's results are perhaps the most interesting, as with the aid of an irrigated compressorium he was able to follow the cleavages of a compressed egg through several divisions. His results showed clearly that, as long as the pressure was exerted by two flat plates, each cleavage plane was at right angles to the plates, whilst the mitotic figure lay in a plane parallel to the plates. It is worth noting that the two plates of the compressorium were fixed during the whole of Ziegler's experiments, and consequently there was no force exerted by the weight of the cover-slip. Ziegler was inclined to regard each cell as a fluid drop and consequently it is a little difficult to see why the pressure within the drop should vary from one plane to another. There can be no doubt whatever that the cleavage planes are at right angles to the plane of the externally applied pressure, but it does not follow that the mechanical pressure of the plates is in any way transmitted directly to the mitotic figure. Since the cytoplasmic matrix of the cell is liquid, the pressure at all points must be equal even if the surface of the cell is subjected to lateral compression. If, however, we are prepared to regard the two growing asters as regions of elastic rigidity, then they will be subjected to pressure as soon as they come into contact with the periphery of the cell; until then they will not be subjected to compression. If one axis of the cell is longer than another, then polar compression of the asters can be delayed by a rotation of their axis into the longest axis of the cell, just as occurs in the normal egg of the nematode. A similar orientation often occurs in normal echinoderm eggs (see fig. 106). If this is sound doctrine, Pflüger's law of cleavage acquires a real meaning; without such an assumption the phenomena of cleavage under pressure seem curiously irrational. A disturbance of the subsequent cleavage planes is not a response to a change in intracellular pressure but to a change in the shape of the cell whereby one axis becomes longer than either of the other two.

There can be little doubt that the application of an external pressure has proved by far the most efficient method of influencing the direction of a cleavage furrow. The facts clearly show that the direction of cleavage of a compressed cell is the result of a movement on the part of the astral axis; in other words, the whole mitotic apparatus moves bodily through the cell until it comes to lie in the longest axis of the cell. This movement could only occur if the pressure exerted on the asters (presumably by the cell surface) is greater than the viscous resistance of the cytoplasm. If this resistance were relatively great, a cell could divide in a plane parallel to instead of at right angles to its longest axis, and this may possibly sometimes

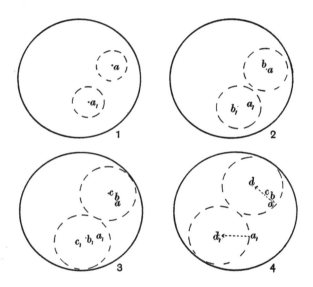

Fig. 106. Movement of two asymmetrically situated asters in an *Echinus* egg. In 1, the centres of the asters are at a, a_1; in 2, they lie at b, b_1; in 3, they lie at c, c_1; in 4, the diameter of each aster is equal to the radius of the egg, so that the centres of the asters (d, d_1) must lie on a diameter of the egg and produce symmetrical cleavage.

be the case in nature. Wilson (1892) and Conklin (1898) have given examples of such anomalous cleavages. It may be remembered that in echinoderm eggs, which are the favourite object of compression, the spindle with the asters can be moved through the cell by centrifugal force, and that there is independent evidence to support the view that the cytoplasmic viscosity is of a low order. It would be interesting to know whether the cytoplasmic viscosity of the mesoblast cells described by Wilson and by Conklin is of a distinctly higher order of magnitude.

The so-called effect of gravity on the direction of cleavage planes

The directions of the early cleavages during the segmentation of an egg are not infrequently defined in terms of inclination to gravity. Thus, in frogs' eggs or in sea urchin eggs the first two cleavages are often described as vertical, whereas the third cleavage plane is horizontal. This nomenclature is convenient, but misleading, since the force of gravity exerts no direct effect on the planes of cleavage either during their formation or afterwards. In the frog's egg the accumulation of heavy yolk at one pole, with a consequent accumulation of cytoplasm at the other, impresses on the undivided egg a definite visible polarity, and a definite orientation in respect to gravity. The whole egg is only in equilibrium with gravity when the flat disc of cytoplasm lies vertically over the yolk; two out of the three rectangular axes of the cytoplasm are equal in length and lie horizontally; the third axis is shorter than the others and is vertical. By dividing the cytoplasm into equal divisions at right angles to its longest axis, the first two cleavage planes are naturally meridional and vertical, whereas the third cleavage plane is more or less equatorial and horizontal. Both Hertwig (1893) and Pflüger (1884) clearly recognised that gravity exerted no direct action on the orientation of the spindle itself, for, as shown by Kathariner (1901), the first two cleavages in eggs rotating on a clinostat retain their orientation in respect to the organised polarity of the egg and exhibit no orientation in respect to the centrifugal force.

Giglio-Tos (1926) and his associates have recently maintained that the first cleavage furrow of echinoderm eggs (species unnamed) is always inclined at an angle of 45° to the vertical and that this is due to the orientation of the cleavage spindle prior to division. These authors maintain that the two asters are free to move on each other and within the cell and that, when the asters are fully formed and equal in size, their position of mechanical equilibrium is reached when the mitotic axis is inclined at an angle of 45° to the vertical.

The observations of the author (Gray, 1927) do not confirm any of the conclusions of Giglio-Tos, but indicate that the effect of gravity on the fertilised eggs of *Echinus esculentus* and *E. miliaris* is of an entirely different nature.

A large number of observations with both *Echinus esculentus* and *E. miliaris* leave no doubt that the direction of the first three cleavages obeys quite strictly the Hertwig-Pflüger law and that the

234   CELL DIVISION

axis joining the centres of the two asters (mitotic axis) can lie in any plane relative to gravity. The asters having taken up their position at the poles of the nucleus maintain the orientation thus acquired until the egg begins to show signs of cleavage furrows. Further, an egg can be rotated so as to bring the mitotic axis into any desired position, and this position is stable. It is only when cleavage begins

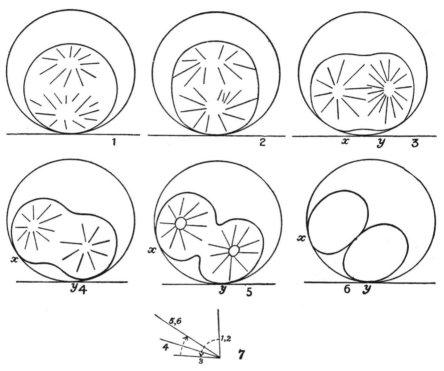

Fig. 107. Orientation of a cleaving egg of *Echinus* in which the astral axis was originally vertical.

that gravity exerts any affect on the orientation of the system. This is most readily observed in an egg whose mitotic axis is vertical as in fig. 107. As the egg elongates upwards (fig. 107, 2) it soon becomes unstable and falls on to one side (fig. 107, 3). This movement is clearly due to gravity, and the egg orientates itself so as to bring its centre of gravity to the lowest possible position. In this way the mitotic axis becomes more or less horizontal (fig. 107, 3). Having reached this position the egg continues to elongate, and its long axis

CELL DIVISION 235

may either continue to remain horizontal or it may tilt upwards as in fig. 107, 4–6. Both types of movement are obviously induced by the accommodation of the egg to the confines of the fertilisation membrane; as the egg elongates, the two points of contact (fig. 107, x and y) move apart forming a longer and longer arc.

When the first cleavage is completed, the long axis of the egg may therefore lie in any position from the horizontal (fig. 107, 3) up to the maximum inclination of about 35° (fig. 107, 6). The first cleavage plane is eventually therefore either vertical or deviates from the vertical by an angle not exceeding 35°. It is clear that these phenomena are due to the fact that on the initiation of cleavage the egg ceases to be spherical and, were it not for the presence of the fertilisation membrane, the egg would only be in equilibrium with gravity when its long axis was horizontal, and the cleavage furrow vertical. As the egg must accommodate itself to the confines of the fertilisation membrane, the egg can be in equilibrium as long as the inclination of its long axis does not exceed a value which depends on the forces exerted at the points of contact with the membrane.

As far as the eggs of *Echinus* are concerned, therefore, the effect of gravity on the orientation of the early cleavage planes is only of an indirect nature, and is based on the fact that in the segmenting eggs the whole system tends to reach an equilibrium position with its centre of gravity in the lowest possible position.

Internal factors which determine the direction of cleavage planes

If Hertwig's law were strictly obeyed by all spherical egg cells (with uniformly distributed yolk and equal cleavage), it follows that only one pattern of cell cleavage planes would be possible. The arrangement is that exemplified by *Echinus* eggs, and is known as the *orthoradial* type of cleavage characteristic of *Echinus, Amphioxus, Synapta, Antedon,* and *Sycandra* (Conklin, 1897). In all these cases the long axis of the third cleavage amphiaster is meridional and at right angles to that of the previous cleavage plane, so that the third cleavage cuts the egg equatorially leaving the two daughter cells of each cleavage in the same meridian of the egg (see fig. 108 *A*). As pointed out by Conklin, orthoradial cleavage is uncommon and even in cases where the early cleavage planes conform in this way to the Hertwig-Pflüger law the later cleavages show a contrary arrangement. By far the most frequent type of cleavage pattern exhibited by spherical eggs during cleavage is the spiral pattern seen in

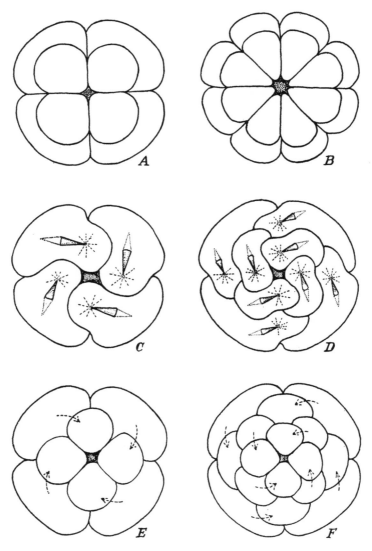

Fig. 108. Orthoradial (*A* and *B*) and spiral cleavage (*C–F*). In orthoradial cleavage the third mitotic spindle is at right angles to the surface of the paper so that the two products of division lie vertically over each other (*A*). The fourth cleavage is at right-angles to the third (*B*). In spiral cleavage the third mitotic spindle is displaced as shown in *C*, so that each micromere lies between two macromeres. If (as in *C* and *E*) the displacement of the mitotic axis is to the right, the third displacement at the next division is to the left (as in *F*). (From Korschelt and Heider.)

CELL DIVISION

molluscs, platode worms and annelids. In these instances the axes of the dividing cells do not coincide with the plane of the long axis of the original undivided cells, but are inclined at an angle to it. The final result of this displacement of the cleavage axis leads to the arrangement (in a four-celled and eight-celled stage respectively) shown in fig. 108 C and E. It can be seen that the two products of cleavage (in figs. 108 C–E, 109) do not lie in the same radius of the egg, but one of them is displaced so as to lie in the furrow between two adjacent cells of the other quartet. The term *spiral* cleavage was applied by Wilson to this arrangement to signify the fact that the products of division lie on a curved radius of the egg for if this curve were produced, it would form a spiral about the egg axis. A typical example of spiral cleavage is provided by the eggs of *Crepidula* described by Conklin (1897). The first cleavage divides the egg

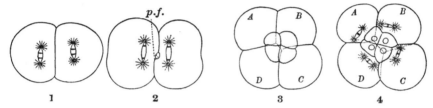

Fig. 109. Typical spiral cleavage of *Crepidula*. The development of the polar furrow (*p.f.*) is seen in 2. Note the displacement of the micromeres in 3. (After Conklin.)

equally into two blastomeres. These cells are at first nearly spherical and touch each other only over a comparatively small area of their surface, although later on they become more closely pressed together and each cell becomes an almost complete hemisphere (fig. 109, 1). At the close of the first cleavage the nuclei and their asters lie directly opposite each other, but, as soon as the blastomeres begin to flatten against each other, the mitotic axes begin to rotate in the direction of the hands of a clock and this direction is often constant in all the eggs of the species. As the time for the second division approaches the two spindles are no longer absolutely parallel to each other, for (when the egg is viewed from one side) they are inclined at an angle to each other; consequently the second cleavage planes do not meet at the centre of the egg but a *polar furrow* is formed (fig. 109, 2 *p.f.*). Polar furrows are essentially typical of spiral cleavage, although they can readily be induced in orthoradial cleavage by experimental means (Gray, 1924 and fig. 112). The axes of the amphiasters of the third cleavage of *Crepidula* are at first rather variable in their orientation in respect to the axes of the cells, although their inner ends are at a higher level than their outer ends and the axes may be radial. As the process of cleavage becomes more complete the spindles, whatever be their original orientation, rapidly begin to show a rotation towards the right-hand side, and after the division wall between the dividing cells has appeared the process of rotation is continued by the blastomeres themselves. In this way each micromere comes to lie in the furrow between two macromeres and to alternate with the macromeres in position (fig. 108 *D–F*). It is to be noted that if in one cleavage the resultant blastomeres are rotated to the right, then at the next division the rotation is to the left; each successive division is, in other words, alternately clockwise and anti-clockwise.

The account of spiral cleavage given above closely follows that of Conklin for *Crepidula*. In contrast to orthoradial division, spiral cleavage appears to exhibit three main features. (i) The displacement of the mitotic axes in respect to the radius of the egg and in respect to each other, (ii) the formation of polar furrows, (iii) the displacement of individual blastomeres in respect to the egg radius.

It has been known for many years that the final result of spiral cleavage leads directly to a geometrical arrangement of cells, which is mechanically stable in that each cell exhibits a minimum surface area to its neighbours and to the environment. Fig. 110 shows that polar furrows and a radial displacement of individual units are as characteristic of soap bubbles as they are of dividing eggs. We have, therefore, to consider how far the arrangement of the cells in *Crepidula* is due to purely mechanical principles, and how far they are the result of biological activity. There can be no question that the rotation of blastomeres (from the orthoradial to the spiral arrangement) reduces the free surface of the individual cells, or that the cells are held together by forces which tend to reduce such surfaces to a minimum (see p. 254); at the same time, from a mechanical point of view there is no reason why rotation should invariably be to the right or to the left, both are equally effective, and one would expect that a given group of egg cells would tend to show both types of rotation with equal frequency. In gasteropod molluscs this does not appear to be the case for the direction of rotation of the cleavage axis appears to be a fixed and hereditary characteristic (for any particular cleavage). For this reason both

CELL DIVISION

Conklin (1897) and Wilson (1892) concluded that spiral cleavage cannot be solely the result of purely mechanical causes which operate at the moment of cleavage. Since the final result produces a geometrical pattern conforming to the law of minimum surface,

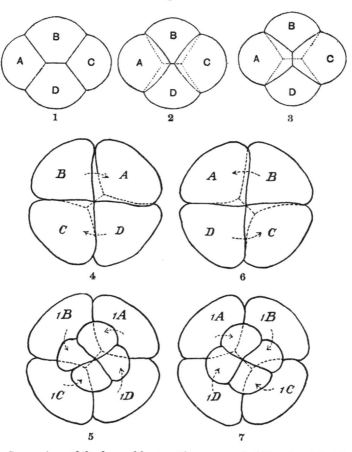

Fig. 110. Comparison of the form of four contiguous soap bubbles, 1–3 (after Robert) with typical spiral cleavage of living eggs. 4 and 5, A dexiotropic division followed by a leiotropic division. 6 and 7, A leiotropic division followed by a dexiotropic division. (After Korschelt and Heider.)

Conklin (1897) concludes that 'We must find the ultimate cause of this anti-clockwise (or clockwise) rotation, not in such external conditions which are, however, incidentally fulfilled but in those more complex internal conditions which direct the course of ontogeny and which in our ignorance we call the coördinating force or hereditary tendency' (p. 80).

This conclusion is, however, somewhat weakened by the recent work of Morgan and Tyler (1930). In the mollusc *Cumingia* the third cleavage is always dexiotropic, so that by the classical rule of alternate displacements the second cleavage should always be leiotropic. In practice, however, the direction of rotation at the

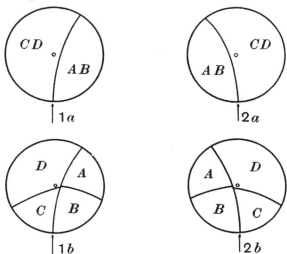

Fig. 111. The position of the cleavage planes of the egg of *Cumingia* with respect to the entrance point of the spermatozoon. In 1a the first cleavage passes to the right of the pole (as marked by the polar bodies which are uppermost), and the blastomere *AB* lies to the right of the point of entry of the sperm. In 2a the situation is reversed and *AB* lies to the left of the point of entry. The second cleavages are shown in 1b and 2b respectively, and are leiotropic and dexiotropic respectively. (From Morgan and Tyler.)

second cleavage can be either clockwise or anti-clockwise according to whether the first cleavage furrow passes to the right or to the left of the polar axis of the cell.

According to Morgan and Tyler the direction of the spindle axes during the second division show no sign of spiral orientation; a fact which differs from Conklin's observations on *Crepidula*. Further in *Cumingia* either the first or the second cleavage plane may become the median plane of the body and this can only be determined after the third cleavage has occurred. In *Nereis*, where the third cleavage is also dexiotropic, only one configuration of cells is found in the four-celled stage, viz. that illustrated in fig. 111, 1b.

In attempting to analyse the nature of the forces which are responsible for the form and position of individual cleavage planes, it is useful to distinguish between two different processes. Firstly, the mechanical principles which control the form and position of the cells after the process of cleavage is complete and secondly, those factors which determine the orientation of the cells during cleavage. These two principles may or may not be the same (see p. 254).

There can be no doubt that if a so-called 'spiral' pattern has not been the result of normal cleavage, it can readily be brought about by suitably applied external pressure. This is the case, for example, in *Echinus* when the egg is exposed to superficial pressure after the first two cleavages are complete; it also occurs normally in a definite percentage of eggs belonging to species whose natural cleavage pattern is orthoradial, e.g. in *Amphioxus*. Spiral patterns of this type are clearly the result of movements executed by fully formed blastomeres and are probably strictly comparable to the movements of soap bubbles or other mechanical systems. When an *Echinus* egg is subjected to pressure before cleavage has occurred, however, it is significant to note that the spiral pattern which results is not reached by way of an orthoradial stage, but is acquired by a gradual orientation of the dividing cells in a way strikingly similar to that observed during segmentation of a normally spiral type such as *Nereis* or *Crepidula* (see fig. 112). Since the spiral pattern can be produced by artificial pressure, and since the final result is obviously in conformity with the law of minimal reaction to external pressure, it seems reasonable to suspect that where it occurs normally it does so in response to an externally applied pressure which owes its origin to the mechanical environment of the blastomeres. At the same time this mechanical environment may be the result of definite ontogenic factors, and it is useful to bear in mind that mechanical conditions may be the result and not the cause of biological activity. This point of view is very clearly expressed by Wilson (1892), Conklin (1897), and by F. R. Lillie (1895); two striking quotations from Lillie are perhaps admissible: 'Almost every detail of the cleavage of the ovum of *Unio* can be shewn to possess some differential significance. The first division is unequal. Why? Because the anlage of the immense shell-gland is found in one of the cells. The apical pole cells divide very slowly and irregularly, lagging behind the other cells. Why? Because the

242   CELL DIVISION

formation of apical organs is delayed to a late stage of development. The second generation of ectomeres is composed of very large cells. Why? Because they form early and voluminous organs (larval mantle). The left member of this generation is larger than the right. Why? Because it contains the larval mesoblast.... One can thus go over every detail of the cleavage, and knowing the fate of the cells, can explain all the irregularities and peculiarities exhibited' (pp. 38,

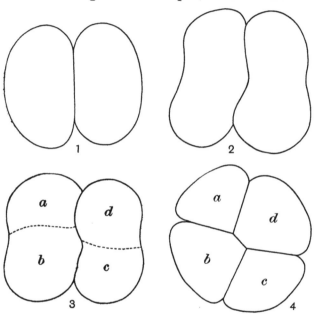

Fig. 112. An egg of *Echinus* segmenting under pressure. The natural orthoradial cleavage is replaced by the spiral type. Compare 1 and 2 with fig. 109, 1 and 2. The polar furrow develops as a displacement of the interface between the first two blastomeres. The dotted lines in 3 indicate the position of the two new vertical cleavage planes seen in 4.

39). In the same paper, Lillie discusses the orientation of the mitotic spindle and subsequent cleavage planes; he concludes that no mechanical explanation will suffice. 'Let us look for a moment at the cleavages of X (First somatoblast). The first position of the spindle is on its left side; the second position on the right side; the third in the middle line towards the apical pole; the fourth in the middle line towards the vegetative pole. In none of these cases does the spindle occupy more than a fraction of the diameter of the blastomere in question. The nucleus has been wandering through

CELL DIVISION

the cytoplasm from one side to the other, from the front to the back, stopping at various stations, and giving off a cell at each one. Finally the nucleus stops in the centre of the cell and a perfectly bilateral spindle (the fifth) is formed. Why does it stop there? Is it because its environment has changed? If so, the change is such as to elude the closest scrutiny. In fact the cell is a builder which lays one stone here, another there, each of which is placed with reference to future development' (p. 46).

The facts of orthoradial and of spiral cleavage lead to a curious situation. In orthoradial cleavage the division planes are in strict accordance with Pflüger's Law, but result in an arrangement of cells which fails to conform to the law of minimal surface area. In spiral cleavage, on the other hand, there is an apparent deviation from Pflüger's Law, but the final result gives a geometrically stable system. Were it not for the fact that the spiral displacements of mitotic axes is believed to begin before the cells begin to elongate, one would suspect that a mechanical solution to these anomalies would not be difficult to locate.

The phenomena of *unequal* cleavage require some consideration. All observers are agreed that unequal cleavage is associated with an asymmetrical arrangement of the mitotic figure. The most striking cases of unequal cleavage are provided by the formation of polar bodies. Conklin (1924) calculates that the polar bodies of *Fulgar* are less than one-millionth of the volume of the egg. In *Crepidula* the maturation spindle can first be observed towards the centre of the egg where, according to Conklin, the two asters appear to be of the same size. Gradually the whole mitotic system migrates to the periphery of the cell and as it does so it shortens in length, and the two asters are no longer equal in size. Conklin believes that the spindle moves passively under the influence of unknown forces, and that the inequality of division is not so much due to an inequality in the size of the asters, as to an inequality in the distribution of the protoplasm which is controlled by each aster. On the other hand, according to Lillie (1901), an inequality of asters can be detected in the eggs of *Nereis* long before the mitotic system migrates to the periphery. A further investigation of the intracellular movements of maturation spindles might throw considerable light on the forces which orientate the asters, but the facts described by Conklin suggest that the rate of growth of an aster is determined in part by the distribution of cytoplasm, rather than *vice versa*.

REFERENCES

Burrows, T. Montrose (1927). 'The mechanism of cell division.' *Amer. Journ. Anat.* 39, *83*.

Bütschli, O. (1900). 'Bemerkungen über Plasmaströmung bei der Zellteilung.' *Arch. f. Entw. Mech.* 10, *53*.

Chambers, R. (1917). 'Micro-dissection studies. II. The cell aster; a reversible gelation phenomenon.' *Journ. Exp. Zool.* 23, *483*.

—— (1919). 'Changes in protoplasmic consistency and their relation to cell division.' *Journ. Gen. Physiol.* 2, *49*.

—— (1924). 'Physical structure of protoplasm.' *General Cytology.* (Chicago.)

Conklin, E. G. (1897). 'The embryology of *Crepidula*, a contribution to the cell-lineage and early development of some marine gasteropods.' *Journ. Morph.* 13, *1*.

—— (1898). 'Cleavage and Differentiation.' *Biolog. Lectures, Wood's Hole.* (Boston.)

—— (1924). *General Cytology.* (Chicago.)

Driesch, H. (1892). 'Entwicklungsmechanisches.' *Anat. Anz.* 7, *584*.

—— (1893 a). 'Zur Verlagerung der Blastomeren des Echinideneies.' *Anat. Anz.* 8, *348*.

—— (1893 b). 'Entwicklungsmechanische Studien. IV.' *Zeit. wiss. Zool.* 55, *1*.

—— (1895). 'Entwicklungsmechanische Studien. VIII.' *Mitt. zool. Sta. Neapel*, 11, *221*.

von Erlanger, R. (1897). 'Beobachtungen über die Befruchtung und ersten Teilungen an den lebenden Eiern kleiner Nematoden.' *Biol. Zentralb.* 17, *152, 339*.

Farr, C. H. (1918). 'Cell-division by furrowing in *Magnolia*.' *Amer. Journ. Bot.* 5, *379*.

Giglio-Tos, E. (1926). 'Die Wirkung der Schwerkraft auf die Richtung der ersten Furchungsspindel im Ei des Seeigels.' *Zeit. wiss. Biol.* Abt. D, 107, *186*.

Gray, J. (1924). 'The mechanism of cell-division. I. The forces which control the form and cleavage of the eggs of *Echinus esculentus*.' *Proc. Camb. Philos. Soc. Biol. Series*, 1, *166*.

—— (1925). 'The mechanism of cell-division. II. Oxygen consumption during cleavage.' *Proc. Camb. Philos. Soc. Biol. Series*, 1, *225*.

—— (1927 a). 'The mechanism of cell-division. III. The relationship between cell-division and growth in segmenting eggs.' *Brit. Journ. Exp. Biol.* 4, *313*.

—— (1927 b). 'The mechanism of cell-division. IV. The effect of gravity on the eggs of *Echinus*.' *Brit. Journ. Exp. Biol.* 5, *102*.

Heilbrunn, L. V. (1921). 'Protoplasmic viscosity changes during mitosis.' *Journ. Exp. Zool.* 34, *417*.

Hertwig, O. (1893). 'Ueber den Werth der ersten Furchungszellen für die Organbildung des Embryo.' *Arch. f. mikr. Anat.* 42, *662*.

Just, E. E. (1912). 'The relation of the first cleavage plane to the entrance point of the sperm.' *Biol. Bull.* 22, *239*.

CELL DIVISION

JUST, E. E. (1922). 'Studies of cell division. I. The effect of dilute seawater on the fertilised eggs of *Echinarachnius parma* during the cleavage cycle.' *Amer. Journ. Physiol.* 61, 505.

KATHARINER, L. (1901). 'Ueber die bedingte Unabhängigkeit der Entwicklung des polar differenzierten Eies von Schwerkraft.' *Arch. Entw. Mech.* 12, 597.

KORSCHELT, E. (1888). 'Zur Bildung des mittleren Keimblatts bei den Echinodermen.' *Zoolog. Jahrb.* Anat. Hefte, 3, 653.

LAMBERT, R. A. and HANES, F. N. (1913). 'Beobachtungen an Gewebskulturen in vitro.' *Virchow's Archiv*, 211, 89.

LEWIS, F. T. (1923). 'The typical shape of polyhedral cells in vegetable parenchyma and the restoration of that shape following cell division.' *Proc. Amer. Acad. Sci.* 58, 537.

—— (1926). 'The effect of cell division on the shape and size of hexagonal cells.' *Anat. Record*, 33, 331.

—— (1928). 'The correlation between cell division and the shapes and size of prismatic cells in the epidermis of *Cucumis*. *Anat. Record*, 38, 341.

LILLIE, F. R. (1895). 'The embryology of the Unionidae.' *Journ. Morph.* 10, 1.

—— (1901). 'The organisation of the egg of *Unio*, based on a study of its maturation, fertilization and cleavage.' *Journ. Morph.* 17, 227.

LOEB, J. (1895). 'Untersuchungen über die physiologischen Wirkungen des Sauerstoffmangels.' *Arch. f. ges. Physiol.* 62, 249.

LYON, E. P. (1902). 'Effects of potassium cyanide and of lack of oxygen upon the fertilized eggs and the embryos of the sea-urchin (*Arbacia punctulata*).' *Amer. Journ. Physiol.* 7, 56.

—— (1904 a). 'Rhythms of CO_2 production during cleavage.' *Science*, 19, 350.

—— (1904 b). 'Rhythms of susceptibility and of carbon dioxide production in cleavage.' *Amer. Journ. Physiol.* 11, 52.

MCCLENDON, J. F. (1912). 'A note on the dynamics of cell division. A reply to Robertson.' *Arch. f. Entw. Mech.* 34, 263.

—— (1913). 'The laws of surface tension and their applicability to living cells and cell division.' *Arch. f. Entw. Mech.* 37, 233.

MAST, S. O. and ROOT, F. M. (1916). 'Observations on amoeba feeding on rotifers, nematodes, and ciliates and their bearing on the surface tension theory.' *Journ. Exp. Zool.* 21, 33.

MATHEWS, A. P. (1907). 'A contribution to the chemistry of cell-division, maturation, and fertilisation.' *Amer. Journ. Physiol.* 18, 87.

MEYERHOF, O. (1911). 'Untersuchungen über die Wärmetönung der vitalen Oxydationsvorgänge im Seeigelei. I–III.' *Biochem. Zeit.* 35, 246.

MORGAN, T. H. (1893). 'Experimental studies on echinoderm eggs.' *Anat. Anz.* 9, 141.

—— (1896). 'The production of artificial astrospheres.' *Arch. f. Entw. Mech.* 3, 339.

—— (1899). 'The action of salt-solutions on the fertilised and unfertilised eggs of *Arbacia*.' *Arch. f. Entw. Mech.* 8, 448.

—— (1900). 'Further studies on the action of salt-solutions and other agents on the eggs of *Arbacia*.' *Arch. f. Entw. Mech.* 10, 489.

—— (1910). 'The effects of altering the position of the cleavage planes in eggs with precocious specification.' *Arch. f. Entw. Mech.* 29, 205.

MORGAN, T. H. and BORING, A. (1903). 'The relation of the first plane of cleavage and the grey crescent to the meridian plane of the embryo of the frog.' *Arch. f. Entw. Mech.* 16, *680*.

MORGAN, T. H. and TYLER, A. (1930). 'The point of entrance of the spermatozoon in relation to the orientation of the embryo in eggs with spiral cleavage.' *Biol. Bull.* 58, *59*.

PFLÜGER, E. (1883). 'Ueber den Einfluss der Schwerkraft auf die Teilung der Zellen.' *Pflüger's Archiv*, 31, *34*.

—— (1884). 'Über die Einwirkung der Schwerkraft und anderer Bedingungen auf die Richtung der Zelltheilung.' *Pflüger's Archiv*, 34, *607*.

ROBERTSON, T. B. (1909). 'Note on the chemical mechanics of cell-division.' *Arch. f. Entw. Mech.* 27, *29*.

—— (1911). 'Further remarks on the chemical mechanics of cell division.' *Arch. f. Entw. Mech.* 32, *308*.

—— (1913). 'Further explanatory remarks concerning the chemical mechanics of cell-division.' *Arch. f. Entw. Mech.* 35, *692*.

ROGERS, C. G. and COLE, K. S. (1925). 'Heat production by the eggs of *Arbacia punctulata* during fertilization and early cleavage.' *Biol. Bull.* 49, *338*.

ROUX, W. (1885). 'Beiträge zur Entwicklungsmechanik. III. Ueber die Bestimmung der Hauptrichtungen des Froschembryo im Ei und über die erste Teilung des Froscheies.' *Breslau ärztl. Zeit.*

—— (1887). 'Beiträge zur Entwickelungsmechanik des Embryo. IV. Die Richtungsbestimmung der Medianebene des Froschembryo durch die Copulationsrichtung des Eikernes und Spermakernes.' *Arch. mikr. Anat.* 29, *157*.

—— (1893). 'Ueber Mosaikarbeit und neuere Entwicklungshypothesen.' *Anat. Hefte*, 1, Abt. II, *227*.

—— (1895). *Gesammelte Abhandlungen über Entwicklungsmechanik der Organismen.* Band 2. (Leipzig.)

—— (1897). 'Ueber die Bedeutung "geringer" Verschiedenheiten der relativen Grösse der Furchungszellen für den Charakter des Furchungsschemas.' *Arch. f. Entw. Mech.* 4, *1*.

SEELIGER, O. (1893). 'Studien zur Entwicklungsgeschichte der Crinoiden.' *Zoolog. Jahrb. Anat.* Heft, 6, *161*.

SPEK, J. (1918). 'Oberflächenspannungdifferenzen als eine Ursache der Zellteilung.' *Arch. f. Entw. Mech.* 44, *1*.

STRANGEWAYS, T. S. P. (1922). 'Observations on the changes seen in living cells during growth and division.' *Proc. Roy. Soc.* B, 94, *137*.

VLÈS, F. (1922). 'Sur les variations des ions H' au voisinage des œufs en division.' *Comptes Rendus*, *643*.

—— (1926). 'Les tensions de surface et les déformations de l'œuf d'Oursin.' *Arch. de Physique biol.* 4, *263*.

WARBURG, O. (1908). 'Beobachtungen über die Oxydationsprozesse im Seeigelei.' *Zeit. f. physiol. Chem.* 57, *1*.

—— (1910). 'Ueber die Oxydationen in lebenden Zellen nach Versuchen am Seeigelei.' *Zeit. f. physiol. Chem.* 66, *305*.

—— (1914). 'Beiträge zur Physiologie der Zelle insbesondere über die Oxydationsgeschwindigkeit in Zellen.' *Ergeb. d. Physiol.* 14, *253*.

WETZEL, G. (1926). 'Zur entwicklungsmechanischen Analyse des Einfaches prismatischen Epithels.' *Arch. f. Entw. Mech.* 107, *177*.

WILSON, E. B. (1892). 'The cell lineage of *Nereis*. A contribution to the cytogeny of the annelid body.' *Journ. Morph.* 6, *361*.
—— (1901 a). 'Experimental studies in cytology. I. A cytological study of artificial parthenogenesis in sea-urchin eggs.' *Arch. f. Entw. Mech.* 12, *529*.
—— (1901 b). 'Experimental studies in cytology. II. Some phenomena of fertilisation and cell division in etherized eggs. III. The effect on cleavage of artificial obliteration of the first cleavage furrow.' *Arch. f. Entw. Mech.* 13, *353*.
WILSON, E. B. and MATHEWS, A. P. (1895). 'Maturation, fertilisation and polarity in the echinoderm egg.' *Journ. Morph.* 10, *319*.
YATSU, N. (1910). 'Experiments on germinal localization in the eggs of *Cerebratulus*.' *Journ. Coll. Sci. Tokyo*, 27, *17*.
ZIEGLER, H. E. (1894). 'Ueber Furchung unter Pressung.' *Anat. Anz.* 9, *132*.
—— (1898 a). 'Experimentelle Untersuchungen über die Zellteilung. I. Die Zerschnürung der Seeigeleier. II. Furchung ohne Chromosomen.' *Arch. f. Entw. Mech.* 6, *249*.
—— (1898 b). 'Experimentelle Untersuchungen über die Zellteilung. III. Die Furchungszellen von *Beroë ovata*.' *Arch. f. Entw. Mech.* 7, *34*.
—— (1903). 'Experimentelle Untersuchungen über die Zellteilung. IV. Die Zellteilung der Furchungszellen von *Beroë* und *Echinus*.' *Arch. f. Entw. Mech.* 16, *155*.

CHAPTER TEN

The Shape of Cells

The shape and form of cells

In the majority of instances, the form of an isolated cell clearly depends upon the particular source from which the cell has been derived, and therefore, to some extent, cell form is a predetermined character which owes its origin to a highly specific type of cell growth or organisation. An isolated muscle fibre, a nerve cell, a red blood corpuscle, and all the Protozoa have their own characteristic forms, which are retained after the death of the cells if suitable killing agents are employed. This diversity of form can readily be correlated with the existence of a solid membrane at the cell surface: a membrane of this type is clearly demonstrable in many cases (see p. 103), and we may safely assume that a given cell maintains its characteristic form by virtue of the rigidity of its surface.

In certain specific cases the form of an isolated cell approximates to that of a sphere, and most analyses of cell form are based on a study of the shapes which such cells assume when subjected to close contact with other similar cells. In analyses of this type there are two distinct problems. Firstly, why is the isolated cell spherical? and secondly, why does it undergo a definite geometrical change in form when in contact with other cells? In just the same way we might consider why an isolated *Paramecium* maintains its characteristic form, and why it changes its form when compressed against its neighbours. Unfortunately the study of the behaviour of spherical cells has been unduly influenced by the assumption that the form of the cells is necessarily due to the operation of forces peculiar to the interface between two liquids (Robert, 1902; Thompson, 1917; Lewis, 1926–8). It would appear, however, that this assumption is contrary to fact. If an isolated living cell is comparable to a homogeneous fluid drop, then when two such cells (immersed in water) come into contact with each other, they should unite together to form one single spherical cell. This is not the case. In order to approximate the cell to a state comparable to a soap bubble, it is necessary to assume that it is essentially a two-phase system composed of a fluid or elastic core, whose surface is covered by a liquid film which is immiscible both with the core and with the surrounding water. When two such drops come into contact, they would automatically unite by their fluid surfaces to form a stable system. This is however not the case with living cells, since two isolated cells show no tendency to form a single system. On the other hand, it can be clearly demonstrated that, like other cells, isolated spherical blastomeres possess solid elastic membranes at their surfaces. It is more probable that the form of a spherical egg owes its characteristic shape to the same forces as are responsible for the form of a typical protozoon. If we attribute the form of a *Paramecium* to differential growth along well defined and specific axes, we can equally well account for the spherical form of an echinoderm egg to growth which tends to be equal in all directions.

If growth occurs in a cell along specific and localised axes, the elastic cell surface will be stretched and will tend to oppose the elongation of these axes, so that presumably such directional growth must sooner or later be accompanied by intercalated growth of the cell membranes at the surface and in this way the elastic tension of the surface will be relieved; alternatively it is possible to suppose that localised growth begins within the solid surface membrane and the distribution of the cytoplasm conforms to the shape of the cell membrane. In either case the final form of the cell depends upon factors which are highly specific, and concerning which we are in complete ignorance.

The whole study of cell form really involves the analysis of two distinct sets of factors—firstly, those intrinsic factors which are intimately associated with specific cell organisation, and secondly, the effects which are superimposed on the natural form of a cell by the mechanical presence of its immediate neighbours and by the external environment. In the following discussion we shall regard all spherical cells as special cases of cell form; this does not indicate that the isolated cell is in any real way comparable to a homogeneous liquid drop.

The form of contiguous spherical cells

For many years it has been known that the shapes of actively dividing or of newly formed cells of all tissues tend to conform to comparatively simple and generalised types, and it is difficult to resist the conclusion that this uniformity of cell form is the direct result of mechanical principles. Since the form of the isolated cell is readily disturbed by the application of mechanical pressure, the suggestion arises that the form of certain cells *in situ* is the result of the mechanical restraint to which the cells are normally exposed. If a spherical cell, enclosed within an elastic membrane, be subjected to unilateral extraneous pressure, the form adopted by the cell will depend on the degree of departure from the spherical state which will enable the cell to generate an elastic force equal to the extraneous pressure; the form occupied by the cell will not depend on its specific nature except in so far as this expression includes the mechanical properties of the cell surface. It is therefore to be expected that wherever spherical cells are compressed together, the geometrical form of the whole system will be of the same type and independent of the specific origin of the cells. The validity of this conclusion is supported by the experiments made by Roux (1897). This author showed that the form occupied by each cell of a segmenting frog's egg is accurately defined by the form occupied by the components of a suitable system of oil drops (in which each drop conformed in size and position to each particular blastomere), the whole of which is subjected to centripetal pressure.

The remarkable accuracy of Roux's models (fig. 113) adds considerable interest to the process by which they were obtained. The procedure involved the preparation of two fluid phases; one of these consisted of olive oil, the other of a mixture of alcohol and water of the same specific gravity as the oil. When a drop of the oil was suspended in the aqueous phase the oil drop assumed a spherical form; when two such drops came into contact they rapidly fused together to form a single spherical drop. In order to obtain a series of spherical drops which would remain as separate entities when in contact, Roux found it necessary to deposit on the surface of each drop a solid elastic membrane which was insoluble in oil and in water. This arrangement was reached by adding to the oil a little oleic acid and to the aqueous phase crystals of calcium acetate. When a drop of the oil came in contact with the surrounding solution, a film of calcium acetate was deposited on the surface of the drop and prevented its fusion with contiguous drops as there was no tendency for the surface membranes to coalesce. To obtain the models of living cells, Roux compressed a series of such oil drops within the rim of a conical wine glass, and by varying the size and number of the drops, and the relative size of the glass container, he showed that

THE SHAPE OF CELLS

most striking models could be obtained of segmenting eggs of different animal types (see fig. 113). If Roux's oil drops are anything more than an analogy to living systems, three facts must be established: (i) each living cell must, when isolated from its neighbours, naturally

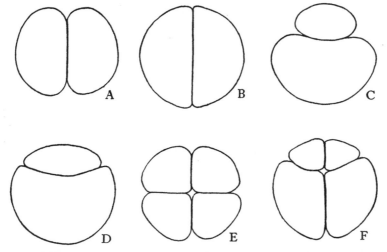

Fig. 113. Oil drops mutually compressed against each other. Note the similarity of A, B and E to orthoradial cleavages of animal eggs. C, D, and F are comparable to unequal cleavages. (From Roux.)

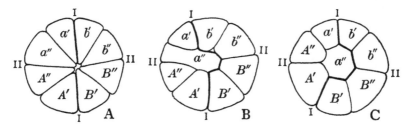

Fig. 114. Three stages in the passage of a large drop of oil (a'') into the centre of a system of smaller drops. Comparable changes of position can be observed when *Echinus* eggs segment under pressure. (From Roux.)

assume a spherical state, (ii) there must exist at the surface of the cell a mechanism whereby two contiguous cells are prevented from complete fusion when subjected to mechanical pressure, (iii) there must exist within the living system a centripetal force compressing all the blastomeres. For strict accuracy, the whole cell system should be confined within a rigid spherical shell. So far, no attempt has been made to determine the truth of these assumptions for the particular living material chiefly concerned in Roux's models, since the blastomeres of the frog are not readily isolated from each other. With other material, however, the evidence supporting the fundamental parallel between the inanimate and animate systems is perhaps convincing. Herbst (1900) showed that when a segmenting echinoderm egg is exposed to artificial sea water which contains no calcium, the individual blastomeres readily separate from each other. From his figures and from subsequent observation (Gray, 1924) it is clear that when isolated in this way each cell is spherical. That the surface of the cell is normally covered with a sticky or adhesive substance which loses its adhesive character when in contact with the normal environment for any length of time has already been shown (Chapter VI). This surface hyaline layer must be present and must enclose the cells in a common investing layer if the cells are to maintain their normal form. It is the removal of this layer which enables the cells to resume their natural spherical form, and it is, therefore, more than probable that it is this layer which exerts the centripetal pressure exerted by the rim of Roux's wine glass. Any environment which alters the mechanical properties of the hyaline layer of the cell alters the degree of mechanical pressure exerted on the individual blastomeres. This can be illustrated with the eggs of *Echinus esculentus* (Gray, 1924). The hyaline membrane of these eggs rapidly loses water and contracts when exposed to acid sea water, consequently when two celled stages of these eggs are treated with acid the pressure exerted by the hyaline layer is increased and the blastomeres are tightly compressed against each other (see fig. 84). By partially removing the hyaline layer from the egg, it can be shown that the compressing effect observed in acid sea water is confined to the area from which the hyaline layer has not been removed.

Similarly if the hyaline layer is lifted away from the egg surface by treatment with hypertonic sea water (Gray, 1924), the blastomeres become spherical (see fig. 83).

It is thus reasonably clear that the system of living echinoderm cells fulfils the fundamental requirements for Roux's models. In other cases it is more likely that an individual blastomere when isolated would only become spherical if the isolation were effected immediately after cell division, and even then only over that region

THE SHAPE OF CELLS

which is directly in contact with the new cleavage interface. Such cases, however, are susceptible to the same type of analysis as applies to echinoderm blastomeres.

The adequacy with which Roux's models reproduce the form of living cells can hardly be questioned, but we are still faced with the fundamental problem of defining the shape of compressed oil drops in such a way as will enable us to feel that the mechanical problem of cell form has been completely solved. Since each individual cell when isolated from its neighbours is assumed to be spherical, it follows that *in situ* its surface area must be greater than when the neighbouring cells are absent. The process of mutual deformation is opposed by the elastic force exerted by each deformed cell and will cease when this force equals that which is pressing the cells together.

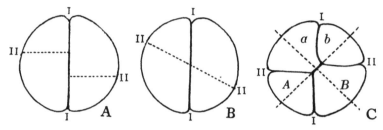

Fig. 115. Two equal compressed oil drops are each divided by an unequal division shown by the dotted lines in *A* or as in *B*. The stable position reached in both cases is shown in *C*. Note the 'polar furrow' in *C* and that the system has two planes of symmetry shown by the dotted lines. (From Roux.)

At the point of equilibrium the surface energy of each cell will be the minimum which is possible under the circumstances, any other condition will be unstable (see also fig. 115). The law of minimum cell surfaces has been known for many years, but it is essential to remember that in its strict and accurate form the law does not in any way define the nature of the forces operating at the cell surface—it simply depends on the existence of free surface energy and this may be of any type. As long as we are dealing with a cell which is completely surrounded by other similar cells, the theoretical form can be deduced from physical data with some degree of certainty, but when we deal with cells, part of whose periphery is not in contact with other cells, the problem becomes much more difficult and much less suitable to geometrical analysis.

The law of minimal surfaces

The newly-formed interface between contiguous cells has that form whereby its surface area is reduced to a minimum. The truth of this statement is clearly illustrated in Thompson's analysis of the segmentation figures of *Erythrotrichia* (fig. 116). To this author we owe a masterly discussion of the whole problem of cell form. Starting with a flat unsegmented disc the first two cleavages divide the disc into four quadrants with the interposition of the small polar furrow necessitated by the fact that four interfaces meeting in a point are physically unstable. The existence of the 'polar' furrow characteristic of the second cleavage is, as pointed out by Thompson (p. 309), the direct consequence of the law of minimal area, for it can

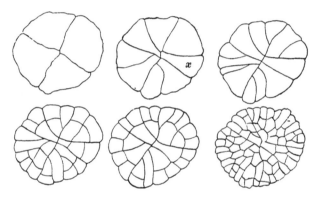

Fig. 116. Segmentation stages of *Erythrotrichia*. (From Thompson.)

be shown that in dividing a closed space into a given number of chambers by partition walls, the least possible area of these partition walls, taken together, can only be attained when they meet together in groups of three at equal angles (see also W. Thomson, 1887).

For the third cleavage there are two possibilities whereby not more than three surfaces shall meet in a point: (i) the third cleavage planes may be periclinal as in fig. 117 B, or (ii) it may be *anticlinal* as in fig. 117 C. Now in order that a quadrant may be divided by an anticlinal partition into two equal parts it is necessary that the circular arc cutting the side of the periphery of the quadrant should include 55° 22′ of the quadrantal arc, and the length of the partition wall is 0·8751 where the radius of the original quadrant is 1·0000. In the

THE SHAPE OF CELLS

case of a periclinal partition the length of the partition wall (for equal cleavage) is 1·111, so that the anticlinal cleavage is the more efficient type. In point of fact, it is this mode of cleavage which characterises the third division of discoidal systems in nature. Subsequent divisions of the two cells of each quadrant also conform to theory to a marked degree. The four-sided cell x (figs. 116 and 117) divides periclinally as one might expect.

If we compare the theoretical arrangement of successive partitions in a discoidal cell as defined by Thompson (see fig. 118), with the arrangement actually found in nature (fig. 116), it is difficult to avoid the conclusion that the law of minimal surfaces is of profound importance in the determination of cell form.

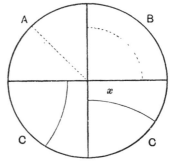

Fig. 117. Alternative cleavages of a quadrant. (From Thompson.)

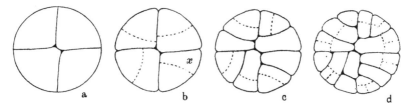

Fig. 118. Theoretical arrangement of partition walls in a discoidal cell. (From Thompson.)

The law of minimum surface as applied to parenchymatous cells

If it be assumed that a given cell when surrounded by other cells (all of the same size and all exerting the same influence on each other) occupies that form in which the law of 'minimal surfaces with no intercellular spaces' is strictly obeyed, then the shape of each cell can be predetermined from geometrical principles. Prior to 1887 it was generally believed that a given space could be completely divided into equal subdivisions (the total surface of which covered a minimum area), when each subdivision had the form of a regular twelve-sided figure or orthic dodecahedron, each facet of which was a regular hexagon. In 1887, however, Kelvin demonstrated that a more stable and more efficient method of equal subdivision of space was presented by a fourteen-sided figure or tetrakaidecahedron (fig. 119); of the fourteen surfaces, six are quadrilateral and eight are hexagonal.

The properties of this figure have been described by Matzke (1927) and by Lewis (1926–8). If each quadrilateral surface has a side of length a, then the side of each hexagonal surface has also a length a. If a section is cut perpendicular to any edge of the solid figure, a hexagonal figure results of which four sides are longer than the remaining two. The two short sides

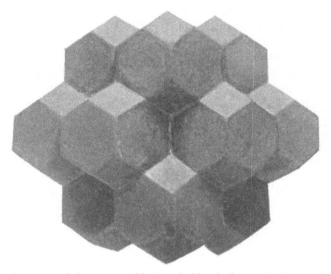

Fig. 119. A group of fourteen orthic tetrakaidecahedra: note the hexagonal and quadrilateral facets. The space enclosed by this group is itself an orthic tetrakaidecahedron. (From Matzke.)

have a length a, whilst the four longer cells have a length $\sqrt{3}a = 1\cdot732a$; the perpendicular distance between two short sides is $2\cdot828a$, whilst the perpendicular distance between two long sides is $2\cdot449a$. The internal angle between two long sides is $109°\ 28'\ 16''$, while that between a long and a short side is $125°\ 15'\ 52''$. The total area of the whole hexagon is $4\sqrt{2}.a$.

If the form of parenchymatous cells conforms to that of a series of orthic tetrakaidecahedra so arranged as to leave no intercellular spaces, then a section cut at right angles to any one cell interface should yield a series of hexagonal figures all of a definite type. That living cells do, in fact, conform to such a system has been assumed on more than one occasion (see D'Arcy Thompson), but only recently have definite data become available (Table XXX). Wetzel (1926)

THE SHAPE OF CELLS

has shown that the pigmented cells of the human retina when viewed as a horizontal section are polygonal in form and possess from four to nine interfaces. More than half of the cells have six interfaces when seen in section. Lewis (1926) has demonstrated the same fact in sections of the pith of *Sambucus canadensis*.

Table XXX

No. of sides	4	5	6	7	8	9	Average
No. of cells: retinal cells	2	106	242	98	3	1	5·993
No. of cells: pith cells	20	251	474	224	30	1	5·996

The cells of *Sambucus* have been examined in detail by Lewis, and from serial sections the number of facets possessed by each cell has been determined.

Table XXXI

No. of cell surfaces	6	7	8	9	10	11	12	13	14	15	16	17	18	19	20
No. of cells	1	1	2	0	2	8	8	21	16	19	10	2	3	6	1

For the hundred cells examined (Table XXXI) the average number of facets per cell was 13·96. It would therefore seem probable that the cell population tends, on the average, to conform to the tetrakaidecahedral form. Wax models of forty-two cells failed, however, to reveal any individual cells with fourteen surfaces. In view of the fact that all the cells are not of exactly the same size, and of the disturbing effect of cell division, this departure from the theoretical result is not altogether surprising. As Lewis points out, the expected result of cell division would be a reduction of the number of facets from fourteen to eleven. Alternatively, Wetzel attributes the irregularity in the polygons seen in section to differential growth; thus in fig. 120 a series of four hexagonal cells might resolve itself into two pentagons and two heptagons if two of the cells increase in size more rapidly than the other two, and thereby displace their neighbours. It must be admitted, however, that there is no general agreement concerning the origin of those cells which exhibit more or less than six sides when viewed in transverse section. Cell division (Lewis, 1926), cell growth (Wetzel, 1926), and cell absorption and fusion (Gräfer, 1919) may well cause irregularities but in no case has the actual process been seen under the microscope.

As pointed out by Lewis, there is definite evidence against the view that the typical form of a parenchymatous cell is that of an *orthic* tetrakaidecahedron. In a figure of this type any section cut transverse to any cell surface yields an irregular hexagon (see p. 256) having four long sides and two short sides. The hexagons seen in actual sections of tissues are, however, variations not of this irregular hexagon but of a regular hexagon, all of whose sides are equal in length. If this be the case, the cells cannot be packed together to form a system with no intercellular spaces.

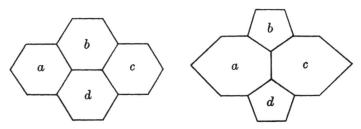

Fig. 120. Diagram to illustrate the transition from hexagonal cross section to pentagonal and heptagonal cross sections by differential growth. The two cells *a* and *c* have displaced the two slower growing cells *b* and *d*. (After Wetzel.)

In any discussion of cell form it is essential to differentiate clearly between the biological facts and the expectations which are based on purely geometrical systems. In view of the natural variations in the size and properties of individual cells, and of the variations in external surroundings to which individual cells may be exposed, it is more remarkable to find such a close parallel between fact and geometrical theory than to find divergencies of a secondary nature.

Limitations of the law of minimal surfaces

In assessing the value of the above facts, it is desirable to remember that a system of living cells differs in many ways from a physical system of soap bubbles, which displays to best advantage the operation of the minimal surface law.

If we are prepared to accept the facts in the form of Errera's Law (as formulated by Thompson), viz. 'A cellular membrane at the moment of its formation tends to assume the form which would be

THE SHAPE OF CELLS

assumed, under the same conditions, by a liquid film destitute of weight'—we come very near to suggesting that these cellular membranes actually possess liquid properties. Both Errera and Hofmeister accepted this view. It is, however, contrary to many well-established facts. As already pointed out, the fundamental law of minimal surfaces does not depend upon the liquid nature of the interfaces, but on the fact that these are the seat of free energy. It seems therefore desirable to restate Errera's Law in a form applicable to modern conceptions of the cell surface. 'A cellular membrane at the moment of its formation tends to assume the form which would be assumed under the same conditions by an elastic membrane destitute of weight.'

There is also, however, a grave difficulty which has so far been overlooked in most theoretical analyses. All diagrammatic representations of cell form assume that the thickness of the cell surface is negligibly small, and that it can be treated as a mathematical surface. Unless the cell membrane is extremely thin and incapable of plastic flow it cannot strictly conform to the ideal state of films destitute of weight. Nor can we neglect this fact without serious consequences. Taking the apparently simple case of two equal and contiguous cells, the ideal state, as defined by Thompson (p. 301) is illustrated by fig. 121; the tensions exerted by all contiguous surfaces (being assumed to be equal to each other) are in equilibrium when each contiguous surface meets its neighbour at an angle of 120°. 'At any point of the boundary of the partition wall, O, the tension exerted by all the films being equal, the angles QOP, ROP, QOR are all equal to 120°. But OQ, OR being tangents, the centres of the two spheres (or circular arcs in the figure) lie on perpendiculars to them; therefore the radii CO, $C'O$ meet at an angle of 60° and COC' is an equilateral triangle. That is to say, the centre of each circle lies on the circumference of the other, the partition lies mid-

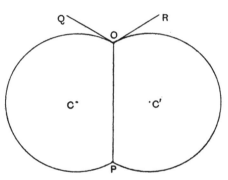

Fig. 121. Position of equilibrium for two contiguous spherical shells each bounded by an elastic membrane whose thickness is either negligibly small or which cannot change by plastic or viscous fluid.

way between the two centres, and the length (i.e. the diameter) of the partition wall, PO, is 1·732 times the radius

$$2 \sin 60° \, r = 1·732 \, r,$$

or 0·866 times the diameter of each of the cells. This gives us then the form of an aggregate of two equal cells under uniform conditions.'
As pointed out elsewhere (Gray, 1924, see also p. 195) this analysis of cell form can be seen to be inadequate by visual observation of any suitable system of living cells. In their natural state the adjacent cells of a two-celled system are not portions of true spheres (fig. 80). The departure from the ideal form is probably due to the fact that the cell surface has a finite thickness.[1] Consider a group of contiguous soap bubbles in which the films have a considerable thickness. In this case the surface energy of the system will not be at its minimum when the free surface occupies the form of two

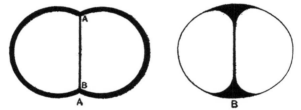

Fig. 122. Two water drops enclosed within a drop of olive oil. A, is unstable; B, is stable. Note the asymmetrical distribution of the oil phase and the compressed form of the water drops in the stable condition. The oil phase is in black.

partial spheres, for the free surface will be still further reduced by a flow of liquid from the poles of the system to the equator, whereby the area in contact with the external medium is still further reduced in area. The new equilibrium is shown in fig. 122B, wherein it can be seen that the two resultant drops are not partial spheres but bear an unmistakable resemblance to the form of living cells (Gray, 1924). It is important to notice that the form of the blastomeres of a sea-urchin is precisely the same as that of a *small* soap bubble, where the thickness of the films is significantly large in comparison with the total area of the bubbles. The tendency for fluid to accumulate at the junction of two liquid surfaces was clearly recognised by Willard Gibbs (1906, p. 290), and the area concerned is sometimes known as Gibbs' ring. Unfortunately the exact form

[1] See footnote, p. 297, Thompson, 1917.

THE SHAPE OF CELLS

of this area has not yet been shown to be susceptible to geometrical or mechanical definition. The departure from the simple theoretical form does not in any way invalidate the law of minimal surfaces, it simply indicates that factors other than those represented in fig. 121 control the form of the minimal surfaces in the region common to two or more surfaces. Thompson's simplified system, whilst applicable to oil films which are negligibly thin compared to the volume of the space they enclose, becomes insufficient when applied to very small drops or to living cells; the 'surface of continuity' becomes in such cases of major importance. There is therefore no true angle of contact between cell surfaces, for at each angle there exists a prismatic accumulation of intercellular substance (see fig. 123). The net result of this disturbing factor was clearly recognised by Thompson in the following paragraph (p. 297): 'We have seen that, at and near the point of contact between our several surfaces, there is a continued balance of forces, carried, so to speak, across the interval; in other words there is a physical continuity between one surface and another. It follows necessarily from this that the surfaces merge one into another by a continuous curve. Whatever be the form of our surfaces and whatever the angle between them, this small intervening surface...is large enough to be a common and conspicuous feature of the microscopy of tissues'.

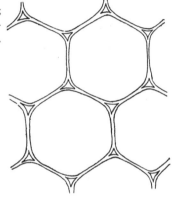

Fig. 123. Section of the parenchyma of maize; note the prismatic interstitial spaces and absence of true angles of intersection. (From Thompson.)

One is inclined to go further than this and admit that in many cases the simpler laws of minimal surfaces are masked with sufficient effect as to render it difficult to define the geometry of cell form with any real accuracy. It will be realised without further comment that the existence of a series of interfaces meeting at a definite angle of 120° is only a theoretical conception.

As far as the evidence goes, it seems fairly clear, however, that the average form of some parenchymatous or epithelial cells approximates with surprising accuracy to the theoretical form of closely packed units which enclose a maximum volume by a minimum of surface. In other cases, however, this is far from being true. The branchial epithelium which covers the gills of *Mytilus* is composed of cells whose outline is greatly wrinkled (fig. 124), and the neighbouring ciliated cells are often rectangular in section. The underlying causes of these anomalies are unknown, but they indicate the danger of applying the principle of minimum cell surfaces over too wide a field.

Finally, it is perhaps permissible to doubt how far future observations of the precise form of prismatic epithelial cells will conform to theoretical expectation, and how far the living cells will be found to vary in form to such a degree as to restrict geometrical analysis to a few selected tissues.

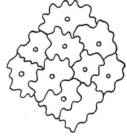

Fig. 124. Outline of cells on branchial epithelium of *Mytilus*. (Diagrammatic.)

Biological conception of cell form

If the form of a cell is strictly controlled by extraneous mechanical forces, it follows that the specific differences seen in the segmentation of different types of spherical eggs must be due to differences in the mechanical surroundings of each cell, rather than to more fundamental differences in the nature of the cells themselves. From this point of view, the segmentation process, as seen in an *Echinus* egg, conforms closely to that of *Arbacia*, not because the two species belong to the same group of animals, but because the mechanical surroundings of each blastomere are closely similar in each case. The segmentation of an *Echinus* egg, from the same point of view, differs from that of *Nereis* because the mechanical surroundings of the blastomeres are different. Similarly, the marked similarity of cleavage pattern seen in Polychaet annelids, gastropod molluscs, and polyclad turbellarias is not so much due to any phylogenetic affinity as to a similarity in the mechanical surroundings of each comparable blastomere. Such an ultra-mechanical conception of segmentation would ascribe the differences in form of all animals to the fact that there comes a time in the segmentation of the egg when a given blastomere divides at a different time or at a different rate in any two given cases. If the mechanical conception of cleavage be extended to include the thesis that the direction of cleavage is itself determined by the mechanical environment of the cell, then the only basis left for differential development lies in inequalities of size of comparable blastomeres or in inequalities of rates of division. In other words, we would be forced to conclude that if the cleavage of a sea urchin's egg could be controlled, so that the size and position of each early blastomere were made to conform to that of a mollusc, the resultant organism would belong to a different phylum to that of the original egg! Putting the mechanistic conception of cell form on one side, it must be admitted that a more comprehensive conception of the development of a living egg is that put forward by F. R. Lillie (1895). '...Each component cell of the organism appears to take up a position and behave in such a manner as clearly foreshadows the final rôle which it will be required to play.' In so doing it must conform to mechanical principles—and if it disobeys Errera's Law, it must do so by definite mechanical means: the cell wall may cease to be the seat of tension energy at the moment of formation, or the cell must prevent this tension from operating in the normal way by an appropriate expenditure of energy.

Grave errors may readily be the result of driving limited data to their logical conclusions. The fact that two equal and contiguous soap bubbles conform to a simple geometrical pattern and that a series of small contiguous soap bubbles tend to approximate to orthic tetrakaidecahedra, does not enable us to define the form of the foam which collects at the surface of a washtub. So with living cells, we can detect unmistakable signs of mechanical forces, but the more delicate forces which are fundamental for the determination of the form of tissues or of whole animals are the result of deep-seated and predetermined characters inherent in the cells, and are not the result solely of those simpler forces to which the cells, once they are formed, are undoubtedly subjected.

The shape of the mammalian red blood corpuscle

In 1919 Hartridge suggested that the characteristic discoidal form of mammalian erythrocytes is an adaptation to the physiological functions of the cell. In order that oxygen should reach the centre simultaneously from all points of the surface, the cell must be either a sphere or an infinitely thin disc. If the red blood cell were spherical, however, it would present a minimum surface per unit volume, and consequently the rate at which oxygen would enter would be reduced to a minimum. In a flat disc, however, oxygen would reach the centre more readily at the periphery than elsewhere. The peculiar form of the mammalian erythrocyte compensates for this by its greater thickness at the periphery—so that oxygen will reach the centre of the cell simultaneously from all parts of its surface. An interesting extension of this conception has been made by Ponder (1925–6), who points out that if a gas starts from a line of equal velocity potential and passes along a line of flow towards one of two adjacent sinks, in doing so it will pass at right angles to all the lines of equal velocity potential which it traverses. If the strength of the two sinks be m_1 and m_2, and if the sinks are separated by a distance a, then the lines of equal velocity potential can be defined by the equation applicable to the equipotential curves of Cayley (1857):

$$\frac{m_1}{r_1} + \frac{m_2}{r_2} = \frac{k}{a},$$

where

$$\frac{1}{r_1} + \frac{1}{r_2} = C.$$

Ponder has shown that if suitable values for the velocity potential (ψ) are selected, the lines of equipotential round the two sinks bear a marked resemblance to the form of an erythrocyte. Such a figure, when rotated about its minor axis, yields a solid approximating to the form of a red blood cell and the equipotential line forming the curve becomes the equivelocity potential surface of the solid of revolution. Gas starting from any point on this surface and moving inwards (from any point on the surface) along lines of flow will reach the circular sink in the same time. Further, any line of equal velocity potential must also be a line of equal gas concentration. If a red blood cell containing no oxygen is exposed to a fluid containing the gas, the surface of the cell will be one of equal gas concentration, and therefore gas will pass across the surface towards the inner parts of the cell to form a series of surfaces of equal gas concentration and of equal velocity potential, and will converge on the circular sink simultaneously from all parts of the cell's surface. Ponder points out that the red blood cell is not quite so rounded at its ends as is the solid of revolution of the curve $\psi = 7.5$, and that its concavity is not quite so deep; further, the volume of a blood cell is about $110\,\mu^3$, whereas that of the theoretical solid of revolution is $196\,\mu^3$. Ponder concludes that, although the form of the cell cannot be rigidly defined by one of the equipotential curves of Cayley, yet the general resemblance between the two systems supports Hartridge's original suggestion. The approximation to the theoretical form indicates that the efficiency of the cell to absorb oxygen is very great compared to other systems of the same volume and very much more efficient than if the cell were spherical.

As long as a mammalian red blood corpuscle is suspended in plasma, it retains its biconcave form more or less indefinitely; if, however, the cells are suspended in isotonic saline (0·85 per cent. NaCl) the form is altered in a curious way as soon as the cells lie within a critical distance of two flat surfaces. If a suspension of corpuscles is enclosed in a thin film between a coverslip and a glass slide, the cells very rapidly lose their typically biconcave shape and become perfect spheres (Ponder, 1929). The reason for this change is obscure, but the rate at which it occurs clearly depends on the distance between the two opposing glass surfaces; if these are very close together the change may be complete within a fraction of a second. Neither the pressure of the coverslip or the use of quartz slides alters the phenomenon. The only factor—apart from the absence of plasma—which is necessary for the cells to acquire the spherical form, is that the surfaces should be such as will be wetted by the saline; if glass is covered with paraffin wax, the normal biconcave form is retained in saline. The whole phenomenon is very obscure, but Ponder is inclined to the belief that it is due to the molecular attraction fields which are known to exist within 10μ of two closely applied surfaces (Hardy and Nottage, 1926).

REFERENCES

CAYLEY, G. (1857). 'On the equipotential curve $\frac{m}{r} + \frac{m'}{r_1} = C$.' *Phil. Mag.* 14, *142*.

ERRERA, L. (1886). 'Sur une condition fondamentale d'équilibre des cellules vivantes.' *C.R. Acad. Sci. Paris*, 103, *822*.

GIBBS, J. WILLARD (1906). *Scientific Papers*. Vol. 1. (London.)

GRÄFER, L. (1914). 'Eine neue Anschauung über physiologische Zellausschaltung.' *Arch. f. Zellforsch.* 12, *387*.

—— (1919). 'Mechanische Betrachtungen und Versuche über Zellform und Zellgrösse.' *Arch. f. Entw. Mech.* 45, *447*.

GRAY, J. (1924). 'The forces which control the form and cleavage of the eggs of *Echinus esculentus*.' *Proc. Camb. Philos. Soc. Biol. Series*, 1, *164*.

HARDY, W. B. and NOTTAGE, M. (1926). 'Studies in adhesion. I.' *Proc. Roy. Soc.* A, 112, *62*.

HARTRIDGE, H. H. (1919). 'Shape of red blood corpuscles.' *Journ. Physiol.* 53, *lxxxi*.

HERBST, C. (1900). 'Über das Auseinandergehen von Furchungs- und Gewebezellen in kalkfreiem Medium.' *Arch. f. Entw. Mech.* 9, *424*.

LEWIS, F. T. (1923). 'The typical shape of polyhedral cells in vegetable parenchyma and the restoration of that shape following cell-division.' *Proc. Amer. Acad. Sci.* 58, *537*.
—— (1926). 'The effect of cell-division on the shape and size of hexagonal cells.' *Anat. Record,* 33, *331*.
—— (1928). 'The correlation between cell-division and the shapes and size of prismatic cells in the epidermis of *Cucumis*.' *Anat. Record,* 38, *341*.
LILLIE, F. R. (1895). 'The embryology of the Unionidae.' *Journ. Morph.* 10, *1*.
MATZKE, E. B. (1927). 'An analysis of the orthic tetrakaidecahedron.' *Bull. Torrey Bot. Club,* 54, *341*.
PONDER, E. (1925–6). 'The shape of the mammalian erythrocyte and its respiratory function.' *Journ. Gen. Physiol.* 9, *197*.
—— (1929). 'On the spherical form of the mammalian erythrocyte.' *Brit. Journ. Exp. Biol.* 6, *387*.
ROBERT, A. (1902). 'Recherches sur le développement des troques.' *Arch. de Zool. expér. et gén.* (3), 10, *269*.
ROUX, W. (1897). 'Ueber die Bedeutung "geringer" Verschiedenheiten der relativen Grösse der Furchungszellen für den Charakter des Furchungsschemas.' *Arch. f. Entw. Mech.* 4, *1*.
*THOMPSON, D'A. W. (1917). *On Growth and Form.* (Cambridge.)
THOMSON, W. (1887). 'On the division of space with minimum partitional area.' *Phil. Mag.* 5th Series, 24, *503*.
WETZEL, G. (1926). 'Zur entwicklungsmechanischen Analyse des Einfaches prismatischen Epithels.' *Arch. f. Entw. Mech.* 107, *177*.

CHAPTER ELEVEN

The Growth of Cells

It is by no means easy to define the process of growth. When we speak of the growth of cartilage within the body of an embryo, at least three concepts are involved. Firstly, there is an increase in the bulk of the whole tissue, which is largely due to the deposition of an intercellular secretion having no obvious vital properties. Secondly, there is a marked increase in the number of cells present. Thirdly, each individual cell undergoes, during each intercleavage period, an increase in size. In order to avoid confusion, it is convenient to restrict the term 'cell growth' to the last of these three phenomena, which is essentially concerned with the formation of new living material.

The growth of a cell must be clearly distinguished from cell division, and for this reason it is convenient to consider at the outset the growth of cells which are not undergoing periodic cleavage. Cells of this type are found in at least two highly differentiated forms—viz. muscle fibres and nerve cells. In the newly-hatched eggs of nematodes the full number of nerve cells and muscle cells of the adult are already present, and as the animal increases in size these cells grow sufficiently rapidly to keep pace with other tissues; no new myoblasts or neuroblasts are formed. It follows that the size of individual cells of this type must be determined by the size of the whole animal or *vice versa*. Analogous facts apply to vertebrate animals. If the myotomes of a young fish embryo are examined, the muscle fibres are seen to originate as small myoblasts which fail to show any trace of striation until the two ends are fixed to the anterior and posterior walls of the myotome. At this stage striations appear and the cell begins to grow. It continues to grow throughout the whole growth cycle of the animal so that it often increases its bulk many thousands of times without undergoing any process of cell division. Unlike the myoblast of the nematodes, however, the growth of the fish's muscle fibre does not keep pace with the growth in girth of the fish, for new fibres are continually being added to the myotomes from fresh undifferentiated myoblasts (see fig. 126 and Table XXXII).

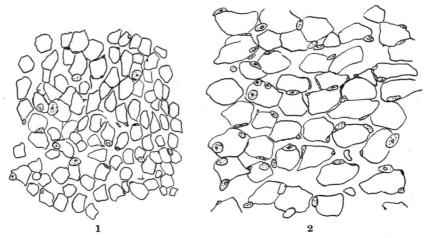

Fig. 125. Camera lucida drawings of transverse sections of somatic muscle fibres of the trout. 1, Fish 1·45 cm. long, cross-sectional body area 1·0 sq. mm. Relative average area of fibre 1·0. 2, Fish 1·65 cm. long, cross-sectional body area 1·90 sq. mm. Relative average area of fibre 2·18. Ratio of cross-sectional body area to relative cross-section of muscle fibre is 1·0 and 0·9 respectively. (From data collected by D. Bhatia.)

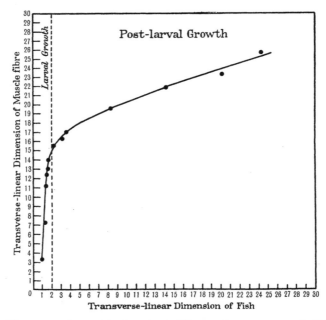

Fig. 126. The graph illustrates that the rapid growth of the somatic muscle fibres during the period of larval life is not maintained at a later stage. During post-larval life the relative rate of growth of the muscle fibre is less than that of the whole body.

Similarly, most if not all the nerve cells of an adult animal are present at an early stage in development. In both muscle and nerve the differentiated cell appears to have lost all faculty for cleavage, although it retains its capacity to grow, and the cessation of cleavage precedes the process whereby these cells begin to exhibit their characteristic and differentiated form. Prior to differentiation, cell division is active.

Table XXXII. Relative area of cross-section of complete fish (*Salmo fario*) and of muscle fibre

Fish	Average muscle fibre
1·0	1·0
2·0	2·1
5·0	4·1
16·8	5·3
69·0	6·5
200·0	8·1
592·0	11·3

In marked contrast to muscle fibres and nerve cells are epithelial cells. The average size of the epithelial cells in the liver, kidney, gut or skin of an adult vertebrate is not markedly larger than that at a much earlier stage in the life history (Illing, 1905). It is not easy to estimate the volume of a tissue cell, but Berezowski's (1910) measurements of cross-sectional area shows clearly that the gut cells of the white mouse rapidly reach their maximum size at an early stage in the growth cycle of the body (see Table XXXIII).

Table XXXIII

Age of mouse	10 days	1 month	2 months	3 months	4 months	5 months
Body volume	4	7	14	20	21	25
Average length of gut cells (μ)	14·3	18·9	20·2	22	21·7	23·2
Average width of gut cells (μ)	4·9	4·9	5·1	4·8	5·1	4·7
Relative cross-sectional area (10 days taken as unity)	1·00	1·32	1·47	1·51	1·58	1·55

THE GROWTH OF CELLS

Probably no hard and fast distinction should be drawn between epithelial cells on the one hand and muscle fibres and nerves on the other, for Plenk (1911) found quite a definite increase in the size of the epithelial cells of amphibians with increasing size of the whole animals.

Table XXXIV

Age of salamander	Volume of cells (μ^3)	
	Oesophageal epithelium	Stomach epithelium
Birth	1458	1242
2 months	3796	2788
12 months	6328	8268

As far as the efficiency of an organ is concerned, there is no very obvious reason why a muscle should increase in size by increasing the size of the existing fibres, whereas the liver should increase in size by increasing the number of its constituent cells.

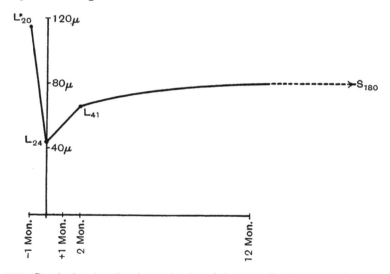

Fig. 127. Graph showing the change in size of the gut cells of *Salamander maculosa* with the age of the animal. The ordinates represent the long axis of the cell. (From Plenk.)

The factors which control the growth of cells are very imperfectly understood. In the case of muscle cells there can be little doubt that mechanical forces play an important part. The size of a fibre varies with its position in the body: it will grow in length wherever it is stretched between two relatively fixed points—a fact well exemplified by the hypertrophic growth of the pregnant mammalian uterus.

The nucleo-cytoplasmic ratio during growth

For many years it has been known that for any given type of epithelial cell, the volume of the cytoplasm tends to bear a fixed ratio to the volume of the nucleus. When, however, a cell is undergoing a series of successive divisions, this nucleo-cytoplasmic ratio is not constant, but varies at different phases of the divisional cycles. For example, Koehler (1912), working with the eggs of *Strongylocentrotus lividus*, found little or no evidence of increase in the total cytoplasmic volume of the whole eggs between the moment of fertilisation and the formation of the gastrula, but the total volume of the nuclei increased about ten times during the same period. If we judge growth by increase in bulk, obviously the growth of the nucleus is of a different order of magnitude to the growth of the cytoplasm.

Table XXXV. Cytoplasmic and nuclear growth in *Strongylocentrotus* eggs (16° C.). (After Koehler.)

No. of cells present	Volume of single nucleus	Volume of cytoplasm in a single cell	Total volume of nuclei	Total volume of cytoplasm	Ratio N/C
2	1395	142625	2790	285250	0·00978
4	742	75078	2968	300312	0·00988
8	607	30324	4856	242592	0·02009
16	547	16659	8754	266544	0·03284
32	366	8230	11722	263370	0·04451
64	332	4197	21235	268576	0·07907
501	63	510	31563	255510	0·12255
607	46	386	27665	234644	0·11917
820	32	263	26294	215396	0·12168
1163	19	242	22121	281790	0·07851

Up to about the ninth cleavage there is a very marked increase in the total volume of the nuclei, whereas the volume of the cytoplasm remains unchanged, and during this period the nucleo-cytoplasmic ratio increases to about ten times its original value. Except for those obtained at low temperatures, Koehler's results suggest that this period of rapidly increasing N/C ratios is followed by a period in which the ratio is fairly steady.

Koehler's results are also of interest when applied to the earlier observations of Popoff (1908) (see also p. 132). Popoff determined the relative volume of the cytoplasm to the nucleus at different stages in the divisional cycles of *Frontonia* and other protozoa. Ten minutes after a vegetative division the volume of the macro-nucleus was found to be approximately 1·3 per cent. of the volume of the cytoplasm: from this period onwards the volume of the cytoplasm was found to increase more rapidly than that of the nucleus, so that after some hours the N/C ratio had fallen to 1·0. Hertwig (1908) pointed out that this process could not go on indefinitely and suggested that it would lead to a state of 'nucleoplasmic strain', which in turn might induce the nucleus to undergo a process of division during which it would increase in volume and thereby re-establish the normal nucleo-cytoplasmic ratio. According to Popoff, the onset of cytoplasmic cleavage is preceded by a phase of very rapid increase in the volume of the nucleus so that, when cleavage occurs, the nucleo-cytoplasmic ratio has regained its normal value. It will be noted that the kinetic cycle of cleavage resembles that in sea urchin eggs in that it is associated with an increase in the volume of the nucleus, whereas the interkinetic cycle is characterised by cytoplasmic growth, a process which is absent in sea urchin eggs, since there is no clearly defined interkinetic period during the early cleavages of the egg cell (Gray, 1927).

Quite how much significance should be attached to observations of nuclear volume is uncertain. If nuclear volume were clearly an index of the amount of an essential nuclear compound the situation would be less indefinite: unfortunately this is not demonstrably true. Masing (1910) and Schackell (1911) failed to obtain any evidence of nuclein synthesis during the period in which the total volume of the nuclei of sea urchin eggs is known to increase by at least ten times. These results have recently been confirmed by Needham and Needham (1930), who concluded that all eggs, except those of terrestrial animals, contain, before fertilisation, all the nuclein constituents requisite for early development. We are forced to assume either that nuclear volume is not an adequate index of nuclear material, or that during development this material gradually passes from the cytoplasm into the nuclei. In both cases it must be admitted that nucleo-cytoplasmic ratios have, at present, no definite chemical significance. For further discussion, reference may be made to Robertson (1923) and Fauré-Fremiet (1925).

Metabolism of growing cells

The physiological state of a growing cell is clearly different from that of a cell which is in equilibrium with its environment—at the same time little or no precise information is available concerning the metabolic activities specific to growing cells. As growth proceeds, two distinct phenomena can be recognised: (a) the synthesis of new living material, (b) a rise in the total metabolism involved by the presence of this new material. As far as can be judged, the syntheses of growth are peculiarly efficient in the sense that little or no dissipation of free energy is involved in the process, so that whether a cell is growing or not the amount of energy set free per unit time per unit of living matter is the same. This conclusion will be considered in detail elsewhere; at the moment it is sufficient to stress the fact that it is by no means easy to define any one chemical or physical property which is peculiar to growing cells—in spite of the fact that the outward and visible signs of growth are themselves obvious. A possible exception is provided by the work of Warburg (1927), who found that young embryonic cells or cancer cells are able to effect anaerobic glycolysis to a much greater extent than the cells of older tissues which are either growing more slowly or not at all. Warburg found that malignant tumour cells produce three to four times more lactic acid per unit of oxygen consumed than do benign tumours, while three to five day old chick embryo tissue in the absence of oxygen produces lactic acid at almost the same rate as malignant tissue, but in the presence of oxygen normal respiration takes place with the formation of very minute amounts of lactic acid. According to Warburg, both growing and non-growing tissues can be classified into four groups according to the nature of their metabolism. (i) Normal resting tissue, characterised by a high rate of respiration and by slight powers of anaerobic glycolysis. (ii) Embryonic tissue with a high rate of respiration and high anaerobic glycolysis, but with low powers of aerobic glycolysis. (iii) Malignant tumour cells with low respiration and high aerobic and anaerobic glycolysis. (iv) Benign tumour cells with less active glycolytic powers than those possessed by malignant tissues. Many of Warburg's conclusions have been confirmed by Murphy and Hawkins (1925), but these authors failed to find any sharp segregation of other tissues into the four categories.

Quite clearly a cell will not grow unless provided with the raw materials with which to build up its essential parts. Each type of cell is to some extent dependent on the presence of specific materials, and these are doubtless of two types. Firstly, raw material for the provision of the free energy required for maintenance; secondly, the raw materials (nitrogenous and otherwise) for cell synthesis. The field of enquiry opened by these conceptions is almost unlimited, and only one or two salient features will here be considered.

In the first place the requisites for growth are at times of a very simple character. Koser and Rettger (1919) have shown that many bacteria are capable of active growth when the only available source of nitrogen is a single amino-acid or even ammonium phosphate; it is more than probable that some types rely solely on free ammonia. Similarly a number of 'typhoid' bacteria can rely on such simple compounds as acetic acid, oxalic acid or glycerol as their source of carbon. The work of Peters (1921) suggests that not only bacteria but also some protozoa are peculiarly modest in the materials required for growth. Between such types on the one hand and the much more specific requirements of mammalian cells on the other, there are doubtless intermediate types of great variety.

In the economy of most animal cells which are undergoing growth, sugars often play an essential rôle. Krontowski and Bronstein (1926) have shown that actively growing cultures of cells *in vitro* absorb sugar from their environment, and Watchorn and Holmes' (1927) work indicates that the sugar is used for maintenance, and is responsible for the production of free energy; in the absence of sugar, the cells fall back on proteins for these purposes with consequent elimination of ammonia and urea.

In addition to the type of material already mentioned, it is necessary to consider two others of a more intricate nature. Some bacteria fail to grow, or only grow to a limited extent, in an otherwise adequate medium if vitamin B is absent (Hosoya and Kuroya, 1923). Similarly, Allen (1914) found that the diatom *Thalassiosira* failed to grow in synthetic media unless a small trace of natural sea water was added; on the other hand, Jameson, Drummond and Coward (1922) obtained a good growth of another genus, *Nitzschia*, in purely synthetic media. Determinations of the vitamin requirements of isolated cells are peculiarly difficult, since it is essential to eliminate the technical errors which can readily creep into the experiments. For example, it does not seem clear whether the 'bios' postulated by Wildiers for the growth of yeast is to be regarded as vitamin B, inositol (Eastcott, 1928), or the hydrogen ion (Darby, 1930). In addition to substances of the vitamin type, cell growth is often dependent on substances peculiar to young or rapidly growing animals—the obvious example of this type being the embryo extract essential for the growth of vertebrate tissues grown *in vitro* and the more problematical 'trephones' described elsewhere (see p. 294).

The efficiency of growth

The efficiency of growth can be expressed in terms of material or in terms of energy. On the former basis it is customary to express the dry weight of new tissue formed as a percentage of the dry weight of the material utilised in its formation. The value so obtained by various authors is found, for ordinary conditions of experiment, to be of the order of 60 per cent. From a practical point of view this figure is of some interest, but theoretically it is not of great significance. The true efficiency of growth is independent of the processes of maintenance peculiar to the tissue, for this only occurs when the new tissue has actually been formed. Over a short period of time the material efficiency is expressed as a ratio

$$\frac{\text{Weight of new tissue formed}}{\text{Weight of raw material used} - \text{Weight of raw material used for maintenance}}.$$

In practice, the weight of raw material used for maintenance has to be calculated from the rate of respiration of the tissue, converting the O_2 consumed, or the CO_2 eliminated, into terms of raw material. Data of this type have been collected from a number of sources and may be illustrated by Table XXXVI, which applies to the developing trout (Gray, 1926). It is clear that after allowing for the 'wastage' of material involved by the maintenance of the living tissue after it is formed, there is no evidence which suggests that the conversion of raw material into new tissue is anything than a highly efficient process.

Returning to the figure of 60 per cent. as an estimate of the gross efficiency of growth, it will be noted that this figure must vary to some extent as time proceeds for an increasingly large percentage of available raw materials will be used up for maintenance leaving, under most experimental conditions, a smaller amount available for growth. Further, the actual value of the efficiency coefficient will vary with any factor which affects differentially the processes of growth and of maintenance.

Table XXXVI

Day of development (after fertilisation)	Dry weight of embryo	Dry weight of yolk equivalent to O_2 consumed	Dry weight of remaining yolk	Total yolk
	(mg.)	(mg.)	(mg.)	(mg.)
46	3·04	1·38	33·21	37·63
50	3·78	1·88	31·57	37·23
53	4·53	2·34	30·75	37·62
57	6·88	3·11	27·68	37·67
60	8·48	3·84	25·01	37·33
68	13·00	6·45	18·53	37·98
71	15·50	7·77	17·63	40·90
75	18·08	9·56	14·76	42·40
78	20·16	10·83	7·79	38·78
80	23·04	11·40	7·38	41·82

Observed average amount of yolk in an unfertilised egg 37·72 mg.

It is, perhaps, more satisfactory to assess the efficiency of cell growth from the point of view of a redistribution of energy rather than of materials, since in some cases a gram of organised tissue may contain a higher energy content than a gram of raw materials. A fairly extensive series of observations are available for the growth of *Aspergillus* from the work of Terroine and Würmser (1922). The mould was grown in a suitable synthetic medium containing glucose. The energy content of the initial and final media, together with that of the mould formed, was determined calorimetrically and expressed in calories. Thus the gross efficiency is

$$\frac{\text{Energy of organised tissue }(T)}{\text{Energy of initial medium }(M_1) - \text{Energy of final medium }(M_2)}.$$

In one experiment the medium originally contained 6·0 calories; it yielded a mycelium containing 3·48 calories, with a final medium containing 0·125 calories; thus the gross efficiency is approximately 60 per cent. Were these figures expressed as grams dry weight of substance the result would be slightly different, since 1 gr. of mycelium is equivalent to 4·8 calories, whereas a gram of glucose contains only 3·76 calories. To reach the true estimate of efficiency it is necessary to allow for the energy dissipated by respiration; knowing the heat produced by the formation of the observed amount of CO_2 from sugar, it was found that 1·97 calories were expended in this way.

Table XXXVII

Energy in original medium	6·00 cal.	Energy in final mycelium	3·48 cal.
		Energy in final medium	0·12 ,,
		Energy equivalent to CO_2 evolved	1·98 ,,
		Unaccounted balance	0·42 ,,
	6·00 cal.		6·00 cal.

The figures show that only 7 per cent. of the total energy of the original medium is unaccounted for and this (0·42 calories) is probably within the region of experimental error. It now remains to consider how far the energy accounted for by respiration can properly be regarded as normal maintenance. From a purely material basis it would seem as though the whole of the heat of respiration can be attributed to this source (see Gray, 1926), but using an indirect method and basing their conclusions on observations of energy, Terroine and Würmser concluded that this is not the case, and that only about one-half of the dissipated energy can be attributed to normal upkeep—and that the true efficiency of growth is thereby reduced to about 72 per cent. when estimated in terms of energy. It is possible that this dissipated energy is stored in the organism as free energy, although the figure (28 per cent.) seems somewhat high. It is worth noting that Terroine and Würmser's analysis only applies to the rather unusual conditions of growth, where the rate of formation of new tissue is constant and independent of the amount of tissue present.

The kinetics of cell growth

When a cell divides into two and each daughter cell acquires the size of the parent cell, we are usually prepared to admit that growth has occurred. Similarly, when a cell increases in size, without any apparent accumulation of secretory products or of excess water, the cell is again exhibiting a process of growth. In actual practice, however, it is exceedingly difficult to put forward a definition of cell growth which will enable us to express the process in a quantitative manner. The final unit of growth must ultimately be expressed in terms of 'growing' material just as the rate of a chemical reaction is expressed in terms of the molecules or ions of each substance involved. For the moment we shall assume that the cell is a natural unit of living matter, and consider how far this conception enables us to express the facts of growth in a reasonable and orderly manner.

Given an actively dividing bacterium, or tissue cell *in vitro*, it seems certain that the cell will grow and reproduce itself at a constant speed as long as the conditions in which it lives are maintained at a satisfactory and constant level. Observations of this type were made by Barber (1908), who isolated single individuals of *Bacillus coli* from a rapidly growing culture, and observed their rate of reproduction at a constant temperature. Table XXXVIII shows that the time which elapsed between one generation and another was approximately constant at 20·0 minutes. Darby's (1930) figures show that the same fact is probably true for protozoa such as *Paramecium*, although in this case the facts are less complete and the degree of variation between individual growth cycles is not yet definable. Using cells grown *in vitro*, Carrel and Ebeling (1921 b) found that fibroblasts in a given culture reproduced themselves once in every forty-eight hours. In this case also, the data are as yet

Table XXXVIII

Generation time in minutes		
18·7	20·2	21·2
19·3	21·2	19·4
20·2	20·0	19·8

incomplete, for it is by no means easy to subject growth *in vitro* to quantitative measurement. Ebeling (1921) has given evidence, however, which indicates that the rate of increase of the surface area of an active culture can be accepted as an index of growth which is accurate to within 6 per cent., and using this method it seems clear that, given a suitable medium, the rate of growth of fibroblasts *in vitro* can be maintained at a high and constant level for very prolonged periods. As long as the conditions of the environment are satisfactory, therefore, there is no inherent tendency for an isolated cell to show a decreasing ability to grow and divide (see, however, Calkins (1926)). Under such circumstances the total number of cells present at any particular multiple of the generation time can theoretically be calculated by a process of compound interest as long as no cell dies or as long as a fixed percentage of the population dies per unit time. If the rate of growth and the frequency of division of each individual derived from the single parent cell were identically the same, it follows that the growth of the culture—as measured in terms of individual cells—would fluctuate between two extremes. At the moment at which all the cells divide the rate of reproduction will be determined by the number of cells (x) present; immediately after division $2x$ cells are present, and there will be no further increase until the next division period is reached. It is a simple matter to calculate the number of cells (x) present at any given time, since $x = e^{kt}$, where k is a constant equal to $\dfrac{\log_e 2}{\text{generation time}}$, and t is the time as calculated to the preceding period of cleavage. In practice, this state of affairs is never completely realised, although it is comparable to the cleavage cycles of the early divisions of a sea urchin's egg (Gray, 1927). Systems of this type do not reveal the rate at which changes are going on within the cell itself, they simply define a hypothetical population in which a definite series of changes are completed within the limits of one generation period, since the unit of growth is the individual cell and the unit of time is one generation period.

In nearly every natural case the conditions of growth (even under optimal conditions) are more complicated. When one bacterium is isolated and its products segregated as a pure culture, the generation time of each cell is not identically the same as that of its neighbours, and consequently at any given moment some cells are dividing, whereas the others are at various intermediate stages of the reproductive cycle. The rate of increase of such a population will be determined by the percentage of cells actually dividing at any instant, and the actual growth of the population can be plotted as a smooth curve (dotted in fig. 128), instead of a series of points restricted to the end of each reproductive period (see fig. 128); the smooth curve is a parabola.

$$\frac{dx}{dt} = kx, \text{ where } k \text{ is equal to } \frac{\log_e 2}{\text{generation time}},$$

or $\qquad x = x_0 e^{kt}$, where x_0 is the number of cells originally present.

Such an analysis only holds good when the *average* generation time of individual cells is the same for all the cells present, although for any given series of filially related cells the generation time may fluctuate from one successive cycle to another. This so-called logarithmic law of growth applies in practice to rapidly growing populations of bacteria (Lane-Claypon, 1909) and to yeast (Slator, 1913; Richards, 1928 *a* and *b*). In cultures of fibroblasts Fischer (1925) states that the average growth rate fluctuates in periodic cycles, so that the law of compound interest will only hold with even approximate accuracy over long periods of time (see p. 293).

The expression $x = x_0 e^{kt}$ is so frequently applied to the kinetics of growth that it is desirable to appreciate its full significance. It implies two things: (i) that all the cells or a fixed percentage of them are always growing and dividing during the whole of their existence,

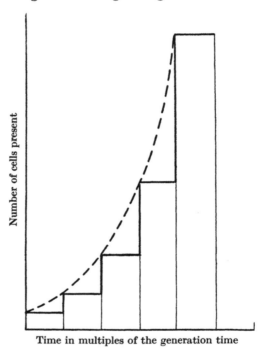

Fig. 128. Graph showing the rate of increase of a cell population in which (*a*) division is synchronous, and the number of cells is suddenly doubled at the end of each generation period, (*b*) division is not synchronous, but a constant proportion of cells are dividing at any given instant.

and (ii) that although the generation times of all the cells present are not all identically the same, yet the variability in this respect remains unchanged; in other words, a cell which divides unduly rapidly must tend to give rise to cells which will divide slowly and *vice versa*. There must be no accumulated tendency to increase the percentage of rapidly dividing cells and to decrease the percentage of slowly dividing units (see p. 311).

It is necessary to remember that in all such systems the units concerned are individual cells, and that the act of cell cleavage is a discontinuous process for the purposes of assessing growth. Hence, although we get a smooth curve which relates cell number to time, the fundamental change is not capable of analysis by such methods. By growth we mean increase in the amount of living material: this obviously goes on in the intercleavage period. If we restrict our data to arbitrary units of whole individual cells all we can say is that, taken as an average of the population, a given cell will produce an amount of new substance equal in bulk to the original cell in a given period of time. Until recently no single cell had been observed at frequent intervals during a period of growth, and our knowledge of the actual rate of intracellular growth was based on indirect analysis. In the case of yeast and bacteria it is possible to estimate the total amount of material present by weighing the cells or by observing their volume (Slator, 1913; Coombs and Stephenson, 1926). Similarly, the respiratory activity can be taken as a measure of the living substance present (Slator, 1913). In both cases the logarithmic law seems to hold with remarkable accuracy. The growth of individual cells has, however, recently been followed by Schmalhausen and Bordzilowskaja (1930), who have measured the rate of increase in size of bacteria and yeast cells. They find that the rate of growth per unit length ($\delta l/\delta t . 1/l$) of *Bacillus megatherium* and of certain other forms remains constant during intercleavage periods of growth, and consequently the course of growth obeys the logarithmic law (see Table XXXIX and fig. 129). This logarithmic increase in size only occurs if the ratio of cell surface to cell volume remains relatively constant during the process of growth. In spherical cells, this is clearly not the case, for as such cells increase in size so the area of cell surface per unit of volume decreases, and at the same time the growth rate declines. Schmalhausen concludes that the absolute rate of growth of cells is controlled by two opposing factors, one of which is directly proportional to the surface area, while the other depends on the volume of the cell.

From the evidence available, it seems fairly safe to assume that a growing cell, of the types mentioned, will produce new material at a constant rate per unit of its own mass if the environment is strictly controlled at a satisfactory and constant level. At the same time this conclusion may not be of general application. Comparatively few types of cells have been investigated by quantitative methods, and within the group of the Protozoa there is considerable evidence to support the view that the growth rate of a population may decline even in a satisfactory medium. In *Uroleptus* and allied forms the growth rate appears to be affected by the intrinsic changes effected by conjugation (see Calkins, 1926, pp. 465–508).

Table XXXIX. *Bacillus megatherium*, 28°–29°. Average cross-section, $1·54\,\mu$

Time in 6-minute intervals	Length in μ (l)	Velocity of growth	Calculated surface	Specific rate of growth ($\delta l/\delta t \cdot 1/l$)	Ratio of surface to volume
0	6·44				
1	6·92	0·48	36·03	0·0120	2·90
2	7·44	0·52	38·44	0·0121	2·88
3	7·96	0·52	40·96	0·0113	2·86
4	8·52	0·56	43·57	0·0113	2·84
5	9·12	0·60	46·37	0·0113	2·82
6	9·80	0·68	49·47	0·0120	2·81
7	10·48	0·68	52·76	0·0112	2·79
8	11·16	0·68	56·05	0·0105	2·78
9	11·96	0·80	59·62	0·0115	2·77
10	12·88	0·92	63·78	0·0123	2·76

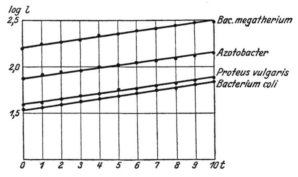

Fig. 129. Logarithmic growth of single bacteria. The ordinates represent the logarithm of the length in microns. (From Schmalhausen and Bordzilowskaja.)

Inhibition of growth

If the number of cells in a given culture is allowed to increase, there comes a time when the compound interest law quite clearly breaks down. Sooner or later the growth rate declines and eventually sinks to zero: after this there is a rapid decline in the number of bacteria present. The factors responsible for this breakdown of the logarithmic law of population growth are not completely known. Graham-Smith (1920) has shown that under certain conditions the amount of foodstuffs available plays an important rôle. In a particular culture medium, Graham-Smith found that *Staphylococcus aureus* reproduces itself until there are approximately ten million organisms present per 0·01 c.c. of medium, and that this figure is independent of the number of bacteria originally inoculated into the medium.

Table XL

No. of bacteria in original inoculum	Maximum no. of bacteria obtained	No. of bacteria in original inoculum	Maximum no. of bacteria obtained
520	9,248,000	34,400	8,720,000
1392	10,606,000	4,300	8,544,000
1784	9,280,000	420	8,496,000
5660	9,872,000	59	7,584,000

There is therefore an upper limit to the density of bacteria which can be obtained in any given culture. This limit is largely dependent on the concentration of nutrient substances in the medium (Table XLI)—see also Penfold and Norris (1912).

Table XLI

Relative concentration of food	Maximum no. of bacteria obtained
100	25,840,000
75	20,300,000
50	15,760,000
25	9,416,000
10	4,273,000

Hand in hand with the decline in growth rate of such cultures there is a decrease in the size of the individual cells, which itself is probably due to a diminution in the concentration of available food (Henrici, 1923 *b*).

The far-reaching effect of food supply upon the growth rate is seen when identical cultures of bacteria are grown at different temperatures (Table XLII). The higher the temperature the more intense are the metabolic processes of the cells, and consequently a higher concentration of food is required to maintain a maximum

THE GROWTH OF CELLS

population. In other words, with identical cultures, the food supply begins to run short sooner at a high temperature than at a low temperature, and the maximum density attainable is correspondingly lower.

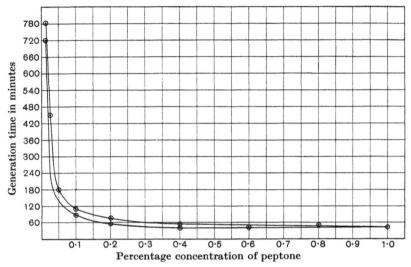

Fig. 130. Graph showing the influence of nutritive material on the rate of growth of bacteria. (From Penfold and Norris.)

Table XLII

Incubation temperature (° C.)	Maximum density of bacteria	Time in days required to reach maximum density
17	18,272,000	8
27	15,164,000	5
37	10,448,000	3

Even when the food supply of a cell culture is very rich, the growth rate may still decline, indicating that other factors are operative in addition to lack of food. Richards has shown that if yeast cultures are subcultured regularly every three hours, the logarithmic law is maintained until a concentration of 70×10^6 cells per c.c. has been reached; failure to subculture results in a decreased growth rate, although abundant food is present (see fig. 131). Richards has shown that the decreased growth rate is largely due to the accumulation of alcohol which begins to inhibit growth at a concentration of about 1 mg. per c.c.

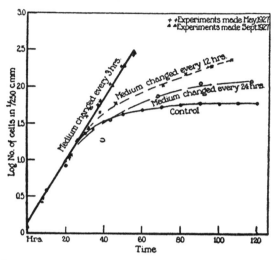

Fig. 131. Graph showing the effect of the removal of toxic products on the growth rate of yeast cells. (From Richards, 1928 b.)

Lag phase of growth

When for any reason an adverse medium effects a markedly depressant action on the growth rate, there is abundant evidence that the effect upon the cells is of a profound nature and that the external environment affects the cell in such a way as to make its recovery in an optimum medium a slow if not impossible process. When cells from an actively growing culture are transferred to a fresh medium—there is no check in the growth rate—the cells continue to grow according to the logarithmic law. If, however, the cells are removed from an ageing culture in which the growth rate has fallen to a low value either from want of food or for any other cause, there is an appreciable period of time during which either no growth occurs or during which the growth rate is lower than that characteristic of later periods of culture. This critical period of depressed growth rate is known as the *lag* phase. The specific underlying causes of the lag in growth rate are obscure (see Penfold, 1914; Buchanan, 1918; Salter, 1919). When the external environment begins to exert a depressant action on growth, a number of cells might be so affected as to be permanently incapable of further growth even when transferred to fresh medium and an apparent lag phase would be observed; on the other hand, the evidence for such a suggestion is not clearly established, for Wilson's (1926) figures suggest that the percentage death rate does not change to a marked degree until the population of the culture is definitely on the down grade. It seems probable that even if only viable cells are considered the lag phase would still persist. The existence of this period is perhaps not surprising, since the affects of adverse conditions on organisms is often not readily and quickly reversible on transference to better conditions.

The general course of the growth of bacterial cultures is well illustrated by Buchanan's diagram (see fig. 132); although the data

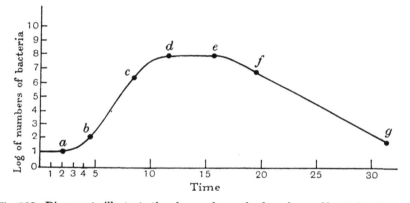

Fig. 132. Diagram to illustrate the phases of growth of a culture of bacteria. $1 \to a$ is the initial stationary phase during which there is no increase in numbers; $a \to b$ is the lag phase during which the rate of growth per unit organism is increasing; $b \to c$ is the logarithmic phase during which the growth rate is constant; $c \to d$ is the phase of negative acceleration during which the growth rate falls to zero; $d \to e$, the number of bacteria present is constant at a maximum value; $e \to f$ is a period of accelerated death; $f \to g$ is the logarithmic death phase. (From Buchanan, 1918.)

for other types of cell are much less extensive, there is no reason to doubt that the general course of growth is essentially similar to that exhibited by bacteria.

Frequent attempts have been made from time to time to express the whole course of unicellular growth in quantitative terms. It is very much open to doubt, however, whether such a procedure is useful in the present state of our knowledge (see Gray, 1929).

One of the most striking conclusions to be derived from the study of growth *in vitro* is that, given suitable conditions, a population of cells, whatever be their properties *in situ*, will continue to grow indefinitely at a steady rate. Fibroblasts, bacteria, yeast, and probably many protozoa, continue to grow and multiply at a rate which is constant within definable limits as long as the external environment is maintained at a satisfactory and constant level. In the case of tissue cells, life under such conditions is clearly not limited to the rate or duration characteristic of corresponding cells *in situ* in the body; Carrel and Ebeling's (1922 a) culture of chicken fibroblasts have already persisted far beyond the normal life of an adult fowl and the amount of tissue to which the original culture has given rise is far in excess of that which would have occurred had the original fibroblasts been grown in their normal positions in the body. To some extent, this is, of course, an artificial comparison, for it might well be argued that a fibroblast, growing in an excess of highly nutrient medium, is provided with an environment very different from the highly concentrated cell population present in an organised body. It is, perhaps, fairer to compare a cell *in situ* with a cell growing in a dense suspension of cells *in vitro*. Here again, however, we must conclude that the behaviour of the two types is curiously similar. Just as the growth rate of a yeast or bacterial suspension tends to decline to zero as the culture becomes older and more concentrated with cells, so a culture of fibroblasts *in vitro* will exhibit the same phenomenon if it is not subcultured or provided with fresh medium. Such comparisons to normal growth *in situ* must not be pushed too far. Almost any disturbance of the external environment will affect the growth rate *in vitro* and from a quantitative point of view it is as yet impossible to distinguish between a variety of factors (lack of food, accumulation of metabolites, lack of adequate growth accelerators)—all of which cause departure from the logarithmic law of growth (see Gray, 1929). We cannot say why growth *in situ* is limited, but the facts of growth *in vitro* clearly demonstrate that the surrounding medium, and the presence or absence of other cells, must play an important rôle.

The parallel between cultures of unicellular organisms and metozoan cells grown *in vitro* is clearly very close. Not only are the two types capable of unlimited logarithmic growth, but when conditions are unfavourable their reactions are closely similar. When a subculture is made from an old culture of bacteria we have seen that there is a marked lag phase before growth is resumed in the fresh medium. Similarly, it has been known for many years that embryonic cells *in vitro* migrate, grow, and divide much more rapidly than cells from older or adult animals; the younger the animal

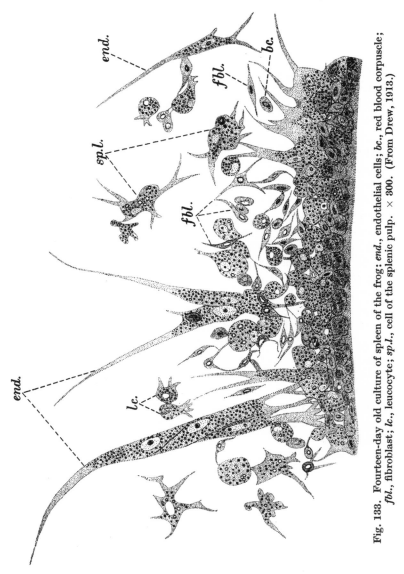

Fig. 133. Fourteen-day old culture of spleen of the frog: *end.*, endothelial cells; *bc.*, red blood corpuscle; *fbl.*, fibroblast; *lc.*, leucocyte; *sp.l.*, cell of the splenic pulp. × 300. (From Drew, 1913.)

from which the original fragment is removed, the more quickly does it show signs of growth *in vitro* (Cohn and Murray, 1925).

Just as in the case of bacteria, there is a latent period before growth begins, and the length of this period is a direct function of the age of the fragment. This may conceivably be due to the accumulation of a growth-inhibiting factor as the organism grows older; until this substance can diffuse away from the cells into the medium, growth would obviously be slow. On the other hand, it may be due to the lack of specific nutritive substances which are only generated in appreciable quantities in the embryo (see p. 294).

Tissue culture

That tissue cells can grow after excision from the body was first demonstrated by Ross Harrison (1910). Small fragments of the central nervous system of a frog, immersed in a hanging drop of lymph, clearly exhibited the outward migration of nerve cells and the subsequent outgrowth of nerve fibres from the cells. This important discovery opened up entirely new methods for the investigation of cell growth and its allied problems, for it was soon shown that the phenomenon is by no means restricted to nerve fibres, but is exhibited by many other types of cells (see Lewis and Lewis, 1924). The technique for successful tissue culture has been described in detail by Strangeways (1924 b). Roughly speaking, success depends upon the isolation of an aseptic fragment of tissue in a suitable nutritive medium. For good results, it is desirable that the growing cells should be provided with a solid surface to which they can attach themselves, and this is usually provided by implanting the fragment in a drop of coagulable plasma, although foreign solids, e.g. glass wool or spider's webs (Harrison, 1914), immersed in a fluid medium will suffice (fig. 134). More recently, Carrel has replaced the 'hanging drop' technique by the 'flask' method, wherein the clot of plasma is attached to the bottom of a small aseptic flask, capable of containing a relatively large amount of fluid medium which can readily be changed from time to time: this obviates the necessity of the constant subculturing necessary in the case of hanging drop cultures.

The growth of the excised fragment is preceded by the outward migration of the cells from their normal position. The migrating cells are mostly derived from the peripheral regions of the fragment and, as already mentioned, movement seems to be restricted to solid surfaces or to the interior of solid media; if a culture is growing in a liquid drop the cells will only migrate when attached to the cover-slip or to the liquid/air film, whereas in a liquid medium containing solid surfaces of fibrin or cotton wool, the cells migrate into the interior of the drop along such surfaces (Harrison, 1914; Fischer, 1925).

It will be noticed that although the phenomenon of migration increases the surface area covered by the fragment, no growth is involved in the sense that there has been an increase in the total

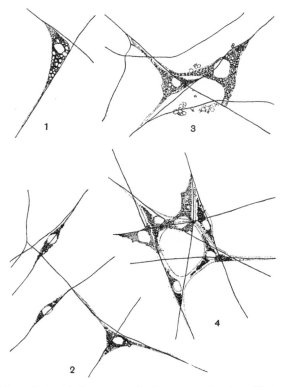

Fig. 134. Cell from the medullary cord of a frog grown *in vitro*. Note attachment of the cells to spider's web. × 300. (Ross Harrison, 1914.)

mass of the living cells present. The mechanism of migration is obscure. Harrison (1911, 1914) regards it as an instance of 'stereotropism' to solid surfaces, whereas Burrows (1913) suggested that the movement away from the periphery of the implant is essentially a chemiotropic movement away from a region of abnormal acidity.

The cells which migrate outwards from the surface of the fragment often show a marked difference in form to that which charac-

THE GROWTH OF CELLS

terises them when *in situ* in the tissue. The characteristic form of migrating cells is illustrated in fig. 135, and is not greatly different

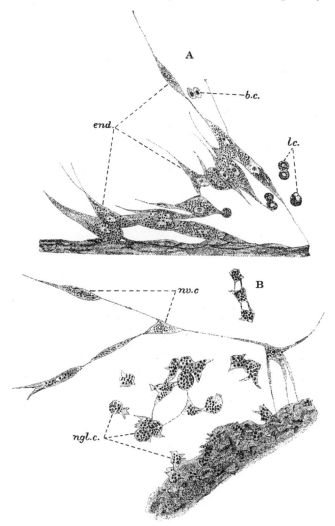

Fig. 135. *A*. Culture of frog's artery in plasma: *b.c.*, red blood corpuscle; *end.*, endothelial cell; *l.c.*, leucocyte. *B*. Culture of frog's cerebrum: *nv.c.*, nerve cell; *ngl.c.*, neuroglia cell. (After Drew.)

for different types of tissue. To some extent the shape of the cells in a culture depends on the nature of the medium in which growth is taking place. Uhlenhuth (1915) found that epithelial cells will grow as a membrane and retain their characteristic form if the growth occurs on the surface of the clot, but if the cells penetrate into the clot itself the cells separate from each other and become spindle-shaped. It is generally admitted that when growing *in vitro* all cells, irrespective of the tissue from which they are derived, tend to

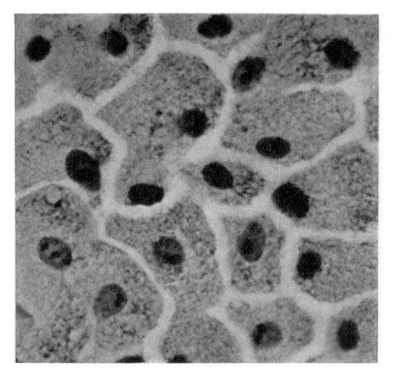

Fig. 136. Stained culture of epithelial cells, two months old, × 1425. (From Fischer, *Journ. Exp. Med.*) Note that the cells have grown as an epithelium and have not dissociated.

acquire a common spindle-shaped form not unlike normal fibroblasts. This fact has been interpreted in two ways. Champy (1913 *a* and *b*) believed that all tissue cells when grown *in vitro* lose their peculiarly differentiated form and all revert to a common embryonic condition; more recently (1921) he has shown that glandular cells when grown *in vitro* do not produce their normal secretory products. On the other hand, Barta (1923) and others claim that no real loss of differentiation occurs, and that the prevalent spindle shape is the result of abnormal cell environments: certainly in very few cases is there reliable evidence to show that one cell type can change into another as a result of growth *in vitro*. On the contrary, there are several examples of 'organised' growth *in vitro*, when the conditions of growth approximate more closely to the normal. Thus A. H. Drew (1923) found that the presence of fibroblasts in a culture of kidney epithelium resulted in the production of kidney tubules, whereas in the absence of fibroblasts unorganised spindle-shaped cells alone were formed.

The migration of cells *in vitro* is both interesting and important, but it is not so fundamentally significant as the fact that migration may be followed by true growth and reproduction.

Having migrated from the periphery of the implant, a limited number of the active cells soon begin to show signs of mitotic division. Considerable caution is required in defining the nature of these cells since most tissue fragments contain cells of more than one type: by a process of subculture, however, it is possible to be reasonably sure that only one type of cell is present. If we regard the active cells of a tissue fragment as a homogeneous population in respect to the period of time which elapses between two successive divisions, we would expect to find that at any given moment there would be approximately the same number of cells undergoing division as at any other moment (see p. 278). Fischer (1925), however, concluded that this is not a true picture of the facts. Using a nine months' old culture of chicken fibroblasts it was found that there were definite periods during which no mitoses were visible; between such periods there were others in which mitoses were relatively frequent. Fischer interprets these observations as an indication that the mitotic activity of one cell can induce a similar activity in another, possibly by means of the protoplasmic bridges which he believes are present between neighbouring cells (see also p. 178). According to Fischer, a single isolated cell will not grow owing to the absence of essential secretions ('desmones') which normally pass from cell to cell; it is claimed that it is the passage of such substances from one cell to another which enables the presence of actively dividing fibroblasts to incite a similar activity in a culture which has hitherto exhibited a much lower rate of growth. The nature of these substances seems problematical, but they may be related to the 'trephones' described by

Carrel (1922). Trephones are the essential principles contained in leucocytes or extracts of leucocytes which stimulate growth in fibroblasts and other tissue cells (see also Carrel and Ebeling, 1923 e).

Media for growth

Only a limited amount of growth *in vitro* will occur if fragments of tissue are isolated in a medium of pure saline (Lewis and Lewis, 1924), and numerous attempts have been made to extend the amount and duration of growth by the addition of nutrient substances to the medium. The presence of plasma is beneficial in that it provides a mechanically suitable medium—it does not, in itself, sustain unlimited growth and may even be toxic if derived from an old animal (Carrel and Ebeling, 1923 a). A very large number of other substances have been investigated but only one—namely, the extract of young embryos—has proved really satisfactory. Using the aqueous extract of crushed chick embryos Ebeling (1922) succeeded in maintaining the growth of chick fibroblasts more or less indefinitely—the same strain having been kept alive in subcultures for at least fifteen years. During this time nearly 2000 generations of cells have occurred, and as far as can be judged there is no diminution in the rate of growth. Of the efficiency of embryo-extracts there can be no doubt and the rate of growth, as judged by the area of new cells formed per unit time per unit of growing tissue, is roughly proportional to the concentration of embryo extract present (Carrel, 1923 b) and inversely proportional to its age. The full significance of this result will be considered later; at the moment it is useful to enquire into the nature of the substance introduced by the extract. We have already seen that for the growth of cell populations there must exist a suitable supply of nitrogenous compounds capable of being built up into new cell structures, and there must also exist a supply of a substance capable of sustaining the respiration of the cells when formed. As far as can be judged by the work of Warburg and Kubowitz (1927) and Watchorn and Holmes (1927), the latter supply can be provided by the presence of glucose. There remains the possibility that embryo extracts provide a supply of nitrogen in a form not available in ordinary plasma. Baker and Carrel (1926–8) attempted to decide this point by dialysing embryo extracts in such a way as to remove amino acids, and found that the residue had lost its power of sustaining unlimited growth; on the other hand, the addition of free amino acids did not raise the efficiency of the residue to its original value as a stimulant of growth. The same authors found that if embryonic tissues are digested by pepsin for a very short period, the resultant medium has very high growth-promoting properties. Prolonged digestion with pepsin or digestion with trypsin fails to give this result, and consequently Baker and Carrel concluded that the efficiency of the embryo extract depends on the presence of substances akin to the proteoses—and not to the smaller amino acid groupings which result from prolonged digestion. Attempts

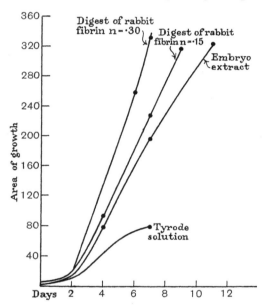

Fig. 137. Graph showing the effect of embryo extract and of fibrin digests on the rate of growth of cells *in vitro*.

to repeat these observations by Willmer (1928) appear to suggest that other factors may also be operative. But, as pointed out by Willmer, the evidence brought forward by Baker and Carrel supports the conclusion that amino acids can provide a source of energy to the cells and that proteoses may provide a source of nitrogen for the formation of new cells.

Amino acids and proteoses cannot, however, be the sole constituents of embryo extracts which influence growth. Carrel, Fischer and others have found that the extract loses its efficiency if heated above 56° C., which suggests that an enzyme is probably involved. According to Carrel, the active principle will not pass through a Chamberland filter and appears to be adsorbed to protein surfaces. On the other hand, Wright (1926) has obtained a highly active growth stimulant which can readily be dialysed through parchment membranes—indicating that such substances need not necessarily have a high molecular weight.

Whatever be the nature of the growth-promoting principle it is present in its most potent form in young embryos and is less obvious in extracts from older individuals. Carrel believes that embryonic extract contains two principles, one which stimulates growth and one which inhibits it. We have already seen that the plasma of young animals sustains a higher rate of growth than does that of older animals. If this difference were due solely to the partial loss of a growth accelerator, the efficiency of older plasma should be increased by concentration. According to Carrel and Ebeling (1921 b, 1923 a) this is not the case, and we have to infer the existence of a depressant factor, which increases in potency as age increases.

So far the culture of cells *in vitro* has been largely restricted to tissues of the higher vertebrates; there are comparatively few observations on invertebrate tissues. Goldschmidt (1917) succeeded in growing the follicular cells of the testes of Lepidoptera, but in most cases invertebrate tissues appear to exhibit only organised growth *in vitro* (Plananians (Murray, 1927); Diptera (Frew, 1928)).

Cell differentiation and metaphasia

There comes a time, sooner or later, when the resultant cells of a segmenting egg can be divided into a series of differentiated categories. This process of cell differentiation is not, as yet, susceptible to physiological analysis, although some of the underlying factors are now known. During the early development of amphibian eggs, in particular, it is possible to state with certainty that the subsequent fate of undifferentiated cells depends upon their position in respect to localised centres, which are capable of organising a series of specific cell types in such a way as to produce a complete embryo. At present, these 'organisers' are purely biological concepts and it is impossible to foretell how far their action on undifferentiated cells will eventually be explicable in physico-chemical terms. We may, however, consider certain minor aspects of cell differentiation. In the first place it is desirable to know how far the process is irreversible or to what extent a particular type of differentiated cell retains the potentiality of conversion into another form. As already mentioned, Champy (1913 a, 1914) believes that when a differentiated tissue cell loses its characteristic place in the body and migrates into an artificial medium, it loses its specific properties and reverts to an undifferentiated embryonic type which, theoretically, should be able to give rise to a variety of other differentiated types. The bulk of opinion is, however, against this view (see Lewis and Lewis, 1924), and most authors believe that although cells grown *in vitro* are very similar in form, yet minor characteristics of a specific nature remain, and that there is no real dedifferentiation to a common embryonic type. On the other hand, Fischer (1925) claims that leucocytes can become fibroblasts, and that fibroblasts can give rise to macrophages.

A number of instances are known where the cells of one tissue, when growing in the body, have given rise to another tissue. According to Adami (1908) such instances of *metaplasia* always obey the rule that epithelial tissues can only be converted into other types of epithelia, and mesoblastic tissues only to other types of mesoblast. It would seem that columnar ciliated epithelial cells can form a squamous epithelium, if subjected to chronic irritation, and the pavement epithelium of the bladder can be converted in a columnar epithelium. Similarly, true bone can be formed in the lungs, or in the walls of arteries (Harvey, 1907). In nearly every case, however, it is exceedingly difficult to make certain that the new tissue has actually arisen from cells which were an integral part of the old tissue, and have not developed from undifferentiated or quiescent cells present in the original tissue in comparatively small numbers. In this connection the work of H. V. Wilson (1907, 1911) is of peculiar interest. Wilson found that if the tissues of the sponge *Microciona* are cut up into small fragments and squeezed through bolting silk, the resultant pulp of cells is capable of reorganising itself into compact tissues or even into new sponge individuals. In the macerated pulp there are cells of three main types: (*a*) dermal cells, (*b*) collar cells, (*c*) amoebocytes. If a group of dermal cells cohere together without the inclusion of either collar cells or amoebocytes, the subsequent mass will form only dermal cells. Similarly, the collar cells can only give rise to collar cells. The amoebocytes, on the other hand, can give rise to either dermal cells or to collar cells. Difficulties arise in the identification of these

298   THE GROWTH OF CELLS

three types of cells in a regenerating mass, for there is a tendency for each type of cell to acquire a common rounded form devoid of any collar. Wilson was inclined to think, however, that dermal cells and collar cells may in some cases actually dedifferentiate to a common type which is totipotent in the sense that, like amoebocytes, they may subsequently give rise to more than one type of differentiated cell. According to Galtsoff (1925), however, the regenerated sponge is derived from two types of cell; the amoebocytes give

Fig. 138. Restitution mass of *Pennaria* (6 days old) showing developing hydranths from a group of artificially separated cells: *op*, perisarc of original mass; *x*, perisarc of outgrowth adherent to glass. (After Wilson.)

rise to the mesenchyme and skeleton, whereas the pinacocytes give rise to the dermal membranes and flagellated chambers (see Table XLIII). For a recent discussion of the whole problem reference should be made to Wilson and Penney (1930). Similar difficulties arise in the case of the restitution bodies of hydoids (fig. 138) (Wilson, 1911).

It is also within the invertebrate kingdom that some of the most striking cases of metaphasia are found. According to Morgan (1904),

THE GROWTH OF CELLS

the muscles of the regenerated crab's claw are derived from the original ectoderm, a fact which vitiates Adami's rule. Similarly, Kroeber (1910) showed that the pharyngeal epithelium of *Allolobophora* can be regenerated from the endoderm of the alimentary canal, although in ontogeny it develops from the ectoderm; a complete survey of such 'totipotent' or 'pleuripotent' systems (see Przibram, 1926) would carry us too far from a consideration of the cell. As already mentioned, the technical difficulties associated with the investigation of cell differentiation are considerable and there are comparatively few instances known where it is safe to conclude

Table XLIII. (From Galtsoff.)

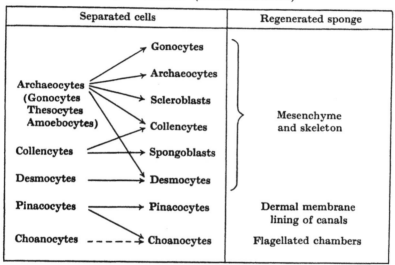

that new growth has not been restricted to previously undifferentiated cells. One of these is provided by the work of G. H. Drew (1911). Drew implanted a small fragment of the ripe ovary of *Pecten* into the adductor muscle of another individual (see fig. 139); a layer of fibroblasts quickly formed round the fragment which itself was then invaded by phagocytes so that it eventually degenerated. After a lapse of about six days no trace of organised ovarian tissue remained, but its site was marked by a cyst which was surrounded by fibroblasts and contained blood corpuscles and other bodies. Three weeks later the innermost fibroblasts gradually changed their shape and formed a cell layer resembling columnar epithelium which later became ciliated. Eventually the whole cyst became lined with well-defined ciliated epithelium; it seems reasonable to suppose that the new epithelium was actually formed from fibroblasts. This case is of peculiar interest for it is hardly conceivable that undifferentiated ciliated cells were originally present in the muscle.

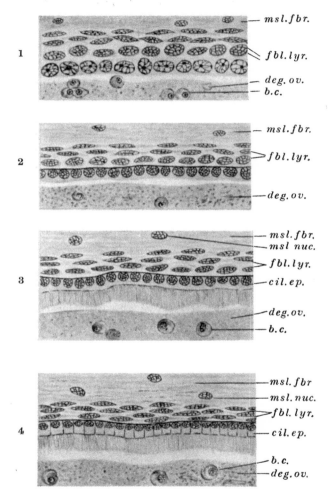

Fig. 139. Formation of ciliated epithelium (*cil. ep.*) from a layer of fibroblasts (*fbl. lyr.*) lining a cyst formed round a piece of ovary (*deg. ov.*) implanted into the adductor muscle of *Pecten*. 1, Portion of cyst wall after 23 days. 2, Cyst wall after 26 days. 3, Cyst wall after 30 days. 4, Cyst wall after 98 days. *b.c.*, blood corpuscles; *msl. fbr.*, muscle fibres; *msl. nuc.*, muscle nuclei. (After Drew, 1911.)

REFERENCES

ADAMI, J. G. (1908). *Principles of Pathology.* (Philadelphia.)
ALLEN, E. J. (1914). 'On the culture of the plankton diatom *Thalassiosira gravida* Cleve, in artificial sea-water.' *Journ. Mar. Biol. Assoc.* 10, *417.*
BAKER, L. E. and CARREL, A. (1925). 'Lipoids as the growth inhibiting factor in serum.' *Journ. Exp. Med.* 42, *143.*
—— (1926 a). 'Effect of the amino acids and dialysable constituents of embryonic tissue juice on the growth of fibroblasts.' *Journ. Exp. Med.* 44, *397.*
—— (1926 b). 'Action on fibroblasts of the protein fraction of embryo tissue extract.' *Journ. Exp. Med.* 44, *387.*
—— (1927). 'Effect of age on serum lipoids and proteins.' *Journ. Exp. Med.* 45, *305.*
—— (1928). 'The effects of digests of pure proteins on cell proliferation.' *Journ. Exp. Med.* 47, *353.*
BARBER, M. A. (1908). 'The rate of multiplication of "*Bacillus coli*" at different temperatures.' *Journ. Inf. Dis.* 5, *379.*
BARNARD, J. E. (1925). 'Cultures from single cells.' *Brit. Journ. Exp. Path.* 6, *39.*
BARTA, E. (1923). 'Some factors regulating the morphology of tissue. Ureter *in vitro.*' *Anat. Record,* 29, *33.*
BEREZOWSKI, A. (1910). 'Studien über die Zellgrösse.' *Arch. f. Zellforsch.* 5, *375.*
BUCHANAN, R. E. (1918). 'Life phases in a bacterial culture.' *Journ. Inf. Dis.* 23, *109.*
BURROWS, M. T. (1913). 'Wound healing *in vitro.*' *Proc. N.Y. Path. Soc.* 13, *131.*
—— (1917 a). 'The oxygen pressure necessary for tissue activity.' *Amer. Journ. Physiol.* 43, *13.*
—— (1917 b). 'Some factors regulating growth.' *Anat. Record,* 11, *335.*
—— (1921). 'The reserve energy of actively growing embryonic tissues.' *Proc. Soc. Exp. Biol. and Med.* 18, *133.*
—— (1923 a). 'Studies on cancer. I. The effect of circulation on the functional activity, migration and growth of tissue cells.' *Proc. Soc. Exp. Biol. and Med.* 21, *94.*
—— (1923 b). 'Studies on cancer. II. The significance of the effect of circulation on the growth of cells.' *Proc. Soc. Exp. Biol. and Med.* 21, *97.*
—— (1923 c). 'Studies on cancer. III. Cellular growth and degeneration in the organism.' *Proc. Soc. Exp. Biol. and Med.* 21, *103.*
BURROWS, M. T. and NEYMANN, C. (1917). 'Studies on the metabolism of cells *in vitro.*' *Journ. Exp. Med.* 25, *93.*
CALKINS, G. N. (1926). *The Biology of the Protozoa.* (New York.)
CARREL, A. (1913). 'Artificial activation of the growth *in vitro* of connective tissue.' *Journ. Exp. Med.* 14, *17.*
—— (1922). 'Growth promoting function of leucocytes.' *Journ. Exp. Med.* 36, *385.*
—— (1923 a). 'A method for the physiological study of tissues *in vitro.*' *Journ. Exp. Med.* 38, *407.*
—— (1923 b). 'Measurement of the inherent growth energy of tissues.' *Journ. Exp. Med.* 38, *521.*

CARREL, A. (1924 a). 'Leucocytic trephones.' *Journ. Amer. Med. Assoc.* 82, *255*.
—— (1924 b). 'Tissue culture and cell physiology.' *Physiol. Review*, 4. *1*.
CARREL, A. and EBELING, A. H. (1921 a). 'Age and multiplication of fibroblasts.' *Journ. Exp. Med.* 34, *599*.
—— (1921 b). 'The multiplication of fibroblasts *in vitro*.' *Journ. Exp. Med.* 34, *317*.
—— (1922 a). 'Heterogenic serum, age, and multiplication of fibroblasts.' *Journ. Exp. Med.* 35, *17*.
—— (1922 b). 'Heat and growth inhibiting action of serum.' *Journ. Exp. Med.* 35, *647*.
—— (1922 c). 'Action of shaken serum on homologous fibroblasts.' *Journ. Exp. Med.* 36, *399*.
—— (1922 d). 'Pure cultures of large mononuclear leucocytes.' *Journ. Exp. Med.* 36, *365*.
—— (1922 e). 'Leucocytic secretions.' *Journ. Exp. Med.* 36, *645*.
—— (1923 a). 'Antagonistic growth principles of serum and their relation to old age.' *Journ. Exp. Med.* 38, *419*.
—— (1923 b). 'Survival and growth of fibroblasts *in vitro*.' *Journ. Exp. Med.* 38, *487*.
—— (1923 c). 'Action of serum on fibroblasts *in vitro*.' *Journ. Exp. Med.* 37, *759*.
—— (1923 d). 'Antagonistic growth-activating and growth-inhibiting principles in serum.' *Journ. Exp. Med.* 37, *653*.
—— (1923 e). 'Tréphones embryonnaires.' *C.R. Soc. Biol.* 89, *1142*.
CHAMPY, C. (1913 a). 'La différenciation des tissus cultivés en dehors de l'organisme.' *Bibliog. anat.* 33, *184*.
—— (1913 b). 'Prolifération atypique des tissus cultivés en dehors de l'organisme.' *C.R. Soc. Biol.* 5, *532*.
—— (1914). 'Quelques résultats de la méthode des cultures de tissus. Le rein.' *Arch. de Zool. expér. et gén.* 54, *307*.
—— (1921). 'Perte de la sécrétion spécifique dans les cellules cultivées *in vitro*.' *C.R. Soc. Biol.* 83, *842*.
COHN, A. E. and MURRAY, H. A. (1925). 'Physiological ontogeny. IV. The negative acceleration of growth with age as demonstrated by tissue cultures.' *Journ. Exp. Med.* 42, *275*.
COOMBS, H. I. and STEPHENSON, M. (1926). 'The gravimetric estimation of bacteria and yeast.' *Biochem. Journ.* 20, *998*.
DARBY, H. H. (1930). 'The experimental production of life cycles in ciliates.' *Journ. Exp. Biol.* 7, *132*.
DREW, A. H. (1913). 'On the culture *in vitro* of some tissues of the adult frog.' *Journ. Path. and Bact.* 17, *581*.
—— (1923). 'Growth and differentiation in tissue cultures.' *Brit. Journ. Exp. Path.* 4, *46*.
DREW, G. H. (1911). 'Experimental metaphasia. I. The formation of columnar ciliated epithelium from fibroblasts in *Pecten*.' *Journ. Exp. Zool.* 10, *349*.
EASTCOTT, E. V. (1928). 'Wildier's Bios. The isolation and identification of "Bios I".' *Journ. Phys. Chem.* 32, *1094*.
EBELING, A. H. (1921). 'Measurement of the growth of tissues *in vitro*.' *Journ. Exp. Med.* 34, *231, 243*.

EBELING, A. H. (1922). 'A ten year old strain of fibroblasts.' *Journ. Exp. Med.* 35, *755.*
—— (1924). 'Cultures pures d'épithélium proliférant *in vitro* depuis 18 mois.' *C.R. Soc. Biol.* 90, *562.*
EBELING, A. H. and FISCHER, A. (1922). 'Mixed cultures of pure strains of fibroblasts and epithelial cells.' *Journ. Exp. Med.* 36, *285.*
FAURÉ-FREMIET, E. (1925). *La cinétique du développement.* (Paris.)
FISCHER, A. (1921). 'Growth of fibroblasts and hydrogen ion concentration of the medium.' *Journ. Exp. Med.* 34, *447.*
—— (1922 a). 'A three year old strain of epithelium.' *Journ. Exp. Med.* 35, *367.*
—— (1922 b). 'A pure strain of cartilage cells *in vitro*.' *Journ. Exp. Med.* 36, *379.*
—— (1923). 'Contributions to the biology of tissue cells. I. The relation of cell crowding to tissue growth *in vitro*.' *Journ. Exp. Med.* 38, *667.*
—— (1925). *Tissue Culture.* (London.)
—— (1925 a). 'A functional study of cell division in cultures of fibroblasts.' *Journ. Cancer Res.* 9, *50.*
—— (1925 b). 'Cytoplasmic growth principles of tissue cells.' *Arch. f. exp. Zellforsch.* 1, *369.*
—— (1926). 'The growth of tissue cells from warm blooded animals at lower temperatures.' *Arch. f. exp. Zellforsch.* 2, *303.*
FREW, J. G. H. (1928). 'A technique for the cultivation of insect tissues.' *Brit. Journ. Exp. Biol.* 6, *1.*
GALTSOFF, P. S. (1925). 'Regeneration after dissociation. (An experimental study on sponges.) II. Histogenesis of *Microciona prolifera*.' *Journ. Exp. Zool.* 42, *223.*
GOLDSCHMIDT, R. (1916). 'Notiz über einige bemerkenswerte Erscheinungen in Gewebskulturen von Insekten.' *Biol. Zentralbl.* 36, *160.*
—— (1917). 'Versuche zur Spermatogenese *in vitro*.' *Arch. f. Zellforsch.* 14, *421.*
GRAHAM-SMITH, G. S. (1920). 'The behaviour of bacteria in fluid cultures as indicated by daily estimates of the numbers of living organisms.' *Journ. Hygiene,* 19, *133.*
GRAY, J. (1926). 'The growth of fish. I. The relationship between embryo and yolk in *Salmo fario*.' *Brit. Journ. Exp. Biol.* 4, *215.*
—— (1927). 'The mechanism of cell-division. III. The relationship between cell-division and growth.' *Brit. Journ. Exp. Biol.* 4, *313.*
—— (1928). 'The growth of fish. II.' *Brit. Journ. Exp. Biol.* 6, *110.*
—— (1929). 'The kinetics of growth.' *Brit. Journ. Exp. Biol.* 6, *248.*
HARRISON, R. G. (1910). 'Experiments upon embryonic tissue isolated in clotted lymph.' *Journ. Exp. Zool.* 9, *799.*
—— (1911). 'The outgrowth of the nerve fibre as a mode of protoplasmic movement.' *Journ. Exp. Zool.* 9, *787.*
—— (1914). 'The reaction of embryonic cells to solid structures.' *Journ. Exp. Zool.* 17, *521.*
HARVEY, W. H. (1907). 'Experimental bone formation in arteries.' *Journ. Med. Research,* 17, *25.*
HEATON, T. B. (1926). 'The nutritive requirements of growing cells.' *Journ. Path. and Bact.* 29, *293.*

HENRICI, A. T. (1923 a). 'Influence of age of the parent culture on the size of cells of *Bacillus megatherium*.' *Proc. Soc. Exp. Biol. Med.* 21, 243.
—— (1923 b). 'Influence of concentration of nutrients on size of cells of *Bacillus megatherium*.' *Proc. Soc. Exp. Biol. Med.* 21, 345.
HERTWIG, R. (1908). 'Ueber neue Probleme der Zellenlehre.' *Arch. f. Zellforsch.* 1, 32.
HOLMES, B. E. and WATCHORN, E. (1927). 'Studies on the metabolism of tissues growing *in vitro*. I. Ammonia and urea production of kidney tissue.' *Biochem. Journ.* 21, 327.
HOLMES, S. J. (1914 a). 'The behaviour of the epidermis of amphibians when cultivated outside the body.' *Journ. Exp. Zool.* 17, 281.
—— (1914 b). 'The life of isolated larval muscle cells.' *Science*, 40, 271.
HOSOYA, S. and KUROYA, M. (1923). 'Water-soluble vitamin and bacterial growth.' *Sci. Reports Govt. Inst. Inf. Dis. Tokyo*, 2, 233.
ILLING, G. (1905). 'Vergleichende histologische Untersuchungen über die Leber der Haussäugetiere.' *Anat. Anz.* 26, 177.
JAMESON, H. L., DRUMMOND, J. C. and COWARD, K. H. (1922). 'Synthesis of vitamin A by a marine diatom (*Nitzschia closterium* W. Sm.) growing in pure culture.' *Biochem. Journ.* 16, 482.
KOEHLER, O. (1912). 'Über die Abhängigkeit der Kernplasmarelation von der Temperatur und vom Reifezustand der Eier.' *Arch. f. Zellforsch.* 8, 272.
KOSER, S. A. and RETTGER, L. F. (1919). 'Studies on bacterial nutrition. The utilisation of nitrogenous compounds of definite chemical composition.' *Journ. Inf. Dis.* 24, 301.
KROEBER, J. (1910). 'An experimental demonstration of the regeneration of the pharynx of *Allolobophora* from endoderm.' *Biol. Bull.* 2, 105.
KRONTOWSKI, A. A. and BRONSTEIN, J. A. (1926). 'Stoffwechselstudien an Gewebekulturen.' *Arch. f. exp. Zellforsch.* 3, 32.
KRONTOWSKI, A. and RADZIMOVSKA, V. V. (1922). 'On the influence of change of concentration of the H˙ to OH′ on the life on the tissue cells of vertebrates.' *Journ. Physiol.* 56, 275.
LAMBERT, R. A. (1913). 'The character of growth *in vitro* with special reference to cell division.' *Journ. Exp. Med.* 17, 499.
LANE-CLAYPON, J. E. (1909). 'Multiplication of bacteria and the influence of temperature and some other conditions thereon.' *Journ. Hygiene*, 9, 239.
LEWIS, W. H. and LEWIS, M. R. (1924). 'Behaviour of cells in tissue culture.' *General Cytology*. (Chicago.)
MASING, E. (1910). 'Ueber das Verhalten der Nucleinsäure bei der Furchung des Seeigeleies.' *Zeit. f. physiol. Chem.* 67, 161.
MORGAN, T. H. (1904). 'Germ layers and regeneration.' *Arch. f. Entw. Mech.* 18, 261.
MURPHY, J. B. and HAWKINS, J. A. (1925). 'Comparative studies on the metabolism of normal and malignant cells.' *Journ. Gen. Physiol.* 8, 115.
MURRAY, M. E. (1927). 'The cultivation of planarian tissues *in vitro*.' *Journ. Exp. Zool.* 47, 467.
NEEDHAM, J. and NEEDHAM, D. M. (1930). 'On phosphorus metabolism in embryonic life. I. Invertebrate eggs.' *Journ. Exp. Biol.* 7, 317.
PENFOLD, W. J. (1914). 'On the nature of bacterial lag.' *Journ. Hygiene*, 14, 215.

PENFOLD, W. J. and NORRIS, D. (1912). 'The relation of concentration of food supply to the generation-time of bacteria.' *Journ. Hygiene*, 12, *527*.
PETERS, R. A. (1921). 'Substances needed for the growth of a pure culture of *Colpidium colpoda*.' *Journ. Physiol.* 55, *1*.
PLENK, H. (1911). 'Ueber Änderungen der Zellgrösse im Zusammenhang mit dem Körperwachstum der Tiere.' *Arb. a. d. Zool. Instit. Wien*, 19, *247*.
POPOFF, M. (1908). 'Experimentelle Zellstudien.' *Arch. f. Zellforsch.* 1, *245*.
PRZIBRAM, H. (1926). 'Transplantation and regeneration: their bearing on developmental mechanics.' *Brit. Journ. Exp. Biol.* 3, *313*.
RICHARDS, O. W. (1928 *a*). 'The growth of the yeast *S. cerevisiae*. I. The growth curve, its mathematical analysis and the effect of temperature on the yeast growth.' *Ann. of Bot.* 42, *271*.
—— (1928 *b*). 'Potentially unlimited multiplication of yeast with constant environment, and the limiting of growth by changing environment.' *Journ. Gen. Physiol.* 11, *525*.
ROBERTSON, T. B. (1923). *The Chemical Basis of Growth and Senescence.* (Philadelphia.)
SALTER, R. C. (1919). 'Observations on the rate of growth of *B. coli*.' *Journ. Inf. Dis.* 24, *260*.
SCHACKELL, L. F. (1911). 'Phosphorus metabolism during early cleavage of the echinoderm egg.' *Science*, 34, *573*.
SCHMALHAUSEN, I. and BORDZILOWSKAJA, N. (1930). 'Das Wachstum niederer Organismen. I. Das individuelle Wachstum der Bakterien und Hefe.' *Zeit. f. wissen. Biol.* Abt. D, 121, *726*.
SLATOR, A. (1913). 'The rate of fermentation by growing yeast cells.' *Biochem. Journ.* 7, *197*.
STEPHENSON, M. (1930). *Bacterial Metabolism.* (London.)
STRANGEWAYS, T. S. P. (1922). 'Observations on the changes seen in living cells during growth and division.' *Proc. Roy. Soc.* B, 94, *137*.
—— (1924 *a*). *Technique of Tissue Culture.* (Cambridge.)
—— (1924 *b*). *Growth in Tissue Cultures.* (Cambridge.)
TERROINE, E. and WÜRMSER, R. (1922). 'L'énergie de croissance. I. Le développement de l'*Aspergillus niger*.' *Bull. Soc. Chim. Biol.* 4, *519*.
UHLENHUTH, E. (1914). 'Cultivation of the skin epithelium of the adult frog *Rana pipiens*.' *Journ. Exp. Med.* 20, *614*.
—— (1915). 'The form of the epithelial cells in cultures of the frog skin and its relation to the consistency of the medium.' *Journ. Exp. Med.* 22, *76*.
—— (1916). 'Changes in pigment epithelial cells and iris pigment cells of *Rana pipiens* induced by changes in environmental conditions.' *Journ. Exp. Med.* 24, *689*.
WARBURG, O. (1927). 'Über die Klassifizierung tierischer Gewebe nach ihrem Stoffwechsel.' *Biochem. Zeit.* 184, *484*.
WARBURG, O. and KUBOWITZ, F. (1927). 'Stoffwechsel wachsender Zellen.' *Biochem. Zeit.* 189, *242*..
WATCHORN, E. and HOLMES, B. E. (1927). 'Studies in the metabolism of tissues growing *in vitro*.' *Biochem. Journ.* 21, *1391*.
WILLMER, E. N. (1928). 'Tissue culture from the standpoint of general physiology.' *Biol. Reviews*, 3, *271*.
WILSON, G. S. (1922). 'The proportion of viable bacteria in young cultures, with especial reference to the technique employed in counting.' *Journ. Bacter.* 7, *405*.

WILSON, G. S. (1926). 'The proportion of viable "Bacilli" in agar cultures of "*B. aertrycke*" (Mutton) with special reference to the change in size of the organisms during growth, and in the opacity to which they give rise.' *Journ. Hygiene*, 25, *150*.

WILSON, H. V. (1907). 'On some phenomena of coalescence and regeneration in sponges.' *Journ. Exp. Zool.* 5, *245*.

—— (1911). 'On the behaviour of the dissociated cells in hydroids, *Alcyonium* and *Asterias*.' *Journ. Exp. Zool.* 11, *280*.

WILSON, H. V. and PENNEY, J. T. (1930). 'The regeneration of sponges (*Microciona*) from dissociated cells.' *Journ. Exp. Zool.* 56, *73*.

WRIGHT, G. P. (1925). 'On the dialysability of the growth activating principle contained in extracts of embryonic juice.' *Journ. Exp. Med.* 43, *591*.

—— (1926). 'Presence of a growth stimulating substance in the yolk of incubated hens' eggs.' *Proc. Soc. Exp. Biol. and Med.* 23, *603*.

CHAPTER TWELVE

Cell Variability

However similar they appear to be at first sight, the units of a biological population are never identically alike: to this rule the cell is no exception. In size, form and activity the individual members of a cell population exhibit differences from each other, and these differences may often play an important part in the history of the whole system.

The variability within a cell population may be of two types. Firstly, there may be a state of *static* variability wherein each cell maintains a particular characteristic at a constant level. For example, the eggs of a trout or of a sea urchin are not all of equal size, but any given egg remains of a constant size, since large eggs do not suddenly become small eggs or *vice versa*. Statistical variation of this type can be defined in quantitative terms if sufficient measurements are available. So far, however, very few observations have been made on individual cell structures and consequently the data of Heiberg (1907) are of some interest. This author measured the diameter of the nucleus in the pancreas and liver cells of young mice and from his data fig. 140 has been constructed. The form of the curve is comparable to that which displays static variation within other types of population and can be expressed algebraically by standard methods (Fisher, 1925). In all such cases we can assign to each member of the population a particular value for the character concerned which will remain constant as time proceeds. Thus one nucleus has a diameter of 6μ and another nucleus has a diameter of 10μ; as far as we know, these values do not change. On the other hand, other characteristics of cell activity may not remain constant in this way; they may vary from time to time, so that a given cell may be highly active at one moment and relatively inactive at a later moment. Such variability is known as *dynamic* variability and may play a very important rôle in biological phenomena. There can be little doubt that neglect of this factor has led to considerable confusion in certain types of cytological experiments.

It may be noted at once that neither static nor dynamic variability within a cell population will seriously invalidate the results of any experiments or observations which are made on systems in which all the members of the population are actively concerned during the whole operation. For example, if we take a million sea urchin eggs *en masse* and measure their total volume or their rate of respiration, we can express the volume or respiration of a single egg by dividing the total volume or activity by the number of eggs. We get, in this way, an average value, which is accurate in so far as an arithmetic mean is a reliable guide to the activity of the whole popu-

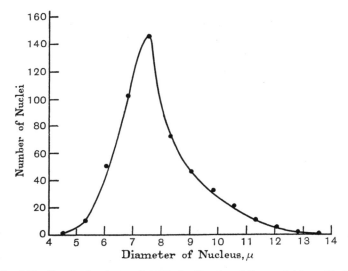

Fig. 140. Graph showing variability in the size of the nuclei in epithelial cells of the mouse. (Data from Heiberg, 1907.)

lation. The larger the number of cells which are employed, the more likely are we to reproduce, in successive experiments, the results obtained by any single observation. So far, cells are similar to any other biological population which exhibits static variability, and we need not worry whether a particular individual cell exhibits variable activity during an experiment; by using large numbers, the chances are in favour of one cell compensating for the changes in another.

Variability introduces a very real difficulty, however, when the conditions of an experiment are such that the units involved are not the same during the whole series of observations. If, for example, we are observing the onset of fatigue in a bundle of muscle fibres, the gradual diminution in contractile response to a stimulus may be due to one of two causes: (i) the response of every fibre may be reduced, (ii) the response of some fibres may disappear altogether, leaving others in a state of relatively normal activity. The same type of problem arises when a muscle or sense organ composed of many cells is exposed to weak stimuli (Hecht and Wolf, 1928). Here the increase in quantitative response to stronger stimuli may either be due to an increased response of every cell or to an increase in the number of cells which actually respond. The variability of response within a population of single cells can be illustrated by the observations of Lund and Logan (1924) on the coalescence of the intracellular vacuoles of *Noctiluca* as a result of an electrical stimulus. From Table XLIV it will be seen that as the intensity of the applied stimulus

Table XLIV

Current intensity in milliamperes	Percentage of cells which responded
20·04	100·0
17·42	100·0
14·42	95·2
10·00	88·2
7·12	87·9
4·90	50·0
4·20	5·7

decreases, so the percentage of cells which respond falls. In a multicellular tissue it would be difficult, if not impossible, to distinguish between a statistical variation of this kind, and a more fundamental change occurring simultaneously in all the cells present. To some extent these difficulties can be overcome by making observations on single cells, and thereby eliminating the static variability of the cell population. Examples of this type are available from several sources—irritability of individual muscle fibres (Jinnaka and Azuma, 1922; Brown and Sichel, 1930), nervous impulses (Adrian and Brock, 1928), exosmosis (Gray, 1921). In many cases, however, it is technically impossible to restrict observations to single cells, and even where this is possible the difficulty sometimes persists (see below). The disturbing influence of static variability upon the observed course of biological processes is so persistent that it is worth while to consider one or two specific instances.

There seems to be only one case in which it is possible to obtain a biological population which is essentially homogeneous in respect to a given character. This fortunate state of affairs is found in rapidly growing cultures of micro-organisms. If we start with two bacteria in a culture, one of which is capable of division once every twenty minutes and the other only once every forty minutes, then after four hours the total number of bacteria will be 4160, of which 4096 are derived from the more active member of the two original organisms; if we now subculture the population there will quickly come a time when the whole population is restricted to descendants of the faster growing bacterium, all others will have been eliminated. In practice the effect of a statistical difference between the growth rates of different types of bacteria or tissue cells is a useful means of obtaining pure cultures. As long as only one type of bacterium is present the rate of growth of a rapidly growing culture follows, very accurately, the simple equation

$$\frac{dx}{dt} = k.x,$$

where x is the number of bacteria present, and where k is a constant whose value is determined by the expression $\frac{\log_e 2}{\tau}$, where τ is the time between each successive reproductive act (see p. 279). This example is of interest for, as pointed out elsewhere (see p. 279), there still exists within the population of growing cells a state of dynamic variation. All the bacteria do not divide at *exactly* the same rate; some bacteria divide after a shorter period of time than others; it is, however, no longer possible to divide the population up into separate permanent categories, since a bacterium which divides rapidly tends to give rise to daughter cells which divide slowly and *vice versa*. In other words, there is no change in the total variability of the population as time goes on. Throughout the whole period of observation there is only one variable, namely, the number of cells present. We must remember however that, since the generation time τ is an average figure which represents the mean about which the whole population fluctuates, the constant k must also represent an average value.

When we are concerned with the reproduction of a mixed population of cells the situation is much more difficult. In this case each type of cell will exhibit dynamic variability about the mean of its own species and not about a mean which is common to all the cells

CELL VARIABILITY

present. The total number of cells formed from each particular species will be ne^{kt}, where n is the number originally present, k is the constant of reproduction of the species, and t is the time. Since the population is heterogeneous, the number of cells N_0 originally present can be divided into categories each possessing a characteristic value of k,

$$N_0 = n_1 + n_2 + n_3 \ldots n_n.$$

After a time t each category will have produced a number of daughter cells—so that the total number in the increased population (N_t) will be

$$N_t = n_1 e^{k_1 t} + n_2 e^{k_2 t} + n_3 e^{k_3 t} \ldots n_n e^{k_n t}.$$

It is obvious that the value of N_t cannot be determined theoretically unless we know the relative values of $n_1, n_2, n_3 \ldots$, or in other words we must know the numbers of the members of each of the original categories together with the value of k characteristic of each. If we were able to stop the course of reproduction at appropriate intervals we would find that a mixed population alters in two respects as time goes on: (i) there is an increase in the total number of cells present, (ii) there is an increase in the percentage of rapidly dividing cells present. The effect of this second change is to alter the *average* generation time for the whole population. We are therefore dealing with a system containing two variables—one of which represents the total number of dividing cells and the other a change in the average generation time. Quite clearly we must stabilise one of these variables before the effect of the other can be determined. Before attempting to do this it is convenient to consider similar populations in which the number of cells is decreasing instead of increasing. A good example of such a system is provided by the reaction between a population of bacteria and a toxic reagent or a disinfectant. Here again we are dealing simultaneously with two separate processes: (i) the reaction between each bacterium and the disinfectant, which leads to the death of the cells, (ii) a change in the variability of the whole population. Only the most sensitive cells are dying during the early stages of the reaction and only the most resistant cells at the end. When we make a graph which correlates the number of surviving bacteria with time, both these changes are involved and both will influence the nature and form of the curve. As a rule the first variable is known as the fundamental reaction. Miss Chick (1908, 1910) and others found that the curve relating the number of bacteria killed with the time of exposure to a disinfectant could be expressed by the equation of a first order reaction and concluded that the fundamental reaction was of a monomolecular nature. The extreme improbability of this conclusion has been pointed out by Brooks (1918). The fundamental condition for a monomolecular reaction is that the same percentage of existing molecules undergoes change at any time during the reaction; it is purely a matter of chance whether a particular molecule undergoes change early in the reaction or late. If all the cells of a population were equally sensitive at all times to the action of the disinfectant, they would all die at precisely the same time. Since the cells do not die simultaneously, there must be individual differences between the cells. Miss Chick (1910) assumed that cyclical changes proceed at the surface of a bacterium whereby its molecules will only react with the disinfectant when they happen to possess a critical energy content, and that the distribution of energy among the individual molecules is of the random type found in inanimate monomolecular systems. Under such circumstances there is no progressive change in variability of the whole system—because both early or late in the reaction the same type of distribution curve will express the population of molecules: there is only one variable, viz. the total number of reacting molecules or cells. In other words the variability of the bacterial population is of a dynamic nature, in that a cell may be sensitive at one moment, and resistant at another. An hypothesis of this type is not very convincing, for we know from other biological systems that a difference in sensitivity between two individual members of a population is often something which is fixed for the life of the animals and, as far as we know, a particular characteristic of this type is not distributed fortuitously to another member of the population. If we assume that a given individual possesses a definite sensitivity (k) to a toxic reagent, and another individual possesses a different and smaller sensitivity (k_1), then the first individual will die before the second, and in a large population the order in which the cells will die will be definitely fixed, and will not be due to chance. In such a system the distribution curve expressing the population in terms of sensitivity will change continuously throughout the process of disinfection for the most sensitive organisms will die before the others. In other words we have again introduced a second and complex variable into the equation which defines the rate of the reaction. As explained above, we cannot proceed further unless one of these two variables can be determined independently; unless we can define the variability of the population of bacteria, we cannot define the nature of the fundamental reaction which occurs between any given bacterium and the molecules of the disinfectant. As pointed out by Brooks, the disinfection/time curves of Chick and others, with their marked similarity to the exponential or monomolecular type, may be explained in either of two ways. (i) We may assume a monomolecular reaction between the molecules of disinfectant and the molecules of the bacteria, if we are prepared to admit a state of dynamic variability in the sensitivity of the latter. (ii) We can assume that the fundamental reaction between the bacterium and the disinfectant proceeds at a uniform rate and that individual cells differ from each other in sensitivity in a particular way. The most probable type of curve relating the number of bacteria with specific degrees of sensitivity is Pearson's (1895) skew curve of limited range,

$$n = n_0 \left(1 + \frac{k}{k_1}\right)^{ak_1} \left(1 - \frac{k}{k_2}\right)^{ak_2},$$

where k_1 and k_2 are the numbers of degrees of resistance beyond the mode possessed by the most fragile and most resistant classes of bacteria. If we put $k_1 = 0$, then the distribution curve becomes

$$n = n_0 \left(1 - \frac{k}{k_2}\right)^{ak_2},$$

and if $\frac{k}{k_2}$ is large the curve relating the number of surviving bacteria with time will approximate to the exponential type. How far either of these explanations are adequate may be doubted, for in the second case the type of distribution curve is highly abnormal and postulates a maximum number of bacteria with a sensitivity at or very near the limit of variation. The immediate point of interest is, however, focussed on the fact that it is impossible to analyse any single series of data in the production of which more than one variable has played its part. The more we try to escape from these difficulties the more unsatisfactory the position seems to become[1]. This is, perhaps, illustrated by the process of decay in the activity of spermatozoa (Gray, 1928). The observed facts can either be harmonised with the conception of a first order reaction occurring within a heterogeneous population of a particular type or as a process of autointoxication

[1] The difficulties have been considerably reduced by the recent work of Ponder (1930).

within a population whose variability in respect to both directions of this reaction gives the same type of curve. Unless we have independent evidence, we can only base our opinions on the intrinsic probability of one or other hypothesis, and not on the goodness of fit of a particular equation. If we attempt to base our conclusions on the goodness of fit of a particular equation we find that we are dealing with equations involving several arbitrary constants. There is the constant peculiar to the fundamental reaction, and there are the constants defining the variability curve—viz. the mean, and the departure from the mean on the two sides. By suitable values a reasonable fit can always be obtained, but the fundamental truth remains that it is on the intrinsic probability of these values that the validity of the whole hypothesis will rest.

Since it is intrinsically improbable that all the cells of a population will be identically alike, it is equally probable that no biological change involving an increase or decrease in the number of active cells present will ever yield a reaction curve comparable to that of a chemical system. Unfortunately we may have to extend this statement to include reactions involving smaller units than cells: if there are units inside the cell which vary among each other and whose properties in this respect are more or less constant during the life of the units, we would be no further forward even if we were able to carry out a series of observations on one single cell. An example of this type is provided by the rate of exosmosis of electrolytes from a single egg cell of the trout (Gray, 1921). As long as the protoplasmic surface of the egg is intact no measurable loss of electrolytes occurs; when the protoplasm dies the loss of electrolytes obeys the simple diffusion laws—so that the rate of loss of electrolytes is determined by the diffusion gradient between the inside and the outside of the cell and by the area of protoplasm which has become permeable. All regions of the egg surface do not become permeable at the same time, however, so that once again there are two variables: (i) the rate at which the surface becomes permeable—which differs in different regions, (ii) the diffusion gradient between the interior of the cell and the external medium. If the means employed to make the protoplasm permeable acts very quickly over the whole surface of the egg the resultant osmosis/time curve becomes concave to the abscissa very soon after the start—but in all other cases the curve clearly reflects the effect of both factors involved (fig. 141 B). When the protoplasm is exposed to a strong toxic reagent, the form of the

CELL VARIABILITY

curve obtained is that due to the effect of the diffusion gradient, since the processes of diffusion are slower than those producing protoplasmic permeability; under these conditions all the protoplasmic units are destroyed at a very early stage, and the fundamental reaction is very rapid compared with that of diffusion. When the toxic agent is weak a considerable time elapses before all the protoplasmic units have been destroyed, and during this period the form of the curve reflects both the variability of the protoplasm and the effect of a falling osmotic gradient between the inside of the egg and the external solution. Similarly, if the fundamental process of disinfection or death of bacteria were extremely rapid, the so-called percentage death/time curve might be in reality

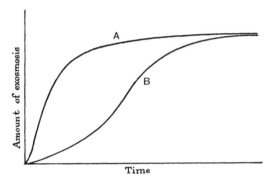

Fig. 141. Curve A shows the rate of exosmosis from a single egg of the trout when exposed to a solution which rapidly destroys the normal impermeability of the protoplasm. Curve B shows the rate of exosmosis from an egg whose impermeability is more slowly destroyed by a less toxic solution.

the mortality curve of the population without reflecting in any obvious way the nature of the fundamental reaction. This was suggested by Loeb and Northrop (1917) for Miss Chick's data. In most cases, however, the observed curve is probably the result of two variables (as pointed out by Brooks), as is quite clearly the case with the data of exosmosis in solutions of weak toxicity.

At this point it is of interest to divert from living cells to inanimate systems. Within the molecular population of a gas the total kinetic energy is not equally distributed among its members. Some molecules will at any given moment be moving very rapidly—others will be moving much more slowly. These facts are expressed in Maxwell's distribution curve. If we were able to mark a single molecule and observe, from time to time, its kinetic energy, we would find that this energy is not fixed and immutable, but varies with time over a range identical with that characteristic of the whole population when measured at any given moment. Such systems, like Miss Chick's conception of bacterial surfaces, are of a dynamic type and play their part in many types of physical change. So far as is known, however,

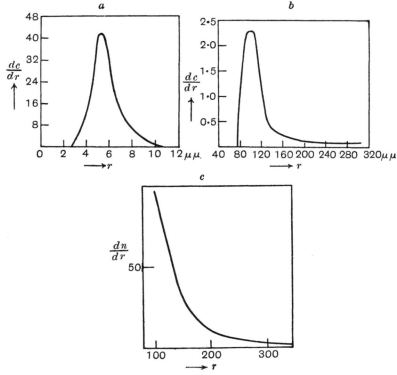

Fig. 142. Distribution curves of size of particles in colloidal suspensions. *a* and *b*, gold sols; *c*, mercury emulsion. (From Svedberg.) Curve *a* is comparable to fig. 140. Curve *b* is similar to that described by Ponder (1930) for the sensitivity of red-blood corpuscles to haemolytic agents. Curve *c* is comparable to those of Chick and others for the death of bacteria in solutions of disinfectants.

the phenomena of *static* variability are only of chemical significance within colloidal systems. Within any given suspension the sizes of the individual particles are not the same, but are distributed on a variability curve of one form or another (fig. 142). The relative number of particles of any given size can be estimated in such systems by a variety of methods (see Svedberg, 1928, pp. 167–183). In many cases the size of a particle does not greatly influence the velocity of chemical reactions in which the particle takes part, but in a few cases there is good evidence to support the view that such a state of static variability within colloidal populations has a very real and profound effect.[1] Svedberg (1928) and others have shown that when a photographic film is exposed to light and subsequently developed, the individual grains of silver halide are either completely reduced to silver and dissolved away or they are not dissolved at all. If, however, potassium bromide is added to the developer there are a certain number of grains which, after development, are only partially dissolved. According to Svedberg the sensitivity to development depends on the presence of one or more 'centres' at which the process of development can occur. These centres are points at which the sensitivity of the grain surface reaches a critically high value; the sensitivity being measured by the least amount of energy required to induce the formation of a reduction centre. When the film is exposed to a stream of α-particles the intensity of incident energy is high enough to produce a centre wherever an α-particle falls on a grain, consequently the process of development follows an exponential curve,

$$P = 100 (1 - e^{-kt}),$$

where P = the percentage of developable grains and k = the number of α-particles falling on one projective area of the halide grain per second. When, on the other hand, the film is exposed to ordinary light the energy of one quantum of light is too small to give a developable centre. It is necessary for a minimum number of quanta to be absorbed within a definitely limited area of the grain surface—and this number varies from one part of the grain to another—consequently the effect of development after varying exposures follows a sigmoid curve which is partly an expression of the variability of the grains of the population and partly of the number of grains present (fig. 143). Systems of this type are admittedly uncommon in physical systems, but it is by no means uncertain that they may prove to be the rule rather than the exception within living systems. Quite clearly the introduction of a state of static variability into a chemical system profoundly modifies the application of the normal laws of mass action: instead of the rate of a reaction being pro-

[1] In this connection it is interesting to note that statistical variations in cell size have recently been held to be responsible for the form of the mortality curve of bacterial populations (Rahn, 1929).

portional to the mass or concentration of the reacting units, it becomes proportional to the mass or concentration of those units which happen to possess a critical degree of statical sensitivity.

We may now consider how far it is possible to escape from some of these difficulties by means of a suggestion made by Ponder (1926).[1] This author was concerned with a population of red blood-corpuscles whose surface possessed varying degrees of resistance to the action of haemolytic agents. The velocity of haemolysis is proportional to the number of molecules of free haemolytic agent (c) and to the number of cells present (n_1) at any given time:

$$\frac{dx}{dt} = k \cdot c \cdot n_1 \qquad \ldots\ldots(ix).$$

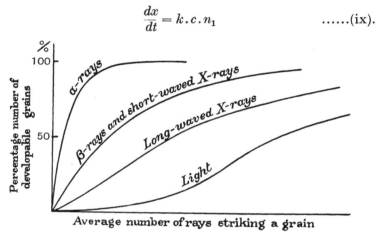

Fig. 143. The effect of radiant energy on a photographic emulsion. (From Svedberg.)

If it be assumed that the variation in sensitivity be expressible by a frequency curve of the type $n = Ae^{-b^2x^2}$, where n is the number of cells broken down by an amount of haemolytic agent x, then as haemolysis proceeds the number of cells haemolysed will be, as with other such systems,

$$n_1 = \int_0^x Ae^{-b^2x^2}\, dx.$$

Expressed as a graph this will result in a sigmoid curve ABC (fig. 144). If we substitute this value for n_1 in equation (ix) we are left with an expression the integration of which is exceedingly unwieldy.

[1] Ponder's recent work (1930) renders this treatment obsolete except as a first approximation.

CELL VARIABILITY 319

Ponder cut the Gordian knot by suggesting that if this curve be replaced by the straight line AB_1C_1, the errors introduced are not greater than others which necessarily creep into experimental procedure. When simplified in this way it is possible to incorporate

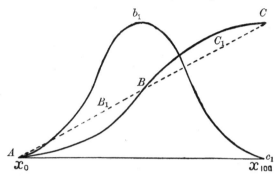

Fig. 144. The curve Ab_1c_1 is the distribution curve of the red blood-corpuscles in respect to their degree of sensitivity (x) to haemolysis. ABC is the integrated curve.

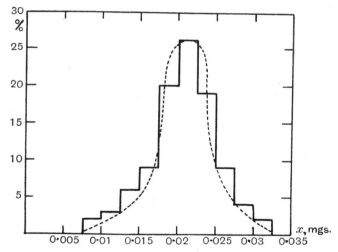

Fig. 145. Frequency polygon illustrating the variation in susceptibility of red blood-corpuscles to cytolysis by saponin. The susceptibility is measured in terms of milligrams of saponin required to produce cytolysis. (Ponder, 1927.)

into the analysis of experimental results a term which will be a quantitative expression of the change in variability of the system, and which at the same time leads to a differential equation which can be integrated to express the progress of the experiment. This pro-

CELL VARIABILITY

cedure is satisfactory as long as we have some external evidence with which to check the arbitrary values assigned to the constants in the initial variability curve.

If such evidence is wanting, Ponder's approximation is liable to be dangerous. An example has recently been quoted elsewhere (Gray, 1929).

An ageing culture of bacteria exhibits a rapidly declining growth rate (see p. 285): if the bacteria are subcultured into a satisfactory medium the full growth rate is eventually resumed, but is preceded by a period of lower growth rate which gradually rises to the full value. Ledingham and Penfold (1914) obtained the following figures.

Table XLV

Time in minutes	No. of bacteria present
0	217·5
45	287
60	345
80	470
100	718
120	1362
150	2535
180	7610

To account for these figures we might make the following hypothesis. At the beginning of the experiment let us imagine that all the bacteria are alive but unable to reproduce, but as time proceeds they gradually recover from the effects of their past history. All the bacteria will not recover at the same rate, some will recover after a short time and some after a longer time; at the end of three hours the culture is known to be growing at a steady rate. If we express the variability of the bacteria by Ponder's approximation, and if the number of actively reproducing bacteria be x, and the number of inactive bacteria be y, then

$$x + y = \frac{217}{180k}(e^{kt} - 1) + 217\left(1 - \frac{t}{180}\right) \quad \ldots\ldots(\text{x}),$$

where k is $\dfrac{\log_e 2}{\text{generation time in minutes}}$.

Giving k the arbitrary value of 0·0285, we get a series of figures which conform very closely to the observed figures. If, however, $k = 0\cdot0285$, then the generation time for the active bacteria should be 24·3 minutes: experimentally, however, k is found to be 0·0385 (= a generation time of 18 minutes), and if we put $k = 0\cdot0385$ in equation (x) we get figures very far removed from the experimental observations.

This example indicates the danger of assigning arbitrary values to constants in any biological equation.

CELL VARIABILITY

Another useful contribution to the theory of static variation within cell populations was made by Yule (1910). If a population of cells is exposed to adverse circumstances (leading to a decrease in their number) for a given number of time units, 1, 2, 3, ..., these periods can be regarded as equivalent to a successive series of incidents each one of which may lead to an elimination of the life of the cell. For any given cell any one of these exposures may be 'effective' or 'non-effective' in that it may damage the cell or leave it uninjured. Let it be supposed that r effective exposures cause the death of the cell, and let the chance of any one exposure being effective on any one individual be p. Such a condition of cumulative death processes is not an unreasonable conception. Under such circumstances the proportion of individual cells which will have received 1, 2, 3, ... effective exposures after the nth exposure will be given by the terms of the binomial expansion

$$q^n, \quad n.q^{n-1}p, \quad \frac{n(n-1)}{1.2}q^{n-2}.p^2 \quad \ldots\ldots(\text{xi}).$$

If r exposures are fatal then the total number of survivors at the end of the nth exposure will be given by the sum of the first r terms of series (xi) and the total proportion surviving at the end of $(n-1)$ exposures will be given by the first r terms of the series (xii)

$$q^{n-1}, \quad (n-1)q^{n-2}p, \quad \frac{(n-1)(n-2)}{1.2}q^{n-3}.p^2 \quad \ldots(\text{xii}).$$

The proportion of the original population which die between the onset of the $(n-1)$th and nth exposure will be the difference between the first r terms of the two above series, and this will reduce to

$$p^r, \quad rp^rq, \quad \frac{r(r+1)}{1.2}(p^rq^2),$$

which is the binomial expansion of $p^r(1-q)^{-r}$. The significance of this result is seen by substituting different values for r. If $r = 1$, one single effective exposure results in death and the series becomes strictly logarithmic, as is apparently the case with bacteria exposed to disinfectants,

$$p.pq.pq^2 \ldots.$$

If however r is greater than 1, the number of deaths per exposure rises gradually to a maximum and then declines as is usually the case. Yule has extended this argument to cover those cases in which death is not necessarily a discontinuous process, but for the present purposes it is sufficient to point out that the wide range of curves shown in fig. 142 can be deduced from Yule's treatment. Unless, therefore, we have reason to suspect that one is more likely to be true than the other, it is quite impossible to foretell what will be the precise effect of statical variation on the process of an intracellular reaction, such as muscle fatigue, or decay in rate of respiration.

As already pointed out problems of statical variation are not restricted to phenomena in which the number of units in a population are undergoing irreversible decay, but are also found in functional systems of a different type. Quite recently Hecht and Wolf (1928) have shown that the resolving power of the bee's eye at different intensities of light can only be interpreted on the assumption that the threshold stimuli for the various ommatidia are not the same but are distributed on a fairly symmetrical distribution curve.

It will readily be admitted that all these attempts to reduce the problem of statical variation to manageable proportions are unsatisfactory. The dominant fact remains that if the members of a cell population carrying out any reaction are themselves liable to become ineffective, there is introduced a factor which cannot adequately be expressed in quantitative terms. There are always two variables: (i) the intensity of the reaction being measured, (ii) the number of units concerned in the reaction at different times; as long as these two variables are unknown in quantity any observed result can only be the result of their combined effects. The only safe and reliable procedure is to design experiments in which static variability plays no part; failing that, some method must be found for measuring the degree of physiological variation by other means, and then applying this value to a particular series of observations. The whole subject is one of great difficulty, but at the same time is of very far-reaching importance. It is significant that problems of a similar nature arise in colloidal suspensions; it may well be that further study of such inanimate systems will be of the greatest value to cell physiology.

REFERENCES

ADRIAN, E. D. and BROCK, D. W. (1928). 'The discharge of impulses in motor nerve fibres. I. Impulses in single fibres of the phrenic nerve.' *Journ. Physiol.* **66**, *81*.

BROOKS, S. C. (1918). 'A theory of the mechanism of disinfection, haemolysis, and similar processes.' *Journ. Gen. Physiol.* **1**, *61*.

Brown, D. E. S. and Sichel, F. J. M. (1930). 'The myogram of the isolated skeletal muscle cell.' *Science*, 72, *17*.
Chick, H. (1908). 'An investigation of the laws of disinfection.' *Journ. Hygiene*, 8, *92*.
—— (1910). 'The process of disinfection by chemical agencies and hot water.' *Journ. Hygiene*, 10, *237*.
Fisher, R. A. (1925). *Statistical Methods for Research Workers*. (Edinburgh.)
Gray, J. (1921). 'Exosmosis from animal cells.' *Journ. Physiol.* 55, *322*.
—— (1928). 'The senescence of spermatozoa.' *Brit. Journ. Exp. Biol.* 5, *345*.
—— (1929). 'The kinetics of growth.' *Brit. Journ. Exp. Biol.* 6, *248*.
Hecht, S. and Wolf, E. (1928). 'The visual acuity of the honey bee.' *Journ. Gen. Physiol.* 12, *727*.
Heiberg, K. A. (1907). 'Über eine erhöhte Grösse der Zellen und deren Teile bei dem ausgewachsenen Organismus verglichen mit dem noch nicht ausgewachsenen.' *Anat. Anz.* 31, *306*.
Jinnaka, S. and Azuma, R. (1922). 'Electric current as a stimulus with respect to its duration and strength.' *Proc. Roy. Soc.* B, 94, *49*.
Ledingham, J. C. G. and Penfold, W. J. (1914). 'Mathematical analysis of the lag-phase in bacterial growth.' *Journ. Hygiene*, 14, *242*.
Loeb, J. and Northrop, J. H. (1917). 'On the influence of food and temperature upon the duration of life.' *Journ. Biol. Chem.* 32, *103*.
Lund, E. J. and Logan, G. A. (1924). 'The relation of the stability of protoplasmic films in *Noctiluca* to the duration and intensity of an applied electric potential.' *Journ. Gen. Physiol.* 7, *461*.
Pearson, K. (1895). 'Skew variation in homogeneous material.' *Phil. Trans. Roy. Soc.* A, 186, *393*.
Ponder, E. (1926). 'The equations applicable to simple haemolytic reactions.' *Proc. Roy. Soc.* B, 100, *199*.
—— (1930) 'The Form of the Frequency Distribution of Red Cell Resistances to Saponin.' *Proc. Roy. Soc.* B, 106, *543*.
Rahn, O. (1929). 'The size of bacteria as the cause of the logarithmic order of death.' *Journ. Gen. Physiol.* 13, *179*.
Richards, O. W. (1928 a). 'Potentially unlimited multiplication of yeast with constant environment and the limiting of growth by changing environment.' *Journ. Gen. Physiol.* 11, *525*.
—— (1928 b). 'Changes in sizes of yeast cells during multiplication.' *Bot. Gazette*, 86, *93*.
Svedberg, T. (1928). *Colloid Chemistry*. 2nd ed. (New York.)
Yule, G. U. (1910). 'On the distribution of deaths with age when the causes of death act cumulatively, and similar frequency distributions.' *Journ. Stat. Soc.* 73, *26*.

CHAPTER THIRTEEN

The Equilibrium between a Living Cell and Water

It is common knowledge that four-fifths of the total weight of living cells consists of water and that in higher vertebrate animals any significant variation in water content may involve far-reaching consequences. Amphibia, however, exhibit a much higher resistance to desiccation, although it is not clear how far the water which can be lost is restricted to the blood and the skin or how far it can be drawn from the cells of the more fundamental tissues; newts can recover from a loss of water equal to half their normal body weight (Gray, 1928). It is, however, within the invertebrate kingdom that organisms exhibit maximum toleration to desiccation; rotifers and nematodes are the outstanding examples (see Davenport, 1897, Vol. 1). The behaviour of the rotifer *Philodina roseata* suggests, however, that the actual loss of water from the living tissues is more apparent than real. When the rotifer is dried in sand on a glass slide it becomes spherical and secretes round itself a gelatinous envelope, within which it remains encysted until water is again available. This capsule is relatively impermeable to water, so that the amount of water within the cells always remains relatively high (Davis, 1873).

Only in a few cases has any attempt been made to correlate the water content of specific types of cell with functional activity. All types of protoplasmic contraction eventually fail if water is removed from the cells beyond a critical level, whilst nuclear and cell division are also sensitive to lack of water (see Chapters VIII and IX).

In the simplest forms of life, e.g. bacteria, the water content of the cells exhibits considerable variation in response to the medium in which the cells are immersed, but in higher animals the normal water content is the result of a more complicated equilibrium. According to Mayer and Schaeffer (1913, 1914), the ability of a cell to hold water depends on the ratio of cholesterol to fatty acids inside the cell, but for an elaboration of this view the original papers must be consulted. For the present purposes, the equilibrium between a living cell and water is chiefly of importance in that it is closely associated with the equilibrium between the cell and the solutes present inside and outside the cell respectively. By observing the passage of water into or out of a cell in different types of environment it is possible to gain some insight into the factors which control this all-important aspect of the cell's activity.

Permeability to water

If living tissues are exposed to a solution whose osmotic pressure is higher than that of their normal environment the cells lose water. If the tissues are replaced in their normal environment they reabsorb the water which had previously been lost. Such facts were first established by Pfeffer (1877) in plant cells in which plasmolysis involves the passage of water from the central vacuole through the protoplasm of the cell. Both qualitatively and quantitatively plant cells behave as though their protoplasmic elements were freely permeable to water, but highly impermeable to the substances in solution. If the cell were not only permeable to water but also to the substances in the outside environment, no water would pass from the cell, since it is the differential pressure of the dissolved substances outside the protoplasm which provides the energy required to move the water. It thus appears that under certain specific conditions the protoplasm of a vacuolated plant cell may be comparable to a membrane such as copper ferrocyanide: it is, in fact, a semipermeable membrane being permeable to water but not to the sugars and other dissolved substances which are capable of exerting an osmotic pressure. The validity of this conception is supported by the following facts. If a membrane of copper ferrocyanide is interposed between a solution of cane sugar and a system of water, the force (P) under which the water tends to pass into the cane sugar is directly proportional to the molecular concentration of the sugar (C) and to the absolute temperature (T),

$$P = kCT,$$

as long as the membrane is completely impermeable to the sugar. The osmotic pressure of all molecules is the same, so that all solutions containing one gram molecule of solvent exert the same osmotic pressure. Overton (1899) found that filaments of *Spirogyra* are just plasmolysed by a 6·0 per cent. solution of cane sugar and remain unplasmolysed by all weaker solutions. The molecular weight of cane sugar is 342 so that a 6 per cent. solution is equivalent to $0.175M$. As long as a living cell of *Spirogyra* is completely impermeable to the molecules of the solute, it will be plasmolysed when the effective concentration of any substance reaches the value of $0.175M$. In the following table are found the results obtained by Overton, who expressed his figures in terms of grams per 100 c.c. and not as gram molecules per litre.

Table XLVI

Substance	Molecular weight	Critical concentration for plasmolysis	
		Observed (%)	Calculated (%)
Cane sugar	342	6.0	—
Mannite	182	3.5	3.19
Dextrose	180	3.3	3.15
Arabinose	150	2.7	2.63
Erythrite	122	2.3	2.14
Asparagin	132	2.5	2.32
Glycocoll	75	1.3	1.32

As pointed out by Höber (1922) the molecular weight of a substance can, in fact, be determined by measuring the concentration necessary for plasmolysis. Thus before the work of de Vries (1888) the molecular weight of raffinose was uncertain: it might have one of three values 396, 594 or 1188. But the cells of *Tradescantia* are plasmolysed by a 3.42 per cent. solution of cane sugar and by 5.96 per cent. raffinose. Since the molecular weight of cane sugar is 342, that of raffinose must be approximately 596.

In Pfeffer's equation $P = kCT$, the molecular concentration C is clearly the reciprocal of the volume occupied by one unit mass of the solute, so that this volume (V) is related to the osmotic pressure by the expression $PV = kT$, which is of the same form as that which applies to the pressure and volume of a gas. In 1877 van't Hoff showed that these two systems are not only similar but are identical, so that k is replaceable by the gas constant R:

$PV = nRT$ where n = number of gram molecules present in unit volume.

In the case of a plant cell the volume occupied by unit mass of the solute is obviously proportional to the volume of the vacuole, so that it is possible to test the relationship $PV = kT$, by measuring the volume of the vacuole when the cell is in equilibrium with plasmolysing solutions of different concentrations. This was done by Höfler (1917) who used cells of filamentous algae.

Table XLVII

Concentration of plasmolysing fluid P	Relative volume of the vacuole V	Isotonic concentration PV
0·30	0·585	0·175
0·35	0·494	0·173
0·45	0·382	0·172
0·60	0·287	0·172

It is seen that van't Hoff's law holds with remarkable accuracy, and that if the cells are to retain their normal volume ($V = 1$), the concentration of the external solution must be approximately $0·175M$.

The osmotic pressure and plasmolysing efficiency of all molecules is the same, but these properties also apply to electrolytic ions. Hence, if a molecule undergoes electrolytic dissociation its plasmolysing efficiency is increased, since each molecule will give rise to two or more ions. If 100 molecules of an electrolyte are dissolved in water and some of these (a) are dissociated to form ions according to the equation (xiii),

$$X_n Y_{n_1} = nX + n_1 Y \qquad \ldots\ldots\text{(xiii)},$$

then the total number of ions present will be $a(n + n_1)$ and the number of undissociated molecules left will be $100 - a$. Hence the total number of molecules and ions will be

$$a(n + n_1) + 100 - a.$$

Each gram molecule of substance will therefore be equivalent, as far as osmotic pressure is concerned, to $\dfrac{a(n + n_1) + 100 - a}{100}$ gram molecules of a non-electrolyte. This fraction is usually known as van't Hoff's factor i, and it can be calculated by a variety of physical methods. If living cells are completely impermeable to a particular type of molecule and to one or more of the ions to which it gives rise, then it should be possible to establish the value of i by observing the concentration of electrolyte required to produce plasmolysis. Observations of this type were made by Fitting (1917) who used the cells of *Tradescantia*.

A consideration of Tables XLVI and XLVIII shows that the plasmolytic efficiency of both non-electrolytes and electrolytes is slightly below the figures which can be calculated theoretically, and this suggests that the surface of the living plant cell is not absolutely impermeable to the substances exerting the osmotic pressure. At the same time the discrepancy is small, and hardly influences the conclusion that to a remarkable degree the protoplasmic elements in a mature plant cell behave as though they possessed true semipermeable properties, in that they are permeable to water but not to molecules of sugars or neutral salts. It is, however, very important to note that the relevant data apply only to experiments of short duration. Probably the protoplasm of plant cells is always permitting the passage of dissolved molecules and ions, but the rate at which this can occur is

Table XLVIII

Salt	van 't Hoff's factor as calculated from		
	Plasmolysis	Freezing point	Electrical conductivity
KNO_3	1·64	1·78	1·83
KCl	1·68	1·84	1·86
K_2SO_4	2·2	2·36	2·38
$NaNO_3$	1·65	1·8	1·83
LiCl	1·73	1·85	1·81
$MgSO_4$	1·01	1·1	1·33
$MgCl_2$	2·39	2·64	2·45
$CaCl_2$	2·37	2·59	2·45

very much less than the rate at which water can pass in or out of the cell—so that it is only in experiments involving considerable periods of time that any discrepancy will be observed which is due to a penetration of the dissolved molecules or ions—assuming always that the protoplasm does not undergo in any experiment a process of irreversible injury. Injury or death rapidly entails a high degree of permeability to sugars and salts (Stiles, 1924).

In being permeable to water the plant cell is not unique; similar properties are displayed by most animal cells. In the latter case, however, there are three important differences from plant cells. (i) There is no large central vacuole, so that changes in volume which result from the passage of water into or out of the cell represent changes in the volume of the protoplast itself. (ii) Most animal cells become instable if the qualitative composition of the surrounding medium differs materially from the normal environment of the cells. These two facts must be borne in mind during any discussion of the osmotic equilibria of animal cells. (iii) Most plant cells are surrounded by a tough cellulose membrane, so that if the cell be immersed in a hypotonic solution water will cease to enter the cell when the osmotic effect of the cell contents is balanced by the hydrostatic pressure of the cellulose wall. In animal cells the hyaline membranes of the cell are often very delicate structures possessing a considerable degree of plasticity, so that their elastic resistance to osmotic forces is probably not very great.

As will soon be evident, the equilibria between animal cells and their aqueous environments are more complex and less well defined than is the case of plants. One of the earliest types of animal cells to be investigated was the red blood-corpuscle of vertebrates. Hamburger (1902) carried out an extensive series of experiments and found that when the osmotic pressure of the medium exceeded a critical value the cells not only changed in appearance but also allowed the haemoglobin to diffuse out of the cell. In this condition, of course, the cell is no longer normal, so that although Hamburger's observations were of interest they were not of great value as a measure of the properties of normal cells. That red blood-corpuscles lose water when exposed to hypertonic solutions is illustrated by Hedin's (1897) experiments in which the volume of the corpuscles was measured in a haemocrit after exposure to a constant centrifugal force for a fixed period of time.

Table XLIX

Concentration of KNO_3 in gram molecules per litre	Volume in c.c. of red blood-corpuscles per 100 c.c. of suspension
0·08	48·6
0·10	46·3
0·12	43·2
0·14	41·2
0·16	39·9
0·18	39·4
0·24	38·7
0·30	37·2

A consideration of these figures (Table XLIX) shows at once that the amount of water lost by exposure to hypertonic solutions is very much less than that calculated by the expression $PV = kT$. Either the cells are permeable to potassium nitrate or the water content of the cells is only controlled by osmotic factors to a comparatively small degree. A similar conclusion is reached from a study of the osmotic properties of cells in intact tissues (Siebeck, 1912). Excised kidneys of the frog were weighed before and after exposure to Ringer solutions of varying strength and care was taken that in each case a true equilibrium was attained before the observations were recorded. If it be assumed that in each 100 gr. of excised kidney the dry contents together with the water held by forces other than osmotic is a grams, and if x is the observed weight when in equilibrium with any given solution, then $x - a$ is the amount of water held in the cell by osmotic forces

$$p(x - a) = k, \text{ so that } p(x - a) = p_1(x_1 - a).$$

Table L

Concentration of Ringer solution p	Relative weight of excised kidney x	Dry weight + non-osmotic fraction a
2·0	79	58
1·0	100	—
0·75	111	68
0·5	127	73
0·25	174	75

From Table L it is evident that after allowing for the dry weight of the kidney (= approximately 15 gr.) there must be at least 40 gr. of water in every 100 gr. of normal tissue which is not held in the cell by simple osmotic forces. This conclusion was originally reached by Overton (1902). Calculations of this type obviously assume that no appreciable mechanical pressure is exerted on the protoplasm by the cell membranes. For example, if a cell is surrounded by a tough and rigid envelope and the cell is then placed in a hypotonic medium, no water can enter the cell since the volume of the latter cannot increase: the osmotic pressure between the outside solution and the protoplasm will be balanced by the mechanical pressure of the envelope. Similarly, if such a cell is placed in a hypertonic solution, loss of water will be inhibited unless the envelope is elastic and can accommodate itself to the reduced volume without any significant changes in energy, or unless the protoplasm can separate itself from the envelope as is the case of plant cells.

When animal cells are immersed in solutions of different osmotic pressure there is general agreement that the resultant changes in volume differ perceptibly from those which would be expected from a system bounded by a true semipermeable membrane, within which all the water was osmotically active. This is clearly the case with kidney cells and with sea urchin eggs (Lücke and McCutcheon, 1927). It does not follow, however, that Overton's explanation is correct. As shown by Hamburger, animal cells readily lose their normal osmotic properties in abnormal environments, and any such change will obviously affect the power of the cell to imbibe or lose water. The recent observations of A. V. Hill (1930) indicate that this factor is, in fact, responsible for the apparent failure to account for the amount of water in muscle cells on the assumption that it is all 'osmotically active'. Instead of observing the changes in weight of a single muscle in a hypertonic or hypotonic medium, Hill compared the changes in weight of two similar muscles in hypertonic and hypotonic solutions respectively; in this way it is possible to compensate, to some extent, for irreversible changes in permeability to solutes, since they will be more or less the same in both lots of tissue. Since the method is of general application it may be considered in some detail. Two gastrocnemii muscles of the frog are taken, having initial weights of unity and which are in equilibrium with Ringer solution of concentration R. Let a be the solid plus 'bound' water fraction, and x be the osmotically active fraction, then $(1 - a - x)$ is the osmotically inactive water fraction. One muscle is then brought into equilibrium with hypertonic Ringer solution $(1 + r) R$, and the other with a hypotonic solution $(1 - r) R$. If the permeability of the muscle cells change in any way x will become $x(1 + \delta)$ where δ may be positive or negative. If $(1 - a - x)$ changes it may become $(1 - a - x)(1 + \delta')$. If A is the final weight of the muscle in the hypotonic solution and B is that of the muscle in hypertonic solution, then

$$A = a + (1 - a - x)(1 + \delta') + \frac{x(1 + \delta)}{1 - r},$$

$$B = a + (1 - a - x)(1 + \delta') + \frac{x(1 + \delta)}{1 + r}.$$

Hence
$$A - B = \frac{x(1 + \delta) 2r}{1 - r^2},$$

and the final osmotically active fraction $x(1 + \delta) = \dfrac{1 - r^2 (A - B)}{2r}$.

When analysed in this way, the amount of water in a muscle cell which is not osmotically active is found to be extremely small, and it looks as though Overton's results were due to irreversible changes caused in the muscle during the experiments or to the failure to reach a true position of equilibrium. It is of interest to note that Hill's data are derived from experiments in which the cells remained alive during the whole series of observations.

We have already seen that a plant cell comes into equilibrium with its surrounding medium when the osmotic force exerted by the cell sap against the external solution is balanced by the hydrostatic pressure of the cellulose wall. In animal cells there is no cellulose wall, and it seems doubtful whether the surface hyaline membranes could exert an elastic force capable of opposing a significant difference in osmotic pressure between the inside and the outside of the cell. It is therefore surprising to find that the cells of animals living in a fresh water environment do not burst or swell continuously. Either the effective difference between the osmotic pressure inside and outside is very slight, or the cell must have some means of excreting water from its interior. Many years ago Hartog (1889) suggested that the contractile vacuoles of Protozoa represented such a mechanism, and Day (1927) has recently reported an increase in the rate of excretion of water by *Amoeba* on transfer to pure distilled water. It is known that salts affect the frequency of contraction of these vacuoles and Grüber (1889) and others reported the absence of vacuoles in marine types. The more recent observations of Adolph (1926) make it doubtful, however, whether the total amount of water excreted by an *Amoeba* is strictly a function of the osmotic force exerted between the external medium and the interior of the cell, since the rate of excretion of water appears to be independent of body surface or volume. The absolute rate of excretion by the vacuole of *Amoeba* appears to be one cubic millimetre of water per square millimetre of body surface every 11–42 minutes, and from 4 to 30 hours are required for the excretion of a volume of water equal to that of the volume of the body.

Before considering the water equilibrium of the cell in greater detail it is useful to remember that the water content of a cell might conceivably be affected by circumstances other than the *number* of molecules or ions in the external medium. We have seen that the water content will change if there is a difference between the osmotic pressure inside and outside the cell which is not compensated by mechanical or other forces. As long as the internal osmotic pressure is maintained by non-electrolytic molecules or by the ions of strong electrolytes, the interior of the cell may legitimately be compared to a sugar solution enclosed within a membrane of copper ferrocyanide. If, however, the osmotic pressure inside the living cell is maintained wholly or to a significant degree by the ions of a weak electrolyte it will be sensitive to any treatment which alters the degree of ionisation of such systems. The importance of this point can be appreciated from the following facts. Powdered gelatin when washed in water absorbs a certain amount of water; if we wash this damp gelatin with sodium chloride which is faintly alkaline no marked change will occur, but if we now rinse the powder with distilled water the particles rapidly absorb a large amount of water which will be liberated again by the addition of sodium chloride to the medium. The nature of these phenomena are now fairly well defined. The amount of water which is taken up by the gelatin is determined by the degree to which the gelatin is ionised in the form of sodium ions and gelatinate ions; when sodium chloride is added the number of free gelatinate ions is decreased and consequently the amount of water held by the particles is reduced. Proctor and Wilson (1916) have shown that the addition of neutral salts to a gelatin system reduces the amount of water which is absorbed in a way which is qualitatively and quantitatively in harmony with a system conforming to a Donnan equilibrium (see p. 399). Fundamentally the processes of swelling of gelatin and the uptake of water by a system surrounded by a copper ferrocyanide membrane are the same; the only difference is that in the former case the gelatin is impermeable to gelatinate ions and permeable to all the others, whereas in the latter case the impermeability is extended to much smaller molecules and ions.

It is clear, therefore, that when we maintain that the surface of a living cell is comparable to a copper ferrocyanide membrane we must base our views on stronger evidence than the plasmolysing action of neutral salts. On the other hand it is equally unsatisfactory to maintain, for similar reasons, that the cell is essentially equivalent to a particle of gelatin (Lapicque, 1924). There are a few significant facts which enable us to differentiate between these two possibilities. As shown by Proctor and Wilson the water content of a fragment of gelatin depends on two factors: (a) the amount of ionised gelatin present, (b) the nature of the salts in the surrounding medium. If a fragment of swollen sodium gelatinate is exposed to a weakly acid solution the fragment rapidly loses water owing to a reduction in the amount of ionised gelatin. If, however, a living cell is exposed to an acid medium, and care is taken to ensure that the acid penetrates into the cell, no detectable change in water content can usually be observed as long as the cell is alive. This fact alone indicates quite clearly that the water content of normal cells is not controlled by the same factors as control the swelling of gelatin: there must be some mechanism other than the presence of intracellular proteins

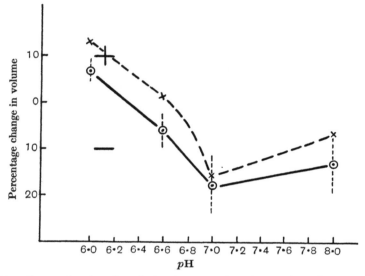

Fig. 146. Curves showing the relation between the volume of an *Amoeba* and the hydrogen-ion concentration of the medium. The ordinates represent the percentage loss or gain in volume. (From Chalkley.)

which limits the entrance and exit of water. The inability of acid to alter the water content of normal cells has been observed by Beutner (1912, 1913), Lücke and McCutcheon (1926, 1927) and others. Again, non-electrolytes influence the swelling of gelatin to a negligible extent, and yet they readily plasmolyse living cells. It should be mentioned, perhaps, that Chalkley (1929) has recently reported changes in the water content of living *Amoebae* in response to changes in the hydrogen ion concentration of their surrounding media (fig. 146).

Post-mortem swelling

The osmotic properties of the cell surface undergo profound and rapid alteration when the cell dies. When death occurs the factors controlling the water content of the cell change and become precisely similar to those of a gelatin system. This fact is often overlooked and merits perhaps a few illustrative examples.

Beutner (1912, 1913) showed that as long as a muscle retains its power of responding to stimuli its water content varied solely with the osmotic pressure of the surrounding medium, although the latter might be abnormally acid. After irritability is lost and the individual fibres begin to die, the water content of the cells increases owing to the acidity of the medium and, as with gelatin, the amount of increase depends on the concentration of neutral salts present. It is a simple matter to state that a cell begins to exhibit imbibitional swelling as soon as it begins to undergo the irreversible processes leading to death: it is more difficult to decide just when these changes begin. In a tissue, or in a population of cells, they do not begin in all the cells at the same time (see Chapter XII), so that we cannot say definitely whether imbibitional swelling is restricted to dead and dying units or whether it is equally distributed throughout all the cells. These difficulties can, however, be overcome to a large extent by using single cells. For this purpose the unfertilised eggs of the trout *Salmo fario* provide very useful material. As long as the cells are alive they are yellow and translucent; when they are dead they are white and opaque. Structurally, each egg consists of a thin protoplasmic film enclosing a clear solution of globulin dissolved in neutral salts; the whole system is enclosed within a tough translucent egg-shell or membrane. The salts which are present in the yolk are largely sodium and calcium chlorides and are present in a concentration equivalent to a depression of the freezing point $-0.45°$ C. As long as the cell is healthy no appreciable quantity of electrolytes diffuses out into the surrounding water, and this state is maintained for many days or even weeks; the moment the cell becomes unhealthy the exit of electrolytes can be detected by a rise in the electrical conductivity of the surrounding medium (Gray, 1921). This test is of extreme delicacy and can be demonstrated with considerable ease. As soon as sufficient electrolytes have left the cell, the globulin in the yolk is precipitated and it is for this reason that the cells become opaque; at no time, however, does the egg membrane (the shell) become permeable to colloids. It can readily be shown that the water content of the dead egg follows the same laws as an equal quantity of globulin enclosed within a parchment membrane, and behaves as a system which strictly obeys the Donnan equilibrium. As long as the cell is alive, however, the water

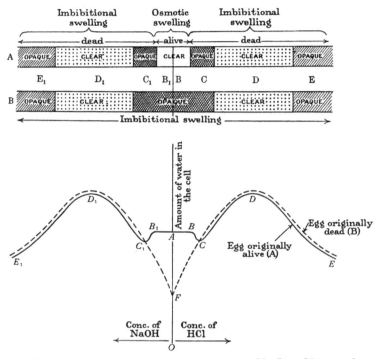

Fig. 147. Diagram illustrating the effect of hydrogen and hydroxyl ions on the uptake of water by (A) living and (B) dead cells. Starting with a living egg of *Salmo fario* the amount of water remains unaffected by the hydrogen ions in the medium until the protoplasmic membrane is destroyed, whereupon the intracellular electrolytes diffuse outwards and the cell becomes opaque: as the acid or alkali diffuses inwards in sufficient concentration to dissolve the globulin the egg becomes transparent and swells: finally when the acid or alkali is present in excess the egg again loses water and becomes opaque (salt action of the excess alkali or acid).

content can only be changed by altering the *number* of osmotically-active ions outside the egg—and during this process the water content is independent of the hydrogen ion concentration of the medium.

The change from osmotic to imbibitional swelling, which attends the onset of surface injury, is of very general application. It can be seen in the curves given by Miss Jordan Lloyd (1916) describing the effect of acids and bases on the water content of frogs' muscles, although this author does not refer to the fact, and concluded, on the other hand, that 'the osmotic phenomena of muscle can be fully explained without assuming the presence of a semipermeable membrane round the muscle fibres'.

If a muscle is exposed to dilute alkalies it first swells, then loses water, and finally swells again; corresponding to these changes are visible changes in appearance. During the first phase the muscle is milky white, in the second it is opaque, in the third it is glassy and transparent. During the two later phases there can be no reasonable doubt that the muscle is dead; during the first phase it is probably alive. From the evidence already presented the whole curve could be explained as follows. During the initial phase the external fluid

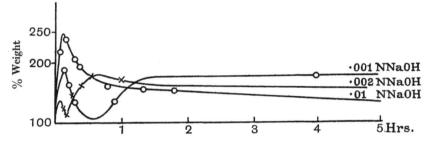

Fig. 148. Graph showing changes in weight of a frog's muscle when immersed in dilute solutions of sodium hydroxide. Note the initial 'osmotic' rise followed by the fall due to exosmosis of salts from the cells. Finally an 'imbibitional' rise occurs. (Fig. from Jordan Lloyd.)

is hypotonic and therefore the cell absorbs water osmotically; after a time, however, the alkali destroys the surface thereby causing a loss of semipermeability; a loss of water then occurs owing to the diffusion outwards of the neutral salts normally present in the cell; finally the alkali diffuses inwards and ionises the proteins which promptly absorb water. Essentially the same phenomena are observed with acids.

The kinetics of osmotic equilibria

The rate at which water passes into living animal cells from hypotonic solutions has been investigated by R. S. Lillie (1916) and more recently by McCutcheon and Lücke (1926). The materials used were the eggs of sea urchins and the amount of water entering the cell in a given time was estimated by measuring the volume of the spherical eggs. As defined by Lillie the rate of entry of water may depend on three main factors. (i) The driving force of osmosis, which will be equal to the difference between the osmotic pressure of the external solution on the one hand, and the osmotic pressure inside the egg together with the elasticity or cohesive strength of the cell surface on the other. (ii) The frictional resistance to the passage of water through the egg surface. (iii) The area of the cell surface. In practice, the volume of an unfertilised egg which is swelling in hypotonic sea water can be determined by an empirical equation, identical with that of a reaction of the first order

$$k = \frac{1}{t} \ln \frac{V_{eq} - V_0}{V_{eq} - V_t} \quad \ldots\ldots(\text{xiv}),$$

where V_{eq} is the volume of the egg in osmotic equilibrium with the hypotonic solution, V_t is its volume at time t, and V_0 the initial volume of the egg. k is a constant.

Lillie's data and those of McCutcheon and Lücke harmonise remarkably well with equation (xiv). It is, however, by no means clear that they can be equally well harmonised with first principles (Northrop, 1927 c). Lillie made two assumptions: (i) that the elastic and cohesive forces of the cell membrane are negligible, (ii) that the change in surface area during swelling can be ignored. In such circumstances the rate of swelling will be proportional to the difference between the osmotic pressure of the cell contents and that of the surrounding medium

$$\frac{dV}{dt} = k(P_t - P_m),$$

where P_t is the osmotic pressure inside the egg and P_m the osmotic pressure of the medium. If the egg is a perfect osmotic system

$$P_t V_t = P_m V_{eq} = k_1,$$

when V_t is the volume at time t, V_{eq} is the volume when in equilibrium with the surrounding medium. Hence

$$\frac{dV}{dt} = k\left(\frac{k_1}{V_t} - \frac{k_1}{V_{eu}}\right) = K\left(\frac{V_{eq} - V_t}{V_{eq} V_t}\right),$$

$$t = \frac{V_{eq}}{K} \int_0^t \frac{V_t}{V_{eq} - V_t} dV,$$

$$K = \frac{V_{eq}}{t}\left(V_0 - V_t - V_{eq} \log_e \frac{V_{eq} - V_0}{V_{eq} - V_t}\right) \quad \ldots\ldots(\text{xv}),$$

where V_0 is the original volume of the egg.

A LIVING CELL AND WATER

Table LI. Changes in volume of unfertilised eggs of *Arbacia*. (From Lillie.)

Units of volume = $10^5 \mu^3$; $V_0 = 21\cdot3$; $V_{eq} = 40\cdot4$; $V_{eq} - V_0 = 19\cdot1$			
t Time in minutes in 40 % sea water	V_t	$\dfrac{V_{eq} - V_0}{V_{eq} - V_t}$	k $k = \dfrac{1}{t} \log_{10} \dfrac{V_{eq} - V_0}{V_{eq} - V_t}$
1	22·9	1·1	41
2	24·5	1·2	40
3	26·1	1·4	49
4	27·9	1·6	51.
5	29·2	1·7	46
6	30·7	1·9	47
7	31·3	2·1	46
8	32·0	2·3	45
9	32·7	2·5	44
10	33·6	2·8	45
11	34·6	3·3	46
12	35·1	3·8	48
13	35·8	4·2	48
14	36·3	4·7	49
			Av. 46

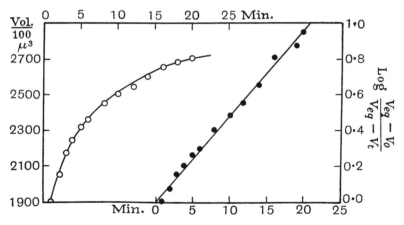

Fig. 149. Graph showing the rate of swelling of unfertilised sea urchin eggs in 60 per cent. sea water at 24·8°. On the left side volumes are plotted against times. On the right side $\log \dfrac{V_{eq} - V_0}{V_{eq} - V_t}$ is plotted against time. (From McCutcheon and Lücke.)

340 THE EQUILIBRIUM BETWEEN

Conversion of K into absolute units gives

$$C = \frac{h \cdot V_{eq}}{S \cdot P_0 t}\left(V_0 - V_t - V_{eq}\log_e \frac{V_{eq} - V_0}{V_{eq} - V_t}\right) \quad \ldots\ldots(\text{xvi}),$$

where C is the c.c. of water which will pass through one square centimetre of the egg surface 1 cm. thick in 1 hour under a pressure of 1 mm. Hg, S = surface area of the egg, P_0 the osmotic constant of the outside solution, and h the thickness of the membrane (Northrop, 1927 c).

Using equation (xvi) Northrop has calculated the value of C/h for Lillie's data—see Table LII.

Table LII. (From Northrop.)

| $V_{eq} = 4.0 \times 10^{-7}$ c.c.; | | | | $V_0 = 2.1 \times 10^{-7}$ c.c.; | | |
| $S = 1.7 \times 10^{-5}$ sq. cm. | | | | $P_0 = 4.6 \times 10^{-3}$ mm. Hg | | |

	Fertilised egg			Artificially parthenogenetic eggs		
$V \times 10^7$ c.c.	t hours	Km	$C/h \times 10^5$	t hours	Km	$C/h \times 10^5$
2·5	0·0080	12·9	3·52	0·0095	10·9	2·98
2·7	0·0130	12·7	3·62	0·0160	10·3	2·97
2·9	0·0200	11·9	3·57	0·0240	10·3	2·97
3·1	0·0290	11·2	3·52	0·0350	9·3	2·91
3·5	0·0550	10·5	3·66	0·0700	8·4	2·92

It will be noted that the values of C/h are more uniform than is the value of the first order reaction constant (Km) as calculated from Lillie's formula.

If we assume that the rate of entry of water is proportional to the surface area of the cell and inversely proportional to the thickness of the surface membrane, and that at all times the volume of the membrane itself remains constant, then

$$\frac{dV}{dt} = 25 C_2 V^{\frac{4}{3}}\left(\frac{P_0}{V} - \frac{P_0}{V_{eq}}\right),$$

or

$$C_2 = \frac{V_{eq}^{\frac{2}{3}}}{25 P_0 t}\left[1.15 \log \frac{V_{eq}^{\frac{2}{3}} + V_{eq}^{\frac{1}{3}} V^{\frac{1}{3}} + V^{\frac{2}{3}}}{(V^{\frac{1}{3}} - V_{eq}^{\frac{1}{3}})^2} + \sqrt{3}\tan^{-1}\frac{2 V^{\frac{1}{3}} + V_{eq}^{\frac{1}{3}}}{V_{eq}^{\frac{1}{3}}\sqrt{3}} - \text{const.}\right].$$

Northrop has calculated the value of the constant C_2 from Lücke and McCutcheon's data; at 11°C. it has a value of 1·45 and at 20·5°C. it is 2·60. If the sea urchin's egg is a perfect osmotic system as postulated by these analyses, the absolute value of the permeability constant should be independent of the osmotic pressures used for changing the volume of the egg. Unfortunately this is not altogether the case. Lücke and McCutcheon's constant varies very

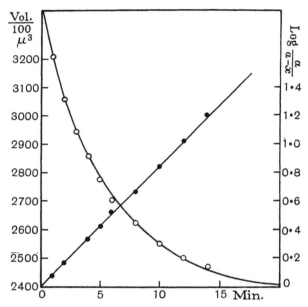

Fig. 150. Figure illustrating exosmosis of water from unfertilised eggs of *Arbacia*. The eggs were exposed to 60 per cent. sea water prior to placing them in normal sea water at 15° C. The curve shows the rate of loss of water on replacement in normal sea water. The straight line shows the logarithm of $\dfrac{a}{a-x}$ plotted against time, where a is the initial volume and x the amount of water which has left the eggs at time t. (From McCutcheon and Lücke.)

greatly with different concentrations of hypotonic sea water and even Northrop's more accurate formula breaks down when the concentration of sea water falls below 60 per cent.

The value of Northrop's analysis does not lie in the ability of his equations to fit the observed data, but in the fact that it is an exact parallel to that applicable to the diffusion of water through an inanimate collodion sac (Northrop, 1927 a and b) and into particles of ionised gelatin (Northrop and Kunitz, 1927). We can therefore compare with some degree of accuracy the relative permeability of living and non-living membranes: the value C/h gives a basis of comparison. Northrop's figures show that the membrane on the surface of a sea urchin's egg is less than 1/1000th times as permeable as a collodion membrane of the same thickness. The *apparent* high degree of permeability is due of course to the extreme thinness of the egg membrane and the high ratio between surface and volume of the egg.

Although the kinetics of swelling of an echinoderm egg and a particle of gelatin follow the same law, there is an essential difference in the way in which the driving force of osmosis is generated in the two cases. In the egg it is due to ions or molecules whose concentrations are not altered by factors influencing the degree of ionisation of proteins; in gelatin on the other hand the osmotic pressure is largely due to protein systems. The facts show that once an osmotic gradient is set up between the interior of an egg or of a particle of gelatin, the equilibrium volume is reached by a flow of water which obeys the same laws in the two cases.

It may well be that different types of cell may generate their osmotic pressures in different ways. Lücke and McCutcheon's (1926) data indicate quite clearly that the volume of echinoderm eggs does not fluctuate with changes in the pH of the medium or of the cell interior. Red blood corpuscles, however, behave differently and in this case it looks as though ionised proteins are responsible for a measurable amount of internal osmotic pressure within the cell (see p. 362).

Before leaving the problem of the osmotic equilibria of animal cells two points should be noticed. (i) In all available analyses the cohesive properties and elasticity of the cell surface have been ignored. (ii) The kinetics of swelling exhibit a very high temperature coefficient ($\mu = 16,000$, McCutcheon and Lücke, 1926). (iii) The equilibria reached in solutions of different osmotic pressures show considerable deviations from the gas laws, although the rate at which the equilibria are reached are explicable without serious deviation from these laws. These three facts arouse the suspicion that there may be other factors involved in addition to those incorporated into Northrop's analysis and which under certain circumstances might seriously invalidate the equations of swelling.

Secretion of water

If the surface of a cell may be regarded as a simple semipermeable membrane then, on interposing a layer of cells between two solutions of the same solute in different concentrations, water should invariably pass from the dilute solution into the concentrated solution, and the rate of transference should be determined by the factors considered by Northrop. It seems clear, however, that there are outstanding deviations to this rule. Reid (1890) showed that the ability of water to pass through the skin of a frog does not simply depend upon the difference in the osmotic pressures of the solutions with which the skin is in contact. When a frog's skin was bathed in normal Ringer solution on both sides, and 5 per cent. glucose was added to the solution in contact with the morphologically inner surface of the skin, 1·83 cubic mm. of solution passed through each square millimetre of the skin in 24 hours: when, however, the same concentration of glucose was in contact with the outer surface of the skin only 0·83 cubic mm. of fluid passed to the inner side. There thus appears to be a difference in the rate at which water will pass under an osmotic force from one side of the skin to the other. This difference is largely a specific property of the healthy living skin, for it is sensitive to the presence of anaesthetics and to those irreversible changes which accompany the death of the cells. Reid concluded that the living skin is able to absorb water from its outer surface and to secrete it on its inner surface, much as a gland cell will secrete fluid at one pole. If therefore an osmotic gradient is established between the two surfaces, the rate of water passing through the skin will depend upon whether the direction of osmotic flow is with or against the flow of water produced by the processes of active secretion. The polarisation of the frog's skin in respect to water transference has been confirmed by Maxwell (1913) and by Adolph (1925 a). Under normal conditions a frog's skin appears to be in osmotic equilibrium when lymph or Ringer is in contact with the inner side and with tap water on the outer side. Adolph found, however, that little or no disturbance in the distribution of water occurred if the tap water on the outside was replaced by Ringer, and concluded that as long as Ringer solution was on the inside of the skin the outer solution could be varied considerably without increasing or decreasing the rate at which water passed through. When, however, tap water is placed in contact with the inner surface, water will pass out to all solutions except tap water. Adolph concluded that under normal conditions, i.e. with lymph on the inside and water on the outside, the natural tendency for the water to pass inwards was opposed by forces in the skin which worked in the opposite direction, i.e. a force which tends to pump water toward the outer surface. Some of Adolph's results are difficult to harmonise with those of Reid and of Maxwell, for both these authors observed an active pumping of water from the outside inwards. Putting these discrepancies on one side, it seems doubtful whether the passage of water across a frog's skin can be accounted for by normal osmotic gradients. The question arises, what is the nature of the forces producing anomalous osmosis? Reid attributed these forces to the active metabolism of the living cells. Adolph goes further and suggests that the forces are electrostatic in nature, since the skin is electrically polarised (Hashida, 1922) and the direction of flow of water tends to be in the direction of the negative side of the skin. On this view the anomalous behaviour of the frog's skin is attributed to those forces which appear to be concerned with anomalous osmosis through inanimate charged membranes, where the flow of water is always towards the negative side; this indicates that the water is itself positively charged.

It is important to bear in mind two points. (i) Within physico-chemical systems anomalous osmosis has only been observed in dilute solutions and where the membrane has been exposed to polyvalent ions; in more concentrated solutions anomalous osmosis becomes negligible. (ii) The theory of anomalous osmosis is by no means clearly defined. If water is to pass *continuously* from a stronger solution to a weaker, the membrane must be doing work; precisely how a collodion membrane coated with protein does this work is not too clear (see Freundlich, 1926); the fact remains that anomalous osmosis occurs. We are apparently dealing with two distinct sources of potential: there is the P.D. between the two sides of the membrane, and the P.D. between the surface of the pores in the membrane and the fluid flowing through the pores. If the positively charged water moves in the direction of the negative side of the membrane, the membrane potential will fall unless a current of electricity is flowing from one side of the membrane to the other. Until the whole situation is considerably clearer it seems a little risky to assume that the behaviour of the frog's skin justifies a very close parallel. Since living frogs' skins behave differently to dead skins, it seems certain that the energy expended by the living skin is not derived from the same source as that which produces anomalous diffusion through a collodion protein membrane (Adolph). In this respect it becomes of importance to know just how far a system Ringer/skin/Ringer is capable of transporting water continuously from one side to the other.

In all cases of anomalous diffusion there must be a difference in the concentration of ions on the two sides of the charged membrane. According to Adolph no water passes through a living frog's skin unless there is a difference in ionic concentration on the two sides—but this disagrees with Reid's data where, with Ringer on both sides, water passed to a sugar solution more easily in one direction than another: it is difficult to see why a non-electrolyte should affect the membrane potential or why there should be a membrane potential in the system Ringer/skin/Ringer, if the potential is solely dependent on a difference in concentration of ions on the two sides (see however p. 395). The only conclusion available seems to be that, by active metabolism, the skin performs work in driving water through the skin—and that it can do this more readily in one direction than in another. It would seem that the subject would prove to be a fruitful line of research—for if the skin is expending energy for the maintenance of an anomalous distribution of water, the permeability of the skin might change in the absence of O_2, and transport of water might continue irrespective of a difference in concentration of ions on the two sides of the membrane. Anomalous diffusion is not restricted to the frog's skin, it occurs to a marked degree in the gut and the kidney[1]. It would be beyond the scope of this book to consider these tissues in any detail: a useful summary of the facts of intestinal absorption is provided by Goldschmidt (1921). It seems clear that both the gut wall and the kidney can effect a transference of water from a hypertonic solution to one which is more dilute, and as in the case of the frog's skin there are two possibilities: (i) the transfer is effected by an active metabolic process of secretion in which the requisite energy is derived from the chemical energy within the cells, or (ii) the energy is derived from (a) an electrostatic source which itself depends on a difference of ions on the two sides of the membrane or (b) a difference in hydrostatic pressure.

It should be noted that although these latter considerations modify to some extent the applicability of the conclusions reached by Lücke and McCutcheon and by Northrop they do not invalidate them. Other things being equal, the rate at which a cell will come into osmotic equilibrium with a surrounding medium will depend on the net value of the differences in osmotic pressure on the two sides. As long as the vital or electrostatic forces remain constant, the curves which illustrate the rate at which equilibrium is reached

[1] For a recent account of the rôles played by the skin and by the kidneys in maintaining the water content of Amphibia, see Adolph (1930).

will always be of the same type; it is only the final equilibrium value which will be affected, and this may account for the discrepancy found between the observed equilibria with those calculated for solutions of varying concentrations.

REFERENCES

ADOLPH, E. F. (1925 a). 'The passage of water through the skin of the frog, in the relation between diffusion and permeability.' *Amer. Journ. Physiol.* 73, *85*.

—— (1925 b). 'Electrostatic forces in the diffusion of water through collodion membranes between solutions of mixed electrolytes.' *Journ. Biol. Chem.* 64, *339*.

—— (1926). 'The metabolism of water in *Amoeba* as measured in the contractile vacuole.' *Journ. Exp. Zool.* 44, *355*.

—— (1930). 'Living Water.' *Quart. Rev. of Biol.* 5, *51*.

BEUTNER, R. (1912). 'Unterscheidung kolloidaler und osmotischer Schwellung beim Muskel.' *Biochem. Zeit.* 39, *280*.

—— (1913). 'Einige weitere Versuche betreffend osmotische und kolloidale Quellung des Muskels.' *Biochem. Zeit.* 48, *217*.

CHALKLEY, H. W. (1929). 'Changes in water content of *Amoeba* in relation to changes in its protoplasmic structure.' *Physiol. Zool.* 2, *535*.

DAVENPORT, C. B. (1897). *Experimental Morphology.* (New York.)

DAVIS, H. (1873). 'A new *Callidina*: with the result of experiments on the desiccation of rotifers.' *Monthly Micr. Journ.* 9, *201*.

DAY, H. C. (1927). 'The formation of contractile vacuoles in *Amoeba proteus*.' *Journ. Morph. and Physiol.* 44, *363*.

FAURÉ-FREMIET, E. (1923). 'Propriétés osmotiques de l'œuf de *Sabellaria alveolata*.' *C.R. Soc. Biol.* 88, *1028*.

FITTING, H. (1917). 'Untersuchungen über isotonische Koeffizienten und ihre Nutzen für Permeabilitätsbestimmungen.' *Jahrb. f. wiss. Bot.* 57, *553*.

FREUNDLICH, H. (1926). *Colloid and Capillary Chemistry.* (Eng. transl.) (London.)

GOLDSCHMIDT, S. (1921). 'On the mechanism of absorption from the intestine.' *Physiol. Rev.* 1, *421*.

GRAY, J. (1921). 'Exosmosis of electrolytes from animal cells.' *Journ. Physiol.* 55, *322*.

—— (1928). 'The rôle of water in the evolution of the terrestrial vertebrates.' *Brit. Journ. Exp. Biol.* 6, *26*.

GRÜBER, A. (1889). 'Biologische Studien an Protozoen.' *Biol. Zentralb.* 9, *14*.

HAMBURGER, H. J. (1902). *Osmotischer Druck und Ionenlehre in den medicinischen Wissenschaften.* (Wiesbaden.)

HARTOG, M. (1889). 'Preliminary note on the functions and homologies of the contractile vacuole in plants and animals.' *Ann. Mag. Nat. Hist.* (6), 3, *64*.

HASHIDA, K. (1922). 'Untersuchungen über das elektromotorische Verhalten der Froschhaut. (i–iii.)' *Journ. Biochem.* 1, *21*.

HEDIN, S. G. (1897). 'Über die Permeabilität der Blutkörperchen.' *Pflüger's Archiv*, 68, *229*.

A LIVING CELL AND WATER

HILL, A. V. (1930). 'The state of water in muscle and blood and the osmotic behaviour of muscle.' *Proc. Roy. Soc.* B, 106, 477.

HÖBER, R. (1922). *Physikalische Chemie der Zelle und der Gewebe.* (Leipzig.)

HÖFLER, K. (1917). 'Die plasmolytisch-volumetrische Methode und ihre Anwendbarkeit zur Messung des osmotischen Wertes lebender Pflanzenzelle.' *Ber. deut. bot. Ges.* 35, 706.

—— (1918 a). 'Permeabilitätsbestimmung nach der plasmometrischen Methode.' *Ber. deut. bot. Ges.* 36, 414.

—— (1918 b). 'Über die Permeabilität der Stengelzellen von *Tradescantia elongata* für Kalisaltpeter.' *Ber. deut. bot. Ges.* 36, 423.

—— (1919). 'Über den zeitlichen Verlauf der Plasmadurchlässigkeit in Salzlösung.' *Ber. deut. bot. Ges.* 37, 304.

KÖPPE, H. (1895). 'Ueber den Quellungsgrad der rothen Blutscheiben durch aequimoleculare Salzlösung und über den osmotischen Druck des Blutplasmas.' *Arch. Anat. u. Physiol.* Physiol. Abt. 154.

LAPICQUE, L. (1924). 'La cellule est-elle enveloppée d'une membrane semiperméable?' *Ann. de Physiol. et de Physiochim. biol.* 1, 85.

LILLIE, R. S. (1916). 'Increase of permeability to water following normal and artificial activation in sea-urchin eggs.' *Amer. Journ. Physiol.* 40, 249.

LLOYD, D. JORDAN (1916). 'The relation of excised muscle to acids, salts, and bases.' *Proc. Roy. Soc.* B, 89, 277.

LÜCKE, B. and McCUTCHEON, M. (1926). 'The effect of hydrogen ion concentration on swelling of cells.' *Journ. Gen. Physiol.* 9, 709.

—— (1927). 'The effect of salt concentration of the medium on the rate of osmosis of water through the membrane of living cells.' *Journ. Gen. Physiol.* 10, 665.

McCUTCHEON, M. and LÜCKE, B. (1926). 'The kinetics of osmotic swelling in living cells.' *Journ. Gen. Physiol.* 9, 697.

—— (1927). 'The kinetics of exosmosis of water from living cells.' *Journ. Gen. Physiol.* 10, 659.

MAXWELL, S. S. (1913). 'On the absorption of water by the skin of the frog.' *Amer. Journ. Physiol.* 32, 286.

MAYER, A. and SCHAEFFER, G. (1913). 'Coefficient lipocytique et imbibition des cellules vivantes par l'eau.' *Comptes Rendus,* 156, 1253.

—— (1914 a). 'Recherches sur les constantes cellulaires. Teneur des cellules en eau. Ier mémoire, discussion théorique. L'eau constante cellulaire.' *Journ. Physiol. et Path. gén.* 16, 1.

—— (1914 b). 'Recherches sur les constantes cellulaires. Teneur des cellules en eau. II. Rapport entre la teneur des cellules en lipoïdes et leur teneur en eau.' *Journ. Physiol. et Path. gén.* 16, 23.

NORTHROP, J. H. (1927 a). 'The kinetics of osmosis.' *Journ. Gen. Physiol.* 10, 883.

—— (1927 b). 'The swelling of iso-electric gelatin in water. I. Equilibrium conditions.' *Journ. Gen. Physiol.* 10, 893.

—— (1927 c). 'Kinetics of the swelling of cells and tissues.' *Journ. Gen. Physiol.* 11, 43.

NORTHROP, J. H. and KUNITZ, M. (1927). 'The swelling of iso-electric gelatin in water.' *Journ. Gen. Physiol.* 10, 905.

OVERTON, E. (1895). 'Über die osmotischen Eigenschaften der lebenden Pflanzen und Tierzellen.' *Vjschr. naturf. Ges. Zürich,* 40, 159.

OVERTON, E. (1899). 'Über die allgemeinen osmotischen Eigenschaften der Zelle, ihre vermuthlichen Ursachen und ihre Bedeutung für die Physiologie.' *Vjschr. naturf. Ges. Zürich*, 44, *88.*
—— (1900). 'Studien über die Aufnahme der Anilinfarben durch die lebende Zelle.' *Jahrb. f. wiss. Bot.* 34, *669.*
—— (1902). 'Beiträge zur allgemeinen Muskel- und Nervenphysiologie.' *Pflüger's Archiv*, 92, *115.*
PFEFFER, W. (1877). *Osmotische Untersuchungen. Studien zur Zellmechanik.* (Leipzig.)
PROCTOR, H. R. and WILSON, J. A. (1916). 'The acid-gelatin equilibrium.' *Journ. Chem. Soc.* 109, *307.*
REID, W. (1890). 'Osmosis experiments with living and dead membranes.' *Journ. Physiol.* 11, *312.*
—— (1901). 'Transport of fluid by certain epithelia.' *Journ. Physiol.* 26, *436.*
SIEBECK, R. (1912). 'Über die osmotische Eigenschaften der Nieren.' *Pflüger's Archiv*, 148, *443.*
STILES, W. (1924). *Permeability.* (London.)
VAN 'T HOFF (1877). 'Die Rolle des osmotischen Druckes in der Analogie zwischen Lösungen und Gasen.' *Zeit. f. physik. Chem.* 1, *481.*
DE VRIES (1888). 'Osmotische Versuche mit lebenden Membranen.' *Zeit. f. physik. Chem.* 2, *415.*
—— (1889). 'Isotonische Koeffizienten einiger Salze.' *Zeit. f. physik. Chem.* 3, *130.*

CHAPTER FOURTEEN

The Permeability of the Cell Surface

The rate at which particular substances pass between the interior of a living cell and its external environment is usually known as the permeability of the cell to those substances. Since an adequate knowledge of cell permeability is perhaps one of the most fundamental needs of cell physiology, it is desirable to eliminate, as far as possible, all ambiguity from the terms employed. The 'permeability of a cell in respect to a given substance' is, in itself, a meaningless expression; it acquires a meaning when the experimental conditions are rigidly defined. Thus, it is illegitimate to speak of the permeability of a sea urchin's egg to water; it is only legitimate to speak of the amount of water which passes into a sea urchin's egg over each square millimetre of surface in unit time from a known environment whose osmotic pressure has a known value. The term 'permeability' involves, therefore, at least four attributes: (i) mass, (ii) area, (iii) time, (iv) concentration and specific nature of the environment.

There can be little doubt that it is the specific nature of the cell surface which enables a living cell to maintain within itself the equilibrium which is essential for life, and it is useful to remember that we are here concerned with a mechanism which is as fundamental and as delicate as that of respiration or any other more obviously vital phenomenon. When we survey all the facts, we are driven to the conclusion that the power possessed by the surface of the cell to allow the passage of some substances and to prevent the passage of others is unique to the living state and is not shared by any inanimate surface or membrane. The cell surface is, in fact, an integral part of the living cell; it may or may not be located within the tangible surface membranes which were considered in a previous chapter (see p. 102).

The data which have been accumulated within recent years are too extensive to submit to reasonably short analysis, but certain outstanding facts are available, and they must be enumerated before any attempt can be made to frame a reasonable hypothesis concerning the structure of cell surfaces.

Permeability to non-electrolytic solutes

As long as disturbing factors are absent it is reasonable to look upon a membrane, living or otherwise, as a sieve through whose pores a substance can pass as long as the diameter of its molecule is not larger than that of the largest pore. This theory in fact applies to various types of inanimate membranes, since Biltz (1910) was able to show that a particular type of collodion membrane would readily permit the passage of all dyes whose molecules did not contain more than about forty-five atoms, whereas dyes with larger molecules only penetrated with much greater difficulty. Presumably only a limited number of pores were large enough to permit the passage of the larger molecules. Similar results have been obtained by Collander (1924), using copper ferrocyanide membranes as well as those of collodion.

Table LIII. Permeability of collodion membranes to crystalloids. (After Collander.)

Substance	Molecular diameter	Relative permeability
Ammonia ...	5·78	100
Formic acid	8·57	86·4
Lactic acid	19·19	30·3
Malic acid	25·27	18·2
Methyl-ethyl-molonic acid	32·98	11·7
Citric acid	36·04	6·0
Cane sugar	70·35	1·6

Clearly anything which tends to clog the pores in a membrane or which tends to enlarge them by effecting a shrinkage of the membrane will affect the permeability in respect to molecules or ions of a critical size. Thus Brown (1915, 1917) was able to alter the permeability of collodion membranes by treatment with a dehydrating agent such as alcohol, and more recently v. Risse (1926 b) and Anselmino (1927) have shown that the pores of a gelatin membrane are altered in diameter by the variation in the degree of swelling of the gelatin which attends a departure from the isoelectric point.

It may be doubted how far pore size is the sole factor which controls the permeability of the cell surface to non-electrolytes and, in any case, it is obvious that the sieve theory in its original form has to be substantially modified; the living cell often exhibits a

PERMEABILITY OF THE CELL SURFACE 351

relative impermeability to very small ions and yet is able to allow larger molecules to pass through fairly readily. Tinker (1916) estimated that the average pore diameter of a copper ferrocyanide membrane is 15–20 $\mu\mu$, whereas the average diameter of the sugar molecules to which it is impermeable is very much less (100 times less). Either the molecules must be hydrated or the pores must be clogged in some way. The significance of these facts will be considered later, but at the moment we may consider the cell surface simply as a sieve-like structure, through the pores of which water can pass with greater freedom than can the molecules of most non-electrolytes.

Since living cells alter their water content in response to alterations in the osmotic pressure of their surrounding media (see Chapter XIII), it follows that the cells must be relatively impermeable to those substances in the media which are exerting the osmotic pressure. It must be remembered, however, that most plasmolytic experiments are of short duration, and the water gained or lost by a cell is measured as soon as the volume or weight of the cell no longer exhibits any change in a given solution. Such experiments do not prove that the cells are completely impermeable to salts or sugar, they only indicate that when the composition of the medium is changed the equilibrium in respect to water is established with much greater rapidity than is that of other substances. It is probable that salts and sugars can pass in and out of living cells with a speed which is of real biological significance, but which is negligibly slow when compared to the speed of movement of water. Complete impermeability to a given solute is probably a rare phenomenon, whereas low permeability is extremely common. If we wish to follow the diffusion of a slowly permeable substance into the cell, we must, therefore, either expose the cell to a relatively high concentration of the solute, or to a lower concentration for a long period of time. Unfortunately experiments of both types are liable to be misinterpreted, for in both cases there is grave danger of injury to the cell and, as we have already seen, this leads to a disorganisation of osmotic relationships.

If the cell surface is to be regarded as an imperfect osmotic membrane, and the cell is immersed in a hypertonic or hypotonic solution of a given solute, the loss or uptake of water will be less than if the surface were completely impermeable. The difference between the theoretical and the observed volume of the cell when the maximum change in volume has occurred will depend on two factors: (i) the relative value of the coefficient of permeability of the solute in the cells (or in the medium) to that of water, (ii) the rate at which the equilibrium volumes are reached. If we plasmolyse a cell rapidly, the minimum volume observed will be very close to that recorded by a perfect osmotic system; if, on the other hand, plasmolysis occurs slowly, the difference between observed and theoretical volume will be greater. If an experiment could be carried on long enough, then equilibrium would be reached when the volume of the cell had returned to the normal value which it possessed at the beginning of the whole experiment. Phenomena of this type are well known in the case of plant cells, although in many cases the secondary uptake of water from hypertonic solutions is due to an irreversible injury of the cells concerned. As far as is known, the rate of penetration of a solute into animal cells has, so far, not been estimated by the rate at which plasmolysis is reversed, but the method was applied by Lepeschkin (1908) to plant cells (see Stiles, 1924, p. 178). The water content of *Spirogyra* cells was reduced by plasmolysis with a solution of sucrose, a substance which penetrates with extreme slowness. The cells were then transferred to an isotonic solution of glycerol whose permeability it was desired to measure. As the glycerol entered, the volume of the cell vacuole increased, and this volume was measured after $\frac{1}{2}$ hour and $2\frac{1}{2}$ hours respectively. The difference between these two volumes gave the volume of glycerol solution which had entered, and since the concentration was known, it is possible to calculate the number of gram molecules entering per hour; by using suitable units of concentration and of cell surface, Lepeschkin estimated that 67–183×10^{-9} gram molecules of glycerol penetrated through each square centimetre of cell surface per hour. More recently both Fitting (1915) and Höfler (1918 a, b, 1919) have estimated the rate of penetration of solutes into plant cells by measuring the rate of de-plasmolysis. As already pointed out, it is essential in such experiments to guard against the danger of irreversible injury to the cells concerned.

In animal cells the permeability of a given solute has seldom been stated in any precise manner. As a rule a substance is regarded as 'permeable' if, within a reasonable time, its presence can be detected inside a cell which has been exposed to a solution containing the solute: if the solute cannot be detected in this way, it is often regarded as 'relatively impermeable'. It must be remembered, however, that if the amount of solute passing into a cell by diffusion is proportional to the difference in concentration inside the cell and in the medium, then the rate of penetration will be affected by any process which tends to immobilise the solute within the cell. This factor probably plays some part in the case of some acid dyes (Ruhland, 1908–1914). A dye which is readily adsorbed to intracellular surfaces will accumulate in a cell more rapidly than one which is not so absorbed, although the coefficient of permeability of the two may be the same.

The permeability of living cells to non-electrolytic substances was extensively investigated by Overton (1895–1900), and his results have been admirably summarised by Jacobs (1924). The ability of an organic molecule to enter a cell appears to depend on whether it is polarised or non-polarised. A non-polar molecule has a structure wherein the electrons are shared by the atoms concerned in such a way as to eliminate intramolecular regions of dissimilar electrical charge. Polar molecules on the other hand possess a distribution of electrons such as will cause marked electrical dissimilarity in different regions of the molecule. The non-polar molecules tend to be more soluble in organic solvents, they do not ionise and are relatively inert: polar molecules tend to dissolve more readily in water, they ionise, and are chemically active. Roughly speaking, non-polar molecules enter living cells very readily: thus the hydrocarbons, acetylene, benzene, xylene, naphthalene were found to penetrate the cell by Overton. If, however, more than one polar group (e.g. OH; COOH; NH_2) is present in a hydrocarbon molecule, the rate of penetration into the cell is very greatly reduced. Thus ethyl alcohol contains no polar group and enters the cell very readily, ethylene glycol will cause temporary plasmolysis, glycerol effects de-plasmolysis very much more slowly, whilst the tetrahydric alcohol, erythritol, is even less able to penetrate the cell (Jacobs, p. 115). Jacobs points out that almost all Overton's results conform to the rule that the ability of any given organic compound to enter a living cell is increased by reducing the number of polar groups within the molecule, and is decreased by increasing the number of these groups.

From a physiological point of view the most important organic solutes are the sugars and the amino acids. Both these substances, as we might expect from their chemical constitution, appear to enter with very great difficulty, and yet it is quite certain that they normally enter in sufficient quantities to provide for the maintenance of life. How far the failure to detect their entry under experimental conditions is due to the extreme slowness at which they enter, or how far the normal process of penetration is essentially inhibited by the experimental procedures hitherto adopted remains obscure.

As in the case of water, so we must consider the possibility of an active secretory mechanism capable of maintaining a difference in the distribution of a solute on the two sides of the cell surface. Apparently little evidence is available in respect to non-electrolytes, although it is probable that the walls of the gut and the cells of the kidney are capable of transporting non-electrolytes by forces other than those concerned in normal diffusion gradients. There is also the related problem of an 'irreciprocal' permeability to such substances recently investigated by Wertheimer (1923–4).

Permeability to electrolytes

The relative impermeability of cells to such non-electrolytes as sugar might possibly be explained by the assumption that the cell surface is a porous structure and that most of the pores are not sufficiently large to allow the passage of comparatively large molecules. In the case of electrolytes, however, this simple hypothesis does not cover the facts, since it will be shown that very often a molecule can pass in or out of a cell, whereas its constituent ions are unable to do so. Osterhout (1925) has recently shown that molecules of sulphuretted hydrogen can pass into plant cells with much greater rapidity than either the hydrogen ions or sulphide ions. This important fact was established by exposing *Valonia* cells to sea water containing a constant total amount of H_2S at different pH. At about pH 5 practically the whole of the H_2S was undissociated; at pH 10 it was all ionised as $H^{\cdot}$ and HS'. It was found that the total amount of H_2S present in the sap was strictly proportional to the amount of ionised H_2S present in the external sea water (see fig. 151).

Similarly, when the total amount of undissociated H_2S in the sea water remained constant, the total sulphide in the sap was independent of pH changes in the external solution.

Essentially similar results were obtained by Osterhout and Dorcas (1925) using CO_2 (fig. 152), and it seems quite clear that little or no CO_2 enters or leaves living cells of *Valonia* except in the form of undissociated carbonic acid (see also Jacques and Osterhout, 1930). It seems probable that all feebly ionised acids may behave in this manner and that the ability of undissociated molecules of electrolytes to penetrate living cells may be fairly widely spread. Miss Irwin (1926 c) has shown that the rate of penetration of cresyl blue into the vacuole of *Nitella* is proportional to the concentration of undissociated molecules of the dye in the external solution (see also p. 373). There is, however, an important difference between the process of penetration by cresyl blue and by CO_2. In the latter case the temperature coefficient suggests that diffusion is the essential factor concerned (Osterhout and Dorcas, 1925), whereas in the case of the dye the

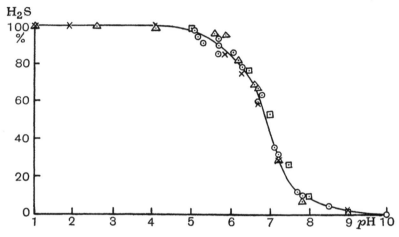

Fig. 151. Curve showing the relation between the total amount of intracellular H_2S and the pH of an external medium containing sulphides. (From Osterhout, 1925.) The figure shows that the total sulphide in the sap of *Valonia* corresponds with the undissociated H_2S in the external solution. The concentration of total sulphide ($H_2S + HS' + S''$) in the sap (⊙) is expressed as a percentage of the total sulphide in the outside solution. The values for the concentration of undissociated H_2S in the sea water as calculated from the dissociation constant (▣) and as determined from the vapour tension (△) and from the rate of evaporation (×) at various pH values of the external solution are expressed as a percentage of the corresponding values in the range pH 1–3, where all the H_2S is regarded as undissociated.

temperature coefficient is surprisingly high, $Q_{10} = 4 \cdot 9$ (Irwin, 1925 a), and suggests that the dye undergoes chemical combination as soon as it enters the cell.

Just as the molecules of a weak acid appear to penetrate readily into a cell, so they will leave a cell with equal facility, and the facts which apply to acids apply equally well to alkalies. It is for this reason that incompletely ionised acids such as carbonic acid, and weakly ionised bases such as ammonia, have a more rapid and profound effect upon intracellular processes than an equal concentration of strong acid or base; similarly a cell will rapidly recover from the effects of a weak acid or alkali, and take a longer time for recovery from the effects of the more completely ionised acids or alkalies (Gray, 1922). That weak acids and weak alkalies penetrate readily into living cells was demonstrated by Newton Harvey (1911) by staining cells with neutral red and exposing them to solutions containing the substance to be investigated. For example, the eggs

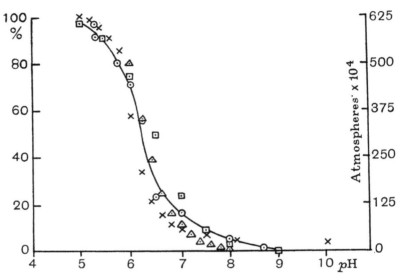

Fig. 152. Graph showing that the total CO_2 in the cell sap of *Valonia* corresponds approximately to the undissociated H_2CO_3 (including CO_2) in the sea water outside. The total CO_2 in the sap (⊙) is expressed as a percentage of that in the sea water outside. The percentage of undissociated H_2CO_3 (including free CO_2) as calculated from the dissociation constant is shown by the symbol ⊡. The partial pressure of free CO_2 in the sea water as determined by McClendon is shown by the symbol △ and is expressed in the units shown to the right of the figure. The concentration of H_2CO_3 in sea water, expressed as a percentage of that found at pH 3 (where H_2CO_3 is regarded as undissociated), is shown by the symbol ×. (From Osterhout and Dorcas, 1925.)

of sea urchins when stained by the dye remain red even if there is sufficient hydroxyl ions in the surrounding sea water to cause rapid disintegration of the egg surface. If, however, the alkalinity of normal sea water is only very slightly varied by the addition of ammonia, the eggs very rapidly turn yellow. On replacing the cells in normal sea water they rapidly regain their red colour (see also p. 85). This type of experiment illustrates two important facts. (i) Ammonia in some form or other is capable of entering into living

PERMEABILITY OF THE CELL SURFACE

cells, and when inside it dissociates to form free hydroxyl ions. (ii) That ammonia is capable of leaving the cell as readily as it enters. Similar facts can be demonstrated with acids. Strong acids only penetrate cells with difficulty, and having entered they only diffuse away with equal difficulty. Weak acids such as carbon dioxide or butyric acid enter cells with much greater freedom, and at the same time they can readily diffuse away. These facts suggest that alkalies and acids do not pass in and out of the living cell in the ionised condition but in the form of undissociated molecules.

Permeability to ions

Since nearly all living cells adjust their water content in response to changes in the osmotic pressure of an external medium containing variable concentrations of fully ionised electrolytes (see p. 328), it follows that the cells must be relatively impermeable to either cations or anions, or to both. That ions cannot pass freely into or out of a cell is directly associated with the fact that all living cells are very poor conductors of electricity (*Fundulus* eggs, Brown (1905);

Table LIV. Illustrating the relatively low conductivity of living blood corpuscles. (From Bugarszky and Tangl.)

Species	Relative conductivity of plasma and corpuscles	
	Plasma ($10^4 \lambda$)	Corpuscles ($10^4 \lambda$)
Horse	105	1·63
Dog	113	1·70
Cat	125	2·20

Table LV. Change in electrical resistance of *Laminaria* tissue on exposure to isotonic NaCl solutions. (From Osterhout, 1914.)

Time (mins.)	Resistance (ohms)
0	980 cells alive
20	745 ,,
40	590 ,,
60	495 ,,
100	395 ,,
150	345 ,,
200	320 ⎫ cells dead
300	320 ⎭ (resistance of sea water 320 ohms)

red blood-corpuscles, Stewart (1899); sea urchin eggs, McClendon (1910); *Laminaria* tissue, Osterhout (1922)). If both anions and cations were able to pass freely through a cell when exposed to an electric field, the conductivity would not differ greatly from that of the surrounding medium: as long as the cell is alive, however, the conductivity is low (Bugarszky and Tangl, 1897) and only rises to that of the surrounding medium when the cell dies (Osterhout, 1914; Shearer, 1919).

The effect of death on the electrical conductivity of bacteria was investigated by Shearer (1919), who found that when *B. coli* are exposed to isotonic NaCl solution the conductivity of a dense suspension rose from a comparatively low figure to that characteristic of the surrounding medium (see fig. 153). An exposure to such solutions of NaCl is fatal to

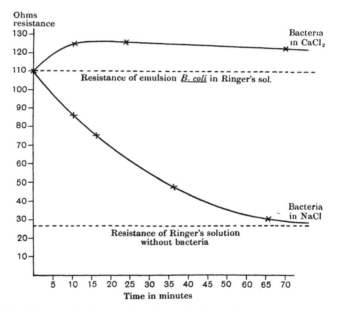

Fig. 153. Graph showing the change in the resistance of bacteria in a solution of pure sodium chloride. (From Shearer.)

the cells after two hours. These results confirm those of Osterhout, but according to Green and Larson (1922) and to Zoond (1927) the conductivity of dead bacteria is no higher than that of the living cells, and the observed changes in the conductivity of a suspension are due to the exosmosis of salts from the cells.

Impermeability of the Cell Surface of Ions

That the barrier to the passage of ions is confined to the periphery of the cell is proved by the work of Höber (1910–13) and later workers who have shown that the resistance to an electric current is limited to the surface layers of the cell only, and that ions can move with comparative freedom within the main body of the cell. The methods employed for these observations are complicated, but in view of the theoretical significance of the results it is desirable to give some indication of the principles involved. Höber's (1910) first method was based on the principle that if an electrical conductor is interposed between the plates of a condenser, the capacity of the condenser is increased and a current is allowed to flow through the circuit. By suspending red blood-corpuscles in isotonic sugar solution and placing the suspension between the plates of a condenser, Höber found that the conductivity of the cells was equivalent to a 0·1 or 0·01M solution of potassium chloride, a much higher value than that obtained when the conductivity was measured by the Kohlrausch bridge: further this high value was unaffected by laking the cells. By another method, based on the ability of the cells to damp the oscillations of a high frequency current, Höber (1912) estimated the internal conductivity of the cells as approximately equal to that of the external serum. Höber (1913) also found that the resistance of red blood-cells to the passage of a current (impedance) decreased with a rise in the frequency of the alternating current which was employed and that at very high frequencies it fell to the low value characteristic of haemolysed cells when measured with low frequencies. Such changes in impedance might be expected to occur if the cell circuits were comparable to that shown in fig. 154, where C is a condenser, r is a resistance equivalent to the resistance of the cell surface, and R is a resistance comparable to that of the cell interior. To the direct current the impedance of the whole circuit is $r + R$, but to currents of infinite frequency the impedance is R. Thus using a variety of methods for the determination of R, Höber found that the internal conductivity of red blood-corpuscles was equivalent to that of a 0·4 per cent. solution of NaCl, whereas the conductivity of the whole cell was only equivalent to 0·02 per cent. NaCl. More recently Fricke and

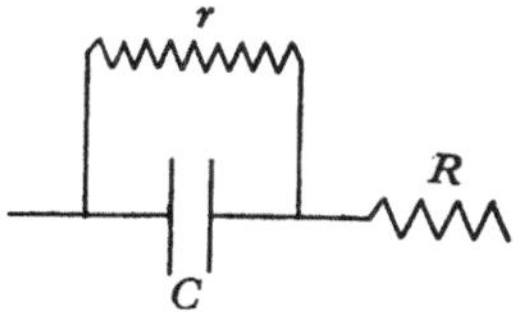

Fig. 154.

PERMEABILITY OF THE CELL SURFACE

Morse (1925) and McClendon (1926, 1927) have found that the resistance and the capacity of cell suspensions are independent of the frequency of the current used as long as that frequency is low; above a critical frequency (ω) they both decrease—a result which confirms Höber's original observations. Fricke and Morse have shown that the resistance ($R\omega$) to a current of ω frequency and the capacity C_ω satisfy the following equations:

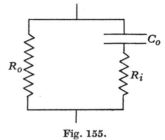

Fig. 155.

$$\frac{1}{R_\omega} = \frac{1}{R_0} + \frac{1}{R_i}\left(\frac{C_0^2 \omega^2 R_i^2}{1 + C_0^2 \omega^2 R_i^2}\right) \quad \ldots\ldots(\text{xvii}),$$

$$C_\omega = \frac{C_0}{1 + C_0^2 \omega^2 R_i^2} \quad \ldots\ldots(\text{xviii}),$$

where C_0 is the static capacity of the corpuscle, and where R_0 is the resistance of the external medium and R_i is the resistance of the interior of the cells; these equations can be derived from first principles if the living cell forms a circuit equivalent to fig. 155.

Table LVI. *Capacity and resistance of ox blood for varying frequencies.* (From Fricke and Morse, 1925.)

Cycles per second	Capacity μμf		Resistance	
	Obs.	Calc.	Obs.	Calc.
87.000	146	146	191	191
833,000	130	135	181	183
1·17 × 10⁶	118	126	174	178
1·52	106	115	168	172
2·04	90	98	159	163
3·04	68	71	148	151
3·82	60	55	144	144
4·52	39	43	138	140
0·00	—	—	(124)	126

The resistance to currents of infinite frequency can be calculated by extrapolation, and the value of R_∞ so obtained gives the resistance which would have obtained at low frequencies if the membranes of the cells were absent:

$$\frac{1}{R_\infty} = \frac{1}{R_0} + \frac{1}{R_i}.$$

PERMEABILITY OF THE CELL SURFACE

From this the internal resistance R_i can be calculated from an equation (see Fricke and Morse) which relates the resistance of the external fluid to that of the whole corpuscle for any given suspension. In this way Fricke and Morse concluded that the specific resistance of the inside of red blood-corpuscles was 3·5 times that of the specific resistance of serum. Cole (1928) by similar means found that the internal resistance of *Arbacia* eggs was 3·6 times that of normal sea water.

From observations of the capacity (C_0) it is possible to determine the approximate thickness of the membrane which is opposing the migration of ions through the cell. Assuming a dielectric constant of 3 (= that of olive oil), Fricke (1925) estimates this thickness as $3·3 \times 10^{-7}$ cm., which is comparable to 20–30 carbon atoms and is therefore equivalent to a monomolecular layer of a serum protein molecule. Using somewhat analogous methods, McClendon (1926) estimated the thickness of the surface membrane at 2–3 carbon atoms only.

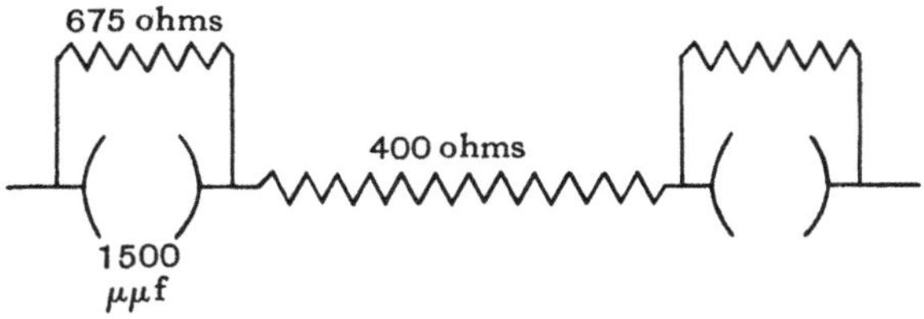

Fig. 156.

McClendon's (1926) conception of the red blood-corpuscle is illustrated in fig. 156.

The two surfaces of the cell membrane represent two plates of a condenser of 1500 micro-microfarads capacity: these are short circuited by a parallel resistance of 675 ohms, indicating that the cell membrane represents a leaking condenser. The cell interior represents a resistance of 400 ohms in parallel with the condenser. The resistance of the two very thin surface membranes is thus three times as great as that of the whole cell interior. We may therefore conclude that the barrier to the diffusion of ions is restricted to the surface layers of the red blood-cell and of sea urchin eggs, a conclusion which is in harmony with the conception of a peripheral semi-permeable membrane already reached from other data. The observations of electrical impedance do not, however, give us any information concerning the nature of the ions to which the cell surface is impermeable: the impermeability may be in respect to cations or anions or both.

Permeability to Anions or Cations

These facts must be considered with reference to others. If red blood-corpuscles are suspended in a solution of sodium chloride and CO_2 is blown through the suspension, the external medium becomes alkaline and at the same time the chlorine content of the cells increases. If instead of being suspended in sodium chloride, the cells are immersed in isotonic sugar solution, no alteration occurs in the alkalinity of the solution when CO_2 is blown through. Höber (1922) interpreted these results as follows. Carbon dioxide passes into the cells in the form of undissociated molecules of carbonic acid, but thereupon ionises to form bicarbonate ions and hydrogen ions. The former are assumed to be capable of rapid diffusion from the cell, but for every bicarbonate ion leaving the cell a chlorine ion must enter. This explanation is based on the assumption that the cell is impermeable to cations. The equilibria reached by blood corpuscles suspended in media of varying composition have recently been discussed in detail by van Slyke (1926), who has gathered together much evidence in support of the view that not only are the cells impermeable to metallic cations, but also that the composition and concentration of the intracellular anions (chiefly HCO_3' and Cl') are determined by the concentration of the intracellular cations (chiefly potassium) and by the concentration of ionised protein (chiefly haemoglobin). The whole system conforms, in fact, to the requirements of a modified Donnan equilibrium and the distribution of electrolytes and water thereby also receives a quantitative explanation. The amount of water in the cells is determined by the number of intracellular ions and these are partly those of haemoglobin; it follows that, unlike other cell systems, any alteration in the pH of the external medium should induce an alteration in the volume of the cells. This is shown to be the case by the following observations of Warburg (1922) (Table LVII).

The behaviour of red blood-corpuscles strongly suggests that their surface layers are impermeable to cations, but permeable to anions of the type of chlorine or bicarbonate. At the same time the evidence is far from complete. Höber's observations on the redistribution of hydrogen ions can receive an equally convincing explanation without postulating a permeability to anions. If blood cells are exposed to an electric field they will pass to the positive pole which would be the case if they were normally surrounded by a layer of cations. In solutions of sodium chloride these cations will be sodium, but in the presence of hydrogen ions (from the CO_2 in the medium) there may well be an interchange of $Na^{\cdot}$ and $H^{\cdot}$, which will increase the alkalinity of the external medium. That other types of cell (e.g. trout eggs) actually exhibit such surface reactions with hydrogen ions there can be no doubt (Gray, 1920), but it may be

Table LVII

pH	Volume of blood cells in percentage of their volume at pH 6·5	
	Obs.	Calc.
6·50	100·0	100
6·80	95·2	96
7·00	92·7	93
7·20	90·7	91
7·40	89·1	88
7·60	87·8	85
7·80	86·3	81

Table LVIII

Tissue	% water (w)	% Cl (c)	pH (calc.)	pH (obs.)	P.D. (calc.)	P.D. (obs.)	$\dfrac{y}{a}$
Red blood-corpuscles	63·6	0·178	7·17	7·23	14·2	—	0·59
Brain (white matter)	70·0	0·155	7·06	—	20·7	17–28	0·45
Brain (grey matter)	81·5	0·115	6·87	—	32·7	—	0·30
Smooth muscle	80·0	0·112	6·87	—	32·9	—	0·29
Liver	84·0	0·096	6·78	6·40–7·04	38·3	—	0·24
Striped muscle	76·0	0·061	6·62	6·02–6·91	47·8	40–80	0·168

that different types of cell behave in a different manner. Höber's hypothesis has been criticised by Rohonyi and Lorant (1916), since these authors report that the anionic exchange between the corpuscles and external medium is not dependent on the integrity of the cell membrane.

It is interesting, however, to note that if van Slyke's analysis can be applied to mammalian cells in general a reasonable explanation is forthcoming for the potential differences which exist between the interior and exterior of different types of living cells (Haldane, 1925).

Haldane's calculations are based on four main assumptions: (i) the cell surface is permeable to water, (ii) it is impermeable to all cations, (iii) it is permeable to Cl' and HCO_3' and that these are the only inorganic anions present in significant quantities, (iv) it is impermeable to colloids.

If a be the molar concentration of salt outside the cell, and y be the molar concentration of anions inside the cell, then the potential difference across the cell surface (E) is $\frac{RT}{F} \log_e \frac{a}{y}$ volts, so that at $38°C$. $E = 0.0617 \log_{10} \frac{a}{y}$ volts. But y can be calculated from the percentage of water in the cell (w) and from the percentage of chlorine content (c); if the ratio of $\frac{Cl' + HCO_3'}{Cl'}$ ($= 1.23$) holds for all tissues and if for isotonic saline a is equal to 0.170, then $\log_{10} \frac{a}{y} = \log_{10} w - \log_{10} c - 2.32$.

But the cH inside the cell is to that outside as a is to y, therefore pH of the tissue = pH of the medium $- \log_{10} w + \log_{10} c + 2.32$, and the injury potential $= 61.7 [\log_{10} w - \log_{10} c - 2.32]$ millivolts.

It is, however, very difficult to believe that all cells are completely impermeable to cations, for under these circumstances such ions could only exert an effect on the cells by operating on their surface. As mentioned elsewhere, most cells respond actively and rapidly to any serious change in their cationic environment. For example, the heart is markedly affected by the ions of hydrogen, potassium, and calcium and if these ions cannot enter the cell they must operate solely at the surface of the tissue. If a heart is perfused with an acid Ringer solution it tends to stop beating in the relaxed condition, and if the excess of hydrogen ions are removed the beat is resumed. If the acid acts solely on the surface of the cells, one would expect to find that the efficiency of any particular solution would depend solely on the pH of the perfusion fluid and would not be increased if for any reason the acid were able to penetrate into the cells. This, however, is not the case. Philippson (1913) showed that the physiological action of an acid which can be shown to penetrate rapidly into the cells is higher than that of an acid which penetrates more slowly, although the pH of both perfusion fluids is the same. Similar facts apply to other tissues, e.g. ciliated cells (Gray, 1922); further, the respiration of cells is much more powerfully reduced by the penetration of acids into the cells than by the application of acids to their surface (Gray, 1924).

Again, if the physiological action of hydrogen ions is solely due to the changes they induce at the surfaces of cells, then these changes should be readily reversible by the addition of hydroxyl ions to the external medium. It is, however, found that the effects of hydrogen ions are much more readily reversible by the addition of hydroxyl ions to the cell interior than by their addition to the external medium. A concrete case is provided by ciliated cells. If such cells are stained with neutral red it can be shown that they resemble many others in that ammonia is capable of producing rapid changes in the alkalinity of the cell interior, whereas this cannot be effected during life by adding sodium hydroxide to the external medium. Yet, if ciliated cells are rendered inactive by the presence of hydrogen ions in the external fluid, the cells are much more readily activated by ammonia than by sodium hydroxide (Gray, 1922). Similarly, although the length of time that the cilia of *Mytilus* will beat in acidified sea water depends to some extent on the pH, it depends to a larger extent on the carbon dioxide tension of the external medium (Haywood, 1925).

These and other facts seem to suggest that the impermeability of the cell surface to cations as exemplified by mammalian red blood-corpuscles cannot readily be extended to other types of tissue.

Permeability to specific ions

So far we have not attempted to consider in any detail the specific relationship between any given ion and its power of penetrating the plasma membrane. To embark on this problem would be a formidable undertaking. In dealing with non-electrolytes many of the facts seem to harmonise with the view that there is a definite relationship between the size of a molecule and the ease with which it can penetrate certain types of membrane: the smaller the molecule the higher is the percentage of pores through which the former can pass. If we attempt to estimate the average pore size of living cells by noting the size of a non-electrolytic molecule which will pass in fairly freely, we must conclude that all the pores are probably of much wider diameter than that of the average electrolytic ion. In order to retain the sieve theory it is necessary to postulate a mechanical or forcible blocking of these pores. Various possibilities appear to exist. The pores may be partially blocked by an adsorbed film of water, or the walls of the pores may offer an electrostatic opposition to ions of opposite sign, or the size of the pores may be affected by adsorption of water by the membrane. These problems belong at the moment to the realm of physical chemistry (see also p. 397). In the present discussion it is desirable to emphasise a fact which is often overlooked in drawing conclusions from physical analogies. In the cell we are dealing with a mechanism which can establish a definite equilibrium between the interior of the cell and its external medium. When a blood corpuscle is immersed in normal serum, or when *Valonia* is in sea water, we must assume that there is strict chemical equilibrium between the amount of potassium inside the cells and the amount in the external sea water. By postulating a membrane which is only very slightly permeable to potassium, we cannot reach a state of equilibrium comparable to that of the normal cell, where the concentration of potassium is much higher than that of the external medium. A physical membrane may alter the *rate* at which equilibrium is effected after a change occurs in the solutions on its two sides, but it cannot— except in so far as it is absolutely impermeable to some active ion —alter the final equilibrium position. It is therefore necessary to bear in mind two concepts: (i) the mechanism responsible for the establishment of an equilibrium condition between the concentration of a specific ion inside the cell and in its normal external medium, (ii) the mechanism which comes into play when any disturbance from this equilibrium takes place. Once again we have to consider how far the normal concentration of an ion inside a cell is maintained by the active processes going on in the cell—which processes may be comparable to those which are obviously operative in the vertebrate kidney—or how far the concentration in the cell is determined by a relatively simple concentration equilibrium on the two sides of the cell membrane.

It is convenient to start from an equilibrium concerning which there is seldom any doubt—viz. the equilibrium which exists between the concentration of a given ion inside the cell with its concentration in the normal environment in which the cell lives. In nearly every case the observed ratios indicate that free diffusion of potassium ions does not occur between the cell and its medium: at the same time the equilibrium position is seldom the same for any two given types of cell—even where the environment is identical. A well-known example (see Table LIX) is provided by red blood-corpuscles (Abderhalden, 1898). An even more startling example is that given by Cooper and Blinks (1928) for *Valonia* and *Halicystis* (fig. 157). For any given ion, therefore, there may be marked differences in the equilibrium concentration in different although allied types of cell. At first sight, the heaping up of an ion inside a cell might be due in

PERMEABILITY OF THE CELL SURFACE 367

part to its immobilisation within the cell (see p. 373), but this cannot be true for metallic ions such as sodium or potassium, for the conductivity of the cell interior is such that most of the electrolytes must be freely ionised.

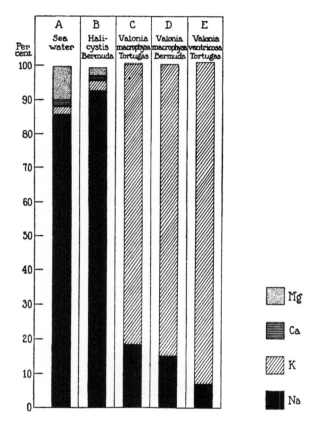

Fig. 157. Chart showing the molar concentration, expressed as per cent. of halide, of the chief elements in the saps of *Valonia ventricosa*, *V. macrophysa*, and *Halicystis* compared with sea water. (From Cooper and Blinks.)

If the sieve theory is to cover facts of this type, it must be capable of applying an effective block to one type of cation and not to another. In this connection the recent work of Michaelis and his collaborators (1926) is of interest, since it shows that membranes of a specific type may exert a marked effect on the relative rates of diffusion of different ions, e.g. potassium and sodium. Michaelis' membranes do not account, however, for a state of equilibrium comparable to that shown in Tables LIX and LX. It will be noted that the cell selects not only one cation in preference to another, but also one anion in preference to other anions; for example, many marine cells have a very high content of phosphate ions, although the concentration in sea water is extremely low. The only conclusion which can be drawn at present is that the cell membrane must be saturated with each ion independently or that the permeability of the membrane in the two directions must be different.

Table LIX

Parts per 1000	Horse		Pig		Dog		Cat	
	Serum	Cor-puscles	Serum	Cor-puscles	Serum	Cor-puscles	Serum	Cor-puscles
Sodium	4·396	—	4·251	—	4·278	2·839	4·439	2·705
Potassium	0·259	4·130	0·270	4·957	0·245	0·273	0·262	0·258

Table LX

Molar concentration of sap as percentage of halide	Sea water (Bermuda)	Halicystis	Valonia macrophysa
Cl and Br	100·00	100·00	100·00
K	2·15	2·58	86·24
Na	85·87	92·80	15·08
Ca	2·05	1·36	0·288
Mg	9·74	2·49	(Trace?)
SO_4	6·26	(Trace?)	(Trace?)

It will also be noted that each type of ion must be considered as a separate entity and it is not permissible to speak of permeability to 'cations' or to 'anions', we must consider each metal or radicle separately. It must be admitted that a consideration of the normal salt content of cells does not suggest any obvious mechanism whereby the equilibrium concentrations of intra- and extra-cellular electrolytes is established. That the equilibrium is probably of a dynamic nature is shown by the fact that growing cells maintain their normal electrolyte concentration, so that the cell must be able to absorb electrolytes if the equilibrium is upset by active growth.

Many attempts have been made to set up within the cell a new equilibrium in response to a change in the external environment.

PERMEABILITY OF THE CELL SURFACE

Cells are placed in an environment differing quantitatively or qualitatively from the normal, and changes in the electrolyte concentration of the cell interior or of the environment are measured. As already pointed out, experiments of this type lead to definite results when incompletely ionised electrolytes are employed and suggest that ions seldom enter the cell. There are, however, a number of observations which suggest that an ion in the external medium can in some cases be exchanged for another within the cell bearing the same charge. Hamburger (1909) reported an exchange of calcium and other ions across the membrane of the red-blood corpuscle, although in this case it is not certain that the exchange was not limited to the surface of the cell. On the other hand Osterhout (1909) demonstrated the penetration of $Ca^{..}$ into the cells of the root-hairs of *Dianthus*, and M. M. Brooks (1922) has shown that $Sr^{..}$ can penetrate into the sap of *Nitella*. Concerning anions, similar phenomena have been observed (Höber, van Slyke). In nearly all experiments of this type there is always some doubt as to how far the phenomena observed are typical of the normal plasma membranes. This doubt is least obvious when dealing with the penetration of hydrogen or hydroxyl ions, since the concentration of these ions can be measured with great accuracy within the limits capable of supporting life. We have already seen that the undissociated molecules of acids enter cells with greater readiness than their derivative ions: at the same time the hydrogen ion must penetrate to some extent. For example, if only undissociated carbonic acid penetrated from acid sea water, the effect on a cell ought to be independent of the pH and solely dependent on CO_2 tension, which is not the case (Haywood, 1925). All the evidence suggests that when exposed to a strong acid, hydrogen ions are slowly set free inside the cell, and are slowly removed on returning the cells to their normal environment (Gray, 1922) and that corresponding phenomena are applicable to bases. Unfortunately most of the observations on the penetration of acids and bases into living cells (Harvey, 1914; Crozier, 1916 a–c) have been performed with solutions of the same molecular strength and not with solutions of the same hydrogen ion concentration. It is possible that the penetration of molecules and of ions is controlled by two separate mechanisms. In the case of molecules we may be dealing with a true equilibrium between the concentration on the two sides of the plasma membrane, whereas the movement of ions may be restricted to a process of exchange, whereby an ion of a given type can only enter if another bearing the same charge leaves the cell. That this does not always occur is shown by the observations of Cooper, Dorcas and Osterhout (1928), who failed to note any rapid exchange of anions or cations by *Valonia*, although when the cells are growing the uptake of salts is measurable. Under favourable circumstances *Valonia* absorbs only 3×10^{-8} gram molecules per hour per square centimetre, and this movement was *against* a concentration gradient of 40 : 1.

There is so much evidence to show that the cell surface is not equally freely permeable to both anions and cations, that one tends to overlook the fact that nevertheless a growing cell must be in a position to absorb both types. How far the recent work of Mond (1927) throws light on this difficulty remains to be seen. Mond has shown that if red blood-corpuscles are on the acid side of their isoelectric point their chlorine content goes up and their potassium content goes down, whilst on the alkaline side the relative concentrations are reversed. This is interpreted by Mond to mean that the permeability of the membrane changes the isoelectric point: this may be true, but there is just the possibility that the observed changes are due directly to the amphoteric nature of the cell envelope. It would be of interest to see whether there is any evidence of a fluctuating change in the pH of the cell interior during cell growth—but the experimental technique would not be easy.

We may conclude that, so far as is known, the living cell exerts a selective influence on the substances which it absorbs of a much more intricate nature than is at present capable of resolution into physico-chemical terms. It should not be forgotten that although the osmotic activity of the interior of a cell may, under certain circumstances, be related to that of the external environment, yet the nature of the active molecules responsible may be qualitatively different on the two sides of the cell surface.

Since fat-soluble substances appear to enter living cells more readily than substances which are not thus soluble (Overton, 1895–1900; Meyer, 1899), it is possible that metallic ions enter the cell in the form of fat-soluble compounds. Little if any direct evidence is available, but there is definite evidence that metallic radicles move in this form from one part of the body to another. For example the crab, *Carcinus moenas*, calcifies its integument, after moulting, by the liberation of calcium salts stored in the liver (Paul and Sharpe, 1916). The calcium is stored in the

PERMEABILITY OF THE CELL SURFACE

liver cells as calcium phosphate, but is set free into the blood as calcium formate and calcium butyrate; in this form it reaches the superficial ectoderm and is there converted into calcium carbonate. It is possible that a similar transformation from water-soluble to the fat-soluble condition occurs during the absorption of metallic compounds by the intestine of vertebrates.

Permeability to dyes

As pointed out by Jacobs (1924), dyes form a unique class of substances for the study of cell permeability. Their penetration can be readily observed and their physical and chemical properties are varied: dyes may be acidic or basic, colloidal or crystalloid, and soluble or insoluble in non-aqueous media.

Table LXI

Dye	Basic or acidic	Solubility in cholesterol	Permeability of living cells
Methylene green	Basic	0	+++
Thionin	,,	0	+++
Malacite green	,,	0	+++
Bismarck brown	,,	0	+++
Victoria blue	,,	+++	0
Basle blue	,,	+++	0
Diazin green	,,	+++	+
Victoria blue	,,	+++	+
Cyanosin	Acidic	+++	0
Bengal rose	,,	++	0 (+)
Oxamin maroon	,,	++	0

The first attempt to analyse the nature of the plasma membrane by measuring its permeability to dyes was made by Overton (1900). Using a number of anilin dyes of various types, he found that all those which entered living cells were absorbed by fat soluble substances, and were also absorbed by lecithin and cholesterol. This result confirmed Overton in his conception of the plasma membrane as a fatty membrane or as one composed of substances allied to fats. From this point of view all substances soluble in fats should enter living cells, all substances not soluble in fats should be unable to enter. That this generalised statement is untrue became apparent when more dyes were investigated. Ruhland (1908–13) showed that there are a number of dyes (e.g. methylene green and thionin) which enter cells readily but which will not dissolve in cholesterol: similarly there are a number of 'fat-soluble' dyes (cyanosin, Bengal rose) which will not enter the living cell (see Table LXI).

Similar exceptions to Overton's rule were found by Höber (1909) and by Garmus (1912). Ruhland finally rejected Overton's hypothesis and suggested that the essential factor which controls the rate of penetration of dyes is their coefficient of diffusion through a gelatinous matrix, which in turn depends upon the size of the molecules. Höber's (1922) conclusions are somewhat different. He points out that the ability of dyes to enter a plant cell is often less than its ability to enter an animal cell, and that this is particularly obvious when the fat-soluble dye is in the colloidal state: it is therefore possible that the failure to enter the plant cell may be due to the inability of a molecular aggregate to penetrate the cellulose wall. If this conclusion is justified, one grave objection to Overton's hypothesis disappears, but it is still necessary to account for the fact that methylene green and thionin (both insoluble in lipoids) will enter both animal and plant cells. Hertz (1922) showed that the entrance of these dyes into *Opalina* is unaffected by the presence of anaesthetics, whereas the entrance of fat-soluble dyes was inhibited. Höber, therefore, suggests that lipoid-soluble and non-lipoid-soluble dyes enter by a different route or by a different mechanism. Similarly Nierenstein (1920) found that an adequate correlation between the ability of basic and acid dyes to enter *Paramecium* and their partition coefficients between a fat-soluble substance and water only exists when the fat solvent (e.g. olive oil) contains both a base and an acid (e.g. diamylamine and oleic acid). So far as these facts can be summarised, they seem to suggest that most cell surfaces are permeable to those dyes which can either (i) dissolve in fats, or (ii) react with some constituent of the membrane, provided that in all cases the molecular aggregates of the dyes are not beyond a certain maximum size (Höber and Kempner, 1908; Höber and Chassin, 1908).

It will be obvious that the ability of a cell to absorb a dye is not a measure of the ability of the dye to penetrate the plasma membrane. Unless a dye is accumulated by the cell it is by no means easy to determine whether it has entered the cell from a dilute solution or not; further, if a dye is accumulated in a non-diffusible state within the cell the effective concentration gradient across the plasma membrane will be maintained for a long time, whereas if the dye is not being so accumulated the gradient will fall rapidly as the dye enters. As a general rule, basic dyes are accumulated by cells, and in the case of plant cells Ruhland attributed this to a reaction with tannic acid. The recent work of Miss Irwin (1922–8) has done much to clear up the situation in respect to one particular dye—viz. cresyl blue. In the first place, the dye clearly accumulates in the sap of vegetable cells—such as of *Valonia* or *Nitella*. Miss Irwin (1925 b) has shown that cresyl blue in the form of the free base diffuses readily in and out (1926 b) of the sap of *Nitella*, and can therefore readily pass through the plasma membrane. When the dye is present as a salt, however, the rate of diffusion is extremely slow. Since the sap (pH 5·5) is normally more acid than sea water (pH 8·0), free base from

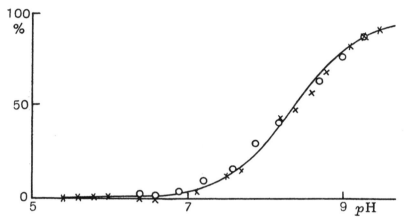

Fig. 158. Curve showing the percentage of undissociated molecules of brilliant cresyl blue at different pH values. The ordinates represent the percentage and the abscissae the pH values. Symbol O represents the data from rates of penetration of the dye into *Nitella*. Symbol × represents the data from the distribution of the dye between chloroform and water. The curve as drawn represents the calculation made from the equation

$$a = \frac{1}{1 + \frac{OH'}{k}}, \text{ where } k = 10^{-5 \cdot 6}. \text{ (From Irwin, 1926.)}$$

the sea water passes into the cells and on reaching the vacuole is ionised and therefore removed from the sphere of outward diffusion. The amount of dye within the vacuole depends therefore on the concentration of free base in the surrounding sea water and on the pH of the sap (see fig. 158). The rate of penetration of the dye depends, however, on factors which affect the protoplasm of the cell—e.g. sodium ions (1926 d, 1927 a, b). The close parallel between the factors involved during the absorption of cresyl blue and those involved during the absorption of H_2S or CO_2 is obvious.

If dyes are only capable of penetrating in the form of the free base, it follows that dyes which can only exist as salts (over the range of pH compatible with life) should not enter living cells. A strongly basic dye of this type is methylene blue. From a spectroscopic analysis of the dye found in the sap of uninjured cells of *Valonia* Miss Irwin (1926 *d*) concludes that methylene blue does not accumulate as such, but only in the form of a less basic derivative trimethyl-thionin. This conclusion has, however, been criticised by M. M. Brooks (1927), who claims that methylene blue penetrates *Valonia* and *Nitella* as such.

In a recent paper, Miss Irwin (1928 *b*) has suggested a mechanism which might account for the observed distribution of a dye between an external medium of sea water and the sap of a vegetable cell. The apparatus consists of a horizontal glass tube with three vertical arms. To the left arm is added a solution of the dye in sea water. To the central arm is added chloroform (representing the non-aqueous layer of the living cell) until it fills the horizontal portions and the lower part of each upright tube. Upon the chloroform in the right arm is poured the sap, artificial or natural, from the living vacuole of *Valonia*. Each phase is kept stirred at a constant rate. Using cresyl blue, azure blue, basic fuchsin, toluidine blue and similar dyes, the dye collects in the 'vacuole' phase just as in the normal cell. Similarly dyes which do not collect in the vacuole of the normal cell are not absorbed by the 'sap' phase of the model. It is interesting to note that some dyes which are readily soluble in chloroform, e.g. crystal violet, do not penetrate into the 'sap' phase. Miss Irwin concludes that the accumulation of a dye in the 'sap' phase depends on the two partition coefficients,

$$K_e = \frac{\text{concentration of dye in the non-aqueous phase}}{\text{concentration of dye in the external solution}},$$

$$K_r = \frac{\text{concentration of dye in the non-aqueous phase}}{\text{concentration of dye in the aqueous sap}}.$$

If K_e is high the dye rapidly accumulates in the non-aqueous phase (chloroform or protoplasm) and if K_r is low the dye will rapidly diffuse from the non-aqueous layer into the sap phase. By replacing the protoplasmic layer of *Valonia* by chloroform, an equilibrium in respect to dyes essentially similar to that of the normal cell can be obtained (Irwin, 1928 *b*). It would appear, however, that during the process of penetration into the interior of the cell, the dye must take part in a chemical change, otherwise it is difficult to account for the high temperature coefficient characteristic of the process (Irwin, 1925 a).

In other cases the absorption of dyes appears to be influenced by factors other than those controlling the degree of dissociation of the molecules. M. M. Brooks (1926) found that the rate of absorption of di-brom-phenol was markedly affected by light. Mond (1924) has recently reported a case of so-called 'irreciprocal' permeability of the gut wall to cyanin; how far this may prove to be explicable by Miss Irwin's hypothesis remains obscure.

In drawing comparisons between these results and those obtained with animal cells, it is advisable to remember that a large aqueous vacuole is not present in animal cells, and that the accumulation of a dye by a vegetable cell is no indication that it will also accumulate in an animal cell, although it is of interest to note that various authors claim that intra-vitam dyes accumulate in 'vacuoles' of the animal cell. It is, however, probable that the surface layer of the animal cell is a separate phase from the cell interior, and since the latter is miscible with water, the whole system is not so radically different to that of plant cells as might appear to be the case at first sight.

Methods of determining the permeability to solutes

In addition to the methods already described, the permeability of a cell to solutes can be determined by a variety of methods. These have been summarised by Stiles (1924) and by Jacobs (1924). If a cell is exposed to the solute in question, its passage into the cell can be determined by an analysis of the cell contents or of the surrounding medium. Thus Janse (1887) exposed *Spirogyra* cells to a solution of KNO_3, and detected the passage of nitrates into the sap by expressing the latter into a solution of diphenylamine and observing the blue colour reaction. A similar method has been more recently employed by Osterhout (1922b), who used the cells of *Nitella* and obtained characteristic crystals on expressing the sap into a solution of nitron in 10 per cent. acetic acid. Using the same material Irwin (1923) detected the passage of chlorine ions into the cells by testing the sap with silver nitrate. Miss Irwin made the significant observation that the rate of accumulation of chlorine in the sap was not increased by increasing the concentration of chlorides in the external medium. Instead of employing chemical tests for intracellular composition, it is possible to make use of characteristic emission spectra: in this way M. M. Brooks (1922) observed the penetration of lithium, caesium and strontium into the cells of *Valonia*. As a rule, direct chemical or physical analysis of cell contents is only satisfactory in the case of large vegetable cells, although comparable methods can be and have been employed in the case of small animal cells, e.g. red blood-corpuscles (Kozawa, 1914).

The analysis of the external medium as opposed to that of the cell contents has been used as an estimate of permeability from time to time. Demoussy (1900) determined the rate at which potassium and calcium disappeared from a medium in contact with the tissues of wheat and maize. The uptake of hydrogen ions by plant tissues was estimated electrometrically by Stiles and Jørgensen (1915 a), and by animal cells by Gray (1920), who used both colorimetric and electrometric methods. The great objection to these methods lies in the fact that we have no direct knowledge that the substances in question actually penetrate as such into the interior of the cell—they may be adsorbed to the surface.

REFERENCES

References marked with an asterisk contain either a general discussion or an extensive bibliography

ABDERHALDEN, E. (1898). 'Zur quantitativen vergleichenden Analyse des Blutes.' *Zeit. f. physiol. Chem.* 25, 67.

ANSELMINO, K. J. (1927). 'Untersuchungen über d. Durchlässigkeit künstlicher kolloidaler Membranen.' *Biochem. Zeit.* 192, 390.

BILTZ, W. (1910). 'Über die Dialysierbarkeit der Farbstoffe.' *Gedenkboek v. Bemmelen*, p. 108. (Te Helder.)

BROOKS, M. M. (1922). 'The penetration of cations into living cells.' *Journ. Gen. Physiol.* 4, 347.

—— (1925). 'The permeability of protoplasm to ions.' *Amer. Journ. Physiol.* 76, 116.

—— (1926). 'The effects of pH, light and other factors on the penetration of 2-6-dibromo-phenol-indophenol and other dyes into a living cell.' *Amer. Journ. Physiol.* 76, 190.

—— (1927). 'Studies on the permeability of living cells. IX. Does methylene blue itself penetrate?' *Univ. Calif. Publ. Zool.* 31, 79.

—— (1929). 'Factors affecting penetration of methylene blue and trimethyl thionine into living cells.' *Proc. Soc. Exp. Biol. and Med.* 26, 290.

BROWN, O. H. (1905). 'Permeability of the egg of *Fundulus*.' *Amer. Journ. Physiol.* 14, 354.

BROWN, W. (1915). 'On the preparation of collodion membranes of differential permeability.' *Biochem. Journ.* 9, 591.

—— (1917). 'Further contributions to the technique of preparing membranes for dialysis.' *Biochem. Journ.* 11, 40.

BUGARSZKY, S. and TANGL, F. (1897). 'Eine Methode zur Bestimmung des relativen Mucus der Blutkörperchen und des Plasmas.' *Zentralb. f. Physiol.* 11, *297*.
COLE, K. S. (1928). 'Electric impedance of suspensions of *Arbacia* eggs.' *Journ. Gen. Physiol.* 12, *37*.
COLLANDER, R. (1921). 'Über die Permeabilität pflanzlicher Protoplasten für Sulphosäurefarbstoffe.' *Jahrb. f. wiss. Bot.* 60, *354*.
—— (1924). 'Über die Durchlässigkeit der Kupferferrocyanidniederschlagsmembran für Nichtelektrolyte.' *Kolloidchem. Beihefte*, 19, *72*.
COOPER, W. C. (Junr) and BLINKS, L. R. (1928). 'The cell sap of *Valonia* and *Halicystis*.' *Science*, 68, *164*.
COOPER, W. C., DORCAS, M. J. and OSTERHOUT, W. J. V. (1928). 'The penetration of strong electrolytes.' *Journ. Gen. Physiol.* 12, *427*.
CROZIER, W. J. (1916 a). 'Cell penetration by acids.' *Journ. Biol. Chem.* 24, *255*.
—— (1916 b). 'Cell penetration by acids. II. Further observations on the blue pigment of *Chromodoris zebra*.' *Journ. Biol. Chem.* 26, *217*.
—— (1916 c). 'Cell penetration by acids. III. Data on some additional acids.' *Journ. Biol. Chem.* 26, *225*.
DEMOUSSY, E. (1900). 'Absorption élective de quelques éléments minéraux par les plantes.' *C.R. Acad. Sci.* 127, *970*.
FITTING, H. (1915). 'Untersuchungen über die Aufnahme von Salzen in die lebende Zelle.' *Jahrb. f. wiss. Bot.* 56, *1*.
—— (1917). 'Untersuchungen über isotonische Koeffizienten und ihre Nutzen für Permeabilitätsbestimmungen.' *Jahrb. f. wiss. Bot.* 57, *553*.
FRICKE, H. (1925). 'The electric capacity of suspensions with special reference to blood.' *Journ. Gen. Physiol.* 9, *137*.
FRICKE, H. and MORSE, S. (1925). 'The electric resistance and capacity of blood for frequencies between 800 and $4\frac{1}{2}$ million cycles.' *Journ. Gen. Physiol.* 9, *153*.
GARMUS, A. (1912). 'Die Permeabilität und das Scheidevermögen der Drüsenzellen für Farbstoffe und eine neue Methode vitaler Beobachtung.' *Zeit. f. Biol.* 58, *185*.
*GELLHORN, E. (1929). *Das Permeabilitätsproblem*. (Berlin.)
GRAY, J. (1920). 'The relation of the animal cell to electrolytes. II. The adsorption of hydrogen ions by living cells.' *Journ. Physiol.* 54, *68*.
—— (1922). 'The mechanism of ciliary movement.' *Proc. Roy. Soc.* B, 93, *104*.
—— (1924). 'The mechanism of ciliary movement. IV. The relation of ciliary activity to oxygen consumption.' *Proc. Roy. Soc.* B, 96, *95*.
GREEN, R. G. and LARSON, W. P. (1922). 'Conductivity of bacterial cells.' *Journ. Inf. Dis.* 30, *550*.
HALDANE, J. B. S. (1925). 'On the origin of the potential differences between the interior and exterior of cells.' *Proc. Camb. Philos. Soc. Biol. Series*, 1, *243*.
HAMBURGER, H. J. (1909). 'Über den Durchtritt von Ca-Ionen durch die Blutkörperchen und dessen Bedingungen.' *Zeit. f. physik. Chem.* 69, *663*.
HARVEY, E. N. (1911). 'Studies on the permeability of cells.' *Journ. Exp. Zool.* 10, *507*.
—— (1914). 'The permeability of cells for acids.' *Intern. Zeit. f. phys.-chem. Biol.* 1, *463*.

HAYWOOD, C. (1925). 'The relative importance of pH and carbon dioxide tension in determining the cessation of ciliary movement in acidified sea water.' *Journ. Gen. Physiol.* 7, *693*.

HERTZ, W. (1922). 'Die Vitalfärbung von *Opalina ranarum* mit Säurefarbstoffen und ihre Beeinflussung durch Narcoticum.' *Pflüger's Archiv*, 196, *444*.

HÖBER, R. (1909). 'Die Durchlässigkeit der Zellen für Farbstoffe.' *Biochem. Zeit.* 20, *56*.

—— (1910). 'Eine Methode die elektrische Leitfähigkeit im Innern von Zellen zu messen.' *Pflüger's Archiv*, 133, *237*.

—— (1912). 'Ein zweites Verfahren die Leitfähigkeit im Innern von Zellen zu messen.' *Pflüger's Archiv*, 148, *189*.

—— (1913). 'Messungen der inneren Leitfähigkeit von Zellen. Dritte Mitteilung.' *Pflüger's Archiv*, 150, *15*.

—— (1914). 'Zur physikalischen Chemie der Vitalfärbung.' *Biochem. Zeit.* 67, *420*.

*—— (1922). *Physikalische Chemie der Zelle und der Gewebe.* (Leipzig.)

HÖBER, R. and CHASSIN, S. (1908). 'Die Farbstoffe als Kolloide und ihre Verhalten in der Niere vom Frosch.' *Kolloid-Zeit.* 3, *76*.

HÖBER, R. and KEMPNER, F. (1908). 'Beobachtungen über Farbstoffausscheidung durch die Nieren.' *Biochem. Zeit.* 11, *105*.

HÖFLER, K. (1918 a). 'Permeabilitätsbestimmung nach der plasmometrischen Methode.' *Ber. deut. bot. Ges.* 36, *414*.

—— (1918 b). 'Ueber die Permeabilität der Stengelzellen von *Tradescantia elongata* für Kalisaltpeter.' *Ber. deut. bot. Ges.* 36, *423*.

—— (1919). 'Über den zeitlichen Verlauf der Plasmadurchlässigkeit in Salzlösung. I.' *Ber. deut. bot. Ges.* 37, *314*.

IRWIN, M. (1922 a). 'The permeability of living cells to dyes as affected by hydrogen ion concentration.' *Journ. Gen. Physiol.* 5, *323*.

—— (1922 b). 'The penetration of dyes as influenced by hydrogen ion concentration.' *Journ. Gen. Physiol.* 5, *727*.

—— (1923). 'The behaviour of chlorides in the cell sap of *Nitella*.' *Journ. Gen. Physiol.* 5, *427*.

—— (1925 a). 'On the accumulation of dye in *Nitella*.' *Journ. Gen. Physiol.* 8, *147*.

—— (1925 b). 'Mechanism of the accumulation of dye in *Nitella* on the basis of the entrance of the dye as undissociated molecules.' *Journ. Gen. Physiol.* 9, *561*.

—— (1926 a). 'Mechanism of the accumulation of dye in *Nitella* on the basis of the entrance of the dye as undissociated molecules.' *Journ. Gen. Physiol.* 9, *561*.

—— (1926 b). 'Exit of dye from living cells of *Nitella* at different pH values.' *Journ. Gen. Physiol.* 10, *75*.

—— (1926 c). 'The penetration of basic dye into *Nitella* and *Valonia* in the presence of certain acids, buffer mixtures and salts.' *Journ. Gen. Physiol.* 10, *271*.

—— (1926 d). 'Certain effects of salts on the penetration of brilliant cresyl blue into *Nitella*.' *Journ. Gen. Physiol.* 10, *425*.

—— (1926 e). 'On the nature of the dye penetrating the vacuole of *Valonia* from solutions of methylene blue.' *Journ. Gen. Physiol.* 10, *927*.

IRWIN, M. (1927 a). 'The effect of acetate buffer mixtures, acetic acid and sodium acetate on the protoplasm as influencing the rate of penetration of cresyl blue into the vacuole of *Nitella*.' *Journ. Gen. Physiol.* 11, *111*.
—— (1927 b). 'Counteraction of the inhibiting effects of various substances on *Nitella*.' *Journ. Gen. Physiol.* 11, *123*.
—— (1928 a). 'Spectrophotometric studies of penetration. V. Resemblances between the living cell and an artificial system in absorbing methylene blue and trimethyl thionin.' *Journ. Gen. Physiol.* 12, *407*.
—— (1928 b). 'Predicting penetration of dyes into living cells by means of an artificial system.' *Proc. Soc. Exp. Biol. and Med.* 26, *125*.
*JACOBS, M. H. (1924). 'Permeability of the cell to diffusing substances.' Section III. *General Cytology*. (Chicago.)
JACQUES, A. G. and OSTERHOUT, W. J. V. (1930). 'The kinetics of penetration. II. The penetration of CO_2 into *Valonia*.' *Journ. Gen. Physiol.* 13, *301*.
JANSE, J. M. (1887 a). 'Plasmolytische Versuche an Algen.' *Bot. Zentralb.* 32, *21*.
—— (1887 b). 'Die Permeabilität des Protoplasma.' *Verslagen en Meded. der K. Akad. Amsterdam*, 4, *332*.
KOZAWA, S. (1914). 'Beiträge zum arteigenen Verhalten der roten Blutkörperchen. III. Artdifferenzen in der Durchlässigkeit der roten Blutkörperchen.' *Biochem. Zeit.* 60, *231*.
LAPICQUE, L. (1907). 'Polarisation de membrane dans les électrolytes du milieu physiologique.' *C.R. Soc. Biol.* 63, *37*.
LEPESCHKIN, W. W. (1908). 'Über den Turgordruck der vakuolisierten Zellen.' *Ber. deut. bot. Ges.* 26 a, *198*.
—— (1909). 'Über die Permeabilitätsbestimmung der Plasmamembran für gelöste Stoffe.' *Ber. deut. bot. Ges.* 27, *129*.
MCCLENDON, J. F. (1910). 'On the dynamics of cell division. II. Changes in permeability of developing eggs to electrolytes.' *Amer. Journ. Physiol.* 27, *240*.
—— (1926). 'Colloidal properties of the surface of the living cell.' *Colloid Symposium Monograph*, 4, *224*.
—— (1927). 'The permeability and the thickness of the plasma membrane as determined by electric currents of high and low frequency.' *Protoplasma*, 3, *71*.
MEYER, H. H. (1899). 'Zur Theorie der Alkoholnarkose. Welche Eigenschaft der Anaesthetica bedingt ihre narkotische Wirkung?' *Arch. exper. Path. u. Pharm.* 42, *109*.
MICHAELIS, L., ELLSWORTH, R. McL. and WEECH, A. A. (1926). 'Studies on the permeability of membranes. II. Determination of ionic transfer numbers in membranes from concentration chains.' *Journ. Gen. Physiol.* 10, *671*.
MICHAELIS, L. and PERLZWEIG, W. A. (1926). 'Studies on the permeability of membranes. I. Introduction and the diffusion of ions across the dried collodion membrane.' *Journ. Gen. Physiol.* 10, *575*.
MICHAELIS, L. and WEECH, A. A. (1927). 'Studies on the permeability of membranes. IV. Variations of transfer numbers with the dried collodion membrane produced by the electric current.' *Journ. Gen. Physiol.* 11, *147*.
MICHAELIS, L., WEECH, A. A. and YAMATORI, A. (1926). 'Studies on the permeability of membranes. III. Electric transfer experiments with the dried collodion membrane.' *Journ. Gen. Physiol.* 10, *685*.

MOND, R. (1924). 'Untersuchungen an isolierten Dünndarm des Frosches. Ein Beiträg zur Frage der gerichteten Permeabilität und der einseitigen Resistenz tierischer Membranen.' *Pflüger's Archiv*, 206, *172*.
—— (1927). 'Umkehr der Anionenpermeabilität der roten Blutkörperchen in eine elektive Durchlässigkeit für Kationen.' *Pflüger's Archiv*, 217, *618*.
NIERENSTEIN, E. (1920). 'Ueber das Wesen der Vitalfärbung.' *Pflüger's Archiv*, 179, *233*.
OSTERHOUT, W. J. V. (1909). 'On the penetration of inorganic salts into living protoplasm.' *Zeit. f. physik. Chem.* 70, *408*.
—— (1914). 'Chemical dynamics of living protoplasm.' *Science*, 39, *544*.
*—— (1922 a). *Injury, Recovery, and Death.* (Philadelphia.)
—— (1922 b). 'Direct and indirect determinations of permeability.' *Journ. Gen. Physiol.* 4, *275*.
—— (1925). 'Is living protoplasm permeable to ions?' *Journ. Gen. Physiol.* 8, *131*.
OSTERHOUT, W. J. V. and DORCAS, M. J. (1925). 'The penetration of CO_2 into living protoplasm.' *Journ. Gen. Physiol.* 9, *255*.
OVERTON, E. (1895). 'Über die osmotischen Eigenschaften der lebenden Pflanzen und Tierzelle.' *Vjschr. naturf. Ges. Zürich*, 40, *159*.
—— (1899). 'Über die allgemeinen osmotischen Eigenschaften der Zelle, ihre vermuthlichen Ursachen und ihre Bedeutung für die Physiologie.' *Vjschr. naturf. Ges. Zürich*, 44, *88*.
—— (1900). 'Studien über die Aufnahme der Anilinfarben durch die lebende Zelle.' *Jahrb. f. wiss. Bot.* 34, *669*.
PAUL, J. H. and SHARPE, J. S. (1916). 'Studies in calcium metabolism. I. The deposition of lime salts in the integument of decapod crustacea.' *Journ. Physiol.* 50, *183*.
PHILIPPSON, M. (1913). 'L'action physiologique des acides et leur solubilité dans les lipoides.' *Résumés IXme Cong. Internat. Physiol.* Groningen, *136*.
v. RISSE, O. (1926 a). 'Über die Durchlässigkeit von Kollodium und Eiweissmembranen für einige Ampholyte. I. Der Einfluss der H˙ und OH′ Ionenkonzentration.' *Pflüger's Archiv*, 212, *375*.
—— (1926 b). 'Über die Durchlässigkeit von Kollodium und Eiweissmembranen für einige Ampholyte. II. Quellungseinflüsse.' *Pflüger's Archiv*, 213, *685*.
ROHONYI, H. and LORANT, A. (1916). 'Zur Kenntnis der Wirkung von CO_2 und O_2 auf die Elektrolytpermeabilität der roten Blutkörperchen.' *Kolloidchem. Beihefte*, 8, *377*.
RUHLAND, W. (1908). 'Beiträge zur Kenntnis der Permeabilität der Plasmahaut.' *Jahrb. f. wiss. Bot.* 46, *1*.
—— (1912 a). 'Studien über die Aufnahme von Kolloiden durch die pflanzliche Plasmahaut.' *Jahrb. f. wiss. Bot.* 51, *376*.
—— (1912 b). 'Die Plasmahaut als Ultrafilter bei der Kolloidaufnahme.' *Ber. deut. bot. Ges.* 30, *139*.
—— (1913). 'Zur Kritik der Lipoid und der Ultrafiltertheorie der Plasmahaut.' *Biochem. Zeit.* 54, *59*.
—— (1914). 'Weitere Beiträge zur Kolloidchemie und physikalischen Chemie der Zelle.' *Jahrb. f. wiss. Bot.* 54, *391*.
SHEARER, C. (1919). 'Studies on the action of electrolytes on bacteria. I. The action of monovalent and divalent salts on the conductivity of bacterial emulsions.' *Journ. Hygiene*, 18, *337*.

VAN SLYKE, D. D. (1926). *Factors affecting the Distribution of Electrolytes, Water, and Gases in the Animal Body.* (Philadelphia.)

STEWART, G. N. (1899). 'The behaviour of the haemoglobin and electrolytes of the coloured corpuscles when blood is laked.' *Journ. Physiol.* 34, *211*.

—— (1901). 'A contribution to our knowledge of the action of saponin on the blood corpuscles and pus corpuscles.' *Journ. Exp. Med.* 6, *257*.

—— (1909). 'The mechanism of haemolysis with special reference to the relation of electrolytes to cells.' *Journ. Pharm. Exp. Therap.* 1, *49*.

*STILES, W. (1924). *Permeability.* (London.)

STILES, W. and JØRGENSEN, I. (1915 a). 'Studies in permeability. I. The exosmosis of electrolytes as a criterion of antagonistic ion action.' *Ann. of Bot.* 29, *349*.

—— (1915 b). 'Studies in permeability. II. The effect of temperature on the permeability of plant cells to the hydrogen ion.' *Ann. of Bot.* 29, *611*.

TINKER, F. (1916). 'The microscopic structure of semipermeable membranes and the part played by surface forces in osmosis.' *Proc. Roy. Soc.* A, 92, *357*.

—— (1917). 'The relative properties of the copper ferrocyanide membrane.' *Proc. Roy. Soc.* A, 93, *268*.

WARBURG, E. J. (1922). 'Studies on carbonic acid compounds and hydrogen ion activities in blood and salt solutions.' *Biochem. Journ.* 16, *153*.

WERTHEIMER, E. (1923 a). 'Über irreziproke Permeabilität. I.' *Pflüger's Archiv*, 199, *383*.

—— (1923 b). 'Über irreziproke Permeabilität. II.' *Pflüger's Archiv*, 200, *82*.

—— (1923 c). 'Die irreziproke Permeabilität von Ionen und Farbstoffen. III.' *Pflüger's Archiv*, 200, *354*.

—— (1923 d). 'Der Salzeffekt an der lebenden Membran. IV.' *Pflüger's Archiv*, 201, *488*.

—— (1923 e). 'Weitere Studien über die Permeabilität lebender Membranen. V. Über die Kräfte die die Wasserbewegung durch eine lebende Membran bedingen.' *Pflüger's Archiv*, 201, *591*.

—— (1924). 'Weitere Studien über die Permeabilität lebender Membranen. VI. Über die Permeabilität von Säuren u. Basen, Einfluss der Temperatur auf die Permeabilität.' *Pflüger's Archiv*, 203, *542*.

ZOOND, A. (1927). 'The interpretation of changes in electrical resistances accompanying the death of bacterial cells.' *Journ. Bact.* 14, *279*.

CHAPTER FIFTEEN

The Nature of the Cell Surface

The barrier which prevents the free diffusion of solutes between the interior of a living cell and its surrounding medium is clearly located at or near the surface of the cell. Although we have no real knowledge of its essential nature, it is customary to speak of this barrier as though it were a definite morphological structure to which the term *plasma membrane* can appropriately be applied. Material conceptions of the plasma membrane have been put forward on many occasions and frequent attempts have been made to manufacture membranes possessing properties comparable to those of the surface of living cells. As already mentioned (p. 371), Overton and others concluded that *lipoid* substances must be present in the plasma membrane, since substances soluble in such media usually enter the cell more rapidly than any others. This suggestion was supported by the observation of Stewart (1909), who observed that when red blood-corpuscles are killed by 12 per cent. formaldehyde they still retain their impermeability to strong electrolytes: if, however, these cells are exposed to ether, this impermeability is at once destroyed. Similarly a substance such as saponin, which is a strong emulsifying agent for fats, is a powerful cytolytic agent. Additional interest was attached to the lipoid theory of the cell surface by the work of Clowes (1916). It had previously been shown by Osterhout (1911) that the semipermeable properties of *Laminaria* cells are lost if the cells are exposed to any solution not containing divalent metallic ions, and that the loss of semipermeability is partially reversible on the addition of calcium. Clowes (1916) showed that the properties of an emulsion of olive oil and water are markedly altered by the presence or absence of calcium ions. Thus an emulsion of olive oil and water in alkaline NaCl yields an emulsion of oil drops in a continuous phase of water: on the addition of calcium, however, the phases are reversed, and the system is converted into water drops dispersed in a continuous phase of oil. If the cell surface is normally characterised by a continuous oil phase, we can understand how it becomes instable in the absence of calcium. According to Gortner and Grendel (1925) the mammalian red blood-corpuscle is bounded by a bimolecular layer of fatty substances, but Mudd and Mudd (1925) have criticised this conclusion and incline to the view that proteins as well as lipoids must be present at the surface of these cells. The arguments for and against the fatty nature of the plasma membrane have been recently summarised by Gellhorn (1929) and by Steward (1929); the latter author rightly criticises the possibility of determining the nature of the cell surface by direct analyses of substances diffusing from the cell into cytolytic media as were attempted by Hansteen-Cranner (1922).

The most obvious difficulty, encountered by the conception of the plasma membrane as a continuous lipoid film, is the necessity of endowing it with an adequate permeability to water. How far this difficulty is as great as it appears to be at first sight is not clear. If the plasma membrane is extremely thin (see p. 361)—and is of the nature of a monomolecular layer—it might be permeable to water, whereas a thicker layer might not be permeable. That water can pass through a thin film of olive oil is suggested by the older observations of Quincke (1894, 1902) and Bütschli (1894), who found that water was absorbed by olive oil which contained potassium oleate. A permeability to water on the part of a fatty film can of course be obtained if it be assumed, with Czapek (1910), that the film is not continuous but is in the form of a discontinuous emulsion in a water permeable phase. Unless, however, this latter phase be impermeable to electrolytes we are no better off than before and, as Nathansohn (1904 a and b) points out, lecithin in the dispersed state does not accelerate the passage of lipoid-soluble substances.

Within recent years the chief interest in the lipoid theory of the plasma membrane has centred on its application to the theory of bioelectric currents. If the protoplasmic surface is immiscible with water, it is possible to draw—as Beutner did—a fairly close parallel between electromotive forces at inanimate phase boundaries and those characteristic of living cells (see p. 393).

The possibility of a *protein* composition for the plasma membrane was first considered by Pfeffer (1900), and the suggestion was later supported by Robertson (1908) and by Lepeschkin (1910–11). It can hardly be doubted that the conclusions reached by these authors were hardly justified by the evidence available (see Blackman, 1912), but certain outstanding phenomena must be considered. Firstly, as pointed out by Robertson, proteins readily form precipitation membranes (see p. 77): the trouble lies, however, in the fact that such membranes are usually readily permeable to strong electrolytes and in this respect exhibit few of the selective properties of the living cell surface. It is doubtless true, as pointed out by Lepeschkin, that semipermeability is suddenly lost when the proteins of the cell are coagulated by heat, but this does not prove that the plasma membrane is related to the proteins except in so far as it may lie in a protein matrix. It may be remembered that the instability of the plasma membrane in the absence of calcium ions (as analysed by Osterhout) can be explained by the dissolution of a protein matrix just as readily as by the dispersion of a lipoid emulsion (see p. 107). That the plasma membrane is not of a simple protein type is shown by the fact that it can and does possess definite osmotic properties. If proteins play any part in the mechanism of the plasma membrane, they must act as a matrix in the pores of which is deposited a film capable of supporting an osmotic pressure; such a film might consist of calcium phosphate and be comparable to the films examined by Meigs (1915) and Lapicque (1907). Such membranes are, it is true, impermeable to many salts, but in other respects they are unsatisfactory. More recent work raises the possibility that if the pores of a protein membrane are small enough they could be blocked by a film of adsorbed molecules of the solute or even by a water film (see p. 351).

It must be admitted that all attempts to manufacture artificial membranes with the requisite properties have so far failed, but if Fricke's estimate of the thickness of the plasma membrane (see p. 361) is of the right order of magnitude, all attempts to reproduce it artificially have been of a rough and ready nature. For a discussion of the various theories concerning the composition of the plasma membrane reference may be made to Höber (1922), Stiles (1924), and Gellhorn (1929). It is, however, worth noting that most, if not all the attempts to reproduce the properties of the cell surface, have tended to overlook the fundamental fact that the living cell is bounded by a surface or a membrane which is capable of generating energy or of transforming energy which is supplied to it in an appropriate form (see p. 21). To suggest that lipoids or proteins are responsible for the observed properties of this surface is equivalent to saying that these bodies are responsible for the properties of protoplasm as a whole. The fundamental problem in both cases is the same, viz. to discover the precise orientation of the constituent parts of the surface, and so discover how the latter can function as a dynamic unit. We may look upon the cell surface from two points of view. Firstly, we may regard it as endowed with life in the sense that it can derive a store of energy from oxidative and other changes, and use this energy for specific purposes. Alternatively, we may regard the surface membrane as a relatively inactive structure only capable of distributing energy when the latter is supplied to it in a particular form; in this case we may hope to find a parallel to the plasma membrane within the realm of inanimate systems.

The cell surface as a living unit

It has been mentioned on more than one occasion that the resistance offered by the cell surface to the diffusion of molecules and ions is greatly decreased when the cell dies. Since this change is irreversible it is difficult to say how far the increased rate of diffusion is the direct result of a decline in the vital activities of the cell, or how far both these phenomena are affected independently by the factor responsible for death; the fact remains, however, that the normal properties of the cell surface are invariably lost after death. If we restrict our attention to unicellular systems any correlation between the 'semipermeability' of the surface and the metabolic activities of the whole cell is at present of a hypothetical nature, but if we are prepared to consider other types of living membranes a number of relevant facts are available.

If a solution of a crystalloid be separated from distilled water by means of a semipermeable membrane, water will diffuse into the solution, and if the solution be confined within an osmometer the process can be made to perform work by lifting a piston or in other ways. The energy for this work is originally present in the solution. If a membrane is interposed between two solutions of equal osmotic pressure, there is no transference of water from one side to the other and no work can be done by the system, since the potential is the same on both sides. In other words, whenever a solution becomes less concentrated it has the capacity of doing work, whenever it becomes concentrated work must be done on the solution. The performance of work implies that two sources of energy are available, one of which is at a higher potential than the other. When a living cell executes a change in the relative concentration of its internal fluids and that of its external environment, where does the requisite energy come from? In inanimate systems no energy source is available if the potential on the two sides of a membrane is identically the same—so that if we interpose a membrane, of any description, between two identical solutions no differential transference of water or solutes can occur and no work can be done. Conversely if a membrane is interposed between two identical solutions and a transference of water or solute spontaneously occurs, the membrane itself must be supplying the requisite energy. We have therefore the possibility of determining whether or not the cell membrane is an active 'living' structure, in the sense that it is the seat of a supply of potential energy which can be so orientated as to do osmotic work. The evidence in respect to the transference of water has already been presented (see p. 343), and although the data are far from complete it certainly looks as though, in some instances, the cell surface is able to effect a transfer of water against or in the absence of any osmotic gradient. When we consider the transfer of dissolved substances the data, also incomplete, give the same conclusion. For example, both Cohnheim (1898, 1899) and Reid (1901) found that dissolved solutes can pass through the gut wall to solutions of identically the same or even stronger concentrations. The same phenomenon occurs in the kidney where substances are not excreted from the blood in accordance with the osmotic gradients, but according to whether the threshold value in the blood is exceeded or not. As far as one can see, the cells selectively absorb substances in such a way as to maintain their internal environment at a constant composition. The respiratory activity of the kidney is clearly associated with its excretory activity (Barcroft and Straub, 1910), which looks as though the requisite energy for excretion were derived from chemical changes of an oxidative nature. Both in the gut and in the kidney the problem is complicated by the presence of hydrostatic forces at the surface of the blood vessels and it is conceivable that this energy might be utilised in some way, although it would not be available at normal cell surfaces. For this reason the experiments of Cohnheim are of peculiar interest. Cohnheim (1901) excised the gut of *Holothuria* and, filling it with sea water, suspended the preparation in a bath of sea water; he reported that after twenty-four hours the fluid within the gut was absorbed by the cells and excreted to the outside medium. Obviously such a transference could not be effected by an inert membrane but must involve active work on the part of the cells. According to Cohnheim, anaesthetics caused a cessation in the secretion of sea water. An attempt by the author some years ago to confirm Cohnheim's results led to a different interpretation of the facts. It is true that the gut of *Holothuria nigra* filled with sea water and immersed in sea water loses weight after some hours, but in all the cases observed no change in weight occurred as long as the gut was healthy and maintained its normal muscular tone. Loss of water appeared to run parallel to a loss of muscular tone and the latter rapidly leads to disintegration of the gut wall; it will be remembered that sea water is not normally in contact with the external wall of the gut and that Cohnheim's experiments were therefore performed on tissue in an abnormal environment; the published work of Oomen (1926) confirms this view. It would be interesting to repeat these experiments using coelomic fluid instead of sea water.

Rather more adequate evidence of a vital activity on the part of cells which are engaged in the establishment of osmotic and electrolytic equilibria is provided by the work of Heidenhain (1894) and Reid (1902), both of whom observed the transference of water and solutes across a gut membrane having identical media on both sides. Further investigation of these facts is much to be desired, for if substantiated they cannot fail to clarify our conceptions of the mechanism whereby a cell comes into equilibrium with its external environment. As noted above, the great difficulty associated with such experiments is the fact that when a living membrane is exposed to the same solution on both sides, the conditions are highly abnormal, and the time factor becomes of very considerable importance; as will be shown later, it is not difficult to find inanimate membranes which will for a short time effect a transfer of water or ions from isopotential solutions. The essential point is whether this occurs to a significant degree for a prolonged period. It will be remembered that both Maxwell (1913) and Adolph (1925) failed to obtain any transfer of water through a frog's skin when both surfaces were bathed in Ringer.

It is, however, possible to attack the problem from another angle. We know that the cell surface separates a relatively concentrated solution from one which is more dilute without any apparent means of opposing the osmotic force. If it does this in the way in which we have suggested, viz. by setting free a store of chemical energy and converting it into such a form as will provide a force equal and opposite to the osmotic force, then it is likely that this store of energy will ultimately be associated with oxidative reactions. The greater the osmotic force the greater must be the opposing force and the greater the output of energy and the more intense the oxidative processes. We would therefore expect to find (i) that if the oxidative processes are upset salts should begin to leak out of the cell, (ii) that if the osmotic gradient on the two sides of the cell is increased or diminished the respiratory activity of the cell should exhibit corresponding changes. Data of this type are not at present available, but if respiration and cell permeability are closely associated with each other it is surprising to think that it has not been recorded hitherto. The possible relationship between metabolic activity and cell permeability has recently been discussed by Straub (1929). The nearest approach to pertinent data is, however, provided by Lund (1926-8) and by Mond (1924-7). It is known that the E.M.F. across a frog's skin depends on the nature of the ions with which the two surfaces are in contact—and is, to some extent, a measure of the ions which can or cannot diffuse from one side to the other. Lund has shown that the E.M.F. of the skin is markedly affected by changes in the oxygen tension of the medium (fig. 159), or by the presence of cyanides. Lund suggests that the bioelectric current is essentially a measure of differences in the intensity of metabolic activity at the two points examined (see fig. 160).

It will be noted that Lund's experiments do not bear directly on the problem concerning the extent to which a cell can perform work to maintain an anomalous distribution of electrolytes between its interior and its external environment: on the other hand they indicate that the source of a P.D. across the cell membrane may be traced to the liberation of metabolites of an electrolytic nature. The only direct proof, that the maintenance of the normal distribution of electrolytes between a cell and its medium involves the performance of osmotic work, would be to show that any appropriate alteration in the electrolytic content of the external medium is reflected by a corresponding alteration in the metabolic activity of the cell.

Alternatively, some indication of the vital activities of the plasma membrane is available from the observation of Osterhout, Damon and Jacques (1927) on the P.D. across the protoplasmic film of *Valonia* cells (see p. 23). If a cell can exhibit for a prolonged period a constant and measurable P.D. across its membrane when the same solution is on both sides—then, as long as the cell obeys the second

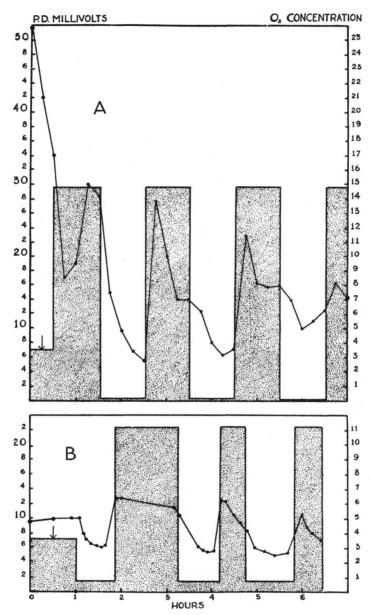

Fig. 159. Diagram showing the reversible effect of different concentrations of oxygen on the magnitude of the electric polarity of two different frogs' skins. Skin *A* has a greater inherent electric polarity than skin *B*. Shaded areas indicate concentration and duration of exposure to oxygen. Arrows indicate oxygen concentration at air saturation. (From Lund.)

law of thermodynamics, the membrane must be generating energy —and in so doing may be regarded as a living unit.

We may approach the problem from still another point of view. If an inanimate membrane is interposed between two solutions of unequal concentration then, whatever be the nature of the membrane, its final effect will be independent of the orientation of the former. Thus if a membrane has side A in contact with solution of concentration C_1 and side B in contact with a similar solution of concentration C_2, the final position of equilibrium in respect to the distribution of ions will be the same as when side A is in contact with solution C_2, and side B in contact with solution C_1. If the membrane is fairly thick, or if ions only adjust themselves to new positions very slowly, the transition period before equilibrium is established may be different in the two cases, but the final equilibrium condition will

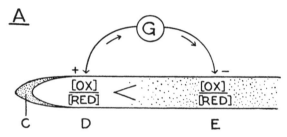

Fig. 160. Diagram of a root tip illustrating electrical polarity due to differences in intensity of respiratory activity. C, root cap; D, region of active cell division; E, region of lower oxidative intensity. (From Lund.)

be the same—in other words there cannot exist a condition of irreciprocal permeability in which substances can move through the membrane more easily in one direction than in the other. It is claimed by various authors that irreciprocal permeability is a specific characteristic of living membranes. This has recently been upheld by Wertheimer (1923-5) using frog's skin: he found that sugar, ions, and gases pass more readily in one direction than in another, although the path of easiest penetration varies for different types of substance. It is, however, very important to differentiate between true irreciprocal permeability and the apparent irreciprocity which is produced when the diffusing substance is removed from the sphere of action on one side of the membrane, as in the cases of carbon dioxide and cresyl blue investigated by Osterhout and by Miss Irwin respectively (see p. 373). Mond (1924) found that the dye cyanosin accumulates rapidly in the gut and does not readily pass out again—but how far this may possibly be due to a change in the degree of its ionisation is not quite clear.

It may be argued that evidence derived from the absorptive properties of the gut or surface skin may not be applicable to cells in general; whilst this is true, it does not invalidate the general argument that the normal plasma membrane may be a generator of energy or be capable of exerting its effects by the absorption of energy which is stored in other parts of the cell. In this respect it is interesting to note that plant cells appear to vary in their ability to absorb electrolytes in accordance with the energy of incident light. M. M. Brooks (1926). Tröndle (1910) reported similar data.

The absorption of salts by somatic cells seems clearly defined in the case of growing cells, for in such cases the percentage of ash does not fall with an increase in the size of the cells; it is possible that in such cases the mechanism of absorption is not closely related to the mechanism which is responsible for the equilibrium conditions during intergrowth periods.

It may be frankly admitted that the conception of the cell surface as an electrical accumulator, which is kept charged by metabolic activity, is not too clearly defined: there are, however, sufficient facts to merit serious attention, particularly in view of its theoretical importance and of the facts to be considered later.

Physical conceptions of the cell membrane

From a physical point of view we may ignore, to some extent, the chemical nature of the membrane and simply consider how far any known physical system exhibits properties comparable to the surface of the cell. We may first consider the maintenance of electrolytic equilibrium between the inside of the cell and the normal external environment. We have seen that the concentration of potassium inside a *Valonia* cell or inside certain red blood-corpuscles is very much higher than that in the external solution. If such systems are in strict thermodynamical equilibrium, then the cell surface must be impermeable to potassium ions or the potassium must be in the unionised condition inside the cells. The latter possibility can be eliminated—since the high conductivity of the interior of the cell makes it very improbable that there is any significant amount of potassium in the unionised condition within the cell. The only alternative is that the cell must be *absolutely* impermeable to potassium ions. If the cell surface were very slightly permeable to both potassium and to chlorine, exosmosis must occur, although it might occur very slowly. If the cell were slightly permeable to potassium ions and not to chlorine, then the ratio of potassium to sodium in the cell would sooner or later become the same as that in the external medium. The essential feature in all such systems is that the final position of equilibrium is independent of the *rate* at which ions diffuse through the membrane. We have seen (in the discussion of the equilibrium distribution of CO_2 or dyes between *Valonia* and its external medium), that the cell must possess a mechanism whereby the substances which diffuse inwards are prevented from diffusing outwards, or in other words, that the cell acts as a trap from which there is no escape. In the case of weak acids and dyes, the substances on entering the cell are converted into a form in which they are unable to diffuse across the membrane in an outward direction: in such cases the substances enter in the undissociated state and are prevented from outward diffusion by electrolytic dissociation; but this explanation is not available for metallic salts. It is important to bear in mind that the cell surface appears to establish a specific equilibrium in respect to individual types of ions and maintains these specific equilibria as long as the cell is alive. When a cell increases in size the different ions enter the cell in just the right proportions, although these proportions differ for different types of cell, e.g. sodium and potassium in *Valonia* and *Halicystis*. Phenomena of this kind are so far unknown in inanimate systems, but find their parallel in the activities of the kidney. Only two alternatives seem possible—either (1) the non-growing cell is absolutely impermeable to cations, but when growth is taking place selective absorption occurs, or (2) the cell is normally permeable and *continuous* selective absorption is taking place. In either case we require an explanation of selective absorption. It will be remembered that in the case of the red blood-corpuscle, the theory of van Slyke (1926) postulates a complete impermeability to potassium.

Instead of considering the equilibrium condition between the concentration of ions inside the cell and in the external medium, we may consider the evidence to be derived from bioelectric phenomena, since the orientation and magnitude of the observed potentials must ultimately depend on a heterogeneous distribution of cations or anions. This point of view is fraught with considerable difficulty and

THE NATURE OF THE CELL SURFACE

danger to the biologist and more properly belongs, at the moment, to the realm of the physical chemist; it is, however, desirable to see the type of evidence at our disposal. It will be recalled that two distinct conceptions of the cell surface are associated with the names of Overton and of Pfeffer respectively.

Overton's results led him to the conclusion that the cell surface is essentially immiscible with water, and is of a fatty nature. It has already been mentioned (p. 22) that, under certain well-defined conditions, an electric current can be made to flow between two unlike aqueous solutions when these are separated by an oil phase of suitable composition. Further, the differences in potential between the two sides of such an oil phase are comparable in magnitude to those observed in living cells. Phase-boundary potentials of this type have already been considered in reference to protoplasmic structure (see p. 24), but for a study of cell permeability a more detailed analysis is desirable, even at the expense of some repetition of the facts.

For a full account of the electrical properties of inanimate membranes and surfaces the monograph of Michaelis (1926) should be consulted; the following description of a typical phase-boundary system is derived from this source.

If we start with two liquids immiscible with each other but each capable of containing a common ion, the junction of the two solutions will be the seat of a phase boundary potential. If we connect the two solutions by a metallic wire as an external circuit, no current will flow because the two solutions are in strict chemical equilibrium. The potential difference will simply depend on the ratio of the solubility of the individual type of ion in the two phases: since this ratio is fixed, no current can flow when the two solutions form a circuit. On the other hand the bounding surface can act as an electrode reversible for the common ion. If therefore we arrange a suitable chain (see fig. 161), composed of two aqueous phases separated by a non-aqueous phase, a current will flow through the system

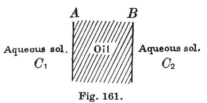

Fig. 161.

when part of a suitable circuit because the potential at A is not the same as at B, and there is no equilibrium in respect to this ion within the oil phase itself. Each phase boundary (A and B) will in fact act as a reversible electrode in respect to any ion which is common to all the liquid phases present. The maximum E.M.F. will be $RT \ln \dfrac{C_1}{C_2} + K$, where K depends on the nature of the ion. The direction of the current will depend upon the nature of the ion common to all phases. Thus, in the following system, used by Beutner (see fig. 161), both the aqueous phases and the oil phase are capable of containing sodium ions $C_1 > C_2$:

| Water + NaCl | Oil + salicylic acid | Water + NaCl |
| C_1 | | C_2 |

As soon as the sodium chloride solutions come into contact with the oil, a trace of NaCl passes from the water into the oil and a trace of salicylic acid into the aqueous phases. In the oil phase double decomposition can take place,

$$\text{NaCl + salicylic acid = Na salicylate + HCl;}$$

whereas in the aqueous phase this does not occur owing to the very low ionisation of salicylic acid. Hence, in the oil phase there are salicylate ions, Na˙ and H˙; but the distribution of these ions in the oil is not uniform, for each side of the oil phase is in equilibrium with a different concentration of sodium ions in the adjacent aqueous phase. Under such conditions the concentration of sodium ions at the interface between the oil and the more concentrated NaCl is greater than at the other interface and the former interface is consequently electropositive. Alternatively, systems having chlorine instead of sodium as a common ion, yield E.M.F.'s in which the more concentrated aqueous phase is electronegative to the more dilute phase.

Aqueous phase (1)	Oil	Aqueous phase (2)
Na˙ (C_1) Cl′ (C_1)	Aniline ions Cl⁻ (C_1') Cl′ (C_2')	Na˙ (C_2) Cl′ (C_2)

If a phase boundary is to exhibit an E.M.F., it is essential that the non-aqueous phase should have an affinity for anions greater or less than that for cations: otherwise the phase boundary potential is zero. The application of these principles to the elucidation of bio-electric phenomena was first made by Beutner (1913, 1920), see p. 22. If the cell surface is the seat of a phase boundary potential, then, on interposing a cell or tissue between two solutions of an electrolyte in unequal concentrations, a galvanic current should flow between the two solutions when the circuit is completed:

Solution 1 | Cell | Solution 2

Using the uninjured surface of an apple, Beutner obtained well-defined E.M.F.'s whose magnitude approached the maximal value characteristic of the oil/water systems previously investigated; in each case the dilute solution was electro-positive (see Table LXII).

THE NATURE OF THE CELL SURFACE

Further experiments indicated, however, that the living system is more complicated than the simplest types of phase-boundary systems. When an apple was subjected to superficial injury and both the injured and uninjured surfaces were in contact with $M/50$ KCl, an E.M.F. of approximately 40 millivolts was observed—this being the normal current of injury. When, however, the amount of tissue between the aqueous phases was decreased the E.M.F. fell rapidly to 10 millivolts. This result can only be explained by assuming that there is a difference between the outer and inner layers of the apple tissue and that when this heterogeneity is reduced, by removing the peripheral layers, so the E.M.F. of the system declines; it will be remembered that in a phase-boundary system the *sine qua non* for an E.M.F. is a heterogeneous distribution

Table LXII

			E.M.F.
$M/10$ NaCl and	$M/10$	NaCl	0
$M/10$,,	,, $M/50$	,,	0·029–0·024 volts
$M/50$,,	,, $M/250$	,,	0·042–0·036 ,,
$M/250$,,	,, $M/1250$	,,	0·041–0·038 ,,

of ions throughout the non-aqueous medium. As pointed out by Michaelis (1926) the apple system represents a chain comparable to

1.	2.	3.	4.
Salt solution	Nitrobenzene poor in picric acid	Nitrobenzene rich in picric acid	Salt solution

Within recent years the biological application of phase-boundary potentials has been developed by Osterhout, some of whose results have already been considered. As mentioned in Chapter III, Osterhout has demonstrated the presence of a persistent E.M.F. in the system,

Valonia sap | Protoplasm | *Valonia* sap,

and has pointed out that if we assume the E.M.F. to be that of a phase-boundary system, then protoplasm must represent a heterogeneous system in respect to an ion common to the sap and the protoplasm. It is clear, however, that any system of this nature cannot be in strict equilibrium if a current is flowing spontaneously through an external circuit; in some manner the membrane must be capable of doing work. If the observed E.M.F. is maintained for long periods in a steady state there must be a supply of free energy available within the membrane—and in this respect we must look upon the membrane as an accumulator—or in other words as a living system. On the other hand, if the E.M.F. is of a transitory nature, it is quite possible that the observed potential differences are of a purely physical nature and only exist during the period of adjustment to a new equilibrium condition where the E.M.F. of the system is zero; until the concentrations of ions at the two interfaces have become equal to each other there will be an E.M.F. It must be remembered, however, that the cell surface is extremely thin and that the total amount of potential energy stored in the form of heterogeneously distributed ions is probably very small.

Beutner's conclusions have not been uncriticised, and to some extent they lost part of their significance when Michaelis and Fujita (1925 a) showed that it is improbable that the living tissues of the apple are responsible for the observed E.M.F.'s. The conception of phase-boundary potentials is, however, valuable not only as an indication of the true state of the protoplasmic/aqueous solution interface, but because it is essentially concerned with a system in which one type of ion (anion or cation) is unable to pass through the membrane. For example, in the system

| Aqueous solution | Oil + acid | Aqueous solution |
| C_1 | | C_2 |

there will be no leakage of ions through the oil phase as long as no electric circuit is established, and as we have seen this state of absolute impermeability is characteristic of the living cell.

Beutner's conception of protoplasm as a water immiscible phase was, however, criticised by Rohonyi (1914, 1922), who pointed out that the essential feature of the oil/water chains does not lie in the immiscibility of the two adjacent phases, but in the fact that only one type of ion—cation or anion—is able to react or be absorbed by the membrane. Membranes possessing this characteristic are not restricted to oils; they may consist of precipitation membranes of the copper ferrocyanide type investigated many years ago by Pfeffer.

Such systems were also examined by Beutner (Michaelis, 1926, p. 215). If a gelatin cup be filled with a solution of potassium ferrocyanide and immersed in a solution of copper sulphate, a current will flow through an external circuit between the sulphate and the ferrocyanide. The copper solution is positive and the E.M.F. is approximately 100 millivolts. If KCl be added to the copper sulphate solution the E.M.F. is changed, whereas it is independent of the concentration of the copper sulphate. The magnitude of the change effected by different concentrations of KCl in the system is such as to indicate that the membrane is behaving as though it were able to absorb potassium ions but not chlorine ions, just as is the case with an oil phase containing an organic acid. Whether we look on the cell surface as a water immiscible fluid or as a precipitation membrane is therefore immaterial, as long as they both show the same differential attraction for cations as opposed to anions. The real difficulty arises in that copper ferrocyanide membranes are permeable to many types of substances to which the cell is not.

Within recent years much attention has been paid by physico-chemists to the properties of protein membranes at various degrees of departure from their isoelectric point. Many such membranes undoubtedly exhibit a differential permeability to cations and anions—and the degree of this difference varies with the pH of the membrane. In this respect the work of Mond and Hoffmann (1928 a and b) and others may be consulted. It must be remembered, however, that all known protein membranes have a much higher capacity for conducting electricity than do copper ferrocyanide membranes, oil films, or the surfaces of living cells. In other words although protein membranes may show a marked difference in respect to their permeability to anions or cations, yet unless the pores in the membrane are extremely small both these types of ion diffuse through the membrane to a very significant extent, and this process of diffusion goes on whether the system is part of an electric circuit or not. In the case of living cells and copper ferrocyanide membranes on the other hand, the leakage of salts to distilled water is very small, although the observed E.M.F.'s may be the same for all these systems. This point is of real significance when we remember that the membrane must not only be the seat of an E.M.F., but must also account for the permanent maintenance of different concentrations of molecules and ions on its two sides.

Until recent years the intensity and persistence of the bioelectric current of injury seemed to eliminate the possibility of attributing these phenomena to the diffusion potentials investigated by Ostwald and others. Michaelis (1926) has, however, revived the conception of diffusion potentials in a modified form. If we interpose between two solutions of KCl ($N/10$ and $N/100$) a perfectly inert membrane, the initial E.M.F. ought to be 0·4 millivolts, this low value being due to the small difference between the migration velocity of potassium ions (64·67) and chlorine ions (65·44). Michaelis found, however, that if a dried collodion membrane is interposed between two such solutions much higher E.M.F.'s are observed. To account for this fact it is assumed that the migration velocity of ions within membranes having very small pores is markedly different to their velocity in water, and that the velocities of different ions are affected by the membrane to a different extent. Thus in an extreme case the velocity of migration of the anion may become zero and the E.M.F. between two solutions of concentrations 10 : 1 will become approximately 57 millivolts; the system will resemble a diphasic system where the non-aqueous phase reacts only with cations. The essential character of Michaelis' theory lies in the possibility of having an infinite variety of conditions between a normal diffusion potential on the one hand (when the relative velocity of migration of different ions is the same as in water) and a maximum value (when the velocity of the anions or cations is reduced to zero) on the other. In the case of phase-boundary potentials (for all dilute solutions) the anions play no part—and the only factor which need be considered is the distribution of the cations within the non-aqueous phase. How far the two theories are strictly alternative to one another or mutually exclusive remains to be seen. It would, however, be interesting to know something of the properties of an aqueous-oil-aqueous system in which the oil contains both a base and an acid radicle which are insoluble in water—for it will be remembered that Nierenstein found that such a system acts reversibly to both acid and basic dyes.

If we look upon the current of injury as equivalent to a phase-boundary current or as due to a specialised type of diffusion potential, we undoubtedly gain some insight into one aspect of biological activity, but we are still a long way from understanding the fundamental nature of the cell surface. The facts must be considered as a whole, and it is perhaps useful to summarise from a biological point of view the main requirements of any theory of cell permeability which is based on the electrical properties of the surface membranes. Firstly, the membrane has an exceedingly high resistance to electric currents which are not of a high frequency of alternation. Secondly, it separates two fluids which differ from each other in two respects—one is more concentrated in electrolytes than the other, and the relative concentrations of any given ion in the system may be substantially different on the two sides of the membrane. Thirdly, the membrane is the seat of a potential, the magnitude of which can only be accounted for on the assumption that the velocity of specific cations through the membrane is relatively very great in comparison with the velocity of their associated anions.

At present no known inanimate system has these properties: at the same time there is no sharp division of theory involved by the alternative suggestion of the plasma membrane as a 'living' unit. Fundamentally the biological structure with its specific properties will no doubt conform to a physico-chemical mechanism—even if it involves a negation of the second law of thermodynamics—the only point of issue at the moment is the source of the energy which is responsible for the maintenance of osmotic equilibria in the cell and of the mechanism whereby this energy is orientated and used in the manner requisite for life. In other words the plasma membrane is part of a dynamic system whose machinery is of a type not yet demonstrable outside the living cell.

The rôle of the Donnan equilibrium in cell physiology

The anomalous distribution of ions which is usually associated with the name of Donnan has played a considerable part in modern theories of cell permeability.

If on one side (1) of a membrane there be a salt NaR which is wholly or partially ionised, and on the other side (2) there be another salt (NaCl) with a common ion, then if the membrane is impermeable to the unionised salt NaR and to the ion R′, but permeable to the molecules and derivative ions of sodium chloride, the system will come into equilibrium when the product of the concentrations of diffusible ions on the two sides of the membrane is the same, i.e. when

$$[Na]_1[Cl]_1 = [Na]_2[Cl]_2 = k\,[NaCl],$$

because the concentration of undissociated sodium chloride must be the same on the two sides. Since, however, $[Na]_2 = [Cl]_2$,

$$[Na]_1[Cl]_1 = [Cl]_2^2,$$

and since
$$[Na]_1 = [R] + [Cl]_1,$$
$$([R] + [Cl]_1)\,[Cl]_1 = [Cl]_2^2,$$

in other words, the concentration of chlorine ions and therefore of sodium ions on side (1) must be less than on side (2).

Assuming complete dissociation of NaR and NaCl, and equal volumes of solution on the two sides of the membrane, and if C_1 is the initial concentration of NaR and C_2 the initial concentration of NaCl, the equilibrium distribution can be defined by the following diagram in which x is the amount of Na· or Cl′ which has moved across the membrane:

I			II	
Na·	R′	Cl′	Na	Cl′
$C_1 + x$	C_1	x	$C_2 - x$	$C_2 - x$

$$(C_1 + x)\,x = (C_2 - x)^2,$$

or

$$\frac{C_2 - x}{x} = \frac{C_1 + C_2}{C_2}.$$

Taking arbitrary values of C_2 and C_1 we get the following series of equilibria.

Table LXIII

Initial concentration of NaR	Initial concentration of NaCl	Ratio of $\dfrac{C_2 - x}{x}$ = distribution of NaCl between II and I
0·01	1·0	1·01
0·1	1·0	1·1
1·0	1·0	2·0
1·0	0·1	11·0
1·0	0·01	99·0

If the concentration of the ions R′ is high compared to that of the diffusible ions on the other side of the membrane, the membrane will appear to be impermeable to NaCl.

It will be noted that there is a difference in chlorine concentration between the two sides of the membrane and consequently if chlorine electrodes are in contact with the two solutions they will be at a different potential. No current, however, can be produced by such a system, since it is in a state of true equilibrium, and consequently Donnan equilibria cannot be responsible for the electromotive properties of living cells. It is, nevertheless, important to decide how far such systems can account for the anomalous distribution of electrolytes between the cell and its external medium. Except in the case of secretory cells, the colloidal substances within a cell are not free to diffuse into the surrounding medium even if the cell be dead (see

p. 335). As long as there are within the cells proteins or other electrolytes of high molecular weight there will always be a Donnan effect if the cell is immersed in an aqueous medium. At the same time this effect can only affect the final equilibrium in such a way as to keep the concentration of freely diffusible ions inside the cell at a lower level than their concentration in the external medium; the cells of most fresh-water organisms exhibit precisely the opposite phenomenon, for the concentration of freely diffusible electrolytes inside the cell is greater than that in the external environment. In order that a Donnan equilibrium should be sufficiently powerful to account for the distribution of electrolytes within such cells, it is essential to postulate that the cell surface shall also be impermeable to other ions (e.g. potassium) in addition to those characteristic of protein and other substances of high molecular weight. An example of this type has already been mentioned in connection with the osmotic properties of the red blood-corpuscles (van Slyke, 1926). It will be noted, however, that although such postulates have their uses, we are still faced with the problem of how the potassium got into the cell in the first instance.

Functional significance of permeability

From time to time a number of cases have been quoted which suggest that marked changes in permeability occur during the life of a cell and that these changes are associated with specific changes in normal activity. Examples of this type will be found in the text-books of Bayliss (1924) and R. S. Lillie (1923). How far some of the evidence is free from criticism is doubtful. It may be admitted that bioelectric phenomena are, in themselves, definite evidence to show that a redistribution of ions has occurred at the cell surface, but it is still uncertain how far the mechanism responsible for this redistribution is also involved in the maintenance of the equilibrium between the electrolytes in the interior of the cell and those in the external medium. R. S. Lillie (1909), Bernstein (1912), and others have urged the possibility that the essential act of cell stimulation involves a localised and transitory increase in permeability to ions. A direct test of this hypothesis was attempted by McClendon (1912, 1929), who observed an increase in the electrical conductivity of a tetanised muscle which was accompanied by an exosmosis of electrolytes from the fibres. The relationship which exists between a contracting muscle and diffusible ions has also been investigated by Mitchell and

Wilson (1921), who found that when frog's muscles are stimulated to contract under conditions which do not involve irreversible stages of fatigue, the cells lose no more potassium than is attributable to a potassium-free medium; on the other hand, it is only during contraction that a muscle will absorb rubidium or caesium. The interpretation of these facts is not easy, but they indicate that the act of contraction alters in some way the diffusibility of electrolytic ions. Parallel to McClendon's results with excited muscle fibres are the observations of Blackman and Paine (1918) on the exosmosis from the pulvinus of *Mimosa* during stimulation.

In accordance with a generalised conception of cell stimulation R. S. Lillie suggested that the act of fertilisation, like the act of stimulating a nerve, involved a temporary increase in the permeability of the cell surface to ions. In 1910 McClendon showed that a dense suspension of the eggs of sea urchins conducts electricity more readily after fertilisation and this fact was confirmed by Gray (1916). In view of the results obtained by the use of more modern methods, these conclusions must be accepted with some caution. Cole (1928), using very high frequency currents, was unable to detect any measurable increase in the internal conductivity of *Arbacia* eggs after fertilisation, although the effect of fertilisation seemed to stabilise the resistance at a level about three times that of normal sea water. Other attempts to associate fertilisation with a change in surface permeability were made by Bachman and Runnström (1912), who used the eggs of amphibia. Attempts by the author to detect such changes in fish eggs failed (Gray, 1920).

One of the clearest cases of altered permeability during the normal life of the cell is that described by R. S. Lillie (1916) for the diffusion of water into sea urchin eggs before and after fertilisation. Water passes into the eggs of *Arbacia* from hypotonic sea water considerably more rapidly after fertilisation than before; since the act of fertilisation profoundly alters the physical properties of the cortex of the egg an altered permeability is perhaps not surprising.

For the relationship between the inferred changes in permeability which accompany stimulation, reference may be made to Keith Lucas (1910), Hill (1910), and R. S. Lillie (1923).

It is by no means easy to co-ordinate the facts which bear on the structure and properties of the cell surface, and it is equally difficult to reach a definite standpoint from which to review our present knowledge. The problems involved are, however, of fundamental importance. Until we know the factors which control the interchange of substances between a cell and its environment, we cannot hope to form an adequate conception of the way in which an organism will respond to a change in its external environment, and until the true nature of the bioelectric current is established, our knowledge of protoplasmic transmission must remain obscure. It is perhaps unfortunate that so much attention has been focussed on the cells of the higher vertebrates, for valuable data might well be derived from a study of the cells of those animals whose internal fluids are known to change in response to changes in environment; in most invertebrates the osmotic pressure of the internal fluids, unlike those of higher vertebrates, changes in sympathy with changes in the external medium. Similarly, the osmotic pressure inside the cells of moulds often bears a definite relationship to that of their medium, and if the cells of a salmon are permeable to water, it is curious to find that the animal can plunge from fresh water to sea water and *vice versa* without any obvious change in the volume of its cells or in the freezing point of its blood.

REFERENCES

References marked with an asterisk contain either a general discussion or an extensive bibliography

ADOLPH, E. F. (1925). 'The passage of water through the skin of the frog, in the relation between diffusion and permeability.' *Amer. Journ. Physiol.* 73, 85.

BACHMAN, E. L. and RUNNSTRÖM, J. (1912). 'Der osmotische Druck während der Embryonalentwicklung von *Rana temporaria*.' *Pflüger's Archiv*, 144, 287.

BARCROFT, J. and STRAUB, H. (1910). 'The secretion of urine.' *Journ. Physiol.* 41, 145.

BAYLISS, W. M. (1924). *Principles of General Physiology*. 4th ed. (London.)

BERNSTEIN, J. (1912). *Elektrobiologie*. (Brunswick.)

BEUTNER, R. (1913). 'Neue Erscheinungen der Elektrizitätserregung, welche einige bioelektrische Phänomene erklären.' *Zeit. f. Elektrochem.* 19, 319.

—— (1920). *Die Entstehung elektr. Ströme in lebenden Geweben.* (Stuttgart.)

BLACKMAN, F. F. (1912). 'The Plasmatic Membrane and its Organisation.' *New Phytol.* 11, 180.

BLACKMAN, V. H. and PAINE, S. G. (1918). 'Studies in the permeability of the pulvinus of *Mimosa pudica*.' *Ann. Bot.* 32, 69.

BROOKS, M. M. (1926). 'Studies on the permeability of living cells. VII. The effects of light on different wave lengths on the penetration of 2-6-dibromophenol indophenol into *Valonia*.' *Protoplasma*, 1, 305.

BROOKS, M. M. (1927). 'Studies on the permeability of living cells. VIII. The effect of chlorides upon the penetration of Dahlia into *Nitella*.' *Protoplasma*, 2, *420*.

BÜTSCHLI, O. (1894). *Investigations on Microscopic Foams and on Protoplasm*. (Eng. transl.) (London.)

CLOWES, G. H. A. (1916). 'Protoplasmic equilibrium. I. Action of antagonistic electrolytes on emulsions and living cells.' *Journ. Phys. Chem.* 20, *407*.

COHNHEIM, O. (1898). 'Über Dünndarmresorption.' *Zeit. f. Biol.* 36, *129*.

—— (1899). 'Versuche am isolierten, überlebenden Dünndarm.' *Zeit. f. Biol.* 38, *419*.

—— (1901). 'Versuche über Resorption, Verdauung und Stoffwechsel von Echinodermen.' *Zeit. f. physiol. Chem.* 33, *9*.

COLE, K. S. (1928). 'Electric impedance of suspensions of *Arbacia* eggs.' *Journ. Gen. Physiol.* 12, *37*.

CZAPEK, F. (1910). 'Ueber die Oberflächenspannung und den Lipoidgehalt der Plasmahaut in lebenden Pflanzenzellen.' *Ber. deut. bot. Ges.* 28, *480*.

DONNAN, F. G. (1924). 'The theory of membrane equilibria.' *Chem. Rev.* 1, *73*.

FUJITA, A. (1925 a). 'Untersuchungen über elektrische Erscheinungen und Ionendurchlässigkeit von Membranen. I. Die Potentialdifferenz an der Apfelschale.' *Biochem. Zeit.* 158, *11*.

—— (1925 b). 'Untersuchungen über elektrische Erscheinungen und Ionendurchlässigkeit von Membranen.' *Biochem. Zeit.* 159, *370*; 162, *245*.

—— (1926). 'Untersuchungen über elektrische Erscheinungen und Ionendurchlässigkeit von Membranen. VIII.' *Biochem. Zeit.* 170, *18*.

*GELLHORN, E. (1929). *Das Permeabilitätsproblem*. (Berlin.)

GORTNER, E. and GRENDEL, F. (1925). 'On bimolecular layers of lipoids on the chromocytes of the blood.' *Journ. Exp. Med.* 41, *439*.

GRAY, J. (1916). 'The electrical conductivity of echinoderm eggs and its bearing on the problems of fertilisation and artificial parthenogenesis.' *Phil. Trans. Roy. Soc.* B, 207, *481*.

—— (1920). 'The relation of the animal cell to electrolytes. I. A physiological study of the egg of the trout.' *Journ. Physiol.* 53, *308*.

HANSTEEN-CRANNER, B. (1922). 'Zur Biochemie und Physiologie der Grenzschichten lebender Pflanzenzellen.' *Meldinger fra Norges Landbrukshøiskole*, 2, *1*.

HEIDENHAIN, R. (1894). 'Neue Versuche über die Aufsaugung im Dünndarm.' *Pflüger's Archiv*, 56, *579*.

HILL, A. V. (1910). 'A new mathematical treatment of changes of ionic concentration in muscle and nerve under the action of electric currents with a theory as to their mode of excitation.' *Journ. Physiol.* 40, *190*.

HÖBER, R. (1922). *Physikalische Chemie der Zelle und der Gewebe*. (Leipzig.)

LAPICQUE, L. (1907). 'Polarisation de membrane dans les électrolytes du milieu physiologique.' *C.R. Soc. Biol.* 63, *37*.

LEPESCHKIN, W. W. (1910–11). 'Zur Kenntnis der Plasmamembran.' *Ber. deut. bot. Ges.* 28, *91*; 28, *383*; 29, *181*; 29, *247*.

LILLIE, R. S. (1909). 'The general biological significance of changes in permeability.' *Biol. Bull.* 17, *188*.

—— (1916). 'Increase of permeability to water following normal and artificial activation in sea-urchin eggs.' *Amer. Journ. Physiol.* 40, *249*.

*LILLIE, R. S. (1923). *Protoplasmic Action and Nervous Action*. (Chicago.)
LUCAS, K. (1910). 'An analysis of changes and differences in the excitatory process of muscles and nerves based on the physical theory of excitation.' *Journ. Physiol.* 40, *225*.
LUND, E. J. (1926). 'The electrical polarity of *Obelia* and frog's skin and its reversible inhibition by cyanide, ether, and chloroform.' *Journ. Exp. Zool.* 44, *383*.
—— (1928 *a*). 'Relation between continuous bio-electric currents and cell respiration. II.' *Journ. Exp. Zool.* 51, *265*.
—— (1928 *b*). 'Relation between continuous bio-electric currents and cell respiration. III. Effects of concentration of oxygen on cell polarity in the frog's skin.' *Journ. Exp. Zool.* 51, *291*.
—— (1928 *c*). 'Relation between continuous bio-electric currents and cell respiration. V. The quantitative relation between E_p and cell oxidation as shewn by the effects of cyanide and oxygen.' *Journ. Exp. Zool.* 51, *327*.
LUND, E. J. and KENYON, W. A. (1927). 'Relation between continuous bio-electric currents and cell respiration. I.' *Journ. Exp. Zool.* 48, *333*.
MCCLENDON, J. F. (1912). 'The increased permeability of striated muscle to ions during contraction.' *Amer. Journ. Physiol.* 29, *302*.
—— (1929). 'Polarization capacity and resistance of salt solutions, agar, erythrocytes, resting and stimulated muscle, and liver measured with a new Wheatstone Bridge designed for electric currents of high and low frequency.' *Protoplasma*, 7, *561*.
MARSH, G. (1928). 'Relation between continuous bio-electric currents and cell respiration. IV. The origin of the electric polarity in the onion root.' *Journ. Exp. Zool.* 51, *309*.
MAXWELL, S. S. (1913). 'On the absorption of water by the skin of the frog.' *Amer. Journ. Physiol.* 32, *286*.
MEIGS, E. B. (1915). 'The osmotic properties of calcium and magnesium phosphates in relation to those of living cells.' *Amer. Journ. Physiol.* 38, *456*.
MICHAELIS, L. (1926). *Hydrogen Ion Concentration*. Vol. 1. Principles of the Theory. (London.)
MICHAELIS, L. and DOKAN, S. (1925). 'Untersuchungen über elektrische Erscheinungen und Ionendurchlässigkeit. VI.' *Biochem. Zeit.* 162, *258*.
MICHAELIS, L. and FUJITA, A. (1925 *a*). 'Untersuchungen über elektrische Erscheinungen und Ionendurchlässigkeit von Membranen. II. Die Permeabilität der Apfelschale.' *Biochem. Zeit.* 158, *28*.
—— (1925 *b*). 'Untersuchungen über elektrische Erscheinungen und Ionendurchlässigkeit von Membranen. IV.' *Biochem. Zeit.* 161, *47*.
—— (1925 *c*). 'Untersuchungen über elektrische Erscheinungen und Ionendurchlässigkeit von Membranen. VII.' *Biochem. Zeit.* 164, *23*.
MITCHELL, P. H. and WILSON, J. W. (1921). 'The selective absorption of potassium by animal cells. I. Conditions controlling absorption and retention of potassium.' *Journ. Gen. Physiol.* 4, *45*.
MOND, R. (1924). 'Untersuchungen zur Theorie der Entstehung der bio-elektrischen Ströme.' *Pflüger's Archiv*, 203, *247*.
—— (1924). 'Untersuchungen am isolierten Dünndarm des Frosches. Ein Beitrag zur Frage der gerichteten Permeabilität und der einseitigen Resistenz tierischer Membranen.' *Pflüger's Archiv*, 206, *172*.

MOND, R. (1927). 'Über die elektromotorischen Kräfte der Magenschleimhaut vom Frosch.' *Pflüger's Archiv*, 215, *468*.
MOND, R. and AMSON, K. (1928). 'Über die Ionenpermeabilität des quergestreiften Muskels.' *Pflüger's Archiv*, 220, *69*.
MOND, R. and HOFFMANN, F. (1928 a). 'Untersuchungen an künstlichen Membranen, die elektiv anionenpermeabel sind.' *Pflüger's Archiv*, 220, *194*.
—— (1928 b). 'Weitere Untersuchungen über die Membranstruktur der roten Blutkörperchen. Die Beziehung zwischen Durchlässigkeit und Molekularvolum.' *Pflüger's Archiv*, 219, *467*.
MUDD, S. and MUDD, E. B. H. (1925). 'On the surface composition of normal and sensitized mammalian blood cells.' *Journ. Exp. Med.* 43, *127*.
NATHANSOHN, A. (1904 a). 'Über die Regulation der Aufnahme anorganischen Salze durch die Knollen von *Dahlia*.' *Jahrb. f. wiss. Bot.* 39, *607*.
—— (1904 b). 'Weitere Mitteilungen über die Regulation der Stoffaufnahme.' *Jahrb. f. wiss. Bot.* 40, *403*.
OOMEN, H. A. P. C. (1926). 'Verdauungsphysiologische Studien an Holothurien.' *Pubbl. Staz. Zool. Napoli*, 7, *215*.
OSTERHOUT, W. J. V. (1911). 'The permeability of living cells to salts in pure and balanced solutions.' *Science*, 34, *187*.
OSTERHOUT, W. J. V., DAMON, E. B. and JACQUES, A. G. (1927). 'Dissimilarity of inner and outer protoplasmic surfaces in *Valonia*.' *Journ. Gen. Physiol.* 11, *193*.
OVERTON, E. (1901). *Studien über Narkose*. (Jena.)
PFEFFER, W. (1900). *The Physiology of Plants*. (Eng. transl.) Vol. 1. (Oxford.)
QUINCKE, G. (1894). 'Über freiwillige Bildung von hohlen Blasen, Schaum, und Myelinformen durch ölsaure Alkalien und verwandte Erscheinungen besonders des Protoplasmas.' *Ann. d. Phys. u. Chem.* N.F. 53, *593*.
—— (1902). 'Über unsichtbare Flüssigkeitsschichten und die Oberflächenspannung flüssiger Niederschlagsmembranen, Zellen, Colloiden, und Gallerten.' *Ann. d. Physik*, 7, *589, 701*.
REID, W. (1901). 'Transport of fluid by certain epithelia.' *Journ. Physiol.* 26, *436*.
—— (1902). 'Intestinal absorption of solutions.' *Journ. Physiol.* 28, *241*.
ROBERTSON, T. B. (1908). 'On the nature of the superficial layer in cells and its relation to their permeability and to the staining of tissues by dyes.' *Journ. Biol. Chem.* 4, *1*.
ROHONYI, H. (1914 a). 'Ionenpermeabilität und Membranpotential.' *Biochem. Zeit.* 66, *231*.
—— (1914 b). 'Zur Theorie der bioelektrischen Ströme.' *Biochem. Zeit.* 66, *248*.
—— (1922). 'Die Entstehung elektrischer Ströme in lebenden Geweben.' *Biochem. Zeit.* 130, *68*.
VAN SLYKE, D. D. (1926). *Factors Affecting the Distribution of Electrolytes, Water and Gases in the Animal Body*. (Philadelphia.)
STEWARD, F. C. (1929). 'Phosphatides in the limiting protoplasmic surface.' *Protoplasma*, 7, *602*.
STEWART, G. N. (1901). 'The conditions that underlie the peculiarities in the behaviour of the coloured blood-corpuscles to certain substances.' *Journ. Physiol.* 26, *470*.

STEWART, G. N. (1909). 'The mechanism of haemolysis with special reference to the relations of electrolytes to cells.' *Journ. Pharm.* 1, *50*.

*STILES, W. (1924). *Permeability.* London.

STRAUB, J. (1929). 'Der Unterschied in osmotischer Konzentration zwischen Eigelb und Eiklar.' *Recueil des travaux chimiques des Pays-bas*, 48, *49*.

TRÖNDLE, A. (1910). 'Der Einfluss des Lichtes auf die Permeabilität des Plasmahaut.' *Jahrb. f. wiss. Bot.* 48, *171*.

WERTHEIMER, E. (1923 a). 'Über irreziproke Permeabilität. I.' *Pflüger's Archiv*, 199, *383*.

—— (1923 b). 'Über irreziproke Permeabilität. II.' *Pflüger's Archiv*, 200, *82*.

—— (1923 c). 'Über irreziproke Permeabilität. III. Die irreziproke Permeabilität von Ionen und Farbstoffen.' *Pflüger's Archiv*, 200, *354*.

—— (1923 d). 'Über irreziproke Permeabilität. IV. Der Salzeffekt an der lebenden Membran.' *Pflüger's Archiv*, 201, *488*.

—— (1923 e). 'Weitere Studien über die Permeabilität lebender Membranen. V. Über die Kräfte, die die Wasserbewegung durch eine lebende Membran bedingen.' *Pflüger's Archiv*, 201, *591*.

—— (1924 a). 'Weitere Studien über die Permeabilität lebender Membranen. VI. Über die Permeabilität von Säuren u. Basen. Einfluss der Temperatur auf die Permeabilität.' *Pflüger's Archiv*, 203, *542*.

—— (1924 b). 'Weitere Untersuchungen an der lebenden Froschhautmembran. VII.' *Pflüger's Archiv*, 206, *162*.

—— (1925 a). 'Weitere Untersuchungen an lebenden Membranen.' *Pflüger's Archiv*, 210, *527*.

—— (1925 b). 'Über die Quellung geschichteter Membranen und ihre Beziehung zur Wasserwanderung.' *Pflüger's Archiv*, 208, *669*.

—— (1925 c). 'Über die irreziproke Permeabilität tierischer Membranen für Gase. Versuche an der Froschhaut und Froschlunge.' *Pflüger's Archiv*, 209, *494*.

CHAPTER SIXTEEN

The Germ Cells

The function of the germ cells is the production of new individuals endowed with the hereditary characters of the parent organisms. Fortunately, our conceptions of this fundamental property is based on a relatively precise knowledge of both morphology and function. The morphology of the germ cells has been admirably described by Wilson (1925) and others, and rightly occupies a central place in the theory of inheritance. The behaviour of the cells as physiological units has also been summarised on more than one occasion. In the present chapter no attempt has been made to cover the whole field, but attention has been focussed on the theoretical significance of a comparatively small number of isolated phenomena. For a complete account of the activity of the germ cells, reference should be made to Wilson (1925), F. R. Lillie (1919), Lillie and Just (1924) and Dalcq (1928).

The spermatozoon

From a morphological point of view a spermatozoon is a unique type of cell. It consists largely of a nucleus together with a typical unit of locomotion—a flagellum. There is little or no cytoplasm, and no visible signs of metabolic reserves. The functions of the two main morphological constituents are beyond doubt—the nucleus is destined to play its peculiar rôle in the transmission of hereditary characters, while the flagellum endows the cell with the requisite powers of active locomotion.

It is not easy to follow the movement of a single spermatozoon in a highly active suspension, but as pointed out elsewhere (p. 472) it is highly probable that the mechanism of movement is essentially similar to that in the flagellum of the sponge (van Trigt, 1919); waves of contraction pass along the tail of the spermatozoon from the head end backwards. As a rule these waves follow a spiral course and pass round the long axis of the tail as well as along it; consequently, there are two forces acting on the medium—a backward thrust which drives the spermatozoon forward and a couple which rotates the cell about its long axis. If we are right in assuming that the nucleus is unlikely to contribute the requisite energy for movement, it follows that the life of the sperm must be strictly limited when moving in a medium which provides no nutrient substances.

From a teleological point of view it is desirable that a spermatozoon should conserve its energies until it reaches the medium in which fertilisation occurs, and in practice this can be accepted as a working rule although a precise formulation of the facts is not easy. A preliminary instance of the phenomena may be of use. The spermatozoa of the trout *Salmo fario* are immobile in the testicular plasma, but become extremely active on the addition of water; the period of activity is, however, extremely short and at 10° C. is limited to two or three minutes (Gray, 1920 b). Similarly, Kölliker (1856) concluded that the sperm of mammals is immobile within the male tract until it is activated by the secretions of the secondary male glands prior to insemination. More recently, however, Young (1929 a) has investigated the conditions in the guinea-pig and concludes that there is no specific activating principle present in the epididymal secretion; according to Moore (1928) if an activating element is present it is derived from the testis itself. It is by no means easy to determine whether spermatozoa are normally motile within the body of the male parent, for the necessary technique usually involves an alteration in the CO_2 and O_2 tension in the neighbourhood of the cells. It is, however, significant to note that Hammond and Asdell (1926) have shown that the total effective life of rabbit sperm is not more than 30 hours in the body of the female, although it may be as long as 38 days in the body of the male. That some spermatozoa can retain their vitality for prolonged periods is also clear in the case of bees and in some fish. Van Oordt (1928) reports that spermatozoa can live within the female tract of the fish *Xiphophorus helleri* for at least 10 months; whether the same fact holds good for the bat is uncertain (Hartman and Cuyler, 1927). The point at issue is not, however, whether spermatozoa can retain their vitality for long periods but whether they can exist for long periods in a state of active locomotion.

Among invertebrates the period of active life can be studied with relative ease. Gemmill (1900) appears to have been the first to detect the important effect played by the density of the suspension on the period of active life. He showed clearly that dilution of the sperm of *Echinus* shortened the period of life. A cursory inspection of the contents of a ripe testis of *Echinus miliaris* reveals the fact that the spermatozoa are present in very great numbers, and that when a drop of this suspension is mixed with sea water the spermatozoa at once exhibit intense and active movement. In other species, notably *Echinus esculentus*, dilution of the sperm will not produce immediate movement unless the sperm is absolutely 'mature'; as a rule there is a variable period during which the cells remain more or less motionless in the diluted suspension. The nature of this activation by dilution is obscure, but it leads to Gemmill's important contribution—viz. that a concentrated sperm suspension will maintain its powers of fertilising eggs longer than one which has been diluted with sea water. If we assume that the metabolic reserves of the sperm are strictly limited and that dilution of a suspension stimulates mechanical movement, then the natural inference is that dilution shortens the effective life by allowing the cells to dissipate their available energy more rapidly than under conditions where movement is less active. This view was supported by the experiments of Cohn (1918). Working with comparatively dilute suspensions of *Arbacia* sperm, Cohn confirmed the fact that the more dilute the suspension the more rapidly did it lose its power of fertilising eggs (Table LXIV).

It is well known that acids inhibit all known types of vibratile movement (see p. 453) and, since Cohn found that the more concentrated is the sperm suspension the higher is its hydrogen ion concentration, it seemed feasible to suppose that the preservation of fertilising power in the concentrated condition is due to the effect of the CO_2 produced by the sperm themselves. Cohn found that the rate of loss of fertilising power is decreased if the area of suspension in contact with air is decreased or if the concentration of the CO_2 in the air is increased (Table LXV).

By estimating the output of CO_2 from the changes observed in the pH of the suspensions, Cohn concluded that the total output of energy during the whole life of a spermatozoon is unaffected by the concentration of the suspension. The rate at which this energy is expended is, however, dependent on the hydrogen ion concentration. In normal sea water, or in a very dilute suspension, the spermatozoa are very active and, since they are thus expending their energy relatively quickly, they soon lose their power of active movement and of fertilising eggs. In more concentrated suspensions the CO_2 generated by the cells raises the hydrogen ion concentration and this

THE GERM CELLS

slows down the rate of expenditure of energy, thereby increasing the length of their effective life.

That accumulated CO_2 will depress the activity of spermatozoa there can be no doubt, but it would seem that the concentration effect

Table LXIV

Age of sperm in hours	Concentration of sperm suspension in percentage of undiluted sperm in 100 c.c. sea water			
	4	1	0·5	0·25
	Percentage of eggs fertilised when 1 drop of sperm was added to eggs in 10 c.c. sea water			
14·2	100	98	67	10
23·6	100	98	15	0
47·0	100	0	0	0
71·9	98	0	0	0
92·0	85	0	0	0

Table LXV

Age of sperm		Approximate CO_2 tension of the air with which the surface of the suspension is in contact	
		0·3 mm.	0·1 mm.
		Percentage of eggs fertilised when 1 drop of sperm is added to 10 drops of eggs in 10 c.c. sea water	
(hr.)	(min.)		
0	15	100	100
1	30	100	100
7	15	100	100
9	35	100	85
13	35	100	77
22	30	90	19
25	30	92	32
29	00	93	25
32	00	81	1
35	50	20	0

originally observed by Gemmill is more complicated in its action than would appear at first sight from Cohn's observations. If accumulation of CO_2 is the sole means whereby a concentrated suspension prolongs the effective life of a spermatozoon, then removal of CO_2 from the suspension should reduce the effective life to the period characteristic of dilute suspensions. It was found by the author (Gray, 1928 a, b) that the uptake of oxygen by a suspension—and hence presumably the CO_2 production—was governed by a series of factors which does not appear in the analysis given by Cohn. If we take two suspensions of the sperm of *Echinus miliaris*, one of which contains 3 mg. of nitrogen equivalents of sperm (= approximately 0·25 c.c. undiluted

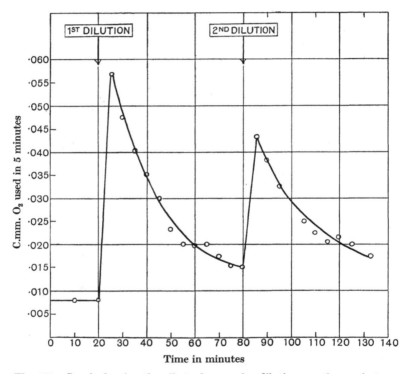

Fig. 162. Graph showing the effect of successive dilutions on the respiratory activity of a suspension of the spermatozoa of *Echinus*. (Gray, 1928 a.)

sperm) in 5 c.c. sea water and compare the rate of oxygen consumption per mg. nitrogen with that of the same amount of sperm in a more concentrated suspension (12 mg. nitrogen per 5 c.c.), the more dilute suspension is found to exhibit a much higher rate of oxygen consumption per unit quantity of sperm than does the more concentrated suspension (fig. 164). It is important to note that these observations were made on suspensions from which the CO_2 was

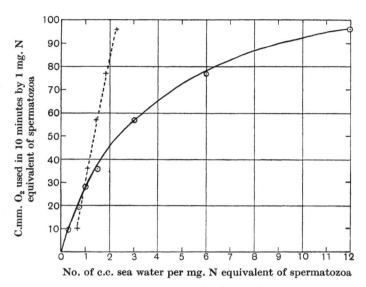

Fig. 163. Graph showing the respiratory activity of a unit amount of the sperm of *Echinus* when diluted with sea water. The dotted line shows the activity plotted against the cube root of the dilution. (Gray, 1928 a.)

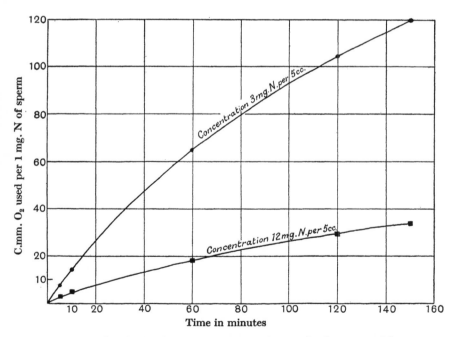

Fig. 164. Graph showing the effect of dilution on the O_2 consumed by a unit quantity of the sperm of *Echinus miliaris*. (Gray, 1928 a.)

being constantly removed by agitation with CO_2 free air. Attempts to raise the level of the respiratory activity of concentrated suspensions to that characteristic of those which are more dilute have invariably failed—and it would appear that a complete activation of individual spermatozoa is only possible in a comparatively dilute suspension where each cell has an adequate space in which to execute its mechanical movements. It should be noted that this factor is not likely to operate in very dilute suspensions such as were investigated by Cohn, but is very obvious in suspensions such as are normally found in the cavities of the male genital organs. In the case of *Echinus* the greater mechanical activity of sperm which follows dilution can usually be detected under the microscope and the gradual loss of fertilising power after dilution is roughly parallel to the decline in mechanical activity. How far the dilution factor affects the duration of active life in other cases is difficult to say, but it may play a part in the conditions investigated by Walton (1927) in mammals. Walton found that there was a significant decline in the percentage fertility of rabbits after artificial insemination with diluted semen. This might conceivably be comparable to the state of affairs in the trout or *Echinus* where dilution causes premature activity and results in an exhaustion of locomotory powers before the sperm reaches the egg. As Walton points out, conditions in the mammalia are complicated; in dilute suspensions the total number of sperm reaching an egg is probably small and dilution has to be effected by the addition of an artificial medium which may injure the cells. Direct observations on the effect of dilution on the period of active life of mammalian sperm have recently been provided by Young (1929 b) for the guinea pig (see Table LXVI); it will be noted that the effect is substantially the same as on echinoid sperm.

The survival of spermatozoa in dilute suspensions is clearly associated with the ability or non-ability of the cells to obtain energy from external sources. When spermatozoa are shed into an environment of sea or fresh water, they must presumably rely on intracellular sources of energy; as long as they are in the body fluids of the male or female parents they may, like other cells, be capable of obtaining energy from external sources. If such an absorption were possible from the medium in which fertilisation occurs, there is no obvious reason why spermatozoa should not possess an active life comparable to that of other cells; at any rate, the period of life would not be limited by the total energy available. Unfortunately, very little accurate information appears to exist concerning the mobility of sperm within the male genital system. Hammond and Asdell (1926) have shown that as long as they are within the male tract the spermatozoa of the rabbit will retain their fertilising power for as long as 38 days, whereas after injection into the body of the female the power of fertilising eggs is lost after 30 hours. The loss of fertilising power in the female tract might be due to one or more causes. If the spermatozoa are more active in the female than in the male, they may exhaust their supplies of energy and so fail to be sufficiently active to fertilise the eggs; on the other hand, we can look upon the external conditions in the female tract as a relatively toxic environment which sooner or later destroys the sperm,

Table LXVI. *Effect of dilution on duration of motion of guinea-pig sperm (46° C.) (From Young)*

Exp. no.	Parts of Locke's solution to undiluted sperm suspension							
	Un-diluted	1:1	2:1	3:1	6:1	7:1	13:1	24:1
	Min.	Min.	Min.	Min.	Min.	Min.	Min.	Min.
1	105	105	—	105	—	70	—	—
2	100	120	—	110	—	60	—	25
3	105	—	85	—	55	—	—	—
4	90	—	60	—	28	—	25	—
5	80	—	70	—	50	—	30	—
6	70	—	40	—	30	—	20	—

although they may still possess ample supplies of energy for movement. It is interesting to note that a similar analysis is applicable to fish spermatozoa (Huxley, 1930; Walton, 1930 b).

Reverting to the spermatozoa of invertebrates, it will be recalled that Cohn's results suggested that motile life is limited by the amount of energy available within each individual cell. Attempts to confirm this conclusion by the author (Gray, 1928 b) failed to give clearly defined results. Using suspensions of *Echinus* sperm from which the CO_2 generated was constantly removed, the rate of decay in the activity of the suspension (as measured by the rate of consumption of oxygen) failed to conform to any simple quantitative law. In the first place, it was found impossible to estimate the total oxygen used by any given suspension, since a small but measurable oxygen consumption persisted up to a point where other oxidative changes of a post-mortem nature obviously set in. This fact, recently confirmed by the data of Carter (1930), makes it impossible to give any clear experimental proof of Cohn's main conclusion: presumably, the amount of energy expended by a spermatozoon is limited, but its amount has yet to be measured quantitatively. The natural and irreversible decay in the activity of a sperm suspension may be due to a variety of causes (Gray, 1928 b), and the facts are probably best described in terms of the statistical variability in the viability

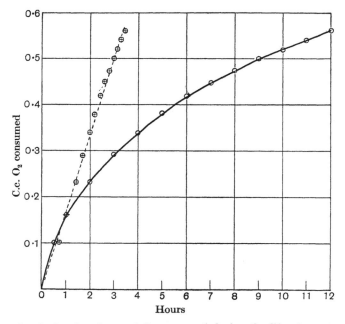

Fig. 165. Graph showing the total O_2 consumed during the life of a suspension of spermatozoa of *Echinus miliaris*. The dotted line shows c.c. O_2 plotted against the square root of the age of the suspension.

of individual units without any reference to the immediate causes which lead to the death of these units (Gray, 1931). It is of interest to note that Young (1929 b) concludes that mammalian spermatozoa actually undergo a process of irreversible decay whilst still within the epididymis.

The survival time of sperm suspensions is more than an academic problem, for in the case of mammals and possibly also fish, it is often of economic importance to effect artificial insemination where a natural mating is either impossible or impracticable. Walton (1930 *a*) has shown that if the spermatozoa are removed from the *vas deferens* of a male rabbit under aseptic conditions in such a way as to prevent contamination with air, the survival time of the suspensions is largely determined by the temperature. At body temperature (40° C.) the maximum survival time (as estimated by fertilising power) is about 13 hours. Above this temperature, the spermatozoa are rapidly destroyed—whereas lowering the temperature to 15° C. prolongs the life to 7 days: below 15° C. the survival time again falls (see fig. 166). It is interesting to note that as drawn from the *vas deferens* the sperm was found to be motile *prior* to being injected into the female (see p. 409).

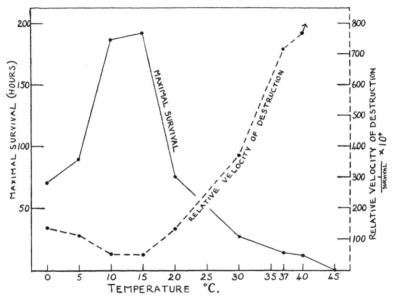

Fig. 166. Maximal survival times and velocity of destruction of the spermatozoa of the rabbit as functions of the temperature. (From Walton.)

Hammond's (1930) experiments on the rabbit differed from those of Walton in that the original spermatozoa were taken from the vagina of a freshly impregnated doe and not from the *vas deferens* of a male. Under such conditions the periods of survival at different temperatures were lower (see Table LXVII).

It is significant that in Hammond's experiments the percentage of sperm found to be motile at room temperature after exposure for variable periods to higher or lower temperatures was found to be a fairly good index of the ability of the suspension to fertilise eggs in the normal way.

To some extent the reactions of spermatozoa to changes in their environment are precisely parallel to those exhibited by other types of cells. As in the case of vibratile cells in general, prolonged movement is dependent on the presence of calcium ions and on a critical minimum concentration of hydroxyl ions (Gray, 1920 a, 1922 b), and in respect to their orientation in an electric field spermatozoa behave as a system of negatively charged particles.

In other respects, however, certain types of sperm show more specific characters. When a sample of sperm drawn from the testis of a ripe *Echinus* is placed in sea water, an immediate outburst of mechanical activity is observed. If the testis is not fully ripe, however, there may

Table LXVII

Temperature	Survival period in hours	
	Sperm from epididymis under paraffin	Sperm from vagina in contact with air
35	14	14
10	168	96
0	60	16

be a significant latent period after dilution before movement reaches its full level. If, however, such sperm is diluted not with fresh sea water but with sea water which has been in contact with ripe unfertilised eggs, the latent period of quiescence is either reduced or abolished altogether. This fact raises the question of how far the natural medium of fertilisation exerts a definite activating influence on the spermatozoa apart from the mechanical effects produced by dilution. Even on sperm which is fully activated in pure sea water (e.g. *Echinus miliaris*), egg secretions exert a definite action, for they maintain the metabolic activity of the cells at its original level, whereas in normal sea water the latter rapidly declines (fig. 167). The nature of the sperm-activating principles in egg secretions of *Echinus* has been investigated by Carter (1930), who finds that to some extent their action is comparable to that exerted by thyroxin and other iodine compounds. On the other hand Clowes and Bachman (1921) found that the activating principle in the egg secretions of *Asterias* and *Echinarachnius* are volatile and that their effects are comparable to those induced by the higher alcohols.

When a suspension of spermatozoa is highly active, it is not uncommon to find that the cells no longer distribute themselves uniformly throughout the medium but tend to group themselves together to form microscopic or macroscopic clusters. This is the

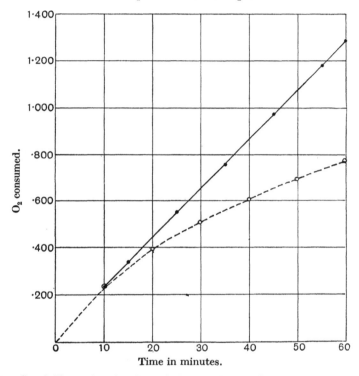

Fig. 167. Graph illustrating the effect of egg secretions on O_2 consumption of *Echinus miliaris* sperm. The unbroken line shows the O_2 consumed by a suspension in the presence of egg secretions; the broken line shows the O_2 consumption of an identical suspension without egg secretions. (After Gray.)

case with the spermatozoa of *Nereis* (F. R. Lillie, 1913). When fresh spermatozoa of *Nereis* are mixed with sea water they rapidly aggregate into clumps of actively moving cells whose size gradually decreases again as the activity of the spermatozoa declines. More than one explanation has been advanced for this phenomenon; Cohn (1918) regards the process of aggregation as the direct result of CO_2 inhibition, whereas Lillie rejects this suggestion. Lillie's explanation is that each spermatozoon exerts a chemotactic effect on its neighbours by virtue of the CO_2 which it liberates; such chemotactic effects are not at all uncommon in flagellate or ciliate protozoa (Fox, 1920; Saunders, 1924). It seems reasonable to suppose that in each case the phenomena are due to an active orientation of the cells to a specific concentration of hydrogen ions (see Gray, 1922 a).

The formation of clusters or aggregates of active spermatozoa readily occurs when the sperm of certain echinoderms is exposed to sea water containing egg secretions (Buller, 1902; F. R. Lillie, 1913). If the sperm are such that they are not highly motile in normal sea water, the first effect of the egg water is the production of a state of intense mechanical activity; within a few seconds clusters of highly motile sperm begin to form in a way which closely resembles the effect of CO_2 on *Nereis* sperm. In some cases the aggregated condition is temporary, for as the sperm begin to be less active the size of the clusters is reduced and the number of isolated feebly moving cells is increased. In other cases, however, aggregation is followed by true agglutination in which the cells adhere firmly to each other after they have ceased to move. The significance to be attached to these reactions must be deferred to a later stage (see p. 433), but it is interesting to note that the reactions of a given type of spermatozoon and egg water are seldom if ever entirely specific. Loeb (1914, 1915) found that the egg water of *Strongylocentrotus franciscanus* will not cause aggregation of the sperm of *S. purpuratus*, whereas the egg water of *S. purpuratus* will agglutinate the sperm of *S. franciscanus*. Just (1919) found that egg water of *Arbacia* will agglutinate the sperm of *Echinarachnius*, whereas the reverse combination is without effect.

The agglutinating principle in egg secretions is a colourless substance of high molecular weight, for it will not pass through a Berkfeld filter and is non-dialysable. According to Glaser (1914 a), it fails to give positive tests for proteins, although Woodward (1918) claims that it is removed from solution by the salting-out action of ammonium sulphate. Whatever be its nature, it appears to differ materially from the activating principle already described. The immediate interest of the sperm agglutinant lies in its possible rôle in the process of fertilisation, for F. R. Lillie's conception of the union between a spermatozoon and an egg is based on the assumption that this union is effected by the mechanism which in some cases can be shown to be responsible for the union between one spermatozoon and another. For this reason Lillie ascribed the name of 'fertilizin' to sperm-agglutinating principles. For a full discussion reference should be made to the original papers or to Lillie and Just (1924).

The egg cell

The egg cells of nearly all the higher animals differ morphologically and physiologically from all the other cells in the body. Typically, the egg is much larger than a somatic cell, and both cytoplasm and nucleus exhibit specific characteristics. The primitive egg cells or *oogonia* can often be recognised at a very early stage in the life history and, as in the case of the spermatogonia, their subsequent development has been studied in great detail (see Wilson, 1925, Chapter IV). For present purposes, it is unnecessary to recapitulate even a summary of the morphological facts, and we may look upon the egg almost exclusively as a physiological unit.

From this point of view an egg is peculiar in two respects. Firstly, it is essentially a totipotent system capable, after fertilisation, of giving rise to a complete new individual. This property is not absolutely specific, for we have seen that in sponges, coelenterates and other forms, aggregates of somatic cells are capable of forming new individuals: as far as is known, however, the ovum (in the metazoa) is the only type of isolated cell which can normally produce a complete organism. Secondly, as the developing egg is often isolated from all external sources of organic material, it must contain within itself all the ingredients requisite for the production of new tissue: we find, in fact, that the cytoplasm of a typical egg contains a much larger percentage of nitrogenous and other deposits than does any other type of cell.

When a young oogonium begins its development, it is usually possible to recognise three distinct phases before the production of a new organism begins. Firstly the cell grows rapidly in size—often reaching a bulk many thousands of times that of typical somatic units; it is during this growth phase that the egg accumulates the *yolk* reserves which will afterwards be used for the manufacture of new cells and tissues. The growth phase is clearly one of great metabolic activity, for there is an increase in the size of the nucleus, and in the bulk of the cytoplasm, as well as a deposition of the yolk. So far our knowledge of these processes is limited to morphological observations, and it is by such means that nuclear growth and yolk formation have been extensively investigated. On the completion of growth, a pause may or may not be interposed in the developmental history before the onset of the second main phase of *maturation*. The most obvious changes effected by maturation are concerned with the nucleus, but there may also be simultaneous changes in the cytoplasm (see p. 95). In some cases maturation is complete before fertilisation, whereas in others maturation is only completed after the spermatozoa has entered the egg.

Nothing is known of the cause which incites the young oogonium to grow or to deposit yolk. In every case, however, there must be an active secretion of nitrogenous bodies. The most usual form of nitrogen storage is the deposition of proteins—which may either be in the solid form (as in amphibian, or elasmobranch eggs) or in the fluid state (many teleostean eggs, e.g. *Salmo*). Gatenby (1922) and Nath (1925) have shown that nitrogenous yolk is not infrequently associated with the extrusion of substances from the nucleus into the cytoplasm. Non-nitrogenous yolk may exist as carbohydrates or as fat. In the eggs of *Ascaris* and some molluscs glycogen is abundant, whereas in other forms (*Nereis*, fish) large oil globules of a fatty nature are very distinct constituents of the cytoplasmic secretions. The chemical changes whereby these secretions subserve the requirements of the developing egg hardly enter into the field of cytology—but reference may be made to Fauré-Fremiet (1925) and to Needham (1931).

Usually the life of a ripe unfertilised egg is considerably longer than that of an active spermatozoon, particularly if the temperature is low or if the egg is deprived of oxygen (Loeb, 1913). Cases are known, however, where the effective life of the ripe egg is very short. Just (1915 b) showed that in order to effect successful artificial insemination of *Platynereis megalops* the eggs and spermatozoa, like those of the trout, must be mixed without previous contamination with sea water. In *Platynereis*, however, it is the eggs which suffer by washing and not the sperm; for if the former are washed in sea water and the water be then removed, no fertilisation is effected by the addition of fresh dry sperm. If the reverse experiment is performed, the previously diluted spermatozoa will fertilise fresh dry eggs after removal of the excess of water. Just regards this loss of fertilising power as due to the loss of an essential 'fertilizin' from the eggs (see p. 421).

The fusion of egg and spermatozoon

There is no evidence which suggests that an animal egg invariably secretes a substance which attracts neighbouring spermatozoa, although in those cases in which the eggs produce a substance which stimulates the male gametes to a higher degree of mechanical activity, a limited amount of chemotactic movement may occur (Dakin and Fordham, 1924; Just, 1930). As far as one can see, contact between sperm and egg is usually fortuitous, so that the chance of fertilisation will depend on the number of spermatozoa and eggs present. This does not imply that the egg is necessarily fertilised by the first spermatozoon with which it comes into contact: probably several collisions occur before the effective spermatozoon reaches the egg. Glaser (1915) observed that this is often the case in *Arbacia* eggs. Quite clearly if a spermatozoon is to fertilise an egg, the two cells must remain in contact with each other after collision. It is generally supposed that the sperm head is adhesive in the sense that after contact the sperm can only be separated from the egg surface by the application of a force which is greater than that which can be exerted by the tail of the spermatozoon. Subsequently, of course, the sperm head is engulfed into the body of the egg cell. The general series of events is not unlike the process of phagocytosis—in which the inanimate particle is replaced by an active spermatozoon. There are, however, two features peculiar to fertilisation; (i) there is a high degree of specificity, (ii) one and only one spermatozoon is usually incorporated into the egg.

If we look upon the opening phase of fertilisation as equivalent to physical adhesion, the problem of specificity resolves itself into a study of specific adhesion between surfaces: at present, this point of view has not led to any very hopeful results, but it is interesting to note that specific adhesion has been observed in the case of red blood-corpuscles (Tait, 1918). At the same time, the attachment of the sperm to the egg initiates far-reaching changes at the egg surface, for within a very short space of time the whole surface is altered in such a way as to prevent adhesion and penetration by other spermatozoa. The physiological nature of these changes is obscure, although they are followed by visible alterations at the periphery of the egg (see below).

In many cases a sperm is able to enter the egg from any point on the egg surface (sea urchins, starfish), even if the egg is surrounded by a relatively tough membrane. Although spermatozoa appear to burrow through such external membranes in considerable numbers, only one effective sperm usually enters the egg cell itself. In other cases, spermatozoa can only enter the egg at one point (the *micropyle*) where there is a definite opening in the external egg membrane (insects, fish). It is interesting to note that in the nemertine *Cerebratulus* the spermatozoon can reach the egg at any point, although a well-defined micropyle is present (Wilson, 1925). The 'burrowing' powers of spermatozoa are often very remarkable; Cragg (1920) describes the migration of the sperm of *Cimex* through chitin. An anomalous method of penetration through egg membranes has recently been described by Chambers (1923) in *Asterias* —where the sperm adheres to one of the fine protoplasmic filaments which protrude from the egg surface, and is passively drawn into the egg by the active contraction of this filament. The description given by Chambers has recently been criticised by Just (1929), who attributes the formation of protoplasmic filaments on the egg surface to abnormal conditions of fertilisation or to moribund egg cells; this conclusion is not accepted by Chambers (1930).

Visible phenomena of fertilisation

The visible phenomena of fertilisation have often been described. Little can be added to the accounts given by Wilson (1925) and by F. R. Lillie (1919), and the following paragraphs are therefore largely restricted to points of theoretical interpretation.

Many invertebrate eggs on contact with an effective spermatozoon exhibit two simultaneous changes in surface structure: (i) at the point of contact with the sperm, the egg surface is protruded to form a small conical projection—the *fertilisation cone*; (2) starting at the point of contact a definite *fertilisation membrane* is rapidly pushed outwards over the whole egg surface. Although these two phenomena occur together they can be regarded both morphologically and physiologically as separate processes.

As far as the fertilisation cone is concerned, it is important to note that, in certain cases, it is formed before the egg cortex is in actual contact with the sperm head (see, however, Just, 1929). Thus in *Nereis*, 'a transparent fertilisation cone arises from the inner wall of the perivitelline space opposite the attached spermatozoon and extends across the space until it touches the membrane at the point of attachment of the spermatozoon. The perforatorium of the spermatozoon pierces the vitelline membrane and becomes imbedded in the cone. These phenomena occupy about fifteen minutes' (F. R. Lillie, 1919, p. 53). It is not easy to visualise the type of physical change responsible for the protrusion of the fertilisation cone, although the behaviour of the egg surface recalls to same extent the formation of food-cups in phagocytosis. Even when the fertilisation cone seems to form after the egg surface and the sperm head are in contact, it is hardly possible to believe that a localised disturbance of surface energy is alone responsible for the formation of the cone—for it must be remembered that the surface of the egg and probably also that of the sperm is of a rigid nature—and this rigidity must be lost before the sperm can be drawn into the fluid interior of the egg (Gray, 1922 a).

Concerning the formation of the fertilisation membrane, a considerable divergence of opinion has existed. There can be no doubt that the membrane in its final form has different properties to any membrane present prior to fertilisation: the fertilisation membrane of a sea urchin's egg is a relatively tough and elastic structure which cannot be identified as such prior to fertilisation. There remain two alternatives: (i) the membrane is an entirely new structure formed subsequently to fertilisation, (ii) it is a modification of a membrane which is present before fertilisation. The first view was held by Harvey (1910, 1914), and is supported by the fact that if ripe sea urchin eggs are shaken into pieces, and the pieces subsequently fertilised—they will all form normal fertilisation membranes. This conception was supported by Loeb (1913), McClendon (1914), and Gray (1922a). It was at one time suggested that the fertilisation membrane is a precipitation membrane due to the interaction of the gelatinous envelope with an oppositely charged exudation from the egg surface. Both McClendon and Gray believed that no fertilisation membrane would form if the gelatinous membrane was removed prior to fertilisation. In point of fact, this conclusion was based on an error in experimental technique, for Hobson (1927) has shown conclusively that the failure to form fertilisation membranes observed by Gray was due, not to the removal of the gelatinous envelope, but to the presence of hydrogen ions which had been used to dissolve this envelope prior to fertilisation. It is clear that the formation of a fertilisation membrane is not dependent on the presence of the gelatinous capsule. According to Kite (1912) and Chambers (1921), it is possible to dissect away from the surface of the unfertilised egg a definite membrane, and if the egg is then fertilised, no fertilisation membrane forms. It looks as though membrane formation represents the elevation of an existing membrane with accompanying changes in its mechanical properties. The extent to which the fertilisation membrane is elevated from the egg surface depends on more than one factor. Loeb (1908 d) showed that the membrane will collapse if proteins are present in the external medium, and from this he concluded that it was impermeable to large molecules and ions. Since the osmotic pressure exerted by solutions of albumen in sea water must be comparatively low, the force required to distort the membrane even in its final tough state must be comparatively small. If, before elevation, there is set free between the membrane and the egg surface a colloid possessing an adequate osmotic pressure, the membrane would be pushed out from the egg surface. Direct evidence of the existence of such a colloid is lacking, but it is significant to note that its existence would account for the fact that the degree of elevation depends on the concentration of hydrogen ions present in the external medium, since the osmotic pressure of a colloid on the alkaline side of its isoelectric point would be depressed by the addition of hydrogen ions. The phenomena observed in *Nereis* (F. R. Lillie, 1911) are somewhat different from those in the sea urchin egg, but are not detrimental to the above interpretation. In *Nereis* the unfertilised egg is surrounded by two layers: (i) a delicate vitelline membrane, (ii) an alveolar layer or *zona radiata*. On fertilisation, fluid from the alveoli of the *zona radiata* passes through the vitelline membrane, and swells up to form a layer of jelly outside the egg. In this way the *zona radiata* is destroyed and remains as the perivitelline space. It looks as though the vitelline membrane were permeable to the colloids liberated by the *zona radiata*, and consequently the vitelline membrane is not lifted away from the egg surface as is the case in sea urchins.

Considerable light was thrown on the mechanism of membrane formation in echinoderm eggs by the discovery that it can be evoked in unfertilised eggs by cytolytic agents such as saponin, soaps, xylol, or chloroform. We know that the elevation of the membrane only occurs in the absence of a critical concentration of hydrogen ions (Gray, 1922 a; Hobson, 1927) and that it will not occur if proteins are present in the external medium. The simplest picture of these facts is provided by the hypothesis that an active spermatozoon or a cytolytic agent alters the cortical layers of the egg in such a way as to liberate a negatively charged colloid whose osmotic pressure is sufficient to overcome the elastic resistance of the membrane lying at the egg surface. It must not be forgotten, however, that during this process the membrane itself undergoes a physical change and becomes thicker and tougher as long as free calcium ions are present (Gray, 1922 a).

The fate of the male nucleus after its absorption into the substance of the egg has been described in detail by F. R. Lillie (1919), and by Wilson (1925), and the essential facts are summarised by Doncaster (1924). From a physiological point of view the movement of the sperm head within the cytoplasm of the egg presents a perplexing problem. The automatic locomotory activity of the sperm is undoubtedly abolished before engulfment, and in most cases the tail does not enter the egg. In order to move against the viscous resistance of the cytoplasm the nucleus must be provided with a definite amount of energy. Various theories concerning the source of this energy have been propounded. Roux (1887) suggested that the two pronuclei attracted each other, whilst others have postulated an amoeboid activity on the part of the nuclei. Alternatively, the male pronucleus has been denied any power of active movement. Chambers (1917) holds that the propulsive unit is the male aster which mechanically forces the two pronuclei into mutual contact (see p. 160).

Physiological effects of fertilisation

In 1895 Loeb showed that fertilised marine eggs fail to develop if deprived of oxygen; and subsequent observations have substantiated his conclusion. Seven years later, the same author (see Loeb, 1913) found that if unfertilised eggs are deprived of oxygen or exposed to a $0.0005M$ solution of KCN in sea water the eggs remained healthy and capable of fertilisation for a much longer time than is the case in normal sea water. Both before fertilisation, and after, the activity of the egg is therefore closely associated with its ability to absorb oxygen. That the nature of this association is altered by the act of fertilisation was first shown by Warburg (1908), who measured the oxygen consumption of *Arbacia* eggs before and after fertilisation; he found that after insemination the oxygen absorbed was six or seven times that prior to fertilisation. Essentially similar results were recorded by Loeb and Wasteneys (1911 a), and more recently Shearer (1922 a) has shown that this significant increase occurs immediately the sperm has effected the visible changes associated with membrane elevation. If we regard the intensity of respiration as a measure of vital activity, the act of fertilisation converts the sea urchin egg from a relatively inert unit into one of intense metabolic change. Before accepting this conclusion, however, it is well to remember that no such change occurs in the eggs of *Asterias* (Loeb and Wasteneys, 1912). Loeb and Wasteneys (1911 *b*) showed that even in *Arbacia* the rate of development of the egg is not directly proportional to the oxygen uptake, for if the temperature is raised, the effect on the rate of oxygen absorption is not the same as on the velocity of segmentation. As already pointed out (p. 30) it is not easy to understand the necessity for the high oxygen uptake by the fertilised egg, unless we postulate membranes or machinery whose existence depends on a constant supply of free energy. At one time Loeb was inclined to associate oxygen uptake with nuclear activity, but as we have seen the evidence against such a view is fairly strong.

Just as the oxygen uptake rises after fertilisation, so the heat production also rises (Meyerhof, 1911; Shearer, 1922 *b*). In both cases it is necessary to distinguish between the large and sudden increase which occurs immediately after fertilisation, with the later and slower increment which occurs when the egg begins to grow (see Chapter XI).

In 1909 R. S. Lillie drew attention to the fact that fertilisation alters the relationship between the egg and its external medium, and suggested that the essential act of fertilisation is accompanied by a temporary excitation of the cell surface, just as a nerve cell exhibits its activity by the existence of an excitation wave of electrical potential. Lillie suggested that the initial action of the sperm produced a temporary condition in which the egg surface was more permeable to ions than it was before fertilisation: this condition was, however, transient and was gradually replaced by one of renewed impermeability. Direct tests of this hypothesis have seldom been attempted, although Miss Hyde (1904) reported that a wave of negative potential passes over the surface of *Fundulus* eggs when fertilisation occurs. Lillie's own data are based on the activating effect of those substances which are known to increase cell permeability, and on the antagonistic effect of substances which act as inhibitors to such changes. If this conception is true, then the initial state of increased permeability must be of much longer duration than is characteristic of excitable tissues such as muscle or nerve, for eggs can be excited to develop by initial treatment which lasts for many minutes before it becomes essential to apply an inhibiting reagent (see p. 439). In 1910 McClendon showed that after fertilisation sea urchin eggs conduct electricity more readily than before fertilisation, and in 1916 Gray attempted to confirm Lillie's hypothesis by direct methods such as McClendon had developed. Any process capable of initiating the cortical changes characteristic of fertilisation resulted in an increased conductivity, but no good evidence was forthcoming which suggested that this condition was transient as Lillie suggested. Using more refined methods Cole (1928) failed to confirm an invariable increase in conductivity such as was described by McClendon and by Gray. These authors were, of course, measuring the conductivity by means of a simple Kohlrausch bridge, whereas Cole employed a much more satisfactory technique and one which did not involve the aggregation of the cells into a closely packed condition. Cole concluded that fertilisation tends to stabilise the permeability of the egg surface to ions rather than to increase it either temporarily or permanently.

On the other hand there can be little doubt that after fertilisation water can pass into an echinoid egg with greater ease than before fertilisation (R. S. Lillie, 1916, 1917). When *Arbacia* eggs are exposed to hypotonic sea water the total amount of water which enters the egg is unaffected by fertilisation, but the rate at which the new equilibrium is reached is considerably higher after fertilisation than before, indicating that a change has occurred in the nature of the cell surface rather than in the amount of osmotically active substances within the cell.

Specificity in fertilisation

To a large degree an effective union between an egg and a spermatozoon only occurs when both gametes are of the same species; and the possibility of producing fertile offspring is often used as a criterion of species. There are, however, many instances known in which the sperm of one species will enter the egg of another species and excite the latter to divide.

Owing to the ease with which their eggs can be artificially inseminated, echinoderms provide peculiarly useful material for the study of cross fertilisation. Using eight different types of echinoids Vernon (1900) found that out of forty-nine different matings, twenty-nine resulted in the production of larvae.

Interspecific hybrids of echinoderms have been described by Shearer, de Morgan, and Fuchs (1913). By using a suitable concentration of sperm of *Echinus esculentus*, *E. acutus* and *E. miliaris* it was found possible to fertilise eggs of any of these forms by the spermatozoa of any other and in every case larvae were obtained. Many of these hybrid larvae, however, failed to metamorphose.

The most complete breakdown of specificity appears to occur in teleost fishes. According to Moenkhaus (1904, 1910), any teleostean egg can be fertilised by the sperm of any other species and Newman (1908–15) described widespread cross fertilisation between different species and genera of these fish. As is general in such experiments, a given species of egg was found to be more readily fertilisable by homogenic sperm than by sperm of a foreign species and the larvae so obtained were more viable.

The ability of two species to cross fertilise is often more or less irreciprocal; for example, the sperm of *E. miliaris* enters the eggs of *E. acutus* with a greater frequency than does the sperm of *E. acutus* into the eggs of *E. miliaris*. Similar observations are well known not only among the generic crosses of echinoderms (Fischel, 1906; Vernon, 1900; Baltzer, 1910; Tennent, 1910) but also in *Amphibia* (Born, 1883, 1886). Fischel (1906) found that the sperm of *Arbacia* will fertilise *Strongylocentrotus* eggs, but the sperm of *Strongylocentrotus* will not fertilise *Arbacia* eggs.

It is significant to note, however, that the block to fertilisation is often not absolute, but can be overcome by a variety of artificial means. Loeb (1903) found that the sperm of *Asterias* will not fertilise the eggs of *Strongylocentrotus* in ordinary sea water, but will do so if the sea water is made slightly hyperalkaline. Similar means of effecting fertilisation have often been employed in other cases with success (see Baltzer, 1910). The ability of a spermatozoon to enter the egg of a different species may also be influenced by the time which has elapsed after the egg has been laid. Thus the fresh eggs of *Hipponoë* or *Toxopneustes* are not fertilisable by the sperm of *Ophiocoma* or *Pentaceros*, whereas cross fertilisation occurs fairly regularly after the eggs have remained in sea water for two or three hours (Tennent, 1910). Similarly the eggs of *Strongylocentrotus* can be fertilised by the sperm of *Mytilus* (Kupelwieser, 1906, 1909), if the eggs are treated with a high concentration of sperm for a prolonged period.

Although ripe sperm will always fertilise the eggs of the same species if the form used has separate sexes, there are a number of cases known where no fertilisation occurs if both gametes are derived from a hermaphrodite individual. The best known case is that of the ascidian *Ciona*. As originally investigated by Castle (1896) and by Morgan (1904) the eggs of *C. intestinalis* appeared to be unfertilised by the sperm of the same individual, although readily fertilisable by sperm from another individual. Fuchs (1914, 1915) working at Naples found, however, that the degree of self-sterility in *Ciona* varied with the quantity of sperm used (see Table LXVIII).

Self-fertilised eggs seldom developed normally. More recently Morgan (1923) found that the barrier to self-fertilisation is located in the membrane which surrounds the egg rather than at the surface of the egg itself. The mature egg is surrounded by a number of

Table LXVIII

	Percentage of eggs fertilised			
	Self-fertilisation		Cross fertilisation	
	5 drops sperm	4 c.c. sperm	5 drops sperm	4 c.c. sperm
Specimen A	0	58	100	100
,, B	0	22	100	100
,, C	12	100	100	100
,, D	0	56	100	100

follicle cells which are bounded on the outer side by a thick membrane. After the egg is laid, it shrinks in size, so that the follicle cells lie freely in a space between the egg and the outer membrane. Morgan found that if the membrane be removed together with the follicle cells, the naked egg is readily fertilisable by the sperm from the same individual and concluded that the barrier to fertilisation is possibly due to the depressant effect of the perivitelline fluid on the activity of the sperm.

The phenomena of hybridisation are of considerable interest in the interpretation of the facts of normal fertilisation. As far as can be seen, there are no general principles which can be used to indicate whether or not a particular cross fertilisation will succeed or not; consequently, it is not easy to see how the union between an egg and a spermatozoon can depend upon the presence or absence of a specific type of substance or molecule. If, on the other hand, a spermatozoon activates an egg by setting up at its surface a critical but non-specific degree of physical potential between the point of contact and the remainder of the egg surface, it is not so difficult to understand how the sperm of a totally different species or phylum can at times activate a foreign egg, and how the possibility of activation may be influenced by changes in external environment (see Gray, 1922 a).

Monospermy and polyspermy

Although a large number of active spermatozoa may burrow through the gelatinous coat of a sea urchin's egg, only one of them enters the cell under normal circumstances. Monospermic fertilisation of this type is characteristic of all small 'alecithal' eggs (sea urchins, worms, molluscs, and mammals). In the case of large and yolky eggs, however, a considerable number of spermatozoa may enter the ovum, although only one is incorporated into the zygote nucleus. The fate of the accessory spermatozoa varies (see Wilson, 1925): they may simply degenerate (insects, urodeles), or as in elasmobranch fish and birds, the accessory sperm nuclei may undergo active mitoses in the peripheral regions of the blastoderm before final resorption.

In sea urchin eggs, polyspermy can readily be effected by experimental means. O. and R. Hertwig (1887) showed that if these eggs are treated with dilute solutions of various drugs (nicotine, strychnine, morphine), or if the temperature is abnormally high, a number of spermatozoa will enter each egg. It is well known that polyspermy frequently occurs if the eggs are allowed to stand in sea water for some hours, or if they are in any way unhealthy. These facts are not only of significance in respect to the physiology of fertilisation, but also illustrate a point of very general significance. We can hardly believe that nicotine, high temperature, and hydrogen ions (Smith and Clowes, 1924) all act on the egg in precisely the same manner —yet as far as polyspermy is concerned their effects are identical: in other words, the response of the cell to an abnormal environment is not necessarily specific—but may be of a generalised nature. We have to look upon the cell as a system whose equilibrium is very easily upset in a variety of ways. Whatever be the point at which a disturbance starts, the end result may sometimes be the same. It is possible to postulate a series of linked or consecutive reactions in which each member is eventually dependent on the activities of the

THE GERM CELLS

others, but we must be careful not to correlate too readily the response of a cell with the specific nature of an applied reagent.

It is important to note that when more than one spermatozoon enters a sea urchin's egg, each male nucleus is capable of mitotic division (see fig. 53). The same phenomenon is observed in frogs' eggs (Brachet, 1911; Herlant, 1911).

It is by no means clear why only one spermatozoon will normally enter small alecithal eggs. Earlier workers suggested that an effective block to other sperm was to be found in the fertilisation membrane, but this can hardly be the case, since Driesch showed that mechanical removal of the fertilisation membrane did not allow additional sperm to enter. More recently Just (1919) has observed that after a sperm has attached itself to the surface of an egg of *Echinarachnius* no other sperm will enter even although it reaches a region of the egg from which the fertilisation membrane has not yet been elevated. It is certain that whatever be the nature of the monospermic mechanism it must come into operation within a very short space of time after effective fertilisation has occurred.

It will be recalled that F. R. Lillie has demonstrated the production of 'fertilizin' from ripe unfertilised eggs of sea urchins and of *Nereis* and has shown that this substance is not produced by fertilised eggs. Lillie regards this substance as an essential requirement for fertilisation, so that if it ceases to be formed as soon as an effective sperm touches the egg, no other sperm will enter if fertilizin has been removed from the sphere of action. From this point of view, monospermy is due to a sudden cessation in the production of fertilizin. For this reaction Lillie put forward a model based on Ehrlich's immunity chain reactions (see fig. 168), in which the removal of fertilizin is effected by an occupation of the spermophile receptor by 'anti-fertilizin' present in the egg, the anti-fertilizin being inactive until one sperm has linked itself to one unit of fertilizin. Lillie's conception has been criticised by Godlewski (1926), who questions the validity of regarding fertilizin as an amboceptor. It will be noted that Lillie's scheme is probably intended to be little more than a pictorial representation of the facts, and is helpful in so far as it leads to a more complete understanding of experimental data. Similar principles to those suggested by Lillie have been applied by Godlewski to sperm antagonism. In 1911 Godlewski found that if the sperm of *Chaetopterus* is mixed with a suspension of spermatozoa of *Sphaerechinus*, the mixture fails to fertilise the eggs of *Sphaer-*

echinus. As far as could be determined, one *Chaetopterus* sperm will totally inhibit the fertilising power of four *Sphaerechinus* spermatozoa, but the effect in part depends on the absolute concentration of both constituents. Godlewski (1926) regards these phenomena as

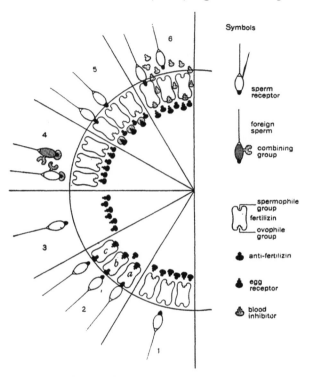

Fig. 168. Diagram illustrating Lillie's analysis of the mechanism of fertilisation. Sector 1, the arrangement of substances immediately prior to fertilisation. Sector 2, the mechanism of normal fertilisation. The sperm receptor unites with the spermophile group of the fertilizin and the egg receptors with the ovophile group of the fertilizin owing to the activation of the latter by the sperm (*a*). Molecules of the anti-fertilizin combine with the spermophile group of the adjacent fertilizin (*b* and *c*) and thus present polyspermy. Sector 3, inhibition of fertilisation by loss of fertilizin. Sector 4, antagonistic sperm action. The sperm receptors are occupied by groups cast off by the foreign spermatozoa. Sector 5, hypothetical blocking of egg receptors. Sector 6, inhibitory action of blood. The ovophile group is occupied by derivatives from the blood. (From F. R. Lillie, 1914.)

analogous to the agglutinating effect of foreign sera on blood corpuscles, but it is not altogether clear that mutual narcosis by CO_2 plays no part, since when dead *Dentalium* sperm is added to *Sphaerechinus* sperm the inhibitory effect is very markedly reduced.

THE GERM CELLS

The production of fertilizin from unfertilised eggs can be clearly demonstrated, as is also the fact that it is not produced by fertilised eggs. If we admit that this substance is essential for fertilisation, the facts can be described in chemical terms without reference to immunological symbols. We can postulate that if one unit of fertilizin reacts with a spermatozoon it will induce a secondary reaction which destroys all the remaining fertilizin in the egg. Since the egg is fertilisable at any point we must postulate a surface layer of fertilizin which is constantly changing by the outward diffusion of fertilizin into the water. Polyspermy can only be prevented by destroying this layer over the whole surface of the egg and there must therefore be a propagated disturbance which starts at the point of entry of the sperm and spreads over the whole egg surface at a rate which is sufficient to complete the destruction of fertilizin before another sperm can touch the egg surface. We can judge this conception in two ways: (i) we can use it as a definite working hypothesis and attempt to isolate the process of fertilizin destruction, and to follow experimentally the propagated wave over the egg surface, or (ii) we can accept it or discard it according to the extent to which it will, as a picture, faithfully portray the whole of the available facts. So long as we are dealing solely with the phenomenon of monospermy the facts can be expressed in at least three ways. We can use the nomenclature of immunology, of chemistry or of physics. In every case we are describing the same facts. In 1922 the author pointed out the difficulty of extending Lillie's original conceptions to the facts of cross fertilisation and artificial parthenogenesis, and urged that these difficulties are diminished if we replace immunological or chemical conceptions by those which are primarily concerned with the distribution of free energy. For example, if we regard the surface of an unfertilised egg as an equipotential surface which is excited at the point of entry of the spermatozoon, then a wave of negative potential will pass over the surface of the egg, if this surface is at all comparable to those of other excitable tissues. If we take a very moderate estimate of the speed of propagation from smooth muscle, then such a wave will complete its circuit over a sea urchin's egg in 0·00001 second. If, on the other hand, the egg is large or if it is subjected to those reagents which reduce the rate of propagation of bioelectric disturbances, the chance of polyspermy is increased. Similarly, if the essential act of fertilisation is to be traced to a localised change in distribution of free energy, it is not surprising to find that artificial fertilisation can be induced by mechanical puncture (see p. 443). Quite recently this conception has been criticised by Just (1928), who prefers to describe the facts in chemical terms. It is obvious that the essential difference between the physical and chemical conceptions of egg activation lies in how far we believe that specific substances must be present at the point of activation; if we are inclined to attribute activation to a particular distribution of specific molecules, then the redistribution of energy becomes of secondary significance. All chemical changes probably involve such a redistribution but a purely chemical conception of fertilisation is not intimately concerned with this fact. On the other hand, if we attribute primary significance to the distribution of energy at the point of contact between the egg and spermatozoon, then the chemical mechanism responsible for this distribution need no longer be specific. We must choose between a physical and a chemical nomenclature solely according to the relative significance which we are inclined to attribute to energy changes as such. If we regard the existence of a critical and localised potential as the essential element of activation, the means whereby this is established are immaterial and egg activation becomes allied to the stimulation of a muscle or a nerve as R. S. Lillie suggested twenty years ago.

The diphasic nature of natural fertilisation

When an egg is fertilised in the ordinary way, a series of events can be traced from the moment the effective spermatozoon reaches the egg until the whole egg embarks on the cleavage cycles which ultimately yield a new organism. The first changes occur at the egg surface and only at a later stage is the whole cell engaged in the processes of nuclear division and cytoplasmic cleavage.

By purely experimental methods it is possible to show that this long cycle of events can be divided into two natural groups, each of which is, to some extent, independent of the other. We know, for example, that the increase in respiratory activity characteristic of the fertilised echinoid egg reaches its full level soon after contact is established between egg and sperm (Shearer, 1922 a); similarly the cortical changes at the egg surface are completed long before the first cleavage cycle begins. By centrifuging the sperm away from the surface of a *Nereis* egg F. R. Lillie (1912 a) enabled us to distinguish those rapid and cortical changes which are the immediate result of fertilisation from those slower and subsequent changes which are dependent on the penetration of the male elements into the cytoplasm of the egg. In *Nereis* the sperm head does not penetrate into the egg until about forty-five minutes after the cortical changes are complete. By applying centrifugal force, Lillie found it possible to dislodge the spermatozoon from the egg surface. After return to normal conditions such eggs complete their maturation, but they do not segment. At the time when the zygote nucleus normally forms its first cleavage spindle, the female pronucleus of the centrifuged egg is clearly visible and appears to be of normal size; the nuclear membrane breaks down and chromosomes are formed in the normal manner; there is, however, in the absence of the sperm nucleus, no sign of spindle or of asters—and the chromosomes lie in a clear area of cytoplasm. Each chromosome splits longitudinally but the halves do not separate from each other. Finally, the chromosomes become scattered and degenerate as chromatic granules. As Lillie says: 'This experiment shows that the fertilisation process may be divided physiologically as well as morphologically into two main phases of external and internal phenomena. The external action is adequate to produce the cortical changes alone, but not the entire series of developmental events'. A similar conclusion may be drawn from the observations of Loeb (1913), who found that the sperm of *Asterias* can excite membrane formation on the eggs of *Strongylocentrotus purpuratus* in hyperalkaline water, although the sperm may not enter the eggs. Such eggs do not segment: if, by chance, the sperm does enter, normal segmentation follows. We can therefore assume that fertilisation effects two changes in the egg: (i) by a modification which starts at the cortex, the metabolism of the egg is increased to its full level: the egg is, in fact, activated; (ii) by penetration into the egg, the sperm nucleus with its attendant aster enables the egg to form the requisite machinery for normal cleavage. Further discussion of these two phases must be deferred until the facts of artificial parthenogenesis have been considered.

Artificial parthenogenesis

Although the main phenomena of artificial parthenogenesis have been known for thirty years they are none the less spectacular and none the less significant. Starting with Mead's observations in 1895 and culminating perhaps in Loeb's later work, we can trace step by step the discovery of a method whereby certain eggs can be induced to develop normally without the intervention of the male gamete. The history of the problem has been given by Loeb (1913) and by Morgan (1927), and is an interesting example of the way in which biological discoveries may be made, although the primary effects of the reagents employed are shrouded in complete mystery and where the methods used are entirely empirical.

The story starts with an attempt by Mead (1895) to excite cell division in unfertilised *Chaetopterus* eggs by treatment with those electrolytes which were known to affect the activity of a muscle fibre. Under normal circumstances the eggs of *Chaetopterus* give off their polar bodies only after fertilisation; Mead found that if the eggs are placed in sea water containing $\frac{1}{4}$–$\frac{1}{2}$ per cent. of potassium chloride the unfertilised eggs not only form their polar bodies but also form a polar lobe which is the natural preliminary to the first cleavage. The addition of potassium chloride to sea water not only altered the concentration of potassium but also altered the osmotic pressure. Shortly afterwards Morgan (1896) showed that if the unfertilised eggs of *Arbacia* are exposed to hypertonic sea water for a limited period, then on transference to normal sea water artificial asters are formed in the cytoplasm and these may result in the formation of a functional amphiaster (see p. 162). In the same year Loeb succeeded in obtaining swimming plutei by exposing the unfertilised eggs of *Arbacia* to 50 c.c. sea water + 50 c.c. $\frac{10}{8}M$ MgCl$_2$ for $1\frac{1}{2}$ to 2 hours and shortly afterwards the same author (Loeb, 1900) showed that the nature of the salt used is of minor importance; the essential factor concerned is osmotic pressure. In two respects the parthenogenetic development of such eggs was abnormal: the eggs developed without the formation of a fertilisation membrane and the larvae failed to swim freely.

During their extensive studies of the effect of drugs on echinoderm mitosis, O. and R. Hertwig (1887) found that when unfertilised sea urchin eggs are exposed to sea water containing chloroform, a typical fertilisation membrane is formed, and Herbst (1893) showed that benzol, toluol, and xylol have a similar effect. Loeb (1905) extended this list by using ethyl acetate, although the ester was found to exhibit a specific difference to chloroform or xylol in that the fertilisation membranes only formed after the eggs had been transferred to normal sea water; when so transferred, the eggs showed incipient signs of segmentation, although no larvae were obtained. Clearly the next step towards effective parthenogenesis consisted in effecting membrane formation and afterwards exposing the same eggs to hypertonic sea water. The eggs so treated developed normally and free swimming larvae were obtained. In all his later observations Loeb replaced the esters of the fatty acids by the acids themselves, so that his final method is as follows. The eggs are placed in 50 c.c. sea water + 2·8 c.c. $N/10$ butyric acid for $1\frac{1}{2}$–4 minutes and are then transferred to 200 c.c. of sea water. After 15–20 minutes they are placed in 50 c.c. sea water + 8 c.c. $2\frac{1}{2}M$ NaCl. After 30–60 minutes they are transferred to clean sea water, and normal development (under favourable circumstances) ensues. The precise details of this technique vary for different species of egg and for different batches of eggs. For British species reference should be made to Shearer and Lloyd (1913).

As already mentioned (see p. 426), membrane formation can be effected by a variety of reagents; saponin, soaps, chloroform, fatty acids, salicylic aldehyde, foreign blood sera, metallic silver (Herbst, 1904), but on the whole saponin or the lower fatty acids have proved to be the most convenient. As far as is known, there are only three substitutes for hypertonic sea water in the second phase of Loeb's technique. These are (i) treatment with cyanides (Loeb, 1913), (ii) anaesthetics (R. S. Lillie, 1913), and (iii) low temperature (Loeb, 1906; Hindle, 1910).

The application of similar methods to eggs, other than those of sea urchins, has led to the production of artificially parthenogenetic larvae in a number of other groups (annelids, molluscs, sipunculids). Curiously enough, if such eggs respond at all, they do so with less difficulty than those of sea urchins. For example, when membranes are artificially produced in such forms as *Polynoë* (Loeb, 1913), *Thalassema* or *Asterias*, normal larvae are obtained without subsequent treatment with hypertonic sea water. Similarly, parthenogenetic larvae can be obtained from the eggs of the molluscs *Lottia* or *Acmaea* by treatment with hypertonic sea water alone. It is quite clear that the two phases of Loeb's final method are by no means universally necessary. An account of parthenogenesis in the various phyla of the animal kingdom will be found in the books of Loeb (1913) and Morgan (1927), and we may confine our attention to the general interpretation of the more outstanding facts.

We have already seen that normal fertilisation can be divided into two series of events: those which result from adhesion between the egg and the spermatozoon, and those which follow the penetration of the sperm into the substance of the eggs. It is at once significant that Loeb's final method of effecting artificial parthenogenesis in sea urchin eggs also involves two processes—one of which induces changes at the surface of the egg precisely similar to those induced by the contact of the egg with an effective spermatozoon. That this morphological parallel has real physiological significance is shown by the fact that in both cases the respiratory activity of the eggs is raised to the same high level (Warburg, 1910), and in both cases there is a marked increase in the ability of the eggs to absorb water from hypotonic solutions (Lillie, R. S., 1916, 1917). It will be remembered that F. R. Lillie succeeded in removing the spermatozoon from the surface of a *Nereis* egg after the cortical changes were completed and showed that the nucleus formed a number of chromosomes but failed to form a normal amphiaster. Much the same phenomena are observed when the eggs of a sea urchin are treated with a membrane-forming substance such as butyric acid. Using *Strongylocentrotus purpuratus* Hindle (1910) showed that eggs exposed to membrane treatment do not divide (unless at low temperatures), but the nucleus undergoes a typical prophase, and forms a large monaster on which lie the scattered chromosomes (fig. 169). More recently, Herlant (1918, 1919) has confirmed this fact in *Paracentrotus lividus*, where the nuclear cycle of the female pronucleus may be repeated several times. It seems reasonable to conclude that just as normal fertilisation can be regarded as a combination of two series of events, so artificial parthenogenesis is divisible into (i) a process operating on the egg surface (which is best induced by treatment with membrane-forming substances), and into (ii) a subsequent process whose distinctive effect is the provision of a normal amphiaster in the place of the single ineffective aster which is the result of the activation process acting by itself. In some way or other hypertonic sea water must provide this amphiaster. It will be recalled that in 1900 Morgan showed that

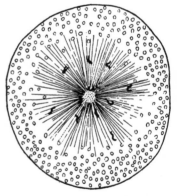

Fig. 169. Monaster in unfertilised egg of *Strongylocentrotus* $2\frac{1}{2}$ hours after treatment with butyric acid in sea water; note the scattered chromosomes. (From Hindle.)

cytasters will form in fertilised eggs if these are subjected to hypertonic sea water and this process was clearly described in *Sphaerechinus* eggs by Wilson (1901). All authors are agreed that hypertonic sea water provides the astral machinery for normal cleavage, but two distinct views are held concerning the way in which this is effected. According to Herlant (1918–19) the final amphiaster is composed of two separate entities—the aster normally associated with the female pronucleus and a second aster which is an artificial product produced *de novo* in the cytoplasm by treatment with hypertonic sea water; this latter aster is comparable to the 'cytasters' originally described by Morgan. If treatment with hypertonic sea water is too severe or too prolonged, more than one effective cytaster may co-operate with the female aster to form a polypolar spindle, and the resultant cleavage is abnormal. Herlant's conclusions have, however, been criticised by Fry (1925). According to Fry, the amphiaster of artificially parthenogenetic eggs is the result of cleavage on the part of the single female aster, while artificial cytasters play no essential rôle; Fry claims that when a cytaster is incorporated into the spindle, the eggs invariably develop abnormally. The present position is therefore uncertain, but it is interesting to note that in normal fertilisation it is undoubtedly the male aster which possesses the power of spontaneous division, whereas the female aster is unable to do so; that the first amphiaster is entirely derived from the male element by no means follows (see p.161).

A study of the sea urchin's egg might justify the conclusion that artificial parthenogenesis can be effected when two reagents are used —one of which activates the egg, and the other completes the machinery for normal cleavage. We must, however, remember that the earlier experiments of Loeb and others show that it is possible to obtain activation without the formation of a normal fertilisation membrane, and in the case of other eggs—notably those of the starfish and *Thalassema* (Lefevre, 1907)—normal development may result from treatment with activating agents only. Further it will be recalled that a limited number of sea urchin eggs can be activated by hypertonic sea water alone. The great variety of effective agents is, in fact, one of the most striking features of artificial parthenogenesis: mechanical agitation (*Asterias*, Mathews, 1901), heat (*Nereis*, Just, 1915 a), saponin (Loeb, 1913), alkalies (Loeb, 1913), and foreign blood sera (*Dendrostoma*, Loeb, 1907, 1908 a, b). As far as we know, the final results are always the same, viz. activation and the production of a normal amphiaster—but the precise chain of events leading up to this result may be different in different cases and it is dangerous to generalise. For example, according to Buchner (1911) in *Asterias glacialis* the amphiaster is derived from the female aster together with that of the second polar body whose nucleus sinks back into the egg and acts as a male pronucleus. Whether this is always the case is doubtful, for Tennent and Hague (1906) previously described the haploid condition of the parthenogenetic eggs of *A. Forbesii*.

Theories concerning the underlying causes of artificial parthenogenesis have been numerous. As is well known, Loeb (1913) looked upon the process of membrane formation in the sea urchin egg as the incipient stages of a cytolysis which, if unchecked, would lead to the destruction of the whole egg. According to Loeb the activation of an unfertilised egg is effected by the introduction into the egg of two substances: (i) a specific cytolysin, which brings about the destruction of the surface layer of the egg, and (ii) a substance which limits or controls the destructive influence of the cytolysin. This scheme is clearly based on the phenomena of artificial parthenogenesis of sea urchin eggs, and encounters difficulties when applied to the facts of normal and cross fertilisation. It is not easy to see why the sperm of *Mytilus* or a shark should be able to introduce a specific lysin into sea urchin eggs, whereas that of a neighbouring species of sea urchin cannot do so—further it is by no means clear that membrane formation and subsequent cytolysis are essentially of the same nature. If sea urchin eggs are placed in a suitable concentration of saponin in sea water, membranes are extruded in the normal way within a few minutes, but there is a long latent period before any visible cytolysis occurs. Further, if hypertonic sea water antagonises cytolysis, it is curious that its effect on the egg can be induced before the egg is subjected to membrane-forming substances. As Loeb himself showed, the 'corrective' treatment with hypertonic sea water is in some cases unnecessary. Although Loeb advocated the conception of specific lysins which destroyed the surface layer of the unfertilised egg, there can be little doubt that he regarded such an effect essentially as a profound change in the physical state of the egg cortex. We have already suggested the advantages to be gained from this point of view: it is more flexible than the hypothesis of specific chemical reactions. Unless we regard the effect of mechanical puncture of an egg as specifically different to that produced by the localised injury of other excitable tissues, there seems some *a priori* evidence in support of the physical hypothesis. Bataillon (1910–11) found that if eggs of the frog are punctured by a fine glass needle in the presence of lymph, a limited number will develop into normal tadpoles. Bataillon regarded this treatment as essentially diphasic—and comparable to Loeb's technique; the puncture was equivalent to surface cytolysis and the action of the lymph to the corrective treatment of hypertonic sea water. Herlant (1911, 1913) regarded the amphiaster (found after puncture in lymph) as a product of two cytasters, which do not develop unless lymph is present. Whatever be the origin of the cleavage centres, it is difficult to resist the conclusion that the initial effect of mechanical puncture is of a physical nature.

REFERENCES

References marked with an asterisk contain either a general discussion or an extensive bibliography

BALTZER, F. (1910). 'Ueber die Beziehung zwischen dem Chromatin und der Entwicklung und der Vererbungsrichtung bei Echinodermenbastarden.' *Archiv f. Zellforsch.* 5, *497*.

BATAILLON, E. (1910). 'L'embryogénèse complète provoquée chez les Amphibiens par figure de l'œuf vierge, larves parthénogénétiques de *Rana fusca*.' *C.R. Acad. Sci. Paris*, 150.

—— (1911). 'La parthénogénèse expérimentale chez *Bufo vulgaria*.' *C.R. Acad. Sci. Paris*, 152.

BORN, G. (1883). 'Beiträge zur Bastardirung der einheimischen Anurenarten.' *Pflüger's Archiv*, 32, *453*.

—— (1886). 'Biologische Untersuchungen. II. Weitere Beiträge zur Bastardirung zwischen den einheimischen Anuren.' *Arch. mikr. Anat.* 27, *192*.

BRACHET, A. (1911). 'Recherches sur l'influence de la polyspermie expérimentale dans le développement de l'œuf de *Rana fusca*.' *Arch. de Zool. expér. et gén.* 6, *1*.

BUCHNER, P. (1911). 'Die Reifung des Seesterneies bei experimenteller Parthenogenese.' *Archiv f. Zellforsch.* 6, *577*.

BULLER, A. H. (1902). 'Is chemotaxis a factor in the fertilization of the eggs of animals?' *Quart. Journ. Micr. Sci.* 56, *145*.

CARTER, G. S. (1930). 'Thyroxine and the oxygen consumption of the spermatozoa of *Echinus miliaris*.' *Journ. Exp. Biol.* 7, *41*.

CASTLE, W. E. (1896). 'The early embryology of *Ciona intestinalis*.' *Bull. Mus. Comp. Zool. Harvard*, 27, *203*.

CHAMBERS, R. (1917). 'Microdissection studies. II. The cell aster: a reversible gelatin phenomenon.' *Journ. Exp. Zool.* 23, *483*.

—— (1921). 'Microdissection studies. III. Some problems in the maturation and fertilisation of the echinoderm egg.' *Biol. Bull.* 41, *318*.

—— (1923). 'The mechanism of the entrance of sperm into the starfish egg.' *Journ. Gen. Physiol.* 5, *821*.

—— (1930). 'The manner of sperm entry in the starfish egg.' *Biol. Bull.* 58, *344*.

CLOWES, G. H. A. and BACHMAN, E. (1921). 'On a volatile sperm-stimulating substance derived from marine eggs.' *Journ. Biol. Chem.* 46, *31*.

CLOWES, G. H. A. and SMITH, H. W. (1923). 'The influence of hydrogen ion concentration on the fertilisation and growth of certain marine eggs.' *Amer. Journ. Physiol.* 64, *144*.

COHN, E. J. (1918). 'Studies in the physiology of spermatozoa.' *Biol. Bull.* 34, *167*.

COLE, K. S. (1928). 'Electrical impedance of suspensions of *Arbacia* eggs.' *Journ. Gen. Physiol.* 12, *37*.

CRAGG, F. W. (1920). 'Further observations on the reproductive system of *Cimex* with special reference to the behaviour of the spermatozoa.' *Ind. Journ. Med. Res.* 8, *32*.

DAKIN, W. J. and FORDHAM, M. G. S. (1924). 'The chemotaxis of spermatozoa and its questioned occurrence in the animal kingdom.' *Brit. Journ. Exp. Biol.* 1, *183*.

*DALCQ, A. (1928). *Les Bases physiologiques de la Fécondation et de la Parthénogénèse*. (Paris.)

*DONCASTER, L. (1924). *An Introduction to the Study of Cytology*. (Cambridge.)

*FAURÉ-FREMIET, E. (1925). *La cinétique du développement*. (Paris.)

FISCHEL, A. (1906). 'Ueber Bastardierungsversuche bei Echinodermen.' *Arch. f. Entw. Mech.* 22, *498*.

FOX, H. M. (1920). 'An investigation into the cause of the spontaneous aggregation of flagellates and into the reactions of flagellates to dissolved oxygen.' *Journ. Gen. Physiol.* 3, *483, 501*.

FRY, H. J. (1925). 'Asters in artificial parthenogenesis. I. Origin of the amphiaster in the eggs of *Echinarachnius parma*.' *Journ. Exp. Zool.* 43, *11*.

FUCHS, H. M. (1914 a). 'On the condition of self-fertilisation in *Ciona*.' *Arch. f. Entw. Mech.* 40, *157*.

—— (1914 b). 'The action of egg-secretions on the fertilizing power of sperm.' *Arch. f. Entw. Mech.* 40, *205*.

—— (1915). 'Studies in the physiology of fertilization.' *Journ. Genetics*, 4, *215*.

GATENBY, J. B. (1922). 'The cytoplasmic inclusions of the germ cells. X. The gametogenesis of *Saccocirrus*.' *Quart. Journ. Micr. Sci.* 66, *1*.

GEMMILL, J. F. (1900). 'On the vitality of the ova and sperm of certain animals.' *Journ. Anat. and Physiol.* 34.

GLASER, O. (1913). 'On inducing development in the sea-urchin (*Arbacia punctulata*), together with considerations on the initiatory effect of fertilisation.' *Science*, 38, *446*.

—— (1914 a). 'A qualitative analysis of the egg secretions of extracts of *Arbacia* and *Asterias*.' *Biol. Bull.* 26, *367*.

—— (1914 b). 'On autoparthenogenesis in *Arbacia* and *Asterias*.' *Biol. Bull.* 26, *387*.

—— (1915). 'Can a single spermatozoon initiate development in *Arbacia*?' *Biol. Bull.* 28, *149*.

—— (1921 a). 'The duality of egg-secretion.' *Amer. Nat.* 55, *368*.

—— (1921 b). 'Fertilization and egg-secretion.' *Biol. Bull.* 41, *63*.

GODLEWSKI, E. (1911). 'Studien über die Entwicklungserregung. II. Antagonismus der Einwirkung des Spermas von verschiedenen Tierklassen.' *Arch. f. Entw. Mech.* 33, *233*.

GODLEWSKI, E. (1926). 'L'inhibition réciproque de l'aptitude à féconder de spermes d'espèces éloignées comme conséquence de l'agglutination des spermatozoïdes.' *Arch. de Biol.* 36, *311*.

GRAY, J. (1915). 'Note on the relation of spermatozoa to electrolytes and its bearing on the problem of fertilisation.' *Quart. Journ. Micr. Sci.* 61, *119*.

—— (1916). 'The electrical conductivity of echinoderm eggs and its bearing on the problems of fertilisation and artificial parthenogenesis.' *Phil. Trans. Roy. Soc.* B, 207, *481*.

—— (1920 a). 'The relation of spermatozoa to certain electrolytes. II.' *Proc. Roy. Soc.* B, 91, *147*.

—— (1920 b). 'The relation of animal cells to electrolytes. I. A physiological study of the egg of the trout.' *Journ. Physiol.* 53, *308*.

—— (1922 a). 'A critical study of the facts of artificial fertilization and normal fertilization.' *Quart. Journ. Micr. Sci.* 66, *419*.

—— (1922 b). 'The mechanism of ciliary movement.' *Proc. Roy. Soc.* B, 93, *104*.

—— (1928 a). 'The effect of dilution on the activity of spermatozoa.' *Brit. Journ. Exp. Biol.* 5, *337*.

—— (1928 b). 'The senescence of spermatozoa.' *Brit. Journ. Exp. Biol.* 5, *345*.

—— (1928 c). 'The effect of egg-secretions on the activity of spermatozoa.' *Brit. Journ. Exp. Biol.* 5, *362*.

—— (1931). 'The senescence of spermatozoa. II.' *Journ. Exp. Biol.* 8, *202*.

HAMMOND, J. (1930). 'The effect of temperature on the survival *in vitro* of rabbit spermatozoa obtained from the vagina.' *Journ. Exp. Biol.* 7, *175*.

HAMMOND, J. and ASDELL, S. A. (1926). 'The vitality of the spermatozoa in the male and female reproductive tracts.' *Brit. Journ. Exp. Biol.* 4, *155*.

HARTMAN, C. and CUYLER, W. K. (1927). 'Is the supposed long life of the bat spermatozoa fact or fable?' *Anat. Record*, 35, *39*.

HARVEY, E. N. (1910). 'The mechanism of membrane formation and other early changes in developing sea-urchin's eggs as bearing on the problem of artificial parthenogenesis.' *Journ. Exp. Zool.* 8, *355*.

—— (1914). 'Is the fertilisation membrane of *Arbacia* eggs a precipitation membrane?' *Biol. Bull.* 27, *237*.

HERBST, C. (1893). 'Über die künstliche Hervorrufung von Dottermembranen an unbefruchteten Seeigeleiern nebst einigen Bemerkungen über die Dotterhautbildung überhaupt.' *Biol. Centrlb.* 13, *14*.

—— (1904). 'Über die künstliche Hervorrufung von Dottermembranen an unbefruchteten Seeigeleiern.' *Mitth. Zool. Sta. Neapel*, 16, *445*.

HERLANT, M. (1911). 'Recherches sur les œufs di- et tri-spermiques de grenouille.' *Arch. de Biol.* 26, *103*.

—— (1913). 'Étude sur les bases cytologiques du mécanisme de la parthénogénèse expérimentale chez les Amphibiens.' *Arch. de Biol.* 28, *505*.

—— (1918). 'Comment agit la solution hypertonique dans la parthénogénèse expérimentale (Méthode de Loeb). I. Origine et signification des asters accessoires.' *Arch. Zool. Exp.* 57, *511*.

—— (1919). 'Comment agit la solution hypertonique dans la parthénogénèse expérimentale (Méthode de Loeb). II. Le mécanisme de la segmentation.' *Arch. Zool. Exp.* 58, *291*.

HERTWIG, O. and R. (1887). 'Ueber den Befruchtungs- und Teilungsvorgang des tierischen Eies unter dem Einfluss äusserer Agentien.' *Jen. Zeit.* 20, N.S. 13, *120, 477.*

HINDLE, E. (1910). 'A cytological study of artificial parthenogenesis in *Strongylocentrotus purpuratus.*' *Arch. f. Entw. Mech.* 31, *145.*

HOBSON, A. D. (1927). 'A study of the fertilisation membrane in the echinoderms.' *Proc. Roy. Soc. Edin.* 47, *94.*

HUXLEY, J. S. (1930). 'The maladaptation of trout spermatozoa to fresh water.' *Nature,* 125, *494.*

HYDE, I. H. (1904). 'Differences in electrical potential in developing eggs.' *Amer. Journ. Physiol.* 12, *241.*

JUST, E. E. (1915 a). 'Initiation of development in *Nereis.*' *Biol. Bull.* 28, *1.*

—— (1915 b). 'An experimental analysis of fertilisation in *Platynereis megalops.*' *Biol. Bull.* 28, *93.*

—— (1919). 'The fertilisation reaction in *Echinarachnius parma.* II. The rôle of fertilizin in straight and cross fertilisation.' *Biol. Bull.* 36, *11.*

—— (1928). 'Initiation of development in *Arbacia.* IV. Some cortical reactions as criteria for optimum fertilisation capacity and their significance for the physiology of development.' *Protoplasma,* 5, *97.*

—— (1929). 'The production of filaments by echinoderm ova as a response to insemination, with special reference to the phenomenon as exhibited by ova of the genus *Asterias.*' *Biol. Bull.* 57, *311.*

—— (1930). 'The present status of the fertilizin theory of fertilization.' *Protoplasma,* 10, *390.*

KITE, G. L. (1912). 'The nature of the fertilization membrane of the egg of the sea-urchin (*Arbacia punctulata*).' *Science,* 36, *562.*

KÖLLIKER, A. (1856). 'Physiologische Studien über die Samenflüssigkeit.' *Zeit. wiss. Zool.* 7, *201.*

KUPELWIESER, H. (1906). 'Versuche ueber Entwicklungserregung und Membranbildung bei Seeigeleiern durch Molluskensperma.' *Biol. Zentralb.* 26, *744.*

—— (1909). 'Entwicklungserregung bei Seeigeleiern durch Molluskensperma.' *Arch. f. Entw. Mech.* 27, *434.*

LEFEVRE, G. (1907). 'Artificial parthenogenesis in *Thalassema mellita.*' *Journ. Exp. Zool.* 4, *91.*

LILLIE, F. R. (1911). 'Studies of fertilisation in *Nereis.* I. The cortical changes in the egg. II. Partial fertilisation.' *Journ. Morph.* 22, *361.*

—— (1912 a). 'Studies of fertilisation in *Nereis.* II. Morphology of the normal fertilisation of *Nereis.* IV. The fertilizing power of portions of the spermatozoon.' *Journ. Exp. Zool.* 12, *413.*

—— (1912 b). 'The penetration of spermatozoon and the origin of the sperm aster in the egg of *Nereis.* On the fertilising power of portions of the spermatozoon.' *Science,* 35, *471.*

—— (1913). 'Studies of fertilisation. V. The behaviour of spermatozoa of *Nereis* and *Arbacia* with special reference to egg-extractives.' *Journ. Exp. Zool.* 14, *515.*

*—— (1919). *Problems of Fertilization.* (Chicago.)

*LILLIE, F. R. and JUST, E. E. (1924). 'Fertilization.' Section VIII. *General Cytology.* (Chicago.)

LILLIE, R. S. (1908). 'Momentary elevation of temperature as a means of producing artificial parthenogenesis in starfish eggs and the conditions of its action.' *Journ. Exp. Zool.* 5, *375*.
—— (1909). 'The general biological significance of changes in the permeability of the surface layer of the plasma membrane of living cells.' *Biol. Bull.* 17, *3*.
—— (1913). 'The physiology of cell division. V. Substitution of anaesthetics for hypertonic sea-water and cyanide in artificial parthenogenesis in starfish eggs.' *Journ. Exp. Zool.* 15, *23*.
—— (1915). 'On the conditions of activation of unfertilised starfish under the influence of high temperatures and fatty acid solutions.' *Biol. Bull.* 28, *260*.
—— (1916). 'Increase of permeability to water following normal and artificial activation in sea-urchin eggs.' *Amer. Journ. Physiol.* 40, *249*.
—— (1917). 'The conditions determining the rate of entrance of water into fertilised and unfertilised *Arbacia* eggs, and the general relation of changes of permeability to activation.' *Amer. Journ. Physiol.* 43, *43*.
LOEB, J. (1895). 'Untersuchungen über die physiologischen Wirkung des Sauerstoffmangels.' *Pflüger's Archiv*, 62, *249*.
—— (1900 a). 'On the artificial production of normal larvae from the unfertilized eggs of the sea-urchin.' *Amer. Journ. Physiol.* 3, *434*.
—— (1900 b). 'Further experiments on artificial parthenogenesis and the nature of the process of fertilization.' *Amer. Journ. Physiol.* 4, *178*.
—— (1903). 'Ueber die Befruchtung von Seeigeleiern durch Seesternsamen?' *Pflüger's Archiv*, 99, *323*.
—— (1905). 'On an improved method of artificial parthenogenesis.' *Univ. Cal. Publ. Physiol.* 2, *83*.
—— (1906). 'Versuche über den chemischen Charakter des Befruchtungsvorgangs.' *Biochem. Zeit.* 1, *183*.
—— (1907). 'Ueber die Hervorrufung der Membranbildung beim Seeigelei durch das Blut gewisser Würmer (Sipunculiden).' *Pflüger's Archiv*, 118, *36*.
—— (1908 a). 'Ueber die Hervorrufung der Membranbildung und Entwicklung beim Seeigelei durch das Blutserum des Kaninchens.' *Pflüger's Archiv*, 122, *196*.
—— (1908 b). 'Weitere Versuche ueber die Entwicklungserregung des Seeigeleies durch das Blutserum von Säugetieren.' *Pflüger's Archiv*, 124, *37*.
—— (1908 c). 'Ueber die Entwicklungserregung unbefruchteter Annelideneier (*Polynoe*) mittels Saponin und Solanin.' *Pflüger's Archiv*, 122, *448*.
—— (1908 d). 'Über die osmotischen Eigenschaften und die Entstehung der Befruchtsmembran beim Seeigelei.' *Arch. f. Entw. Mech.* 26, *82*.
*—— (1913). *Artificial Parthenogenesis and Fertilisation*. (Chicago.)
—— (1914). 'Cluster formation of spermatozoa caused by specific substances from eggs.' *Journ. Exp. Zool.* 17, *123*.
—— (1915). 'On the nature of the conditions which determine or prevent the entrance of the spermatozoön into the egg.' *Amer. Nat.* 49, *257*.
LOEB, J. and WASTENEYS, H. (1911 a). 'Die Beeinflussung der Entwicklung und der Oxydationsvorgänge im Seeigelei.' *Biochem. Zeit.* 37, *410*.

LOEB, J. and WASTENEYS, H. (1911 b). 'Sind die Oxydationsvorgänge die unabhängige Variables in den Lebenserscheinungen?' *Biochem. Zeit.* 36, 345.
—— (1912). 'Die Oxydationsvorgänge im befruchteten und unbefruchteten Seesternei.' *Arch. f. Ent. Mech.* 35, 555.
McCLENDON, J. F. (1910). 'On the dynamics of cell division. II. Changes in permeability of developing eggs to electrolytes.' *Amer. Journ. Phys.* 27, 240.
—— (1914). 'On the nature and formation of the fertilisation membrane of the echinoderm egg.' *Intern. Zeits. f. phys.-chem. Biol.* 1, 163.
MASING, E. (1910). 'Ueber das Verhalten der Nucleinsäme bei der Furchung des Seeigeleies.' *Zeit. f. physiol. Chem.* 67, 671.
MATHEWS, A. P. (1900). 'Some ways of causing mitotic division in unfertilized *Arbacia* eggs.' *Amer. Journ. Physiol.* 4, 343.
—— (1901). 'Artificial parthenogenesis produced by mechanical agitation.' *Amer. Journ. Physiol.* 6, 142.
MEAD, A. D. (1895). 'Some observations on the maturation and fecundation of *Chaetopterus pergamentaceus* (Cuvier).' *Journ. Morph.* 10, 313.
MEYERHOF, O. (1911). 'Untersuchungen über die Wärmetönung der vitalen Oxydationsvorgänge in Eiern. I–III.' *Biochem. Zeit.* 35, 246.
MOENKHAUS, W. J. (1904). 'The development of the hybrids between *Fundulus heteroclitus* and *Menidia notata* with especial reference to the behaviour of the maternal and paternal chromatin.' *Amer. Journ. Anat.* 3, 29.
—— (1910). 'Cross-fertilisation among fishes.' *Proc. Ind. Acad. Sci.* 353.
MOORE, C. R. (1928). 'On the properties of the gonads as controllers of somatic and psychical characters. X. Spermatozoön activity and the testis hormone.' *Journ. Exp. Zool.* 50, 455.
MORGAN, T. H. (1896). 'The production of artificial astrospheres.' *Arch. f. Entw. Mech.* 3, 339.
—— (1900). 'Further studies on the action of salt solutions and of other agents on the egg of *Arbacia*.' *Arch. f. Entw. Mech.* 10, 489.
—— (1904). 'Self-fertilisation induced by artificial means.' *Journ. Exp. Zool.* 1, 135.
—— (1905). 'Some further experiments on self-fertilisation in *Ciona*.' *Biol. Bull.* 8, 313.
—— (1910). 'Cross- and self-fertilisation in *Ciona intestinalis*.' *Arch. f. Entw. Mech.* 30, 206.
—— (1923). 'Removal of the block to self-fertilisation in the ascidian *Ciona*.' *Proc. Nat. Acad. Sci.* 9, 170.
*—— (1927). *Experimental Embryology*. (New York.)
NATH, V. (1925). 'Cell inclusions in the oogenesis of scorpions.' *Proc. Roy. Soc. B*, 98, 44.
NEEDHAM, J. (1931). *Chemical Embryology*. (Cambridge.)
NEWMAN, H. H. (1908). 'The process of heredity as exhibited by the development of *Fundulus* hybrids.' *Journ. Exp. Zool.* 5, 505.
—— (1910). 'Further studies of the process of heredity in *Fundulus* hybrids.' *Journ. Exp. Zool.* 8, 113.
—— (1914). 'Modes of inheritance in teleost hybrids.' *Journ. Exp. Zool.* 16, 447.
—— (1915). 'Development and heredity in heterogenic teleost hybrids.' *Journ. Exp. Zool.* 18, 511.

THE GERM CELLS 449

VAN OORDT, G. J. (1928). 'The duration of life of the spermatozoa in the fertilised female of *Xiphophorus helleri* (Regan).' *Tijdschrift der Ned. Dierkundige Vereen.* 3, 77.

ROUX, W. (1887). 'Die Bestimmung der Medianebene des Froschembryos durch die Copulationsrichtung des Eikernes und des Spermakernes.' *Arch. f. mikr. Anat.* 29, 157.

SAUNDERS, J. T. (1924). 'The effect of the hydrogen ion concentration on the behaviour growth, and occurrence of *Spirostomum*.' *Proc. Camb. Philos. Soc. Biol. Series*, 1, 189.

SHEARER, C. (1922 a). 'On the oxidation process of the echinoderm egg during fertilization.' *Proc. Roy. Soc.* B, 93, 213.

—— (1922 b). 'On the heat production and oxidation processes of the echinoderm egg during fertilization and early development.' *Proc. Roy. Soc.* B, 93, 410.

SHEARER, C. and LLOYD, D. J. (1913). 'On methods of producing artificial parthenogenesis in *Echinus esculentus* and the rearing of the parthenogenetic plutei through metamorphosis.' *Quart. Journ. Micr. Sci.* 58, 523.

SHEARER, C., DE MORGAN, W. and FUCHS, H. M. (1913). 'On the experimental hybridisation of echinoids.' *Phil. Trans. Roy. Soc.* B, 204, 255.

SMITH, H. W. and CLOWES, G. H. A. (1924). 'The influence of hydrogen-ion concentration on unfertilised *Arbacia*, *Asterias* and *Chaetopterus* eggs.' *Biol. Bull.* 47, 304.

TAIT, J. (1918). 'Capillary phenomena observed in blood cells: thigmocytes, phagocytosis, amoeboid movement, differential adhesiveness of corpuscles, emigration of leucocytes.' *Quart. Journ. Exp. Physiol.* 12, 1.

TENNENT, D. H. (1910). 'Echinoderm hybridization.' *Publ.* 132, *Carneg. Instit. Wash.*

TENNENT, D. H. and HAGUE, M. J. (1906). 'Studies on the development of the starfish egg.' *Journ. Exp. Zool.* 3, 517.

VAN TRIGT, H. (1919). *A Contribution to the Physiology of the Freshwater Sponges*. (Leiden.)

VERNON, H. M. (1900). 'Cross fertilization among the Echinoidea.' *Arch. Entw. Mech.* 9, 464.

WALTON, A. (1927). 'The relation between "density" of sperm-suspension and fertility as determined by artificial insemination of rabbits.' *Proc. Roy. Soc.* B, 101, 303.

—— (1930 a). 'The effect of temperature on the survival *in vitro* of rabbit spermatozoa obtained from the *vas deferens*.' *Journ. Exp. Biol.* 7, 201.

—— (1930 b). 'The maladaptation of trout spermatozoa to fresh water.' *Nature*, 125, 564.

WARBURG, O. (1908). 'Beobachtungen über die Oxydationsprocesse im Seeigelei.' *Zeit. f. physiol. Chem.* 57, 1.

—— (1909). 'Über die Oxydationen im Ei. II. Mittheilung.' *Zeit. f. physiol. Chem.* 60, 443.

—— (1910). 'Über die Oxydationen in lebenden Zellen nach Versuchen am Seeigelei.' *Zeit. f. physiol. Chem.* 66, 305.

—— (1914 a). 'Über die Rolle des Eisen in der Atmung des Seeigeleies nebst Bemerkungen über einige durch Eisen beschleunigte Oxidationen.' *Zeit. f. physiol. Chem.* 92, 321.

—— (1914 b). 'Zellstruktur und Oxydationsgeschwindigkeit nach Versuchen am Seeigelei.' *Pflüger's Archiv*, 158, 189.

WARBURG, O. (1915). 'Notizen zur Entwicklungsphysiologie des Seeigeleies.' *Pflüger's Archiv*, 160, *324*.

WILSON, E. B. (1901 a). 'Experimental studies in cytology. I. Cytological study on artificial parthenogenesis in sea-urchin eggs.' *Arch. f. Entw. Mech.* 12, *529*.

—— (1901 b). 'Experimental studies in cytology. II. Some phenomena of fertilisation and cell division in etherized eggs.' *Arch. f. Entw. Mech.* 13, *353*.

*—— (1925). *The Cell in Development and Inheritance*. (New York.)

WOODWARD, A. E. (1918). 'Studies in the physiological significance of certain precipitates from the egg secretions of *Arbacia* and *Asterias*.' *Journ. Exp. Zool.* 26, *459*.

—— (1921). 'The parthenogenetic effect of echinoderm egg-secretions on the eggs of *Nereis limbate*.' *Biol. Bull.* 41, *276*.

WOODWARD, A. E. and HAGUE, F. S. (1917). 'Iodine as a parthenogenetic agent.' *Biol. Bull.* 33, *355*.

YOUNG, W. C. (1929 a). 'A study of the function of the epididymis. I. Is the attainment of the full spermatozoön maturity attributable to some specific action of the epididymal secretion.' *Journ. Morph. and Physiol.* 47, *479*.

—— (1929 b). 'The importance of an ageing process in sperm for the length of the period during which fertilizing capacity is retained by sperm isolated in the epididymis of the guinea-pig.' *Journ. Morph. and Physiol.* 48, *475*.

CHAPTER SEVENTEEN

Contractile Cells

Of the three main types of contractile cells—muscular, vibratile, and amoeboid—the first has naturally received the greatest attention. The essential properties of a contracting muscle are, however, best illustrated by multicellular units in which the behaviour of individual cells can only be observed with great difficulty. Hence although it may be illogical to consider the problems of contractile protoplasm without extensive reference to the muscle cell we shall, in the present chapter, largely restrict attention to vibratile and amoeboid movements for in these cases the cell is the natural and obvious unit of observation.

It may be useful to remember at the outset that the mechanism whereby a cell is capable of altering its form, and thereby performing mechanical work, is unknown even in muscle. There are two general theories. According to the first of these, the essential step in protoplasmic contraction is a localised alteration in the distribution of water within the cell. Since lactate ions increase in number as a result of vertebrate muscular movements, and since there is a large amount of protein present in the cell, it is assumed that at the moment of contraction the affinity of certain protein surfaces for water is altered by the presence of hydrogen ions which are produced with the lactate ions from a carbohydrate precursor. The presence of these hydrogen ions at the protein surface alters the degree of imbibition of this surface and so either attracts water from other parts of the cell or allows water to pass from the protein elsewhere, according to the original charge on the protein. It is this movement of water which is associated with contraction and which is, of course, a process capable of generating very powerful forces. During muscular relaxation the hydrogen ions are removed from the surface by neutralisation and the *status quo ante* is resumed. This theory has been discussed at some length by Ritchie (1928), and can be applied in more than one form to all types of contractile movement. The alternative theory assumes that the hydrogen ions produce free energy by accumulation at a liquid interface, thereby changing its surface tension: whether or not such a system can produce sufficiently large forces to account for powerful muscular contractions is not yet clear (see Hill, 1925, 1926 b). Without considering such theories in any detail, it is perhaps worth noting that, at present, the fundamental mechanism of muscular contraction is just as mysterious as that of any other expression of cell activity. Contractile protoplasm often has no optical structure and it is useful to remember that the concepts of intracellular membranes or of liquid interfaces are little more than morphological models of contractile systems which have replaced the cruder notions of 'fibrillae' and protoplasmic alveoli.

It is not unreasonable to imagine that the underlying mechanism of contraction is essentially similar in all types of movement (see Gray, 1922; Pantin, 1923; Hill, 1926 b). The evidence to support such a view is entirely indirect and is largely based on a similarity of reaction of the three main types of movement to a variety of environmental changes (see Gray, 1924 and Table LXIX). Inferences of this type are clearly not without their dangers—for we may find that although the reactions of a muscle cell, a cilium and a pseudopodium are remarkably alike, yet they are due to cell properties which are not directly associated with contraction and with contraction alone. An example may make this clearer. Muscular, ciliary and amoeboid movement are all inhibited by an excess of hydrogen ions within the cell (Gray, 1922; Pantin, 1923). From this fact alternative inferences may be drawn: (i) the three types of contractile mechanism are directly inhibited by hydrogen ions, or (ii) contractile processes, like other expressions of the cell's activity, are affected by any disturbance of the general electrolytic equilibrium of the cell when this is upset by the presence of additional hydrogen ions. Unless it can be shown that the contractile systems react to specific changes in environment in a way not exhibited by other cell activities, there will always be some doubt concerning the conclusions drawn from observations of this type. A similar problem arises in connection with the rôle of glycogen in muscle and in ciliated cells. In certain types of vertebrate muscle there can be little doubt that this substance is the ultimate source of the energy for movement and that it gives rise to the lactic acid which is ultimately oxidised to carbon dioxide and water. Quite recently, Beutler (1929) has demonstrated the presence of glycogen in the ciliated cells of Actinians and it is tempting to assume that its rôle is comparable to that

Table LXIX. A comparison of the properties of muscular, ciliary and amoeboid movements. (See also Gray, 1924)

Reagent	Cardiac muscle	Coelenterate muscle	Cilia of *Mytilus*	Amoeba *limax*
Temperature	Q_{10} for frog's sinus (1·2°–30·5°) 3·1–1·3. Oxygen consumption proportional to rate of beat		Q_{10} for interval (0°–32·5°) 3·1–1·92. Oxygen consumption proportional to rate of beat. Phases of heat rigor parallel to those of cardiac muscles	Q_{10} for interval (0°–20°) 16·0–2·04
Oxygen	Oxygen is not directly required by the contractile mechanism, but is required for prolonged activity		Same as for heart muscle	Same as for heart muscle
pH	Frog's sinus active from pH 4–9·5. Acids reduce speed. Excess of OH′ reduces rate of relaxation	Acids reduce rate of contractions	Active between pH 5–9·5. Acids reduce speed. Excess of OH′ reduces speed of relaxation	Active between pH 5–8·0. Acids reduce speed of movement
Calcium	Absence of Ca·· causes stoppage in diastole. Ca·· can be partially replaced by OH′	? Relatively insensible to absence of Ca··	Absence of Ca·· causes stoppage in relaxed position. Ca·· can be partially replaced by OH′	Absence of Ca·· causes stoppage. Ca·· can be partially replaced by OH′
Non-electrolytes	Will beat normally in isotonic dextrose for a considerable period			
Osmotic pressure	The amplitude but not the rate is affected in hypertonic solutions	No diminution in rate in hypertonic solutions, but form of pulsation is irregular	The amplitude but not the rate is affected in hypertonic solutions	Form of movement changed by hypertonic solutions
Veratrin	Can replace potassium	[Not known]	Can replace potassium	[Not known]

in vertebrate muscle. It is clear, however, that this conclusion must rest on the proof that the activity of the cilia leads directly to a decrease in the amount of glycogen present (see Gray, 1928, p. 108). In drawing comparisons between muscular and other contractile tissues a second type of difficulty arises. The contraction of a striped muscle fibre involves at least four processes. (i) The process whereby an applied stimulus excites a localised spot on the surface of the fibre. (ii) The process of contraction which occurs at the point of

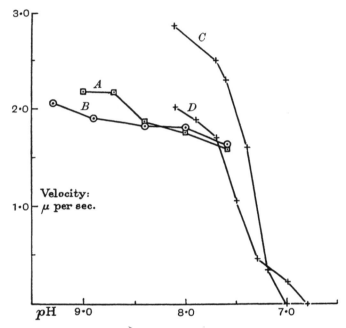

Fig. 170. Effect of hydrogen ions on velocity of movement of *Amoeba limax*. (From Pantin.)

stimulation. (iii) The process whereby a contraction started at one point is transmitted along the fibre. (iv) The process whereby an excited fibre, having carried out a contraction, is able to return to a state in which a second stimulus will initiate a second contraction similar to the first. Even in different types of muscle cell these four phenomena are not equally clearly defined. Thus, in a sinus preparation of the heart, the initiation of contraction occurs in the tissue itself or on its surface; it is, in fact, a spontaneous event. When an attempt is made to compare the whole cycle in a muscle, in a cilium, and in a pseudopodium it is not easy to make sure that a given environmental change is affecting the same point in each of the three types of tissue. The physical nature of excitation, inhibition and conduction are by no means clear; at the same time when we mechanically inhibit a ciliated comb of *Beroë*, and compare it with precisely the same phenomenon in the muscular dorsal fin of *Motella*, it is difficult to believe that the physiological mechanism is different in the two cases, however different the morphological form of the two systems. It is largely a matter of personal opinion how far a working hypothesis is valuable or otherwise and, if we bear in mind the difficulties, there is much to be said for exploring the possibilities of a mechanism common to all types of contractile cells.

Amoeboid movement

The visual phenomena of amoeboid movement have been extensively described by Schaeffer (1920), whose textbook together with the original papers of Pantin (1923–30) includes most of our knowledge of this particular type of contraction. Only a very brief consideration of the main phenomena is given below.

When an amoeba of the *proteus* type is in motion, endoplasmic granules can be seen streaming forward in the direction of the advancing pseudopodium; inclusions within the superficial ectoplasm are, on the other hand, motionless. In the great majority of cases there is no backward current of endoplasm. This fact is of importance for it indicates that, during motion, the ectoplasm is not a rigid inert sheath (see p. 456), but represents a phase which is undergoing a constant change in form and distribution as long as motion persists. When a pseudopodium is being formed the ectoplasm must either be exhibiting plastic flow or must be actively expanding; when a pseudopodium is being withdrawn, the ectoplasm must be flowing in the direction of movement or be in an active state of contraction. So far as is known, the thickness of the ectoplasmic layer does not alter materially when such movements occur, and we must therefore assume that when a pseudopodium is advancing, ectoplasm is constantly being generated from endoplasm. When the pseudopodium is being withdrawn ectoplasm is being converted into endoplasm. In *Amoeba limax* this process of 'endoplasmic/ectoplastic conversion' is somewhat similar, for as shown by Bütschli in 1880, endoplasm is probably converted into ectoplasm only at the anterior end of the organism, whereas the reverse process occurs at the posterior end.

Numerous theories have been advanced as possible explanations of amoeboid movement (see Schaeffer, 1920). Roughly they can be divided into two categories. Firstly, we can follow Rhumbler's (1898) hypothesis which suggested that the cyclical conversion of endoplasm into ectoplasm and *vice versa* is not simply the result of

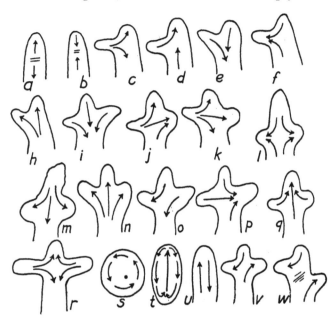

Fig. 171. Diagram illustrating endoplasmic currents in pseudopodia of *Amoebae*. *a* and *b* illustrate opposing currents; *c* to *r* are of common occurrence; *s*, rotational streaming seen in some species only; *t*, fountain currents in *A. blattae*. (From Schaeffer.)

endoplasmic streaming but is actually its cause. Conceptions of this type have been developed by Hyman (1917), Schaeffer (1920) and by Pantin (1923), all of whom look upon the ectoplasm of an *Amoeba* as an actively contractile sheath which generates energy and squeezes the endoplasm forward (see fig. 175). On the other hand, it is conceivable that the source of energy for amoeboid movement lies in the endoplasm rather than in the ectoplasm. If the movement within the endoplasm is entirely passive it should cease when no obvious changes occur in the shape of the *Amoeba*, and the latter is at rest. According to Schaeffer (1920), when an *Amoeba* is suspended in a jelly, so that no locomotion occurs, the endoplasm continues to stream forward, and under these circumstances there is a backward current of ectoplasm which completes the path of circulation. This phenomenon is comparable to that which normally occurs in *A. blattae* (fig. 171 *t*), in *Paramecium* or within a plant cell, although in these cases the ectoplasm itself does not move, and the backward current, like the forward current, is restricted to the endoplasm. Two alternatives seem possible. Either the source of energy in all cases lies in the endoplasm, or the endoplasmic streaming of *A. limax* or *A. proteus* is of a different nature to that in other forms.

Typical endoplasmic streaming or cyclosis was investigated in some detail by Ewart (1903) and more recently by Schaeffer (1920). In plant cells (e.g. of filamentous algae such as *Chara* and *Nitella*) the superficial layer of ectoplasm is rigidly attached to the cellulose wall and takes no part in the movement. As in *Amoeba*, the velocity of streaming is greatest in the central layers of the endoplasm and least at the point of junction of endoplasm with ectoplasm; while at the inner surface of the endoplasm some degree of movement is exhibited by the adjacent layers of the cell sap (see fig. 172). It was these facts which led Ewart to conclude that the energy for movement was developed in the endoplasm. A similar type of movement

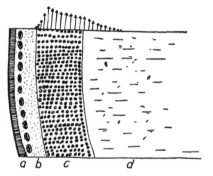

Fig. 172. Section of a cell of *Chara* showing movement of globules in the endoplasm. *a*, cell wall; *b*, ectoplasm; *c*, endoplasm; *d*, cell sap. The arrows indicate by their length the relative velocity of streaming. (From Schaeffer.)

has been described by Schaeffer (1920) in *Frontonia leucas* (fig. 173). This ciliate feeds on filamentous algae. 'As soon as the tip of a filament is well in the mouth and in contact with the endoplasm, streaming begins in the endoplasm in the region of the mouth and takes a direction directly back against the aboral wall, almost, if not quite perpendicular to the longitudinal axis. This stream of endoplasm carries the filament back to the aboral wall, sometimes pushing out the wall a considerable distance. Presently, however, the filament is carried posteriorly along the aboral wall by the streaming protoplasm, which has by this time become rotational, and after reaching the posterior end the filament is brought up along the oral wall. The rotational streaming continues until the entire filament is wound up' (Schaeffer, p. 98). This description of streaming protoplasm suggests that the active centres of movement lie in the endoplasm and that the ectoplasm is passively stretched to accommodate the volume of the ingested filament. It is, however, possible that an active extension of ectoplasm occurs which sets the endoplasm in motion and which sucks the filament into the interior of the cell. On the other hand, the marked similarity of the rotational type of streaming to that seen in plant cells suggests that the ectoplasm may be an inert rather than an actively contractile sheath.

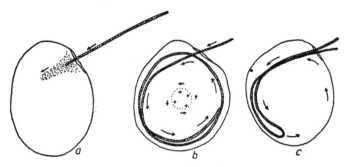

Fig. 173. Ingestion of an algal filament by *Frontonia leucas*. *a*, beginning of ingestion; note endoplasmic streaming in front of the tip of the filament; *b*, almost two complete coils of the alga have been rolled into the *Frontonia* by rotary streaming, note absence of streaming at the centre; *c*, ingestion of a double strand of algal filament. (From Schaeffer.)

It is perhaps worth while remembering that some pseudopodia are purely ectoplasmic and contain no endoplasm—this is the case in some leucocytes and possibly also in *Amoeba radiosa*. Such pseudopodia are markedly contractile. Pseudopodia of this type probably bear some relationship to the surface films characteristic of leucocytes. These films have recently been examined by Fauré-Fremiet (1929); they are often not more than 0.25μ in thickness and may be as thin as 0.1μ. These figures give some conception of the small dimensions of the fundamental units of contraction.

The various theories of amoeboid movement have been summarised by Schaeffer (1920); it is, however, of interest to give some indication of the trend of modern thought. We have seen that Rhumbler attached significance to the process of conversion of endoplasm into ectoplasm and *vice versa*, and that this point of view is maintained by modern authors. Jennings (1904, 1915) on the other hand, denied that this cyclical process proceeds continuously. He observed that a foreign particle attached to the upper surface of *A. proteus* was carried forward to the anterior end as the animal moved, it then passed down to the undersurface where it remained fixed in reference to the substratum until the *Amoeba* had passed over it, after this it was again carried forward to the anterior end. Jennings concluded that an *Amoeba* 'rolled' forward much as though it were a fluid mass enclosed within distensible walls. These results are difficult to harmonise with Dellinger's (1906) observation that pseudopodia can be pushed out independently of a substratum, and with the work of Schaeffer, who has shown that the behaviour of a particle attached to

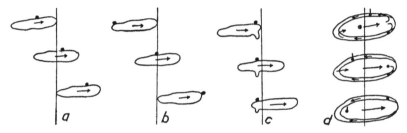

Fig. 174. Motion of *Amoebae* relative to the substratum and to an attached particle. In *a* the particle is moving at the same speed as the organism; in *b* the particle is moving forward over the surface of the *Amoeba*, e.g. *A. discoides*; in *c* the particle is stationary relative to the substratum, and is moving backwards relative to the surface of the organisms; this occurs when attached particles are heavy; *d*, movement of ectoplasm in an *Amoeba* suspended in a jelly. The vertical lines represent a fixed point in the environment. (From Schaeffer.)

the surface by no means always conforms to Jenning's description (see fig. 174). In some cases, it seems that the mechanism responsible for the movement of surface particles bears no real relationship to the mechanism of amoeboid locomotion, for according to Schaeffer (1920) there may exist a peripheral flow to the anterior end, whereas a particle motionless in the underlying ectoplasm may be carried in the opposite direction.

One of the simplest types of amoeboid movement is that displayed by *Amoeba limax*, which can be regarded as obtruding a single persistent pseudopodium at the anterior end. In this type the endoplasm and its included granules stream towards the advancing pseudopodium (fig. 175), whose anterior end consists of clear fluid ectoplasm (Pantin, 1923). The endoplasmic granules flow forward until they reach a position just behind this hyaline cap—there they are checked as though there were a barrier to the entrance of the ectoplasmic area. 'The ectoplasm at the sides of the clear anterior region seems to become more rigid. Since the ectoplasm in the middle and that which is anterior is advancing continuously, this solidification at the sides results in the formation of a tube of gelated ectoplasm. As the advancing pseudopodium continuously adds fresh solid ectoplasm, each portion of this tube, once formed, moves further and further back towards the hind end of the amoeba' (Pantin, 1923, p. 30). Pantin concludes that the tube of gelated ectoplasm contracts continuously, and at the hind end is contracting as fast as the fresh ectoplasm is formed at the anterior end so that there is a constant volume of ectoplasm present during motion. The interpretation

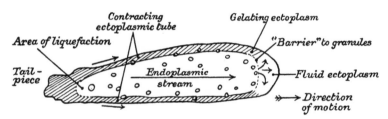

Fig. 175. Diagram illustrating the mechanism of movement of *A. limax*. The mechanical energy is generated in the ectoplasm while the endoplasm flows forward passively. (From Pantin.)

given by Pantin is, as he says, essentially the same as that suggested by Rhumbler, and opposed to those of Jennings or Dellinger: Rhumbler, however, described a reverse current of endoplasm or ectoplasm, whereas according to Pantin there is no posteriorly directed flow of ectoplasm. Relative to the substratum a particle included in the ectoplasmic sheath is motionless; as the endoplasm flows forward, however, the particle comes to lie nearer and nearer to the posterior end of the *Amoeba* (fig. 176), where eventually it becomes incorporated into the endoplasm owing to the conversion of ectoplasm into endoplasm. In *A. limax*, therefore, we may regard endoplasmic flow as the direct result of ectoplasmic contraction, but the whole cycle of events involves (i) the formation of liquid ectoplasm from endoplasm at the anterior end of the animal, (ii) solidification of ectoplasm as it is forced away from the anterior end, (iii) the power of the ectoplasm to contract and force endoplasm forwards, (iv) the liquefication of ectoplasm at the posterior end and subsequent conversion into endoplasm.

If the ectoplasm is actively contractile it is of interest to consider how far this process is comparable to the other types of contractile mechanisms. By staining certain types of *Amoeba* with neutral red, Pantin inferred that there is some indication that the ectoplasm is more acid in the neighbourhood of an active pseudopodium than is the case elsewhere. Whilst it is, admittedly, somewhat difficult to estimate the precise significance of his observation, Pantin concluded

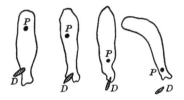 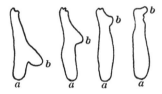

A. Four successive drawings to show migration of particles (*P* and *D*) to the hind end of *Amoeba limax*.

B. Migration of a retracting pseudopodia *b* to the hind end of an advancing *Amoeba*. Direction of movement is towards *a*.

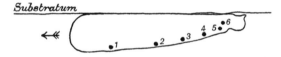

C. Successive positions of a single large particle embedded in the ectoplasm of *Amoeba limax*.

Fig. 176. *A–C*, figures illustrating the progressive changes in position of particles or pseudopodia on an *Amoeba limax*. In *A* and *C* note that as the *Amoebae* moves forward so the attached particles (which are stationary relative to the substrate) come to lie nearer to the posterior end of the organism. (From Pantin.)

tentatively that the normal contraction of ectoplasm is due to the production of an acid: this leads to an imbibition of water with local swelling. In this way a pseudopodium is protruded. In order to account for the subsequent solidification of the ectoplasm it is assumed that the acid raises the surface tension of the ectoplasm, thereby leading to the condensation of any substances which will decrease the free surface energy. As the gelated tube passes back towards the hind end of the *Amoeba*, the acid is neutralised, the imbibed water is lost, and so contraction occurs.

It would be difficult to test this hypothesis by direct observation, but it is of definite interest since it raises the question as to how far all types of protoplasmic motion are of the same fundamental nature (see p. 452). This general aspect of amoeboid movement has been considered by Pantin (1923, 1930 a–c). Roughly speaking, the evidence is based on the reaction of *Amoeba limax* to acids (fig. 170), oxygen, and narcotics. Each of these distinctive reagents, together with others of a minor nature, exert on amoeboid movement effects which are directly comparable with their effects on muscle and on ciliated cells. Pantin (1930 a) has recently shown that a marine *Amoeba* of the *limax* type will continue to move when oxygen is completely absent from its environment, although movement gradually decreases in speed, and ceases after 6–12 hours (fig. 177). The actual cessation of movement

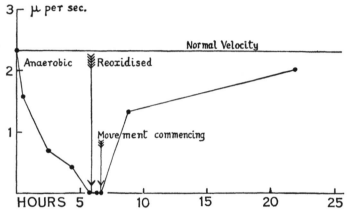

Fig. 177. Velocity of movement of *Amoeba* during anaerobiosis and recovery. (From Pantin.)

is abrupt and is accompanied by morphological changes and by an apparent rise in the viscosity of the protoplasm. This is clearly analogous to the effect of oxygen lack on muscle and on cilia. Pantin concluded that the effect is exerted on the cell either by an exhaustion of an essential precursor substance of unknown nature or, more probably, by the accumulation of products which are normally removed from the cell when oxygen is present. If an *Amoeba* has been subjected to oxygen lack, it will recover if oxygen is again provided in the environment. It is interesting to note that these *Amoeba* are insensitive to changes in oxygen tension until the latter falls below 30–40 mm. Hg (fig. 178) (Pantin, 1930 b), and that the resultant phenomena are to be associated with the pressure of the oxygen

CONTRACTILE CELLS

present rather than with the total amount of oxygen which can diffuse into the cell in a given time. This fact, together with the effect of cyanides and anaesthetics, is explicable by the hypothesis that an essential part of the cycle involves the presence of cytochrome or of a similar body (Pantin, 1930 c). It is clear that the resemblance between the amoeboid cycle and those of muscle and of cilia, is quite well defined, although at present we cannot demonstrate experimentally that amoeboid movement automatically involves an increase in oxygen consumption by the cell nor do we know the nature of the substance which provides the energy for movement.

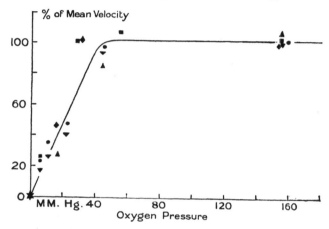

Fig. 178. Percentage mean velocity of *Amoeba* after eight hours in sea-water at various oxygen pressures. (From Pantin.)

Vibratile movement[1]

The small size of even the largest flagella and cilia not only limits the methods available for their study but raises difficult problems of dimensions. Cilia, which appear under the microscope to be moving with extreme rapidity, are actually moving with a very low velocity. It is important to realise that all ciliary movement is essentially slow, although it may produce results superficially similar to those of more rapid muscular movement. In terms of their own dimensions a flagellated spermatozoon is probably moving at approximately the same relative speed as an average fish, but the absolute velocity of the latter is very much greater.

[1] For a fuller account of vibratile movement reference may be made to Gray (1928) from which the following account has been extracted.

One theory after another has been advanced to elucidate the essential mechanism of a cilium, but no convincing correlation has yet been made between the structure as seen under the microscope and what is known of the changes occurring in the living cell. When we remember that really nothing is known of the machinery whereby a muscle develops a tension when it is stimulated, it is not surprising that the ciliary mechanism also remains obscure. The morphological structure of ciliated cells is peculiarly constant no matter what is the organ or the animal in which it is found (fig. 179 *A* and *B*). For the detailed morphology of the ciliary mechanism reference should be made to suitable monographs (Erhard, 1910; Prénant, 1913; Saguchi, 1917).

Table LXX. (See *Tabulae Biologicae*, vol. IV)

Organism	Absolute speed of movement (per sec.)	Relative speed = multiple of organisms' length in 1 sec.
Bacillus subtilis	10 μ–15 μ	5
Spirillum volutans	110 μ	8·5
Euglena sp.	115–235 μ	3–5
Paramecium	1·3 mm.	6
Spirostomum	600 μ	0·6
Halteria saltans	2·3 mm.	77
Amoeba	0·5 μ–2·5 μ	0·02
Alectrion (ciliary movement)	2·5 mm.	0·01
Fish	50–200 cm.	1–4
Spermatozoa (fish)	33–180 μ	2–8
Man (walking)	150 cm.	0·75

In its simplest form the vibratile element shows no optical structure even under polarised light. When viewed by transmitted light or on a dark background both cilia and flagella are optically homogeneous. When viewed under polarised light an apparent doubly refracting property is observed (Engelmann, 1898), but there can be little doubt that the phenomenon is due to diffraction at the surface of the flagellum or cilium and not to its internal structure (MacKinnon and Vlès, 1908).

It has been shown that nearly all the larger cilia occurring in nature can be resolved into a number of smaller units each of which possesses the power of movement as long as it is attached to the cell. According to Taylor (1920) the cirri of *Euplotes patella* are to be regarded as a series of vibratile filaments embedded in a viscous or sticky matrix. On the other hand Carter (1924) showed that the *laterofrontal* cilia on the gills of *Mytilus* are normally composed of a series of triangular plates in close contact with one another, but no matrix appears to be present. After displacement from their

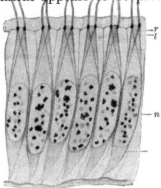

Fig. 179 A. Six laterofrontal cells of the gill of *Mytilus edulis*. n, Nucleus; f, common fibril; r, l, ciliary rootlets. (After Bhatia.)

Fig. 180. Laterofrontal cilium of *Mytilus* distorted by a needle (p). The whole cilium is composed of a series of triangular platelets similar to that seen to the left of the needle. (After Carter.)

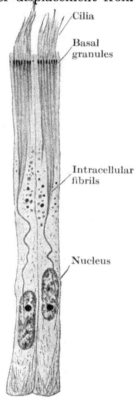

Fig. 179 B. Side view of cells of typhlosole of *Anodon*. (From Gurwitsch.)

normal position individual plates readily regain their normal position and beat in unison. A comparable structure occurs in the 'undulating membrane' (*membranella*) of *Blepharisma* (Chambers and Dawson, 1925). On puncture with a needle (fig. 181) this membrane at once splits along a line running through the puncture and part of the membrane breaks into a series of very fine cilia which

466 CONTRACTILE CELLS

beat out of unison. As soon as the needle is removed from the membrane the row of isolated cilia quickly reforms an apparently homogeneous membrane, the elements of which beat synchronously.

The nature of the ciliary mechanism has long been the subject of morphological discussion, and no serious attempt will be made to gather together the numerous theories which have been put forward. Adequate reviews from the historical and other points of view have been made by Pütter (1903), Erhard (1910), and Prénant (1913).

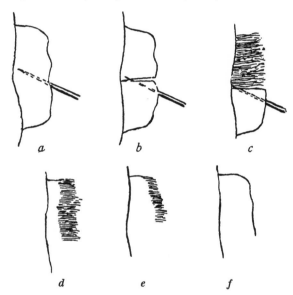

Fig. 181. Microdissection of the membranella of *Blepharisma undulans*. When the membranella is held by the needle, as in *a*, it splits at the point of contact, and the region beyond the needle breaks into a series of very fine cilia which beat out of unison. On removing the needle the cilia rapidly unite again (*d–f*) to form an apparently homogeneous membrane. (After Chambers and Dawson.)

There are, however, a limited number of facts which are of value, and which must eventually be woven together into a complete theory. In the first place, it is certain that the motivity of the cilium is independent of the cell structure as a whole. Peter (1899) was the first to show that when a ciliated protozoon is crushed into quite small fragments the cilia continue to beat as long as they are in organic connection with a fragment of cytoplasm. Erhard (1910) and Gray (1922) have shown that profound changes can occur in the proximal cytoplasm and in the nucleus without entailing a cessation

CONTRACTILE CELLS

of ciliary movement. According to Engelmann (1898) enucleated spermatozoa exhibit active movement. There is, therefore, general agreement that the whole of the essential ciliary mechanism lies at the distal end of the cell. Beyond this point there is marked divergence of opinion. Peter observed that cilia completely isolated from the distal protoplasm are motionless, and since basal granules are always present in this region it may not unreasonably be suspected that they are the kinetic centres of the system. Engelmann (1898) stated that if the tail of the frog's spermatozoon be cut off from the 'middle piece' all movement ceases, whereas if the cut is made between the nucleus and the 'middle piece' active movement continues. A similar state of affairs can be observed in the *lateral* epithelium of the gills of *Mytilus* (see fig. 182) where the cilia can be stripped away from the cells, leaving the basal granules behind; such cilia are always motionless.

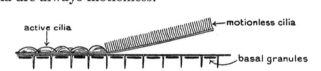

Fig. 182. Excised strip of lateral epithelium of *Mytilus* (diagrammatic). Note that the cilia are active as long as they are in organic communication with the cells.

The views of Peter received considerable support from the so-called 'Henneguy-Lenhossék theory' of the origin of the basal granules. According to these authors the basal granule of a cilium is homologous and sometimes identical with the centrosome of cell division. Just as the centrosome was regarded as the kinetic centre for nuclear division, so in its rôle as basal granule it was the kinetic centre for ciliary movement. It is impossible to discuss in any detail the cytological evidence in support of these conclusions, although it is in some cases remarkably strong. The latest adherents to Henneguy's view are Jordan and Helvestine (1922), who claim that the ciliated cells in the epididymis of the rat divide amitotically because the division centres of the cells are functioning as basal granules (see also p. 164).

The movement of an individual cilium

When observed under the microscope most cilia appear to be moving with very high velocity, and it seems difficult to accept the statement that even the most rapid types seldom attain a velocity of twenty feet per hour. A simple calculation, however, rapidly disposes of any conception of a cilium as a rapidly moving unit. Let the length of a cilium be 10μ and let it oscillate through an amplitude of $180°$ twelve times every second; the total distance travelled in 1 second by the tip of the cilium is therefore

$$12\,[\pi\,(10\,\mu)] = 375\,\mu \text{ (approximately)}.$$

According to Kraft (1890), the velocity of the effective stroke is five times that of the recovery stroke so that during the former phase the tip of the cilium will move with an approximate velocity of 1 mm. a second or 12 feet per hour. Without allowing for changes of velocity during the complete cycle, Bidder estimates the velocity of the tips of the flagella of sponges at 14 feet per hour. 'We forget, as we look through a microscope...that though distance is magnified, time is not magnified' (Bidder, 1923, p. 302). It must not be forgotten, however, that these slow linear velocities are associated with angular velocities of quite a high order; in the first example given the cilium is moving with the angular velocity of a flywheel running at 360 revolutions a minute.

The difficulty of observing the beat of a healthy cilium under the microscope is not so much due to its apparent high linear velocity as to the inability of changing the axis of the human eye at a frequency equal to the frequency with which the cilium changes its direction of movement. This difficulty can be overcome by observation in intermittent light of suitable frequency (Gray, 1930).

Various modifications of vibratile movement have been known for many years (Valentin, 1842), but it is only comparatively recently that any detailed description has been given of the behaviour of those units (excluding flagella) which are known to perform a considerable amount of mechanical work.

The first of the more modern observations are those of Williams (1907). Unfortunately the parent organism is only described as 'an unidentified but common larva of a protobranch mollusc' from Narragansett Bay. The large velar cilia (see fig. 183), by which the larva moves, are arranged along the edges of an overhanging groove, each cilium being somewhat curved with its concave face towards the side of the effective stroke (fig. 183, *A*). This position can be readily observed whenever the animal comes to rest, but the details of the beat can only be seen in larvae of reduced activity (e.g. when under a coverslip). In its resting position (fig. 183, *A*) the cilium lies wholly outside the adjacent groove and the first phase of the beat is such that the base of the cilium moves backwards to fit quite closely into the groove (fig. 183, *B*). This is the end of the backward or preparatory movement. The forward stroke of the cilium now takes place and is divided into two distinct phases: (i) the base of the cilium begins to move upwards and forwards, whilst the overhanging lip of the groove prevents the movement of the cilium as a whole (fig. 183, *C* and *D*); (ii) as the forward movement of the base continues the more distal part of the cilium eventually ceases to be crimped against the edge of the groove, and is suddenly released; it flies forward, with a very rapid effective stroke, far past

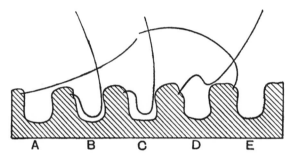

Fig. 183. *A*, Position of rest of velar cilium as described by Williams. *B*, Backward preparatory stroke; note that the cilium fits into the velar groove. *C*, *D*, Opening phases of forward effective stroke; note that the cilium is crimped against the side of the groove. In *D* the cilium has just released itself from the groove and rapidly flies forward to position *E*. From *E* it swings back to the position of rest at *A*. The cilium is therefore more flexible during the backward preparatory stroke than during the forward effective stroke.

its position of rest (fig. 183, *E*). Finally the cilium swings back to its position of rest and the cycle is repeated.

The essential features to note are that (i) during its preparatory stroke the cilium accommodates its form to that of the groove and is obviously very easily flexed, (ii) during the forward stroke it is not so easily flexed but offers considerable resistance to the edge of the groove.

The form of the beat of the frontal and abfrontal cilia on the gills of *Mytilus* was described by the author (Gray, 1922, 1930, fig. 184). In this case the difference in flexibility during the two phases of the beat is very clear. During both the effective and recovery strokes the cilium is moving against the resistance of the water; but during the effective stroke the cilium behaves as a fairly rigid rod fixed at one end to the cell and its tip traces an arc of 90°; during the recovery stroke on the other hand the cilium moves backwards as a limp thread along which a backward stress is passing from base to tip. Both effective and recovery strokes take place in the same plane. These observations were confirmed by Carter (1924) for the large abfrontal cilia on the gills of *Mytilus*, who showed that during the effective beat the whole cilium is rigid throughout its whole length: during the recovery beat the cilium is limp and frequently slips below the microdissecting needle which is used as an obstruction. According to Carter these properties remain for some time

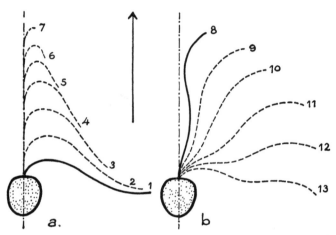

Fig. 185. Simplest type of movement of flagellum of *Monas* during rapid forward movement. *a*, 1–7, Successive stages in preparatory stroke. Note the flexure begins at the base and spreads to the tip. *b*, 8–13, Successive stages in the effective stroke. Note the rigidity of the cilium. The arrow indicates the direction of movement of the organism. (After Krijgsman.)

after the cilium is brought to rest, for they can be observed by moving the cilium artificially through the two phases of its beat.

Recently Krijgsman (1925), using dark ground illumination, has analysed the movements of the flagellum of *Monas*. This protozoon can move forward either slowly or rapidly, it can move backward, and it can move laterally. The simplest movements of the flagellum during forward movement are shown in fig. 185. It will be noted that the form of the flagellum during the two phases of the beat is essentially the same as that of the frontal cilia of *Mytilus*. The recovery stroke (fig. 185, *a*) takes place by means of a bending movement which begins at the base of the flagellum and spreads to its tip.

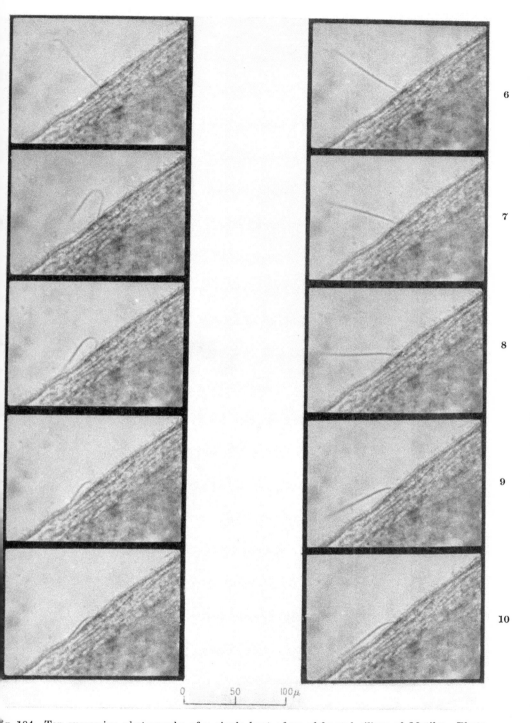

Fig. 184. Ten successive photographs of a single beat of an abfrontal cilium of *Mytilus*. Photographs 1–5 show the form of the recovery stroke; photographs 6–10 show the form of the effective stroke. Interval between each photograph 0·05 second. (Gray 1930.)

In the effective stroke the flagellum moves back as a more or less rigid structure (fig. 185, b).

The observations of van Trigt (1919) show very clearly how the flagellum of *Spongilla* may also exhibit marked variations in the form of movement, and how these variations are associated with varying degrees of activity (fig. 186).

When moving with its normal rapid frequency, the flagellum of *Spongilla* exhibits a series of waves whose amplitude[1] and wave length are both small. These waves pass from the base to the tip of the flagellum with considerable velocity and a current of water is thereby driven forwards from the cell in a direction parallel to the long axis of flagellar movement (fig. 186, a). Soon after isolation from the

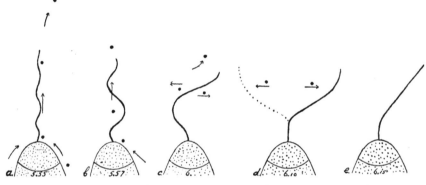

Fig. 186. Successive stages in the reduction in speed of the flagellar movement of a choanocyte of *Spongilla*. The collar is entirely retracted. The arrows indicate the direction of the water currents: the figures indicate the times of observation. a, Rapidly moving flagellum immediately after isolation of cell; b, two minutes later, note increased amplitude of waves; c, five minutes after isolation; d, fifteen minutes after isolation; e, the flagellum at rest. Note the flexed position of rest. (After van Trigt.)

animal, however, the activity and frequency of vibration decrease, and at the same time the amplitude and the wave length of the individual waves increase (fig. 186, b and c). Eventually the wave length and amplitude of the slowly moving flagellum become so large that the undulatory type of movement appears to pass into one of lateral displacement only, as in fig. 186, d; finally the flagellum comes to rest in a more or less straight condition. There can be little

[1] By 'amplitude' is meant the extent of the displacement of any point on the flagellum in a line at right angles to the main axis of the moving flagellum. By 'wave length' is meant the length of flagellum occupied by one complete wave.

doubt that during typical undulatory movement the flagellum forms part of a helix and that the diameter of the helix increases with decrease in the frequency of vibration. Van Trigt states that in some cases there is no doubt that the point of the flagellum during rapid movement travels in an elliptical orbit, but from his description one may infer that the axes of the ellipse are not increased uniformly when the 'amplitude' is increased. This point is of some theoretical importance and will be discussed later.

Of the numerous attempts to analyse the propulsive action of a flagellum, perhaps the best known is that of Bütschli (1889)[1]. His explanation was as follows. Any point (P) on the flagellum carries out a series of lateral movements and in moving from side to side exerts a pressure on the water at right angles to its surface. If this force be PA, it can be resolved into two forces PB and PC. The force PB is directed parallel to the long axis of the organism and tends to drive it through the water cell foremost. The force PC is at right angles to PB and tends to rotate the organism on its own axis. Hensen (1881) analysed the movement of a flagellate spermatozoon as follows. 'If the undulating membrane be examined, waves are seen to travel over its surface from the front to the back. Any one point moves laterally to and fro through a given distance with a force F. This force can be resolved into two components; one force kF is exerted tangentially to the membrane and tends to compress the latter; the other force k_1F is at right angles to the surface of the membrane. k_1F can in turn be resolved into two components, k_2F which exerts a backward pressure on the water and drives the organism forward, and k_3F at right angles to the surface which tends to rotate the organism on its own axis; opposed to this latter force is an equal and opposite force exerted by a point in which the transverse direction of movement is in the opposite phase.'

Both Bütschli and Hensen appear to have overlooked the fact that in order to produce propulsion there must be a force which is always applied to the water in the same direction and which is independent of the phase of lateral movement. There can be little doubt that this condition is satisfied in flagellated organisms not because each particle of the flagellum is moving laterally to and fro but by the transmission of the waves from one end of the flagellum to the other, and because the direction of the transmission is always the

[1] For a recent analysis of the mechanics of vibratile structures, see Ludwig, W. (1930), *Zeit. f. wiss. Biol.* Abt. C, 13, *397*.

same. A stationary wave, as apparently contemplated by Bütschli, could not effect propulsion since the forces acting on the water are equal and opposite during the two phases of movement. If, however, the waves are being transmitted in one direction only as is shown in fig. 187, definite propulsive forces are present which always act in a direction opposite to that of the waves. The propulsive power of the flagellum is equivalent to that which would be produced by projecting along the length of the flagellum a series of 'humps' of the same form as the waves, the velocity of the humps being made equal to the velocity of movement of each wave. If the waves pass from the base of the flagellum to its tip, the organism is driven forward in front of the flagellum; if the waves pass from the tip to

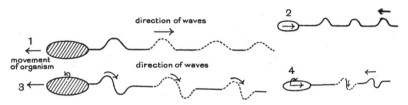

Fig. 187. Diagram illustrating the dependence of the direction of movement of an organism on the direction of the waves which pass along its flagellum. 1, The waves originate at the base of the flagellum and travel straight along it to its tip: the animal moves straight ahead with the flagellum pointing backwards. 2, The waves originate at the tip of the flagellum and pass to its base: the animal moves with the flagellum in front. 3, The waves originate at the base of the flagellum and pass along and round it in a clockwise direction: the animal moves with the flagellum behind and at the same time rotates in an anti-clockwise direction. 4, The waves originate at the tip and pass backwards with an anti-clockwise rotation: the animal moves with the flagellum in front and rotates in a clockwise direction.

the base the organism is drawn forward with the flagellum in front. If the waves pass along the flagellum in one plane there will be no force tending to rotate the animal on its axis: if, however, the waves pass round the flagellum as well as along it the organism will rotate. As the waves rotate, pressure is exerted on the water in such a way as to drive water in the direction in which the waves are travelling, so that the flagellum with the attached cell tends to rotate in the opposite direction, just as when a cilium beats in one plane the movement of the organism is in the direction opposite to that of the effective stroke. If the waves pass clockwise round the flagellum, the organism will be rotated anti-clockwise, as was observed by Reichert (1909) in *Spirillum*. 'The body on account of the right-handed rotation of the flagellum (?), always turns to the left about its long axis.'

In fig. 188 each of the rotatory forces (e.g. x, y) generated by the waves on the flagellum can be resolved into a transverse component (e.g. a, d), and the sum of these causes a rotation of the organism about its longitudinal axis vh.

The cilium and flagellum as actively contractile units

Concerning the essential nature of a cilium there are two radically opposite views. According to Schäfer (1904) the whole of the mechanical energy liberated by the cilium is derived as such from the cell and is transmitted as mechanical energy to be liberated as work by virtue of the elasticity of the cilium. According to Heidenhain (1911), however, the mechanical energy liberated by the cilium is stored as chemical energy in the cilium itself until a disturbance arises in the cell or elsewhere which liberates it as mechanical energy. In other words, we may look on the cilium as a passive unit mechanically operated by the cell, or as an active structure potentially capable of autonomous contraction in all or some of its elements. From a morphological point of view these possibilities were considered by Williams (1907).

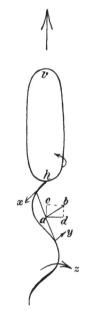

Fig. 188. Diagram illustrating the forces of rotation of *Spirillum*. The arrow z indicates the direction in which the waves travel round the flagellum. (After Reichert.)

The chief arguments in favour of the passive nature of vibratile organs appear to be as follows: (i) The flagella of *Spirillum* are only 0.05μ in diameter (Fuhrmann, 1910), and it is difficult to conceive how any adequately heterogeneous mechanism can be contained in a cross section which approaches ultramicroscopic dimensions. (ii) Many cilia and flagella show no evidence of structural complexity even under the highest magnifications. (iii) Most cilia are motionless when separated from the cell. (iv) Some observers, including Williams (1907), claim to have seen active movements in the body of the cell corresponding in frequency to the movement of the cilia. These and other facts have led to the suggestion that the cilium or flagellum does not generate mechanical energy along its whole length but is merely a means whereby this type of energy, when generated at the distal end of the cell, is transmitted away from its source and is liberated for work. This view was in fact adopted by Peter (1899), Benda (1901), Joseph (1902) and others.

There appear to be two ways in which to test the conception of the flagellum as an inert but flexible rod. Firstly, if a wave can pass down a flagellum at constant speed and without change in form, then the wave must be part of a sine wave (otherwise as it moves along it will resolve itself into a series of sine waves of different wave lengths each travelling with a different velocity, so that the form of the wave will vary at different points along its path). Now in order to produce a sine wave in an elastic rod it is necessary to exert a bending force and a longitudinal thrust, and it is very difficult to see how such forces could be exerted by movements entirely restricted to that part of the flagellum which is in contact with the cell. Secondly, if a vibrating rod is to exhibit no reflected waves which travel back towards their original source, the vibrations must be very highly damped; but if the damping is high, the amplitude of the waves must invariably decrease as they pass from their source, and this is apparently not the case in living flagella.

The evidence, as far as it goes, seems to suggest that a flagellum cannot be regarded as an inert flexible unit operated mechanically by the cell, since waves of varying form pass along its length without change in shape and without apparent change in amplitude.

That movements of a flagellum or cilium are not infrequently accompanied by mechanical movements within the cell is highly probable, but this evidence does not show that the flagellum or cilium is not also an actively contractile element. The contractile properties at the distal end of ciliated cells have recently been investigated by Merton (1924, 1927).

Almost overwhelming evidence in favour of the active nature of the cilium or flagellum is provided by the following facts.

(i) In some cilia there can be no doubt that the angular velocity of the tip is greater than that of the base, so that the whole cilium becomes bent during the effective stroke. It is exceedingly difficult to see how such a movement could be effected by the liberation of kinetic energy at the extreme base of the cilium; on the other hand, the nature of this movement can be readily visualised by the assumption that the effective stroke is due to the development of a bending force along the whole length of the cilium.

(ii) Among the Mastigophora it is not at all uncommon for the flagellum to extend forward in front of the cell and to move the animal through the water by movements restricted to its tip. This type of movement is seen, for example, in *Peranema* and in *Monas*. It hardly seems possible to believe that sufficient kinetic energy can be transmitted along the proximal part of the cilium to induce active movements at the tip. Dr Bidder informs me that he has seen, not infrequently, lateral movement of the extreme tip of a sponge flagellum although the rest of the flagellum remained motionless and straight. These facts together with the observation that sudden bends may occur at any point on a slowly moving flagellum can only be explained by the assumption that kinetic energy can be generated at any point in the flagellum.

(iii) As already mentioned there is no evidence that all waves of distortion passing along a flagellum undergo changes in amplitude. If this could be supported by quantitative measurement, quite conclusive evidence would be available to show that the flagellum is an active unit capable of generating kinetic energy along its length. A moving wave cannot provide the energy for propelling an organism and at the same time pass on with unreduced amplitude unless the energy being lost is continually being replaced as the wave moves along.

(iv) In all tractella, or flagella which draw an organism forward through the water, the waves of distortion pass from the tip of the flagellum to its base. It hardly seems possible to imagine that such movements could be exhibited by an inert filament whose sole source of energy lay in the body of the cell.

We may therefore conclude that a flagellum or a cilium is to be regarded as an active unit (comparable perhaps to a muscle fibre) in that it generates its own mechanical energy. When a stimulated muscle fibre develops a tension, the force of the contraction is directed along the longitudinal axis of the fibre; we do not know how this force is produced, but if it can be produced longitudinally in a muscle fibre, it is not unreasonable to suppose that it can be produced transversely and also as torsion in a flagellum or cilium. Consider, for example, an india-rubber rod. If its ends are compressed it shortens and a longitudinal strain exists between adjacent particles. If it is bent a transverse strain is set up. If one end is fixed and the other is twisted torsional strains arise. We can imagine that in a muscle a longitudinal strain is set up internally and shortening is the result. If a corresponding transverse strain is set up internally in a cilium it will cause bending; and a torsional strain will cause twisting. In other words the essential changes in a muscle fibre and in a flagellum may very likely be the same, although they occur along different axes.

From the evidence already given we may conclude that the mechanical energy for movement is generated along the length of the cilium and is not transmitted as hydrostatic pressure by the cell.

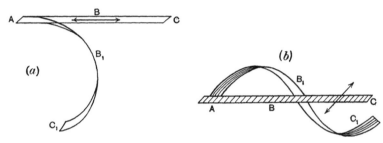

Fig. 189. Figures showing the form of a strip of paper caused to bend by adding water to one side. In (a) the strip ABC has been cut so that the natural axis of bending (indicated by the arrow) is parallel to the longitudinal axis of the strip and the latter bends into an arc of a circle AB_1C_1. In (b) the strip has been cut so that the axis of bending is inclined at an angle to the longitudinal axis of the strip; the latter bends into a helix. Suitable strips can be cut from a sheet of note paper; type (a) is provided by a strip cut parallel to the shorter edge of the paper; type (b) by a strip cut along a diagonal of the sheet.

We know that when water is removed from the cilium by plasmolysis, the cilia of *Mytilus* or of the frog come to rest at the beginning of their preparatory stroke (Gray, 1922), so that there is some slight experimental evidence in favour of regarding the normal stroke as due to a redistribution of water within the cilium. If one side of a straight cilium becomes capable of absorbing more water than the other, the cilium will bend into the arc of a circle whose convex side is the side containing the excess of water. A model can readily be made with a strip of paper (fig. 189). If both sides of the strip are equally dry or equally wet the strip remains flat, but if one side becomes more moist than the other the paper bends so that the damp side is outermost. As the moisture soaks through the paper the strip gradually straightens. We may picture a cilium as in fig. 190. Whilst at rest, water is equally distributed across its whole section, movement occurs because certain elements lying nearer to the side *AB* than to *CD* acquire an affinity for water, thereby causing water to flow from the side *CD* to *AB*; *AB* expands and *CD* contracts—thereby bending the cilium. Relaxation occurs because *AB* loses its increased affinity and the water distributes itself equally over the cross section of the cilium. The essential change in the affinity for water may well be due to a modification in the degree of

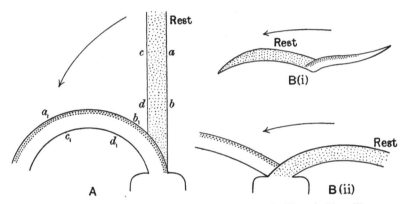

Fig. 190. Hypothetical mechanism of different types of cilia. *A*, The cilium at rest is straight, and the water molecules are uniformly distributed. Contraction occurs by the aggregation of water molecules to one side of the cilium, causing this side to become convex. Relaxation occurs when the water molecules again distribute themselves uniformly. *B* (i), The cilium is at rest at the end of the effective stroke and in this position the water molecules are uniformly distributed. The preparatory stroke is complete when the water is aggregated to one side. *B* (ii), The cilium is at rest at the beginning of the effective stroke and during the effective stroke the water is aggregated to one side.

ionisation of a protein or similar molecule: thus, the localised production of an acid would produce this effect whilst the neutralisation of the acid would return the system to the *status quo*.

The following points in this scheme are noticeable: (i) the general mechanism is that applicable to muscle (see Hill and Hartree, 1920), since an essential feature is a localised change in the distribution of water; (ii) both effective beat and recovery beat are active processes which depend on the rate at which water is taken up by the activated units and the rate at which it is lost; (iii) it can readily be applied to pendular cilia or to typical helical movement, and by postulating a system of localised control on the part of the cell, a limited length or area of a flagellum can be activated whilst the remainder continues at rest.

In comparison with a muscle fibre a ciliated cell generates forces of very small magnitude; even numerous active collar cells of sponges cannot produce a hydrostatic pressure greater than 4 mm. of water (Parker, 1914). Clearly ciliary activity would be quite unable to provide the force required to maintain an active circulation throughout the body of a large animal. At the same time a ciliated epithelium can set in motion relatively heavy bodies. Maxwell (1905) showed that a weight of 5·0 gr. can be moved by 58 sq. m. of the frog's oesophageal epithelium at a velocity of 1 mm. per minute. Roughly speaking 100 sq. m. of epithelium can perform 10 gr. mm. of work per minute, while one cilium of *Paramecium* exerts, according to Jensen (1893), a force of $4·5 \times 10^{-7}$ mg.

The metabolism of ciliated cells

As already mentioned ciliary activity involves the consumption of oxygen—and as far as can be seen the rate at which oxygen is taken up is almost a linear function of the rate at which the cilia will drive a particle over a ciliated surface (Gray, 1923) (fig. 191). In the absence of O_2, movement continues for 30–45 minutes (Gray, 1924), and in this respect ciliary movement is clearly analogous to many types of muscular movement. Although it is almost certain that the energy of movement is derived in the long run from the oxidation of a reserve material in the cells, the nature of this substance is not yet established, although it may be glycogen in the case of *Pecten* tissues (see Gray, 1928, p. 108) and in actinians (Beutler, 1929). That there are at least five steps in the complete ciliary cycle can be inferred from the way in which the ciliated cells of *Mytilus* react to changes in their environment. The detailed evidence for this has been discussed elsewhere (Gray, 1928) and need only be summarised here. (i) Within the cells is a reserve (G) of chemical energy, which may or may not be glycogen. (ii) There is also a derivative of this reserve, the quantity of which derivative (X) determines the period during which the cilia will beat when there is no longer any possibility of obtaining fresh supplies from the ultimate reserve in the cells. (iii) The rate at which the cilia beat depends on the temperature, and on the nature of the cations present in the environment of the cells. (iv) The actual contractile mechanism must be intact, and this is only possible if the osmotic pressure of the environment lies within certain limits. (v) Free O_2 must be available. Many of the facts are summarised in fig. 192, but for a more detailed description reference may be made to Gray (1928).

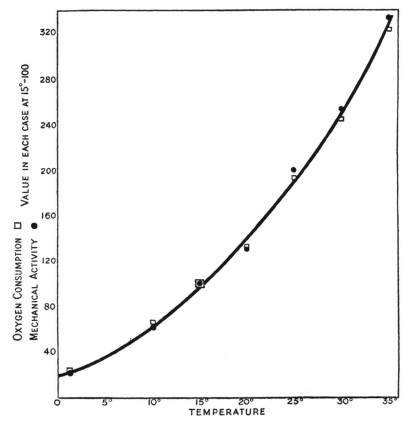

Fig. 191. Diagram showing the effect of temperature on the oxygen consumption and on the mechanical activity of the frontal cilia of *Mytilus*.

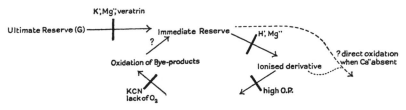

Fig. 192. The bars which cross the arrows indicate the point in the ciliary cycle at which the various reagents exert their characteristic effects.

Excitation and inhibition of cilia

If a piece of frog's epithelium or gill of *Mytilus* be exposed to conditions (e.g. CO_2, KCN, cold) under which movement is extremely slow or just abolished, the cilia can be induced to beat more actively for a short time by mechanical irritation with a needle or other instrument. Kraft (1890) found that a mechanical stimulus thus applied to a localised region of a weak ciliary field induced activity along a band beginning at the spot stimulated and continuing for some distance in the direction of the normal effective beat (fig. 193). The activity never extended far in the opposite direction or laterally from the point of stimulation. It is not clear from these experiments how far the main band of induced activity is the result of mechanical stimulus by the water current set up by the area (1, 2, 3, 4) subjected to direct stimulation.

Mechanical activation of stationary cilia also occurs in Ctenophores. If a complete ciliated comb is removed from *Pleurobrachia*, its cilia exhibit active movement for some time after excision, but after a time the movements gradually subside. From this condition active movement can again be evoked by mechanical agitation, and a series of well-defined beats are observed in which the metachronal waves pass away from the direction of the effective beat.

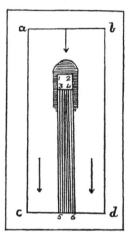

Fig. 193. Transmission of excitation over ciliated epithelium a, b, c, d of the frog's oesophagus. The area 1, 2, 3, 4 was stimulated mechanically. The shaded area shows the secondarily excited area. Note that the transmission is almost entirely restricted to the area 3, 4, 5, 6. The arrows indicate the direction of the effective beat of the cilia. (After Kraft.)

It may be noted that these excitatory effects of stimuli are almost always restricted to tissues in a subnormal condition of activity and it may be doubted whether they ever occur in a normal epithelium *in situ*. Merton (1923 b) states that if a small fragment from the lips of the snail *Physa* be dissected out with the attached nerve the cilia soon come to rest unless the nerve is stimulated. As Carter (1926) points out it is difficult to make sure that the stimulus is conducted to the cilia by the nerve and not in a more direct manner. On the whole the phenomenon of extraneous excitation of cilia recalls the effect of such treatment on moribund or quiescent cardiac muscle: in both cases the healthy tissue is automatically rhythmical, but when this property is lost for any reason it can be induced to reappear for a short time by providing an external stimulus.

Whereas excitatory effects of external stimuli are comparatively rare, it is by no means uncommon to find clear instances of ciliary inhibition in perfectly normal tissues either *in situ* or after excision. Locomotory cilia are almost invariably under the control of the animal and this control is often, if not always, of an inhibitory nature.

The clearest cases of inhibition are those found in the higher invertebrata. The ciliated larvae of annelids, molluscs and polyzoa are all capable of bringing their cilia to rest, and they do so at irregular intervals or when externally stimulated. A typical instance has recently been investigated by Carter (1926) in the velar cilia of nudibranch larvae. During life these cilia show well-marked alternations of rest and activity; if, however, a portion of the velum be excised from the animal, the cilia on this portion cease to show any interruption of activity and beat steadily for many hours. Carter has shown that narcotics, used in such concentrations as are necessary to anaesthetise nerves, abolish the periods of ciliary rest in the intact larva. Similar results were obtained with other types of larvae by Merton (1923 *a*), but Carter was the first to demonstrate the presence of the inhibitory nerve fibres which control the cilia. The power of ciliary inhibition is also lost by the Ctenophore *Beroë* when the animal is exposed to 0·2 per cent. chloral hydrate (Göthlin, 1920). Taken in conjunction with the fact that nerve endings can be traced to intimate association with the ciliated cells (figs. 194, 195), the evidence strongly suggests that the normal periods of rest shown by the velar cilia are due to nervous inhibition.

Another interesting example of 'controlled' locomotory cilia is found in the snail, *Alectrion trivitta* (Copeland, 1919). This animal moves by means of its ciliated foot; when the animal is at rest the cilia are motionless, but progression is entirely due to the activity of the cilia. Copeland found that when a resting animal is stimulated by touching one of its tentacles with a piece of fish meat the proboscis is extended and the pedal cilia begin to beat. At first sight this appears to be a reflex response involving the excitation of the cilia by a motor nerve. The phenomenon is not simple, however, for ciliary movement is greatly reduced or ceases altogether when the proboscis is worked over the surface of the foot as is often the case. On withdrawal of the proboscis the cilia show increased activity. On excising the foot, exactly the same series of events occur as in the Ctenophores. Immediately after excision, the cilia of *Alectrion* are motionless, but after a brief interval short outbursts

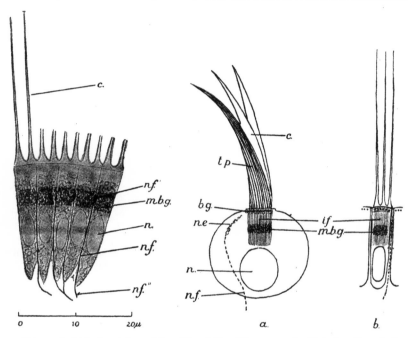

Fig. 194. *Archidoris tuberculata.* Living cells of the velar cilia stained *intra vitam* with methylene blue showing nerve fibrils. *n.f.*, nerve fibrils between the ciliated cells; *m.b.g.*, granules; *n.*, nucleus; *n.f."*, nerve fibrils passing into the tissues of the body; *c.*, cilium. (After Carter.)

Fig. 195. Diagram of ciliated cells of the molluscan velum. *a*, side view of cilium; *b*, end view. Note the compound nature of the cilium. *t.p.*, triangular plates of cilium in side view; *b.g.*, basal granules; *i.f.*, internal fibres; *m.b.g.*, granules which stain with methylene blue; *n.*, nucleus; *n.f.*, nerve fibrils between the ciliated cells; *c.*, cilia. (After Carter.)

of ciliary movement occur simultaneously with muscular twitchings. Gradually these periods of ciliary and muscular activity become more and more frequent, and during the periods of muscular quiescence the cilia no longer cease to beat, although they beat more slowly than during muscular activity. Finally, after some hours, when all muscular movement has ceased, the cilia are found to be beating regularly at a constant rate which continues as long as the preparation remains intact.

In Ctenophores ciliary control is well defined. All these animals possess an aboral 'sense organ' (fig. 196) from which finely ciliated grooves pass to the aboral end of each row of ciliated plates. In *Pleurobrachia* the normal mode of progression is with the mouth forward and with the contractile tentacles extended posteriorly. If the animal is given a slight blow the ciliated plates instantly become motionless, the tentacles contract, and the organism sinks slowly in the water. After a short pause the cilia begin to beat again, the tentacles relax, and progressive movement is resumed. A marked case of mechanical inhibition has been described by R. S. Lillie (1906) in the genus *Eucharis*.

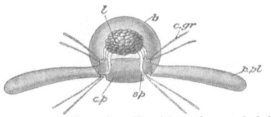

Fig. 196. 'Sense organ' of the Ctenophore *Hormiphora plumosa*. *b*., bell; *c.p.*, ciliated plate; *c.gr.*, ciliated groove leading to locomotory ciliated plates; *l.*, lithites; *p.pl.*, polar plate; *sp.*, spring. (From Parker and Haswell.)

It seems clear that three distinct types of ciliary movement exist: (*a*) cilia which are normally in an active state of movement independent of any obvious external stimulus, (*b*) cilia which are motionless or only feebly active except when a stimulus is applied, (*c*) cilia which are active but can be brought to rest by some type of inhibitory control. The parallel to cardiac muscle is again striking. In both tissues one condition is fairly readily convertible into another except from type (*a*) to (*c*).

Metachronal rhythm

A cursory examination of any active ciliated epithelium reveals the fact that although the cilia may be beating at the same rate they are not beating in the same phase. Any particular cilium is slightly in advance of the cilium behind it in the series and slightly behind the one just in front of it. This regular sequence is known as *metachronism* and is observable in nearly all ciliated epithelia. Since all the cilia lying in the same line across the epithelium beat in approximately the same phase, regular waves of activity can be seen passing over the surface, thereby giving the well-known analogy to the waves which pass over a field of corn when exposed to the breeze. The crests of the metachronal waves are cilia at the zenith of their effective stroke; the troughs of the waves are cilia at or near the beginning of the recovery stroke. The form of the waves has recently been described by the author (Gray, 1930).

The direction of the metachronal waves varies with different tissues. In the case of the frog's epithelium, or the frontal cilia of *Mytilus*, the waves move in the direction of the ciliary current, i.e. in

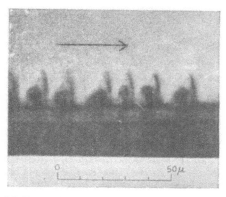

Fig. 197. Lateral epithelium of *Mytilus edulis* gill. The effective beat is in a plane at right angles to the plane of the paper: the metachronal wave passes in the plane of the paper in the direction of the arrow.

the direction of the effective stroke of the cilia. In Ctenophores, however, the waves usually move in exactly the opposite direction. In these animals the effective stroke of the ciliated plates is in an aboral direction, so that the animals move mouth foremost. The metachronal waves, however, pass over the row of cilia from the aboral to the oral end. In the *lateral* epithelium on the gills of *Mytilus* the metachronal wave moves at right angles to the effective stroke (fig. 197).

Although the direction of the metachronal wave thus differs in different tissues, it is remarkably constant in each particular case. In the frog's oesophagus, von Brücke (1916) showed that if portions of the epithelium were excised and replaced after being turned through 180° the original direction of the metachronal wave remained, although diametrically opposed to that of the rest of the epithelium. In *Mytilus* the lateral wave passes in opposite directions on the two sides of the gill filament. It is only in certain Ctenophores that any evidence of wave reversal is found. In *Pleurobrachia* a rapid wave, on reaching the oral end of the ciliated row may be reflected aborally, but according to Parker (1905 b) it seldom passes over more than one-third of the whole row. Verworn (1890) induced reversed waves by stimulating the oral end of the row.

In the case of the Ctenophores there is no doubt that, during life, the activity of the cilia is co-ordinated to some degree by definite 'sense' organs at the aboral end of the animal. At the same time metachronal rhythm is a fundamental property of ciliated epithelium, just as the power of automatic movement is a property of the individual cells. The beautiful rhythm of the *lateral* epithelium of *Mytilus* can be seen in gill fragments, whose proximal and distal ends have both been cut. The problem presented by the facts resolves itself into finding out how automatic units when side by side in a tissue beat in an orderly sequence.

REFERENCES

References marked with an asterisk contain either a general discussion or an extensive bibliography

BENDA, C. (1901). 'Ueber neue Darstellungsmethoden der Zentralkörperchen und die Verwandtschaft der Basalkörper der Cilien mit Zentralkörperchen.' *Verh. physiol. Ges. Berlin*, *147*.

BEUTLER, R. (1929). 'Liefert das Glykogen die Energie für den Flimmerschlag?' *Zeit. f. wiss. Biol.* Abt. C, 10, *440*.

BHATIA, D. (1926). 'The structure of the latero-frontal cells of the gills of *Mytilus edulis*.' *Quart. Journ. Micr. Sci.* 70, *681*.

BIDDER, G. P. (1923). 'The relation of the form of a sponge to its currents.' *Quart. Journ. Micr. Sci.* 67, *293*.

v. BRÜCKE, E. T. (1916). 'Versuche an ausgeschnittenen und nach einer Drehung um 180° reimplantierten Flimmerschleimhaut-Stücken.' *Pflüger's Archiv*, 166, *45*.

BÜTSCHLI, O. (1889). 'Protozoa.' *Broon's Thierreich*. (Leipzig.)

CARTER, G. S. (1924). 'On the structure and movements of the laterofrontal cilia of the gills of *Mytilus*.' *Proc. Roy. Soc.* B, 96, *115*.

—— (1926). 'On the nervous control of the velar cilia of the nudibranch veliger.' *Brit. Journ. Exp. Biol.* 4, *1*.

CHAMBERS, R. and DAWSON, J. A. (1925). 'The structure of the undulating membrane in the ciliate *Blepharisma*.' *Biol. Bull.* 48, *240*.

COPELAND, M. (1919). 'Locomotion in two species of the gastropod genus *Alectrion* with observations on the behaviour of the pedal cilia.' *Biol. Bull.* 37, *126*.

—— (1922). 'Ciliary and muscular locomotion in the gastropod genus *Polinices*.' *Biol. Bull.* 42, *132*.

DELLINGER, O. P. (1906). 'Locomotion of amoebae and allied forms.' *Journ. Exp. Zool.* 13, *337*.
—— (1909). 'The cilium as a key to the structure of contractile protoplasm.' *Journ. Morph.* 20, *170*.
ENGELMANN, T. W. (1898). *Dictionnaire de Physiologie.* (Paris.)
ERHARD, H. (1910). 'Studien über Flimmerzellen.' *Archiv f. Zellforsch.* 4, *309*.
EWART, A. J. (1903). *Protoplasmic Streaming in Plants.* (Oxford.)
FAURÉ-FREMIET, E. (1929). 'Caractères physico-chimiques des choano-leucocytes de quelques invertébrés.' *Protoplasma*, 6, *521*.
FUHRMANN, F. (1910) 'Die Geisseln von *Spirillum volutans.*' *Centralb. f. Bakteriol.* II, 25, *129*.
GÖTHLIN, G. F. (1920). 'Inhibition of ciliary movement in *Beroë cucumis.*' *Journ. Exp. Zool.* 31, *403*.
GRAY, J. (1922). 'The mechanism of ciliary movement.' *Proc. Roy. Soc.* B, 93, *104*.
—— (1923). 'The mechanism of ciliary movement. III. The effect of temperature.' *Proc. Roy. Soc.* B, 95, *6*.
—— (1924). 'The mechanism of ciliary movement. IV. The relation of ciliary activity to oxygen consumption.' *Proc. Roy. Soc.* B, 96, *95*.
*—— (1928). *Ciliary Movement.* (Cambridge.)
—— (1930). 'The mechanism of ciliary movement. VI. Photographic and stroboscopic analysis of ciliary movement.' *Proc. Roy. Soc.* B, 107, *313*.
HEIDENHAIN, M. (1911). *Plasma und Zelle*, 1, 2. (Jena.)
HENSEN, V. (1881). *Hermann's Handbuch der Physiol.* 6, *89*.
HILL, A. V. (1925). 'The surface tension theory of muscular contraction.' *Proc. Roy. Soc.* B, 98, *506*.
—— (1926 a). *Aspects of Biochemistry.* (London.)
—— (1926 b). 'The laws of muscular motion.' *Proc. Roy. Soc.* B, 100, *87*.
HILL, A. V. and HARTREE, W. (1920 a). 'The thermo-elastic properties of muscle.' *Phil. Trans. Roy. Soc.* B, 210, *153*.
—— (1920 b). 'The four phases of heat-production in muscle.' *Journ. Physiol.* 54, *84*.
HYMAN, L. H. (1917). 'Metabolic gradients in ameba and their relation to the mechanism of ameboid movement.' *Journ. Exp. Zool.* 24, *55*.
JENNINGS, H. S. (1904). 'The movements and reactions of *Amoeba*.' *Carn. Instit. Wash. Publ.* 16, *129*.
—— (1915). *Behaviour of the Lower Organisms.* (New York.)
JENSEN, P. (1893). 'Die absolute Kraft einer Flimmerzelle.' *Pflüger's Archiv*, 54, *537*.
JORDAN, A. E. and HELVESTINE, F. (1922). 'Citogenesis in the epididymis of the white rat.' *Anat. Record*, 25, *7*.
JOSEPH, H. (1902). 'Beiträge zur Flimmerzellen und Centrosomenfrage.' *Arb. a. d. Zool. Instit. Wien*, 14, *1*.
KRAFT, H. (1890). 'Zur Physiologie des Flimmerepithels bei Wirbelthieren.' *Pflüger's Archiv*, 47, *196*.
KRIJGSMAN, B. J. (1925). 'Beiträge zum Problem der Geisselbewegung.' *Archiv f. Protist.* 52, *478*.
LILLIE, R. S. (1906). 'The relation between contractility and coagulation of the colloids in the ctenophore swimming-plate.' *Amer. Journ. Physiol.* 16, *117*.

MacKinnon, D. L. and Vlès, F. (1908). 'On the optical properties of contractile organs.' *Journ. Roy. Micr. Sci.* 553.
Maxwell, S. S. (1905). 'The effect of salt solutions on ciliary activity.' *Amer. Journ. Physiol.* 13, *154*.
Merton, H. (1923 a). 'Studien über Flimmerbewegung.' *Pflüger's Archiv,* 198, *1*.
—— (1923 b). '"Willkürliche" Flimmerbewegung bei Metazoen.' *Biol. Zentralb.* 43, *157*.
—— (1924). 'Lebenduntersuchungen an den Zwitterdrüsen der Lungenschnecken.' *Zeit. f. wiss. Biol.* 1, *671*.
—— (1927). 'Zur Frage nach dem motorischen Zentrum der Flimmerzellen.' *Zeit. f. Anat. u. Entwickl.* 83, *222*.
Pantin, C. F. A. (1923). 'On the physiology of amoeboid movement. I.' *Journ. Mar. Biol. Assoc.* 13, *24*.
—— (1924). 'On the physiology of amoeboid movement. II. The effect of temperature.' *Brit. Journ. Exp. Biol.* 1, *519*.
—— (1926 a). 'On the physiology of amoeboid movement. III. The action of calcium.' *Brit. Journ. Exp. Biol.* 3, *275*.
—— (1926 b). 'On the physiology of amoeboid movement. IV. The action of magnesium.' *Brit. Journ. Exp. Biol.* 3, *297*.
—— (1930 a). 'On the physiology of amoeboid movement. V. Anaerobic movement.' *Proc. Roy. Soc.* B, 105, *538*.
—— (1930 b). 'On the physiology of amoeboid movement. VI. The action of oxygen.' *Proc. Roy. Soc.* B, 105, *555*.
—— (1930 c). 'On the physiology of amoeboid movement. VII. The action of anaesthetics.' *Proc. Roy. Soc.* B, 105, *565*.
Parker, G. H. (1896). 'The reactions of *Metridium* to food and other substances.' *Bull. Mus. Comp. Zool. Harvard,* 29, *107*.
—— (1905 a). 'The reversal of ciliary movement in metazoans.' *Amer. Journ. Physiol.* 13, *1*.
—— (1905 b). 'The movements of the swimming-plates in Ctenophores with reference to the theories of ciliary metachronism.' *Journ. Exp. Zool.* 2, *407*.
—— (1914). 'On the strength and volume of water currents produced by sponges.' *Journ. Exp. Zool.* 16, *443*.
—— (1918). *The Elementary Nervous System.* (Philadelphia.)
Peter, K. (1899). 'Das Centrum für die Flimmer- und Geisselbewegung.' *Anat. Anz.* 15, *271*.
*Prénant, A. (1913). 'Les appareils ciliés et leurs dérivés.' *Journ. de l'Anat. et Physiol.* 49, *150, 545*.
*Pütter, A. (1903). 'Die Flimmerbewegung.' *Ergeb. d. Physiol.* 2, 11, *1*.
Reichert, K. (1909). 'Ueber die Sichtbarmachung der Geisseln und die Geisselbewegung der Bakterien.' *Centralb. f. Bakteriol.* i, 51, *14*.
Rhumbler, L. (1898). 'Physikalische Analyse von Lebenserscheinungen der Zelle.' *Arch. f. Entw. Mech.* 7, *103*.
—— (1899). 'Physikalische Analyse von Lebenserscheinungen der Zelle.' *Arch. f. Entw. Mech.* 9, *32*.
—— (1905). 'Zur Theorie der Oberflächenkraft der Amöben.' *Zeit. wiss. Zool.* 83, *1*.
—— (1914). 'Das Protoplasma als physikalisches System.' *Ergeb. d. Physiol.* 14, *474*.

RITCHIE, A. D. (1928). *The Comparative Physiology of Muscular Tissue.* (Cambridge.)
SAGUCHI, S. (1917). 'Studies on ciliated cells.' *Journ. Morph.* 29, *217.*
*SCHAEFFER, A. A. (1920). *Ameboid Movement.* (Princeton.)
SCHÄFER, E. A. (1904). 'Theories of ciliary movement.' *Anat. Anz.* 26, *517.*
TAYLOR, C. V. (1920). 'Demonstration of the function of the neuromotor apparatus in *Euglotes* by the method of micro-dissection.' *Univ. Cal. Publ. Zool.* 19, *403.*
VAN TRIGT, H. (1919). *A Contribution to the Physiology of the Freshwater Sponges.* (Leiden.)
VALENTIN, G. (1842). *Wagner's Handwörterb. der Physiol.* 1, *484.*
VERWORN, M. (1890). 'Studien zur Physiologie der Flimmerbewegung.' *Pflüger's Archiv,* 48, *149.*
WILLIAMS, L. W. (1907). 'The structure of cilia, especially in Gastropods.' *Amer. Nat.* 41, *545.*

CHAPTER EIGHTEEN

Phagocytosis

The power of cells to ingest solid particles is widespread among the animal kingdom, and is often of far-reaching significance. In many protozoa, coelenterates, and molluscs, ingestion is an essential preliminary to digestion; in other cases ingestion may lead to the elimination of waste or foreign particles from the animal's body.

In attempting to analyse the mechanism whereby the particles are absorbed into the interior of the cell, at least three distinct phenomena must be considered. Given a particle situated at some distance from a phagocytic cell, ingestion will only occur if the two units can come into intimate contact with each other; having come into contact, it is necessary for the cell and the particle to adhere to each other to prevent a chance disturbance from separating them again; finally there occurs the actual process of ingestion whereby an adherent particle is drawn into the interior of the phagocyte.

If a cell is to come into contact with a particle situated some distance from it, it follows that either the cell or the particle or both must be in motion. Clearly the chances of collision will depend on the rate at which the two components approach each other and on the distance which they are apart at any given moment. If, for example, there are a large number of particles present in a suspension of leucocytes the chances of a collision between a cell and a particle will be greater than if a smaller number of particles are present. As Fenn (1920 a) has shown, the phagocytic activity of a suspension of cells and inert particles can be expressed by a logarithmic formula

$$k = \frac{1}{t} \log \frac{a}{a-x} \qquad \ldots\ldots(\text{xix}),$$

where a is the number of particles originally present, and $a - x$ is the number which are uningested at time t. These results are obviously in harmony with the view that the same percentage of uningested particles is being taken up by the cells at any given instant. The facts suggest that the rate of ingestion of the inert particles may be proportional to three variables—the number of cells present, the number of particles present, and the chance of collision between a cell and a particle. Equation (xix) will only hold good if the number of active cells is constant during the whole period of observation—in other words a leucocyte which has ingested one or more particles must be just as likely to ingest another particle as a leucocyte which has not yet encountered a particle at all. The proof of this conclusion is presented by McKendrick (1914) as follows. If ϕ be the probability that an individual leucocyte in a culture will ingest a particle within a given time, and if there be V_0 leucocytes present on adding the particles, then ϕV_0 leucocytes will ingest a particle. When this has occurred there will be $V_0 - \phi V_0$ leucocytes

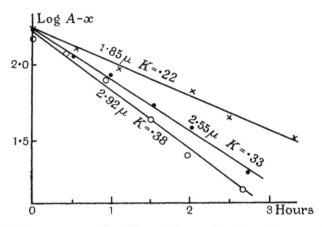

Fig. 198. Ordinates represent logarithms of the number of particles not taken up by leucocytes plotted against the time as abscissa. The slope of the graph (K) is shown for quartz particles of three different sizes. (From Fenn, 1920.)

without ingested particles, and the rate of reduction of this latter class will be given by equation (xx)

$$\frac{dV_0}{dt} = -\phi V_0 \qquad \ldots\ldots(\text{xx}).$$

If a leucocyte which has ingested a particle has an undiminished chance of ingesting a second particle, and after that a third and so on, there will arise a series of classes containing one, two, three, ... x particles. The rate at which any given class will increase will be proportional to the difference between the number of individuals in the class containing one less particle and the number of individuals which already contain the given number. Thus the rate, at which the number of leucocytes containing five particles will increase, is dependent on the number of leucocytes which already contain four particles, and on the number which already contain five—since any one of the latter by ingesting another particle passes out of the five class

$$\frac{dV_x}{dt} = (V_{x-1} - V_x)\phi \qquad \ldots\ldots(\text{xxi}).$$

From these two equations, McKendrick derives a formula by which the expected distribution can be calculated

$$V_x = \frac{V_0 \left[\log_e \frac{a_0}{V_0}\right]^x}{x!} \qquad \ldots\ldots(\text{xxii}),$$

where a_0 is the total number of leucocytes present. The validity of this analysis is shown in Table LXXI.

Table LXXI. Random distribution of bacteria among ingesting leucocytes. (From McKendrick)

No. of bacteria present in a leucocyte x	No. of leucocytes in each class (V_x)			
	Exp. 1		Exp. 2	
	Obs.	Calc.	Obs.	Calc.
0	19	(19)	41	(41)
1	59	58	126	119
2	98	88	154	173
3	88	90	164	168
4	65	68	121	122
5	37	42	62	71
6	17	21	36	34
7	8	9	35	14
8	5	4	5	5
9	2	1	2	2
10	1	—	3	—
11	0	—	1	—
12	1	—	—	—

If we assume that the rate of ingestion of particles can be expressed by equation (xxiii)

$$-\frac{dn}{dt} = k.L.n\phi \qquad \ldots\ldots(\text{xxiii}),$$

where L is the number of leucocytes, and n is the number of un-ingested particles, then k (the index of phagocytosis, as defined by Fenn) will depend on the chance of collision (ϕ) between leucocyte and particle. This in turn will depend on the relative size of the particles and on their density of distribution relative to the leucocytes. The larger the particles the greater are the chances of a collision with a leucocyte. At the same time, if particles and cells are settling in a test tube the chances of a collision will depend upon the rate at which the particles settle relative to the rate of settling of the cells. If a suspension of cells of diameter C and velocity Vc, and of particles of diameter P and velocity Vp is allowed to settle in a test tube, the chances of collision (ϕ) will be proportional to the velocity

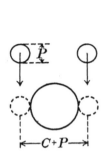
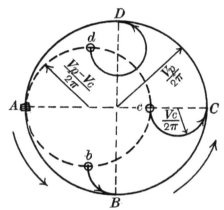

Fig. 199. Diagram illustrating the chance of collision between a particle, P, moving towards a cell, C. (From Fenn, 1920.)

Fig. 200. Diagram illustrating orbits described by a particle at A and cells at b, c and d in a rotating mixture. The figure represents a cross-section of the medium which is rotating to the left. The direction of rotation of the cells and particle is counter-clockwise. Collisions occur at B, C and D. The chance of collision is proportional to the circumference $Abcd$. (From Fenn, 1920.)

of the particles relative to the cells ($Vp - Vc$) and to the square of the sum of their diameters,

$$\phi = (Vp - Vc)(c + P)^2,$$

and this relationship will hold good when the mixtures are not settling under gravity but are slowly rotating on a horizontal drum. Under the latter circumstances the cells and the particles will not settle in a straight line but will each move in circular orbits whose circumference will be V/n, where V is the velocity of movement under gravity and n is the frequency of rotation of the drum (fig. 200). The chance of a collision will be proportional to the circumference $Vp - Vc$, but this will decrease as n increases, and there will come a time when these orbits are so small as to make a collision unlikely; in other words by rotating the suspension more rapidly the amount of phagocytosis should be decreased. To test this conclusion Fenn employed six small tubes (3 mm. in diameter × 1·5 cm. long) sealed at one end. These tubes were filled with suspensions of cells and quartz particles and sealed with paraffin. Three of the tubes were rotated at 0·3 revolutions per minute and three at 19 revolutions per minute. At intervals of about one hour, one tube from each set was removed from the drums and the number of uningested particles determined. It was found that the particles in the slowly rotating tube were ingested 2·5 times more rapidly than those in the rapidly rotating tube: the value k (Table LXXII) being calculated from the theoretical formula.

Table LXXII

Hours	19 revolutions per minute		0·3 revolutions per minute	
	No. of particles not ingested	No. of cells	No. of particles not ingested	No. of cells
0·0	324	76	324	76
0·6	262	76	124	48
1·75	157	—	55	—
2·6	64	53	9	32

Counts refer to volumes of 0·02 c.mm.
$k = 0·22$. $k = 0·52$.

For this particular experiment the average distance between the particles was 37μ. At the slower rate of revolution the diameter of the orbits followed by the cells was 279μ, whereas that of the particles was 462μ. At the more rapid rate of revolution the orbits of the cells were $4·6\mu$, whereas that of the particles was $7·3\mu$. In the latter case the chances of a collision were very small, and were it not for the fact that an ideal distribution of 37μ between particles and cells is statistically improbable, probably no phagocytosis would have occurred, if it is legitimate to neglect the possible effects of Brownian movement and centrifugal force. These observations show that the relative chances of a collision between a cell and a particle can be determined with a satisfactory degree of accuracy, if we know the size and velocity of movement of the particles and cells. Fenn (1920) was able to confirm these results by taking advantage of the fact that small particles settle under gravity more slowly than do larger particles (see Table LXXIII).

These results show fairly clearly that the three variables mentioned on p. 490 each play their part in fixing the rate at which phagocytic cells ingest inert particles of quartz. For any given suspension the rate depends upon the number of cells present, the number of particles present, and on the chance of mutual collision. Using carbon particles instead of quartz the results obtained by Fenn were essentially similar although the value of k sometimes decreased as time proceeded. It is interesting to find that the agreement between the theory for ideally distributed populations holds

Table LXXIII. Comparison of theoretical and experimental rates of phagocytosis by leucocytes of the rat. (From Fenn, 1920)

Average diameter of particles of quartz	$2 \cdot 92\,\mu$	$2 \cdot 55\,\mu$	$1 \cdot 85\,\mu$
Average velocity of settling in cm. per hour	2·81	2·14	1·13
Chances of collision	151·8	61·8	76·5
Relative rates of collision (theoretical)	1·0	0·41	0·51
Average relative value of 'k' as determined experimentally	1·0	0·64	0·61

good for systems in which the particle size and other factors are by no means uniform in all cases.

Since carbon or quartz particles are completely insoluble in water it is not surprising to find that their chances of being absorbed by a leucocyte depend entirely on fortuitous encounter with a cell. In the case of soluble particles this is not the case, and there are several instances known in which the movement of the leucocyte is directed towards a neighbouring particle. A good example of such directive movements is given by Fenn (1922). Using mixtures of quartz and manganese dioxide particles in equal numbers, Fenn found that the frequency of collision between leucocytes and manganese dioxide was twenty-four times as great as with the quartz. It is clear that such anomalous figures are not due to chance. 'In one case (see fig. 201) a leucocyte seemed obviously to be attracted from a distance of perhaps 30 microns toward a group of two manganese and one quartz particles. From its unusually rapid movements the leucocyte appeared to be "in haste" and its path was directly blocked by the quartz particle. Two pseudopods were put out "straddling" this obstruction, each of which promptly ingested a manganese particle. After this feat the leucocyte rounded up and wandered away with its twin load leaving the quartz particle undisturbed. The whole performance could not have lasted over two minutes' (Fenn, 1922, p. 318).

Similar selection was observed for *Penicillium* spores as opposed to quartz, supporting the observation of Commandon (1919) that leucocytes are attracted by starch grains.

If the degree of phagocytosis of inert particles were solely dependent on the chance of collision between particle and cell it

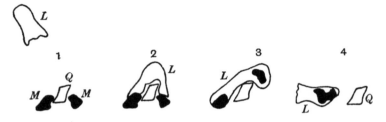

Fig. 201. Diagram of an actual case in which a leucocyte (L) was apparently attracted towards two particles of manganese dioxide (M, M) which it ingested with such precision that it seemed almost 'purposeful', an intervening particle of quartz (Q) being completely neglected. Time about 2 minutes. (From Fenn, 1922.)

follows that carbon particles should be less readily ingested than quartz particles of the same size, since their velocity of movement when settling under gravity is less than that of quartz. In practice, however, this is not the case (Table LXXIV) and Fenn (1920 c) found that carbon particles are ingested three or four times more readily than quartz particles of the same size when allowance is made for the difference in the velocity of movement.

The differences observed in the relative ease of phagocytosis of carbon and quartz particles will be considered in more detail elsewhere (p. 503), at present it is convenient to consider a possible criticism of the method whereby these differences were first established.

When a suspension of leucocytes and particles is actively agitated in a tube, the forcible collisions which occur are hardly comparable to those effected between a moving leucocyte and a stationary par-

PHAGOCYTOSIS

ticle. Fenn (1920 c) therefore investigated the relative ease of phagocytosis of carbon and quartz by enclosing between two coverslips a thin film of a dense suspension of cells containing equal numbers of quartz and carbon particles; after incubation the numbers of uningested particles was determined. Under these conditions the particles were stationary and ingestion occurred by the active movement of the cells. Although under these conditions the cells tend to spread out on the glass surfaces and eventually cease their activities, the technique is a useful one, since the relationship of one cell to different particles can be studied. Thus in three minutes one leuco-

Table LXXIV. Showing the observed and calculated ratios between the index of phagocytosis (k) of carbon and quartz particles. The calculated ratios are based on the assumption that the frequency of ingestion is directly proportional to the chance of fortuitous collision in both cases

Ratio $\dfrac{k \text{ carbon}}{k \text{ quartz}}$	
Observed	Calculated on basis of chance of collision
1·07	0·36
1·0 +	0·36
6·0	1·1
0·87	0·21
2·2	0·19
(1·6	1·9)
(0·77	1·1)

cyte refused one quartz particle and one carbon particle, and ingested three carbon particles. This preference of leucocytes for carbon is illustrated by fig. 202. It is quite clear that the effect of active leucocytes on a mixture of carbon and quartz particles cannot be explained on the basis of chance collisions alone. In this respect Fenn has demonstrated the important fact that although the carbon particles are more readily ingested than quartz, yet nevertheless the number of collisions with leucocytes is the same. A suspension of leucocytes and equal numbers of carbon and quartz particles was prepared at a sufficiently low temperature to keep the cells inactive. Specimens of this culture were then placed on a cool slide and the position of each cell and each particle was noted. The cells were then activated by raising the temperature, and each encounter of a cell with a particle was observed. In one such experiment the field contained 44 carbon particles, 38 quartz particles and 77 leucocytes. After twenty-four minutes of activity the number of observed collisions between leucocytes and quartz was 36, whereas there were 37 between leucocytes and carbon: at the same time only 1 quartz particle had been ingested, whereas 12 carbon particles were ingested in the same period. These observations bring us to the second factor concerned in phagocytosis, namely the ability of the particle to adhere to the surface of the cell after collision had occurred.

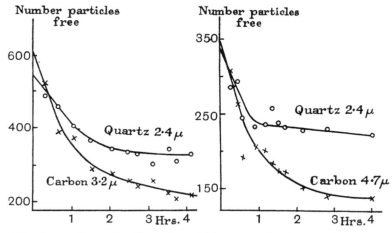

Fig. 202. Curves illustrating the preferential ingestion of carbon particles as opposed to quartz. These data were obtained from suspensions in which the particles were stationary. (From Fenn.)

Actual ingestion is often a comparatively slow process and until it is complete there must exist a mechanism whereby any chance disturbance will not separate the particle from the surface of the cell. Our conception of this mechanism will depend upon whether we are prepared to look upon the cell surface as essentially of a liquid nature. Most workers on phagocytosis have accepted this point of view and it will therefore be considered in some detail.

If a fluid drop (C) (fig. 203) is in contact with a solid surface (G) and with another immiscible fluid P, and the system is in equilibrium, the angle of contact between the drop and the solid surface can be defined in terms of the surface tensions of the three interfaces

$$\cos A = \frac{{}_G T_P - {}_G T_C}{{}_C T_P} \qquad \ldots\ldots\text{(xxiv)}.$$

If $\cos A = 1$, then $_C T_P = {_G T_P} - {_G T_C}$ and the drop C will spread over the surface of G. This type of system clearly conforms to a state of minimum surface energy: if the total surface energy is reduced by spreading, spreading will actually occur. Similar conditions will hold good when the drop C comes into contact with a particle of G instead of with a plane surface. It is well known that many types of leucocytes, when in contact with a glass surface, tend to spread out into thin films after which they degenerate and die. We may look upon this phenomenon from two points of view. We may regard it as a pathological condition in which the cells are losing their natural properties owing to an abnormal environment. On the other hand we may assume that the forces which bind the cells to the glass

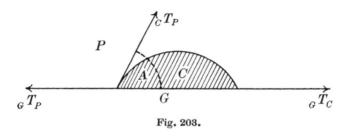

Fig. 203.

are essentially the same as those which bind the cell to a particle of glass during the preliminary phases of ingestion. The latter point of view has been adopted by Tait (1918) and by Ponder (1925) but rejected by Lison (1929). Tait and Gunn (1918) observed that the phagocytic cells of crustacean blood would not ingest fatty particles nor did they tend to spread out on a fatty surface: they would ingest carbon, quartz and glass, and similarly when in contact with surfaces of these substances they spread out to a thin film and then degenerated. The conclusions of Tait and Gunn were as follows. If the cell spreads out on a surface of a compound G, then a particle of G can be ingested by normal cells. If cells will not spread on another compound G_1, then a particle of G_1 may or may not be ingested by normal cells. Tait's conception of ingestion is thus comparable to the spreading of a liquid drop on the surface of a solid or of another immiscible fluid. From this point of view phagocytosis is solely the result of a redistribution of surface energy—and does not involve any active process on the part of the cell. A more elaborate analysis of the theory of adhesion of cells to particles has been developed by

Ponder (1925). If the surface tension at the particle/medium interface be $_GT_P$, that at the cell/medium interface be $_CT_P$, and that at the particle/cell interface be $_GT_C$, then the three phases can remain in contact when the critical angle (θ) between the particle and the free cell surface is defined by equation (xxiv). The values of cos A must lie between -1 and $+1$. If $\frac{_GT_P - _GT_C}{_CT_P} < -1$ the particle would flow round the cell, or if the particle is rigid, no ingestion will occur. If $\frac{_GT_P - _GT_C}{_CT_P}$ is greater than -1 but less than $+1$, then true equilibrium with partial ingestion will occur. If $\frac{_GT_P - _GT_C}{_CT_P}$ is equal to $+1$ or more, then the cell will flow completely round the particle, and digestion will be complete. Ponder also considers the possible effect of electrostatic changes of similar sign at the surfaces of the particle and of the cell and concludes that although work is required to move the particle to within a short distance of the cell surface, there will then be an attractive force which will prevent the particle leaving the surface.

The theories of Tait and of Ponder, like the earlier suggestions of Rhumbler (1898), are open to objections. (i) There is no direct evidence which indicates that the surface of a cell is a fluid—on the other hand there is evidence to show that amoeboid cells are normally coated by a firm pellicle. Unless this pellicle breaks down in the region of contact, the cell (which is assumed to have a fluid interior) will not flow round the particle even if the surface tensions are of the limiting value. (ii) It fails to cover the cases in which ingestion is preceded by the formation of 'food cups'. It is difficult to imagine that the mechanics of food cup formation and of ingestion are fundamentally different. (iii) There is no evidence to show that an alteration in the surface tension of the particle/medium interface affects in any way the phagocytic powers of cells. It might, perhaps, be urged that such evidence would be very difficult to obtain. (iv) In a protein solution all foreign particles are probably coated with a protein film, so that unless this film is destroyed at the point of contact all particles should be ingested with equal ease. (v) Ledingham (1912) found that at low temperatures micro-organisms tend to collect at the periphery of leucocytes but were not ingested. Since surface tension is increased by decreasing the temperature it is difficult to see why cells should fail to ingest at low temperatures if the controlling factor is surface tension.

Tait's observation that *Astacus* leucocytes would not ingest fat droplets is not universally true for phagocytic cells. In attempting to inject various oils into *Amoeba dubia*, Dawson and Belkin (1929) observed that in many cases the drops of oil instead of being injected into the organism were merely brought into intimate contact with the outermost surface of the organism. On removing the pipette, the drop was not dislodged from the surface of the *Amoeba* but continued to remain attached to it, usually assuming the form of a slightly concave hemisphere with the concave face in contact with the *Amoeba*. Whenever this occurred there was 'instant' and typical response on the part of the organism. The protoplasm in contact with the oil extruded a pseudopodium to the tip of which the oil remained in contact while the main body of the amoeba continued to flow into the pseudopodium; the general appearance of the *Amoeba* changed, approximating to the *A. limax* form. Progression always occurred with the 'cap' of oil in front, and the cap might remain attached in this position for some days. If a drop of oil was permitted to remain in contact with an amoeba for several seconds, the amoeba reacted by flowing round and over the oil, forming a normal food cup; when the pipette was removed, complete ingestion of the oil took place. For ingestion to occur the oil had to be supported against the pipette—for if the droplet was ejected into the medium directly in the path of the amoeba, the latter extruded pseudopodia towards it, but digestion failed because the streaming pseudopodium pushed the oil away from it. If the drop was held against the surface of the amoeba ingestion occurred. These observations indicate that the process of adhesion between a cell and an oil surface is distinct from the subsequent process of ingestion into the cell—which is contrary to the conceptions put forward by Tait. It is also interesting to note that all *Amoebae* cannot ingest oil drops in the way described for *Amoeba dubia*. Dawson and Belkin found that *A. proteus* failed to ingest oil and this they correlate with the fact that this species possesses a much tougher surface layer than does *A. dubia*.

As already pointed out it is by no means certain that the spreading of leucocytes on plane surfaces has any real connection with their phagocytic powers. Barikine (1910) and later Fenn (1922-3) have shown that reagents which influence the adhesion of leucocytes to surfaces do not affect phagocytosis of particles in the same way. Fenn determined the percentage of leucocytes which can be washed off a glass slide by a stream of water, thereby getting some indication of the adhesive force between the two units. The results indicate that the factors which influence the adhesion of the cells to various surfaces are not the same as those which influence the phagocytosis of discrete particles. Thus in acid solutions quartz particles are phagocytised more readily than carbon particles, but the stickiness of leucocytes to coal is greater than to glass in both acid and alkali. Again calcium chloride, ethyl alcohol, and iodoform, which were found by Hamburger (1912) to increase phagocytosis of carbon, have no effect or slightly decrease the stickiness of leucocytes to glass. Very significant, perhaps, is the observation that the presence of

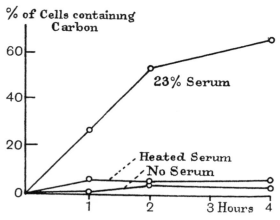

Fig. 204. The presence of serum increases the phagocytosis of carbon particles, but its efficiency is destroyed by heating to 56° C. for 40 mins. (From Fenn.)

serum increases the phagocytosis of solid particles but decreases the stickiness to glass, coal, mica and paraffin. This suggests that the spreading of the leucocyte on a glass surface is an indication that the cell is no longer in its normal condition; the more normal the environment of the cell the less likely it is to undergo pathological change.

It is possible to seek for a physical explanation of the adhesion of cell to particle from an entirely different source to surface tension. We may regard the adhesion as akin to that between discrete inanimate particles in colloidal suspension. As a general rule the stability of discrete particles is at a minimum at the isoelectric point, so that one might infer that phagocytosis would be greatest at the hydrogen ion concentration at which the repulsive effect between particle and cell was at a minimum. The observations of Fenn (1923), however, do not support this view. Phagocytosis for both carbon and quartz is greatest near neutrality, in acid solution quartz is ingested more rapidly than carbon, in alkaline solutions the reverse is the case (see Table LXXV).

Fenn (1922) concludes that the cataphoretic potential between the particles and the medium is not the controlling factor for ingestion, since in the presence of serum both quartz and carbon particles will be coated with plasma proteins and therefore bear the same change. Although neither surface tension nor surface charge appears to give an adequate picture of the selective action of cells for carbon as opposed to quartz, it is interesting to recall Rhumbler's

Table LXXV. Increase in the ratio of ingested carbon to ingested quartz particles with increase of pH

Exp. no.	Serum %	pH	$\dfrac{\% \text{ carbon}}{\% \text{ quartz}}$
1	17	7·2	0·88
		7·6	1·8
2	7	6·4	0·43
		7·3	1·39
		7·6	2·93

experiments with oil and alcohol. Rhumbler made a suspension of coal and quartz particles in oil and sprayed this into 70 per cent. alcohol: the coal particles remained inside the oil drops while the quartz collected on the surface. Similarly, according to Fenn, carbon collects more readily than quartz at the interface between chloroform and water.

As already mentioned it is by no means certain that the concepts of economy of surface energy applicable to a liquid interface can give an adequate picture of the process of ingestion of a particle by a living cell. When a leucocyte spreads on glass the process is irreversible and in this respect such cells are highly abnormal. As far as can be judged, the phagocytic activities of the cells are greatest when in their normal environment for, contrary to the views expressed by Hamburger (1912), Fenn (1920 b) found that the presence of normal blood serum greatly increased the power of leucocytes to ingest carbon and quartz. Similarly Ouweelen (1917) observed a failure of leucocytes to ingest starch grains in the absence of serum.

It is possible to look upon the process of ingestion as akin to amoeboid movement and regard the energy expended as being derived from metabolic sources rather than the surface energy at the cell/medium interface. Such a conception is in harmony with the fact that all cell surfaces exhibit plastic rather than viscous flow—in other words they are of a solid and not of a liquid nature. That the surface of invertebrate leucocytes is of a solid nature is very

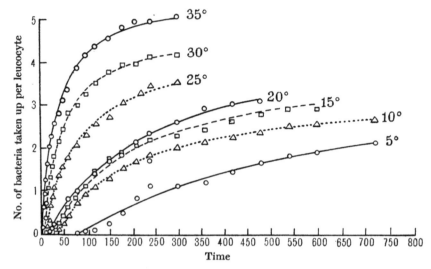

Fig. 205. Curves showing the number of bacteria taken up per leucocyte (ordinates) as a function of the time (abscissae) at different temperatures. Note the shorter latent period, higher maximum and shorter duration of the experiment at higher temperatures. (After Fenn, 1922 a.)

clearly demonstrated by the recent observations of Fauré-Fremiet (1929) (see fig. 31). The high temperature coefficient of ingestion recorded by Fenn, viz. 3·2 ± 0·4 at 27°–35° C. and 14·2 ± 1·7 at 20°–27° C. suggests that a chemical process is involved. If ingestion is the result of the redistribution of surface energy at a liquid interface, such figures are very remarkable: they are not so difficult to explain if the energy required to rupture the cell surface is being derived from metabolic processes inside the cell. The observations of Ledingham (1908) suggest, however, that although the whole cycle of phagocytosis may have a high temperature coefficient, the effect of temperature on the actual process of ingestion is not nearly so obvious. Ledingham found that if a mixture of *Staphylococcus aureus*, serum, and leucocytes was ingested at 18° C., the phagocytic activity of the cells was about one-quarter to one-fifth of that exhibited by the cells of a similar suspension incubated at 37° C. If, however, the bacteria and serum were previously mixed and incubated before the leucocytes were added, then subsequent incubation at 18° produced as much phagocytosis as at 37° C. These observations must, however, be considered in conjunction with the fact, already mentioned, that at still lower temperatures bacteria will adhere to but are not ingested by leucocytes. The amoeboid theory of phagocytosis is difficult to harmonise with the statement that leucocytes incapable of active movement can ingest particles (Friedemann and Schönfeld, 1917).

REFERENCES

BARIKINE, W. (1910). 'Sur le mécanisme de la phagocytose *in vitro*.' *Zeit. f. Immunitätsforsch.* 8, *72*.

COMMANDON, J. (1919). 'Tactisme produit par l'amidon sur les leucocytes. Enrobement du charbon.' *C.R. Soc. Biol.* 82, *1171*.

—— (1919). 'Action de la température sur la vitesse de reptation des leucocytes.' *C.R. Soc. Biol.* 82, *1305*.

DAWSON, J. A. and BELKIN, M. (1929). 'The digestion of oils by *Amoeba proteus*.' *Biol. Bull.* 56, *80*.

FAURÉ-FREMIET, E. (1929). 'Caractères physico-chimiques des choanoleucocytes de quelques invertébrés.' *Protoplasma*, 6, *521*.

FENN, W. O. (1920 a). 'The phagocytosis of solid particles. I. Quartz.' *Journ. Gen. Physiol.* 3, *439*.

—— (1920 b). 'The phagocytosis of solid particles. II. Carbon.' *Journ. Gen. Physiol.* 3, *465*.

—— (1920 c). 'The phagocytosis of solid particles. III. Carbon and quartz.' *Journ. Gen. Physiol.* 3, *575*.

—— (1922 a). 'The temperature coefficient of phagocytosis.' *Journ. Gen. Physiol.* 4, *331*.

—— (1922 b). 'The theoretical response of living cells to contact with solid bodies.' *Journ. Gen. Physiol.* 4, *373*.

—— (1922 c). 'The adhesiveness of leucocytes to solid surfaces.' *Journ. Gen. Physiol.* 5, *143*.

—— (1922 d). 'Effect of the hydrogen ion concentration on the phagocytosis and adhesiveness of leucocytes.' *Journ. Gen. Physiol.* 5, *169*.

—— (1923). 'The phagocytosis of solid particles. IV. Carbon and quartz in solutions of varying acidity.' *Journ. Gen. Physiol.* 5, *311*.

FRIEDEMANN, W. and SCHÖNFELD, A. (1917). 'Über die physikalisch-chemischen Bedingungen der Leucocytenbewegung.' *Biochem. Zeit.* 80, *312*.

HAMBURGER, H. J. (1912). *Physikalische-chemische Untersuchungen über Phagozyten.* (Wiesbaden.)

LEDINGHAM, J. C. G. (1908). 'The influence of temperature on phagocytosis.' *Proc. Roy. Soc.* B, 80, *188.*

—— (1912). 'The mechanism of phagocytosis from the adsorption point of view.' *Journ. Hygiene,* 12, *320.*

LISON, L. (1929). 'La tension superficielle joue-t-elle un rôle dans la genèse et le maintien des lames hyaloplasmiques chez les amibocytes?' *Protoplasma,* 7, *489.*

MCKENDRICK, A. G. (1914). 'Studies on the theory of continuous probabilities with special reference to its bearing on natural phenomena of a progressive nature.' *Proc. London Math. Soc.* Series 2, 13, *401.*

OUWEELEN, J. (1917). 'Über den Einfluss des Serums auf die Phagozytose von Kohle und Amylum. IV.' *Arch. f. ges. Physiol.* 169, *129.*

PONDER, E. (1925). 'A contribution to the theory of phagocytosis.' *Journ. Gen. Physiol.* 9, *827.*

RHUMBLER, L. (1898). 'Physikalische Analyse von Lebenserscheinungen der Zelle. I. Bewegung, Nahrungsaufnahme, Defäkation, Vacuolen-Pulsation und Gehäusebau bei lobosen Rhizopoden.' *Arch. f. Entw. Mech.* 7, *103, 199.*

—— (1914). 'Das Protoplasma als physikalisches System.' *Ergeb. d. Physiol.* 14, *474.*

TAIT, J. (1918). 'Capillary phenomena observed in blood cells: thigmocytes, phagocytosis, amoeboid movement, differential adhesiveness of corpuscles, emigration of leucocytes.' *Quart. Journ. Exp. Physiol.* 12, *1.*

TAIT, J. and GUNN, J. D. (1918). 'The blood of *Astacus fluviatilis*: a study in crustacean blood with special reference to coagulation and phagocytosis.' *Quart. Journ. Exp. Physiol.* 12, *35.*

INDEX OF AUTHORS

Abderhalden, E., 366
Abramson, H. A., 51, 54
Adami, J. G., 297
Adolph, E. F., 332, 343 seq., 387
Adrian, E. D. and Brock, D. W., 309
Albrecht, E., 120
Allen, E. J., 274
Andrews, E. A., 4, 180
Anselmino, K. J., 350

Baas Becking, L. G. M., Bakhingzen, H. v. d. S. and Hotelling, H., 59, 60, 66, 68
Bachman, E. L. and Runnström, J., 402
Baker, L. E. and Carrel, A., 294, 295
Baltzer, F., 175, 430
Barber, M. A., 278
Barcroft, J. and Straub, H., 386
Barikine, W., 501
Baron, M. A., 182
Barta, E., 292
Bataillon, E., 443
Bayliss, W. M., 58, 61, 88, 401
Bělař, K., 122, 145
Bělehrádek, J., 66
Benda, C., 475
Benecke, W., 117
van Beneden, E., 165
Berezowski, A., 269
Bernstein, J., 401
Beutler, R., 452, 479
Beutner, R., 22, 334, 335, 383, 394, 396
Bhatia, D., 268, 465
Bidder, G. P., 468
Biltz, W., 350
Bjerknes, J., 171, 172
Blackman, F. F., 383
Blackman, V. H. and Paine, S. G., 402
Bogue, R. H., 33
Bolles Lee, A., 80
Borsook, H. and MacFadyen, D. A., 36
Boveri, T., 89, 161, 165, 166
Brachet, A., 95, 132, 433
Brooks, M. M., 369, 374–376, 391
Brooks, S. C., 87, 312, 313
Brown, D. E. S. and Sichel, F. J. M., 309
Brown, O. H., 357
Brown, W., 350
v. Brücke, E. T., 485
Buchanan, R. E., 285
— and Fulmer, E. I., 51
Buchner, P., 442
Bugarszky, S. and Tangl, F., 358
Buglia, A., 43
Buller, A. H., 420

Burrows, M. T., 122, 211, 290
Bütschli, O., 71, 73, 74, 169, 170, 383, 455, 472

Calkins, G. N., 278, 282
Cameron, A. T., 9
Cannon, H. G., 152, 153, 172
Carrel, A., 192, 289, 294, 296
— and Ebeling, A. H., 278, 287, 294, 296
Carter, G. S., 115, 416, 418, 465, 470, 481, 482
Castle, W. E., 431
Cayley, G., 264
Cernovodeanu, P. and Henri, H., 51
Chalkley, H. W., 334
Chambers, H. and Scott, G., 177
Chambers, R., 72, 76, 77, 85, 86, 124, 125, 129, 150–154, 156, 158, 160, 195, 206, 209, 212–214, 424–427
— and Dawson, J. A., 465
— and Pollack, H., 85
Chambers, R., Pollack, H. and Hiller, S., 85
Chambers, R. and Rényi, G. S., 120, 122
— and Reznikoff, P., 76
— and Sands, H. C., 126, 149, 156
Champy, C., 122, 292, 297
Chick, H., 311–316
Clowes, G. H. A., 382
— and Bachman, E., 418
Cohn, A. E. and Murray, H. A., 288
Cohn, E. J., 410–414, 419
Cohnheim, O., 386, 387
Cole, K. S., 361, 402, 429
Collander, R., 350
Commandon, J., 496
— and Jolly, J., 120
Conklin, E. G., 89, 93–96, 131, 133, 164, 232, 235–243
Coombs, H. I. and Stephenson, M., 281
Cooper, W. C., Dorcas, M. J., and Osterhout, W. J. V., 370
— and Blinks, L. R., 366
Copeland, M., 482
Coulter, C. B., 49
Cox, L. W. R. and Harris, J. E., 59
Cragg, F. W., 424
Crampton, H. E., 90
Crozier, W. J., 369
Cunningham, R., 62
Czapek, F. 383

INDEX OF AUTHORS

Dakin, W. J. and Fordham, M. G. S., 423
Dalcq, A., 408
Danchakoff, V., 132
Darby, H. H., 275, 278
Davenport, C. B., 324
Davis, H., 324
Dawson, J. A. and Belkin, M., 501
Day, H. C., 332
Dellinger, O. P., 459
Demoussey, E., 376
Dixon, M., 19, 20
— and Elliott, K. A. C., 20
Dobell, C. C., 2
Doncaster, L., 427
— and Gray, J., 175
Donnan, F. G., 49, 399 seq.
Drew, A. H., 293
Drew, G. H., 299
Driesch, H., 2, 88–90, 135, 231
Dustin, A. P., 177

Eastcott, E. V., 275
Ebeling, A. H., 278, 294
Einstein, A., 59
Emerson, R., 20
Engelmann, T. W., 464, 467
Ephrussi, B., 148, 173
Erdmann, R., 134
Erhard, H., 464, 466
v. Erlanger, R., 196, 209, 229
Errera, L., 259
Ewart, A. J., 457

Farr, C. H., 210
Fauré-Fremiet, E., 133, 134, 272, 422, 458, 504
Fenn, W. O., 490–503
Fetter, D., 63
Feulgen, R., 84, 128
Fischel, A., 90, 430
Fischer, A., 33, 42, 178, 279, 290, 293, 297
Fisher, R. A., 307
Fitting, H., 327, 352
Fox, H. M., 420
Freundlich, H., 33, 34, 49, 53, 58, 59, 344
— and Seifriz, W., 69
Frew, J. G. H., 296
Fricke, H., 361, 384
— and Morse, S., 359–361
Friedemann, W. and Schönfeld, A., 505
Fry, H. J., 441
Fuchs, H. M., 431
Fuhrmann, F., 474

Gaidukov, N., 58
Galtsoff, P. S., 106, 107, 115, 298
Garmus, A., 372
Gatenby, J. B., 422

Gelfan, S., 87
Gellhorn, E., 383, 384
Gemmill, J. F., 409–411
Gibbs, J. Willard, 260
Giglio-Tos, E., 233
Glaser, O., 420, 423
Godlewski, E., 133, 433
Goldschmidt, R., 296
Goldschmidt, S., 345
Gortner, E. and Grendel, F., 383
Göthlin, G. F., 482
Gotschlich, E., 43
Gräfer, L., 258
Graham-Smith, G. S., 283
Gray, J., 20, 46, 51, 105, 113, 124, 131, 134, 148, 159–167, 174, 180, 183, 194 seq., 207, 252, 260, 272–279, 286, 309, 313, 320, 324, 335, 363, 369, 376, 402, 409 seq., 420, 426 seq., 452, 468, 477–480
Green, R. G. and Larson, W. P., 358
Gross, R., 120
Grüber, A., 71, 332
Gutherz, S., 177
Gutstein, M., 81
Gurwitsch, A., 181, 182
— and Gurwitsch, L., 182
Guye, C. E., 13–15

Haberlandt, G., 178
Haldane, J. B. S., 363
Hamburger, H. J., 329, 331, 369, 502, 503
Hammond, J., 417
— and Asdell, S. A., 409, 415
Hansteen-Cranner, B., 383
Hardy, W. B., 33, 39, 40, 49, 74, 122, 156
— and Nottage, M., 265
— and Wood, T. B., 106, 109
Hargitt, C. W., 96
Harrison, R. G., 289, 290
Hartman, C. and Cuyler, W. K., 409
Hartog, M., 172, 332
Hartridge, H. H., 263
Harvey, E. N., 356, 369, 425
Harvey, W. H., 297
Hashida, K., 344
Hatschek, E., 65
Hauser, E. A., 60
Haywood, C., 365, 369
Hecht, S. and Wolf, E., 309, 322
Hedin, S. G., 329
Heiberg, K. A., 307
Heidenhain, M., 165, 166, 474
Heidenhain, R., 387
Heilbronn, A., 64, 66
Heilbrunn, L. V., 42–44, 58, 59, 62–66, 68, 206, 209

INDEX OF AUTHORS

Henrici, A. T., 283
Hensen, V., 472
Herbst, C., 107, 252, 438, 439
Herlant, M., 132, 161, 433, 440–443
Hertwig, O., 135, 161, 230, 233, 272, 432, 438
Hertwig, R., 132, 161, 432, 438
Hertz, W., 372
v. Herwenden, M. A., 60
v. Hevesy, G., 49
Hill, A. V., 10, 14, 21, 331, 332, 402, 452
— and Hartree, W., 478
Hindle, E., 439, 440
Hinshelwood, C. N., 14
Höber, R., 326, 359, 360, 362, 363, 369, 372, 384
— and Chassin, S., 372
— and Kempner, F., 372
Hobson, A. D., 425, 426
van 't Hoff, J. H., 327
Höfler, K., 327, 352
Hofmeister, F., 183, 259
Hosoya, S. and Kuroya, M., 274
Howland, R. B., 103
Hunter, O. W., 80
Huxley, J. S., 415
Hyde, I. H., 428
Hyman, L. H., 456

Illing, G., 269
Irwin, M., 355, 373–375, 390
Isawaki, Y., 177

Jacobs, M. H., 353, 371, 375
Jacques, A. G. and Osterhout, W. J. V., 354
Jameson, H. L., Drummond, J. C. and Coward, K. H., 274
Janse, J. M., 375
Jennings, H. S., 459
Jensen, P., 479
Jinnaka, S. and Azuma, R., 309
Jolly, J., 173
Jordan, A. E. and Helvestine, F., 164, 467
Joseph, H., 475
Just, E. E., 195 seq., 211, 212, 227, 228, 420, 422, 424, 433, 436, 441

Karsten, G., 183
Kathariner, L., 233
Keilin, D., 20
Kermack, W. O., McKendrick, A. G. and Ponder, E., 63
Kiesel, A., 79
Kindred, J. E., 164
Kite, G. L., 77, 102, 124, 125
Klein, E., 165
Koehler, O., 271

Kölliker, A., 409
Konopacki, M., 174
Kornfeld, W., 183
Korschelt, E., 222
Koser, S. A. and Rettger, L. F., 274
Kozawa, S., 376
Kraft, H., 481
Krijgsman, B. J., 470
Kroeber, J., 299
Krogh, A., 105
Krontowski, A. A. and Bronstein, J. A., 274
Kruse, W., 80
Kupelwieser, H., 430

Ladenburg, R., 62
Lamb, A. B., 171, 172
Lambert, R. A., 146
— and Hanes, F. M., 192
Lane-Claypon, J. E., 279
Lapicque, L., 333, 384
Leblond, M., 58
Ledingham, J. C. G., 500, 504, 505
— and Penfold, W. J., 320
Leduc, S., 169
Lefevre, G., 441
Lepeschkin, W. W., 352, 383, 384
Levi, G., 146, 147
Lewis, F. T., 223–225, 248, 256–258
Lewis, M. R., 155
Lewis, W. H., 74
— and Lewis, M. R., 120, 122, 146–148, 289, 294, 297
Lillie, F. R., 2, 96, 130, 132, 161, 181, 241–243, 263, 408, 419–427, 433–437, 440
— and Just, E. E., 408, 421
Lillie, R. S., 118, 131, 152, 337 seq., 401, 402, 428, 429, 436, 439, 440, 484
Lison, L., 499
Lloyd, D. Jordan, 336
Loeb, J., 115, 131, 133, 196, 219, 420, 422, 425–427, 430, 437–442
— and Beutner, R., 22–24
— and Northrop, J. H., 315
— and Wasteneys, H., 20, 427, 428
Lucas, K., 402
Lucke, B. and McCutcheon, M., 331, 334, 342, 345
Ludwig, W., 472
Lund, E. J., 27, 388
— and Logan, G. A., 28, 309
Lynch, V., 130
Lyon, E. P., 61, 217

Macallum, A. B., 81–83
McClendon, J. F., 163, 167, 168, 214, 215, 358–361, 401, 402, 425, 429

INDEX OF AUTHORS

McCutcheon, M. and Lucke, B., 337 seq.
MacDonald, J. S., 23
McKendrick, A. G., 491
MacKinnon, D. L. and Vlès, F., 464
Mainx, F., 182
Mann, G., 41, 83
Martens, P., 122, 123, 145, 151
Masing, E., 133, 272
Mast, S. O., 75
— and Root, F. M., 215
Mathews, A. P., 125, 131, 149, 173, 219, 441
Matzke, E. B., 256
Maxwell, S. S., 343, 387, 479
Mayer, A. and Schaeffer, G., 324
Mead, A. D., 437, 438
Meek, C. F. O., 167
Meigs, E. B., 384
Merton, H., 475, 481, 482
Meves, F., 167
Meyer, H. H., 370
Meyerhof, O., 167, 222, 428
Michaelis, L., 367, 393, 396–398
— and Fujita, A., 23, 396
Miescher, H., 126, 128
Minot, C. S., 132
Mitchell, P. H. and Wilson, J. W., 401
Mockeridge, F. A., 128
Moenkhaus, W. J., 430
Mond, R., 27, 370, 375, 388, 390
— and Hoffmann, F., 397
Moore, C. R., 409
Morgan, T. H., 2, 31, 89, 90, 94, 137, 156, 161, 163, 298, 431, 438–440
— and Boring, A., 227
— and Spooner, G. B., 90, 94
— and Tyler, A., 229, 240
Morgulis, S., 183
Morpurgo, B., 183
Mudd, S., 53
— and Mudd, E. B. H., 383
Murphy, J. B. and Hawkins, J. A., 273
Murray, M. E., 296

Nath, V., 422
Nathanson, A., 383
Needham, J., 422
— and Needham, D. M., 85, 272
Newman, H. H., 430
Nierenstein, E., 372, 398
Northrop, J. H., 49, 50, 52, 53, 338 seq., 345
— and Kunitz, M., 341

Oliver, J. and Barnard, L., 51
Oomen, H. A. P. C., 387
v. Oordt, G. J., 409

Osterhout, W. J. V., 24, 28, 118, 130, 354, 357, 358, 369, 375, 382, 384, 390, 395
—, Damon, E. B. and Jacques, A. G., 23, 388
— and Dorcas, M. J., 354, 355
— and Harris, E. S., 24–26

Ouweelen, J., 504
Overton, E., 325, 326, 330–332, 353, 370, 371, 382, 393

Painter, T. S., 161
Pantin, C. F. A., 66–68, 452, 455–463
Parker, G. H., 479, 486
Paul, J. H. and Sharpe, J. S., 370
Pauli, W., 33
Pearson, K., 313
Peebles, F., 124
Penfold, W. J., 285
— and Norris, D., 283
Pentimalli, F., 168
Perrin, J., 59
Peter, K., 129, 466, 467, 475
Peters, R. A., 274
Pfeffer, W., 103, 325, 326, 383, 393, 396
Pflüger, E., 230, 231, 233
Phillipson, M., 364
Plenk, H., 270
Popoff, M., 132, 272
Policard, A., 122
Pollack, H., 80, 81, 86
Polowzow, W., 180
Ponder, E., 264, 265, 313, 318–320, 499
Powis, F., 49
Prénant, A., 167, 464, 466
Proctor, H. R. and Wilson, J. A., 333
Prowazek, S. v., 71
Przibram, H., 299
Pütter, A., 466

Quincke, G., 383

Rahn, O., 317
Reichert, K., 473
Reid, W., 343, 386, 387
Reznikoff, P. and Pollack, H., 87
Rhumbler, L., 171, 460, 500, 503
Richards, O. W., 279, 284
Ringer, S., 45
v. Risse, O., 350
Ritchie, A. D., 451
Robert, A., 248
Robertson, T. B., 134, 214, 216, 272, 383, 384
— and Miyake, K., 110, 113

INDEX OF AUTHORS

Rogers, C. G. and Cole, K. S., 222
Rohonyi, H., 396
— and Lorant, A., 363
Roux, W., 227, 228, 231, 250–253, 427
Ruhland, W., 353, 371, 372
Russ, C., 52

Sachs, J., 43
Saguchi, S., 464
Sakamura, T., 149
Salter, R. C., 285
Saunders, J. T., 420
Scarth, G. W., 58
Schackell, L. F., 133, 272
Schalek, E. and Szegvary, A., 57, 69
Schäfer, E., 474
Schaeffer, A. A., 455–459
Schitz, V., 122
Schmalhausen, I. and Bordzilowskaja, N., 281
Seeliger, O., 222
Seifriz, W., 57, 64, 70, 76, 77, 105
Sharp, L. W., 1
Shearer, C., 18, 156, 358, 427, 428, 436
— and Lloyd, D. Jordan, 439
—, de Morgan, W. and Fuchs, H. M., 430
Shiwago, P., 120
Siebeck, R., 330
Simms, H. S., 47
Slator, A., 279, 281
van Slyke, D. D., 362, 363, 369, 392
— and Bosworth, A. W., 112
— and Winter, O. B., 112
Smith, H. W. and Clowes, G. H. A., 174, 432
v. Smoluchowski, M., 7, 12, 59
Solger, B., 165
Sorin, A. N., 182
Spek, J., 196, 214
Spitzer, W., 131
Stephenson, M., 29
Steward, F. C., 383
Stewart, G. N., 358, 382
Stiles, W., 66, 328, 352, 375, 384
— and Jørgensen, I., 376
Strangeways, T. S. P., 142, 146, 150, 154, 190, 289
— and Canti, R., 41
— and Hopwood, F. L., 174
Strasburger, E., 132
Straub, J., 27, 388
Svedberg, T., 7, 33, 57, 69, 316, 317

Tait, J., 423, 499
— and Gunn, J. D., 499

Taylor, C. V., 464
Tennent, D. H., 430
— and Hogue, M. J., 442
Terroine, E. F. and Würmser, R., 276, 277
—, Würmser, R. and Montané, J., 80
Thompson, D'A. W., 248, 254–261
Thomson, W., 255
Tinker, F., 351
Townsend, C. O., 130, 132
v. Trigt, H., 408, 471, 472
Tröndle, A., 391

Uhlenhuth, E., 292
Unna, P. G. and Tielemann, E. T., 81

Valentin, G., 468
Vernon, H. M., 429, 430
Verworn, M., 69, 130, 486
Vlès, F., 216, 218, 219
Voss, H., 128
de Vries, H., 326

Walton, A., 414, 415, 417
Warburg, E. J., 362
Warburg, O., 19–21, 131, 217, 221, 273, 427, 440
— and Kubowitz, F., 294
Watchorn, E. and Holmes, E. B., 274, 294
Weber, F., 66
Wertheimer, E., 354, 390
Wetzel, G., 224, 256, 257
Whiteside, B., 127
Williams, L. W., 468, 474
Willmer, E. N., 295
Wilson, E. B., 72, 88–90, 135, 136, 152, 161–163, 167, 180, 201, 202, 232, 239, 241, 408, 421, 424, 427, 432, 441
— and Mathews, A. P., 227
Wilson, G. S., 286
Wilson, H. V., 297
— and Penney, J. T., 298
Wilson, J. A., 49
Winslow, C. E. A. and Fleeson, E. H., 47
—, Falk, I. S. and Caulfield, M. F., 52
Woodward, A. E., 420
Wright, G. P., 146, 147, 178, 296

Yatsu, N., 95, 164, 231
Young, W. C., 409, 414, 416
Yule, G. U., 321

Ziegler, H. E., 231
Zimmermann, K. W., 165
Zoond, A., 358

INDEX OF SUBJECTS

Acids, physiological effects of, 364
Acmaea, artificial parthenogenesis of, 439
Actinosphaerium, structure of, 69–71
Aethalium, ectoplasm of, 103
Alectrion, cilia of, 482, 483
Allolobophora, regeneration of, 299
Ammonium oleate, viscosity of, 70
Amoeba, Brownian movement in, 61
— , contractile vacuoles of, 332
— , ingestion of oils by, 501
— , properties of enucleated, 130
— , structure of, 75, 77, 102, 103
— , water content of, 334
Amoeboid movement, 455–463
Amphiaster, 162 seq.
Amphioxus eggs, segmentation of, 235
Anaphase, 165 seq.
Antedon eggs, segmentation of, 235
Arbacia, nuclei of, 127
— , spermatozoa of, 427
Arbacia eggs, artificial parthenogenesis of, 438
— , cell division in, 196 seq.
— , CO_2 production of, 217
— , electrical conductivity of, 361
— , fertilisation of, 423
— , hyaline layer of, 211
— , O_2 consumption of, 427
— , velocity of segmentation, 174
— , viscosity of, 43, 59, 61
— , water content of, 337 seq., 402
Archidoris, ciliated cells of, 483
Archiplasm, 162
Arrhenatherium, nuclei of, 122
Aspergillus, proteins of, 80
Asterias eggs, artificial parthenogenesis of, 439, 441, 442
— , fertilisation of, 424
— , nuclei of, 124, 132
— , structure of, 73
— , velocity of segmentation, 174
Asters, 157 seq., 164
— , rôle during cleavage, 201
Azotobacter, proteins of, 80

Bacillus, growth of, 278, 281
— , size of, 9
Bacteria, agglutination of, 49 seq.

Bacteria, death curves of, 311 seq.
— , electrical conductivity of, 358
— , growth of, 278–286
— , ingestion of, 491, 492
Badhamia, protoplasmic viscosity of, 64
Bee's eye, resolving power of, 322
Beroë, ciliated plates of, 455, 482
Bioelectric currents, 21 seq., 392 seq.
Blepharisma, membranella of, 465
Blood corpuscles, agglutination of, 49 seq.
— , cataphoresis of, 51
— , composition of, 9
— , elasticity of, 105
— , electrolytes in, 368, 370
— , isoelectric point of, 49
— , water content of, 363
Brownian movement, 7, 58
— , inhibition of, 60

Calcium, test for intracellular, 80
Casein, dispersion of, 113
Cataphoresis, 47 seq.
Cell division, 189 seq.
— , anastral, 190
— , astral, 193, 206, 211
— , energetics of, 216
— , epithelial cells, 222 seq.
— unequal, 243
Cell growth, 267 seq.
— , effect of food on, 283
— , effect of metabolites on, 284
— , inhibition of, 282
— , kinetics of, 277
— , lag-phase of, 285
— , media for, 294
Cell membranes, 102 seq.
Cell surfaces, lipoid theory of, 382, 383
— , permeability of, 349 seq., 385 seq.
— , protein theory of, 383
Cell Theory, 1
Cells, differentiation of, 296
— , electrical conductivity of, 357 seq.
— , permeability of, 353 seq.
— , rigidity of, 216 seq.
— , shapes of, 248 seq., 255
— , size of, 269
— , variability of, 307 seq.
— , water equilibrium of, 324 seq.
Centrosome, 155, 164
Cerebratulus eggs, cytoplasm of, 88, 89, 95

INDEX OF SUBJECTS

Cerebratulus eggs, enucleated, 164
Chaetopterus eggs, artificial parthenogenesis in, 438
— , development of, 2, 96
— , fertilisation of, 229
Chlorella, respiration of, 20
Chondrioderma, surface of, 103
Chromosomes, mechanical properties of, 150
Cilia, contractile properties of, 474
— , excitation of, 481
— , inhibition of, 482
— , movement of, 467 seq., 477 seq.
— , velocity of, 468
Ciliated cells, structure of, 466
— , metabolism of, 479
Cimex, spermatozoa of, 424
Ciona, self-sterility of, 431
Cleavage, orthoradial, 235 seq.
— , spiral, 235 seq.
Cleavage planes, effect of gravity on, 233
— , effect of pressure on, 230, 242
— , orientation of, 226, 235
Colloidal particles, size of, 316
Colloids, 34 seq.
— , lyophil, 35
— , lyophob, 34
Contractile cells, 451 seq.
— vacuoles, 332
Crepidula, nuclei of, 131
— , spiral cleavage of, 237, 238, 241
Cryptobranchus, erythrocytes of, 105
Ctenolabrus eggs, effect of O_2 on, 115
Cucumis, cell division in, 225
Cumingia eggs, cleavage planes of, 240
— , viscosity of, 43, 62, 66
Cyclosis, 457, 458
Cynthia, nuclear growth in, 133
Cytasters, 163
Cytochrome, 20
Cytoplasm, differentiation of, 88 seq.

Dendrostoma, artificial parthenogenesis of, 441
Dentalium eggs, development of, 89 seq.
Deplasmolysis, 352
Desmones, 178
Dianthus cells, calcium in, 369
Differentiation, 296
— , without cleavage, 2
Dissosteira, nuclei of, 129 seq., 150 seq.
Donnan equilibrium, 333, 362, 399
Dyes, diffusion of, 371–374

Echinarachnius eggs, consistency of, 105
— , hyaline layer of, 211
— , viscosity of, 64, 105

Echinoderm eggs, optical structure of, 104
— , respiration of, 217 seq.
— , rigidity of, 216 seq.
— , structure of, 104
Echinus eggs, cell division of, 194 seq., 222 seq.
— , effects of gravity on, 233
— , effects of pressure on, 230
— , heat production of, 18
— , hyaline membrane of, 104, 252
— , nuclei of, 124, 125
— , protoplasmic structure of, 73
— , respiration of, 166, 219
— , segmentation of, 254
— , velocity of segmentation of, 146
Echinus, spermatozoa of, 409 seq.
Eggs, development of, 421
— , electrical conductivity of, 402, 429
— , fertilisation of, 423 seq.
— , water content of, 335 seq.
Electrokinesis, 47
Electrolytes, secretion of, 386, 387
Electrostatic charge, effect of ions on, 51
Emulsions, sensitivity of, 317
Endosmosis, electrical, 48, 53
Entropy, 16
Enzymes, oxidising, 19
Errera's Law, 258
Erythrotrichia, segmentation of, 254
Eucharis, cilia of, 484
Euplotes, cirri of, 464
Exosmosis, 314

Fertilisation cone, 424
— , heat production, 428
— membranes, 438 seq., 424 seq.
— , partial, 437
— , physiological effects of, 427
— , respiration and, 428
— , specificity in, 429 seq.
— , theory of, 432 seq.
Fertilizin, 433–435
Feulgen reaction, 84, 128
Fixation, theory of, 39 seq.
Flagella, movement of, 470 seq.
Frog's skin, permeability of, 343–345
Frontonia, cyclosis in, 457
— , division of, 272
Fundulus eggs, bioelectric currents of, 428
— , effect of $Ca^{..}$ lack, 116
— , electrical conductivity of, 428

INDEX OF SUBJECTS

Galleria, mitoses in, 177
Gel, 65, 68 seq.
—, iron oxide, 57, 69
Gelatin, water content of, 333
Gibb's ring, 260
Gold, colloidal, 7, 34
Gromia, structure of protoplasm of, 74
Growth, efficiency of, 275
—, metabolism during, 273
Gut cells, permeability of, 345

Halicystis, sap of, 366, 368
Histriobdella, spindles of, 156
Hyacinth cells, effect of current on, 167
Hyaloplasmic layer, 104, 196 seq.
Hydroids, regeneration of, 298

Illyanassa, cytoplasm of, 90
Imbibitional swelling, 335 seq.
Intercellular bridges, 4
— matrices, 106 seq.
Ions, antagonistic, 47
Iris, protoplasm of, 76
Iron, microchemistry of, 82
Isoelectric point, 36, 49

Kidney, water content of, 330
Kinetics of swelling, 337 seq.

Lag-phase of growth, 320
Laminaria, electrical conductivity of, 357, 382
Limnea, nuclei of, 120
Lottia eggs, artificial parthenogenesis of, 439

Machines, efficiency of, 11
Manganese dioxide, ingestion of, 495, 496
Membranella, structure of, 466
Membranes, sieve theory of, 350, 367
Metabolism of growing cells, 273
Metachronal rhythm, 484 seq.
Metaplasia, 297
Micro-analysis, 81 seq.
Microciona, restitution bodies of, 106, 297
Micro-organisms, metabolism of, 29
Minimum surface, law of, 258 seq.
Mitogenic rays, 181
Mitosis, 141 seq.
—, abnormal, 173 seq.
—, mechanics of, 166 seq.
—, metaphase of, 152 seq.
—, stimulation of, 177
—, velocity of, 145
Monas, flagellum of, 470
Monaster, 161
Monotropa, nuclei of, 130

Motella, vibratile fins of, 455
Mucor, nitrogen content of, 80
Muscles, water content of, 331, 335, 337
Mytilus, 36, 47
—, cilia of, 465–469, 485
Myxomycetes, ectoplasm of, 103

Necrohormones, 177
Nereis eggs, cleavage planes of, 241
—, effect of pressure on, 135
—, fertilisation of, 227, 228, 426, 433
—, mitosis in, 152, 161, 243
—, partial fertilisation of, 437, 440
—, segmentation of, 262
—, viscosity of, 66
Nereis sperm, agglutination of, 419
—, aggregation of, 419
Nitella, bioelectric properties of, 24
—, electrical conductivity of, 87
—, permeability of, 369, 373 seq., 392
Nitzschia, growth of, 274
Noctiluca, electrical stimulation of, 28, 309
Nostoc, nuclear derivatives in, 128, 129
Nucleic acid, 84
Nucleo-cytoplasmic ratio, 132 seq., 271
Nucleus, biological properties of, 126 seq.
—, catalysts of, 134
—, chemistry of, 126 seq.
—, effects of injury on, 129, 150
—, optical structure of, 120 seq.
— and regeneration, 130
— and respiration, 131
—, rôle in development of, 135 seq.
—, synthesis of, 133
—, variability of, 308
—, viscosity of, 125

Oil drops, form of, 250
—, ingestion of, 501
Opalina, permeability of, 372
Osmosis, anomalous, 344
Osmotically inactive water, 331 seq.
Oxidase, indophenol, 21

Paracentrotus eggs, artificial parthenogenesis of, 439
—, respiration of, 219
—, rigidity of, 216 seq.
—, velocity of segmentation, 148
Paramecium, cilia of, 479
—, cleavage of, 215
—, form of, 248, 249
—, viscosity of, 63
Paris, chromosomes of, 151
Parthenogenesis, artificial, 437 seq.

INDEX OF SUBJECTS 515

Particles, hydration of, 35, 60
— , size of colloidal, 316
Pecten, metabolism of cilia, 479
— , metaplastic growth in, 299
Penicillium spores, ingestion of, 496
Peranema, flagellum of, 476
Permeability, coefficients of, 340, 349
— , dyes, 371 seq.
— , electrolytes, 354 seq.
— , functional significance of, 401
— , irreciprocal, 375
— , non-electrolytes, 350 seq.
— , water, 325 seq.
Phagocytosis, 490 seq.
— , effect of serum on, 503
— , effect of temperature on, 504
— , mechanism of, 496
Phosphorus, microchemistry of, 83
Physa, cilia of, 481
Planarians, regeneration of, 2
Plant cells, permeability to water, 325 seq.
Plasmolysis, 325 seq.
Platynereis eggs, fertilisation of, 422
Pleurobrachia, cilia of, 481, 484, 486
Polar furrow, formation of, 237
Polynoë eggs, artificial parthenogenesis of, 439
Polyspermy, 432
Populations, growth of, 310 seq.
Potassium, microchemistry of, 82
Potentials, phase boundary, 22, 383, 393 seq.
Prophase, mitotic, 148
Proteins, ampholytic properties of, 35
— , denatured, 37
— , heat coagulation of, 37
— , reactions with calcium, 45
Protoplasm, chemical composition of, 79
— , electrical conductivity of, 87
— , heat coagulation of, 42 seq.
— , miscibility of, 76 seq.
— , optical properties of, 71
— , pH of, 85
— , streaming of, 457, 458
— , viscosity of, 59 seq., 66 seq.

Quartz, cataphoresis of, 54
— , ingestion of, 493 seq.

Rat, ciliated cells of, 467
Red blood-corpuscles, conductivity of, 357
— , elasticity of, 105
— , electrolytes in, 368, 370
— , form of, 263 seq.

Red blood-corpuscles, water content of, 329, 363
Respiration, inhibition of, 20
— , intracellular, 19
Respiratory ferment, 19
Restitution bodies, 297
Reticularia, composition of, 79
— , protoplasmic viscosity of, 64
Rhizopus, protoplasmic viscosity of, 57
Rotifers, water content of, 324

Salamandra, leucocytes of, 165
Salmo, eggs of, 71, 314, 422
— , muscle fibres of, 269
— , spermatozoa of, 409
Sambucus, form of cells in, 257
Sols, 65, 68 seq.
— , iron oxide, 57
Spermatozoa, 408 seq.
— , agglutination of, 420
— , aggregation of, 419
— , antagonism of, 433
— , movement of, 472
— , respiration of, 410 seq.
— , survival of, 417
Sphaerechinus eggs, cross fertilisation of, 175
— , cytasters in, 441
Spindle, fibres, 155
— , mechanical properties of, 156
Spirillum, movement of, 473
Spirogyra, effect of Ca·· lack on 116, 117
— , permeability of, 375
— , plasmolysis of, 325, 326, 352
— , potassium in, 82
— , protoplasmic viscosity of, 59, 68
Spirostomum, protoplasm of, 71
Sponges, regeneration of, 297 seq.
Spongilla, flagellum of, 471
Staining, differential, 42
Staphylococcus, growth of, 283
— , ingestion of, 505
Statistical laws, 7
Stentor, enucleation of, 130
Stokes' Law, 62
Strongylocentrotus, artificial parthenogenesis of, 439
— , cross fertilisation of, 175
— , cytoplasm of, 89
— , sperm agglutination of, 420
Styela, cytoplasmic differences in, 93 seq.
Suspensions, colloidal, 7
Swelling, post-mortem, 335
Sycandra eggs, segmentation of, 235
Symphoricarpus, protoplasm of, 58
Synapta eggs, segmentation of, 235

INDEX OF SUBJECTS

Tetrakaidecahedron, properties of, 256
Thalassema eggs, artificial parthenogenesis of, 439
Thalassiosira, growth of, 274
Thermodynamics, second law of, 8, 14
Tissue-culture, 289 seq.
Toxopneustes eggs, cytasters in, 163
— , fertilisation of, 227
— , monastral, 161
Tradescantia, chromosomes of, 149
— , plasmolysis of, 326 seq.
— , protoplasm of, 58
— , spindles of, 156
Tripneustes, protoplasm of, 64
Triton, nuclei of, 120

Unio, nuclei of, 120
Uroleptus, growth rate of, 282

Valonia, bioelectric properties of, 23, 24, 376, 395
— , permeability of, 370, 373 seq., 376, 392
— , sap of, 366, 368
Variability, dynamic, 308 seq.
— , static, 307 seq.
Vespertilio, nuclear growth in, 133
Vibratile movement, 463 seq.
Vicia faba, protoplasmic viscosity of, 64
Viscosity, protoplasmic, 59 seq., 66 seq.
Vorticella, cytoplasm of, 82

Water, permeability of cells to, 325 seq.
— , secretion of, 343

Xiphophorus, spermatozoa of, 409